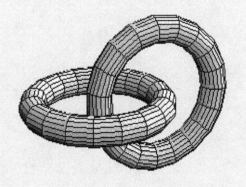

Introduction to the
Variational Calculus

by J.H. Heinbockel
Emeritus Professor of Mathematics
Old Dominion University

Trafford rev. 11/13/2018

 www.trafford.com

North America & international
toll-free: 1 888 232 4444 (USA & Canada)
fax: 812 355 4082

PREFACE

This is an introductory text to the mathematical discipline called the calculus of variations. The material presented is designed for engineers, physicists, mathematicians and other scientists who desire an introductory level text for teaching or for a reference book in this subject area. It is assumed that the readers of this text have completed a series of calculus courses and that they have some knowledge of both ordinary and partial differential equations. The text can be used for an introductory course in the calculus of variations for students at either the upper undergraduate or the beginning graduate level.

Chapter one presents selected material from the subject areas of introductory calculus, advanced calculus and differential equations. Much of the material presented in chapter one can be viewed as background material that is needed for subsequent chapters. Chapter two reviews maximum and minimum problems from the areas of calculus and advanced calculus and then introduces problems associated with optimization from the subject area of operations research. Chapter three introduces variational concepts and notations associated with finding the maximum and minimum values of functionals. A variety of functionals are considered together with the presentation of numerous worked examples and applications to aid in the understanding and utilization of the variational calculus. Chapter four considers the subject area of extreme values associated with functionals which have constraints. Also presented in this chapter are some of the more detailed analysis concepts associated with various types of functionals and boundary conditions. Chapters five and six present selected applications of the calculus of variations using direct methods and approximation or numerical methods to solve problems. The majority of the applications presented are from the traditional subject areas of mechanics, physics and engineering. The appendix A contains units of measurements from the Système International d'Unitès. The appendix B contains the gradient, divergence curl and Laplacian in Cartesian, cylindrical and spherical coordinates. The appendix C contains solutions to selected problems. A more detailed description of the material content can be discerned from an examination of the table of contents.

The equations, figures and examples presented are numbered using the notation of chapter number followed by a dash and then the number associated with the equation, figure or example. There are numerous worked examples and figures throughout the text. The example problems are terminated with the ■ symbol.

J.H. Heinbockel, 2006

Table of Contents

Introduction to the
Variational Calculus

Chapter 1   Preliminary Concepts 1

Functions and derivatives, Rolle's theorem, Mean value theorem, Higher ordered derivatives, Curves in space, Curvilinear coordinates, Integration, First mean value theorem for integrals, Line integrals, Simple closed curves, Green's theorem in the plane, Stokes theorem, Gauss divergence theorem, Differentiation of composite functions, Parametric equations, Function defined by an integral, Implicit functions, Implicit function theorem, System of two equations with four unknowns, System of three equations with five unknowns, Chain rule differentiation for functions of more than one variable, Derivatives of implicit functions, One equation in many variables, System of equations, System of three equations with six unknowns, Transformations, Directional derivatives, Euler's theorem for homogeneous functions, Taylor series, Linear differential equations, Constant coefficients

Exercises 47

Chapter 2   Maxima and Minima 55

Functions of a single real variable, Tests for maximum and minimum values, First derivative test, Second derivative test, Further investigation of critical points, Functions of two variables, Analysis of second directional derivative, Generalization, Derivative test for functions of two variables, Derivative test for functions of three variables, Lagrange multipliers, Generalization of Lagrange multipliers, Mathematical programming, Linear programming, Maxwell Boltzmann distribution

Exercises 87

Chapter 3   Introduction to the Calculus of Variations 95

Functionals, Basic lemma used in the calculus of variations, Notation, General approach,

[f1]: Integrand $\mathbf{f(x, y, y')}$, Invariance under a change of variables,

Parametric representation, The variational notation δ, Other functionals,

[f2]: Integrand $\mathbf{f(x, y, y', y'')}$,

[f3]: Integrand $\mathbf{f(x, y, y', y'', y''', \ldots, y^{(n)})}$,

[f4]: Integrand $\mathbf{f\left(x, y_1, y_1', y_1'', \ldots, y_1^{(n_1)}, y_2, y_2', y_2'', \ldots, y_2^{(n_2)}, \ldots, y_m, y_m', y_m'', \ldots, y_m^{(n_m)}\right)}$,

[f5]: Integrand $\mathbf{f(t, y_1(t), y_2(t), \dot{y}_1(t), \dot{y}_2(t))}$,

[f6]: Integrand $\mathbf{f(t, y_1, y_2, \ldots, y_n, \dot{y}_1, \dot{y}_2, \ldots, \dot{y}_n)}$,

[f7]: Integrand $\mathbf{F\left(x, y, w, \dfrac{\partial w}{\partial x}, \dfrac{\partial w}{\partial y}\right)}$,

[f8]: Integrand $\mathbf{F\left(u, v, x, y, z, \dfrac{\partial x}{\partial u}, \dfrac{\partial x}{\partial v}, \dfrac{\partial y}{\partial u}, \dfrac{\partial y}{\partial v}, \dfrac{\partial z}{\partial u}, \dfrac{\partial z}{\partial v}\right)}$,

[f9]: Integrand $\mathbf{F\left(x, y, z, w, \dfrac{\partial w}{\partial x}, \dfrac{\partial w}{\partial y}, \dfrac{\partial w}{\partial z}\right)}$,

[f10]: Integrand $\mathbf{F(x_1, x_2, \ldots, x_n, w, w_{x_1}, w_{x_2}, \ldots, w_{x_n})}$,

[f11]: Integrand $\mathbf{F(t, x, y, z, u, v, w, u_t, u_x, u_y, u_z, v_t, v_x, v_y, v_z, w_t, w_x, w_y, w_z)}$,

[f12]: Integrand $\mathbf{F\left(x, y, z, \dfrac{\partial z}{\partial x}, \dfrac{\partial z}{\partial y}, \dfrac{\partial^2 z}{\partial x^2}, \dfrac{\partial^2 z}{\partial x \partial y}, \dfrac{\partial^2 z}{\partial y^2}\right)}$

[f13]: Integrand $\mathbf{F\left(x, y, u, v, \dfrac{\partial u}{\partial x}, \dfrac{\partial u}{\partial y}, \dfrac{\partial v}{\partial x}, \dfrac{\partial v}{\partial y}\right)}$,

Differential constraint conditions, Weierstrass criticism

Exercises 131

Chapter 4   Additional Variational Concepts 137

Natural boundary conditions, Natural boundary conditions for other functionals,

More natural boundary conditions, Tests for maxima and minima, The Legendre and

Jacobi analysis, Background material for the Jacobi differential equation,

A more general functional, General variation, Movable boundaries, End points on two

different curves, Free end points, Functional containing end points,

Broken extremal (weak variations), Weierstrass E-function, Euler-Lagrange equation with

constraint conditions, Geodesic curves, Isoperimetric problems, Generalization, Modifying

natural boundary conditions

Exercises 187

Table of Contents

Chapter 5   Applications of the Variational Calculus 195

The brachistrochrone problem, The hanging, chain, rope or cable,
Soap film problem, Hamilton's principle and Euler-Lagrange equations,
Generalization, Nonconservative systems, Lagrangian for spring-mass system,
Hamilton's equation of motion, Inverse square law of attraction, spring-mass
system, Pendulum systems, Navier's equations, Elastic beam theory

Exercises 231

Chapter 6   Additional Applications 243

The vibrating string, Other variational problems similar to the vibrating string,
The vibrating membrane, Generalized brachistrochrone problem, Holonomic
and nonholonomic systems, Eigenvalues and eigenfunctions, Alternate form
for Sturm-Liouville system, Schrödinger equation, Solutions of Schrödinger equation for
the hydrogen atom, Maxwell's equations, Numerical methods, Search techniques,
scaling, Rayleigh-Ritz method for functionals, Galerkin method, Rayleigh-Ritz, Galerkin
and collocation methods for differential equations, Rayleigh-Ritz method and B-splines,
B-splines, Introduction to the finite element method.

Exercises 301

Bibliography 310
Appendix A Units of Measurement 312
Appendix B Gradient, Divergence and Curl 314
Appendix C Solutions to Selected Exercises 318
Index 348

Chapter 1
Preliminary Concepts

The prerequisite mathematical background for the material presented in this text are basic concepts from the subject areas of calculus, advanced calculus, differential equations and partial differential equations. The material selected for presentation in this first chapter is a combination of background preliminaries for review purposes together with an introduction of new concepts associated with the background material. We begin by examining selected fundamentals from the subject area of calculus.

Functions and derivatives

A real function f of a single real variable x is a rule which defines a one-to-one correspondence between elements x in a set A and image elements y in a set B. A real function f of a single variable x can be represented using the notation $y = f(x)$. The set A is called the domain of the function f and elements $x \in A$ are real numbers for which the image element $y = f(x) \in B$ is also a real quantity.

The set of image elements $y \in B$, as x varies over all elements in the domain, constitutes the range of the function f. An element x belonging to the domain of the function is called an independent variable and the corresponding image element y, in the range of f, is called a dependent variable. Functions can be represented graphically by plotting a set of points (x, y) on a Cartesian set of axes. A function $f(x)$ is continuous[‡] at a point x_0 if (i) $f(x_0)$ is defined, and (ii) $\lim_{x \to x_0} f(x)$ exists and (iii) $\lim_{x \to x_0} f(x) = f(x_0)$. A function is called discontinuous at a point if it is not continuous at the point.

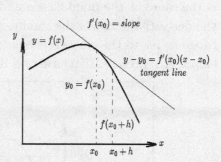

Let $y = f(x)$ represent a function of a single real variable x. The derivative of this function, evaluated at a point x_0 is denoted using the notation $\frac{dy}{dx}\big|_{x=x_0} = f'(x_0)$ and is defined by the limiting process

$$\frac{dy}{dx}\bigg|_{x=x_0} = f'(x_0) = \lim_{h \to 0} \frac{f(x_0 + h) - f(x_0)}{h} \tag{1.1}$$

if this limit exists. The derivative at a point x_0 can also be written as the limiting process

$$\frac{dy}{dx}\bigg|_{x=x_0} = f'(x_0) = \lim_{x \to x_0} \frac{f(x) - f(x_0)}{x - x_0} \tag{1.2}$$

by making the substitution $h = x - x_0$ in the equation (1.1). The derivative $\frac{dy}{dx}$ evaluated at a point x_0 denotes the slope of the tangent line to the curve at the point $(x_0, f(x_0))$ on the

---

[‡] A more formal definition of continuity is that $f(x)$ is continuous for $x = x_0$ if for every small $\epsilon > 0$ there exists a $\delta > 0$ such that $|f(x) - f(x_0)| < \epsilon$, whenever $|x - x_0| < \delta$. A function is called continuous over an interval if it is continuous at all points within the interval.

curve $y = f(x)$. The equation of this tangent line is $y - y_0 = f'(x_0)(x - x_0)$. Higher ordered derivatives are defined as derivatives of lower ordered derivatives. For example, the second derivative $\frac{d^2 y}{dx^2}$ is defined as the derivative of a first derivative or $\frac{d^2 y}{dx^2} = \frac{d}{dx}\left(\frac{dy}{dx}\right)$.

Note that primes ' are used to denote differentiation with respect to the argument of the function. When the number of primes becomes too large we switch to an index in parenthesizes. For example, if $y = y(x)$ denotes a function which has derivatives through the nth order, then these derivatives are denoted

$$\frac{dy}{dx} = y', \quad \frac{d^2 y}{dx^2} = y'', \quad \frac{d^3 y}{dx^3} = y''', \quad \frac{d^4 y}{dx^4} = y^{(4)}, \quad \dots, \quad \frac{d^n y}{dx^n} = y^{(n)} \tag{1.3}$$

A δ-neighborhood of a point x_0 on the x-axis is defined as the set of points x satisfying $\{x \mid |x - x_0| < \delta\}$. This δ-neighborhood can also be written as the set of points x satisfying the inequality $x_0 - \delta < x < x_0 + \delta$.

Rolle's theorem

Let $y = f(x)$ denote a function of a single real variable x, which is continuous and differentiable at all points within an interval $a \leq x \leq b$. The Rolle's theorem, which is associated with continuous functions with continuous derivatives, states that if the height of the curve is the same at the points $x = a$ and $x = b$, then there exists at least one point $\xi \in (a, b)$ where the derivative satisfies the condition $f'(\xi) = 0$. This can be interpreted geometrically that the tangent line to the curve represented by $y = f(x)$, at the point $(\xi, f(\xi))$, is horizontal with zero slope. A representative situation illustrating the Rolle's theorem is given in the figure 1-1.

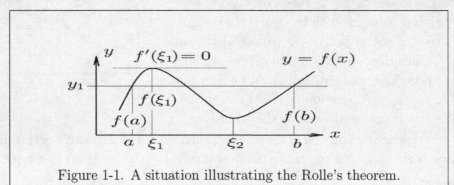

Figure 1-1. A situation illustrating the Rolle's theorem.

Mean value theorem

The mean value theorem is associated with functions $y = f(x)$ which are continuous and differentiable at all points within an interval $a \leq x \leq b$. Recall that a line passing through two different points $(a, f(a))$ and $(b, f(b))$ on a given curve is called a secant line. The mean value theorem states that if a function is continuous, then there exists at least one point $c \in (a, b)$ where the slope of the curve $y = f(x)$ at $x = c$ is the same as the slope of the secant line through the end points $(a, f(a))$ and $(b, f(b))$. The mean value theorem can be written $f'(c) = \frac{f(b) - f(a)}{b - a}$ for some value c satisfying $a < c < b$. A representative situation illustrating the mean value theorem is given in the figure 1-2.

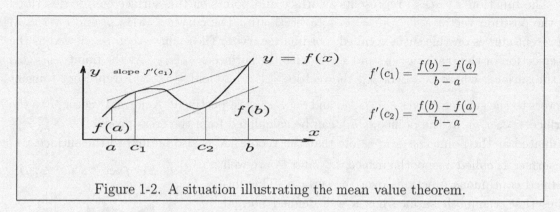

Figure 1-2. A situation illustrating the mean value theorem.

A function f of two real variables x and y is a rule which assigns a one-to-one correspondence between an ordered pair of real numbers (x, y) from a set A and an image element z in a set B. The image element is denoted by the notation $z = f(x, y)$ where (x, y) represents a point in a domain A and z represents a point in the range B. Here x and y are real quantities and f represents a rule for producing a real image point z. The quantities x and y are called the independent variables of the function and z is called the dependent variable of the function.

A δ-neighborhood of a point (x_0, y_0), lying in the x, y-plane, is defined as the set of points (x, y) inside a small circle with center (x_0, y_0) and radius δ. All points (x, y) lying in a δ-neighborhood of the point (x_0, y_0) must satisfy the inequality

$$(x - x_0)^2 + (y - y_0)^2 < \delta^2. \tag{1.4}$$

A function $f(x, y)$ is continuous[‡] at a point (x_0, y_0) if (i) $f(x_0, y_0)$ exists, (ii) $\lim_{\substack{x \to x_0 \\ y \to y_0}} f(x, y)$ exists, and (iii) $\lim_{\substack{x \to x_0 \\ y \to y_0}} f(x, y) = f(x_0, y_0)$ and this limit must exist and be independent of the method that $x \to x_0$ and $y \to y_0$. If $f(x, y)$ is continuous for all points in a region R, then $f(x, y)$ is said to be continuous over the region R. A set of points R is called an open set, if every point $(x, y) \in R$ has some δ-neighborhood which lies entirely within the set R. If, in addition, the set R has the property that any two arbitrary distinct points $(x_0, y_0) \in R$ and $(x_1, y_1) \in R$ can be joined by connected line segments, which lie within the region R, then the set R is called a connected open set or a domain. The boundary of the region R is denoted ∂R and represents a set of points with the following property. A point $(x, y) \in \partial R$ is called a boundary point of the region R if every δ-neighborhood of the point (x, y) contains at least one point within the region R and at least one point not in the region R. A closed region R is one that contains its boundary points.

The partial derivatives of a function $z = z(x, y)$ with respect to x and y are defined

$$\frac{\partial z}{\partial x} = \lim_{\Delta x \to 0} \frac{z(x + \Delta x, y) - z(x, y)}{\Delta x} \quad \text{and} \quad \frac{\partial z}{\partial y} = \lim_{\Delta y \to 0} \frac{z(x, y + \Delta y) - z(x, y)}{\Delta y} \tag{1.5}$$

if these limits exist.

[‡] A more formal definition of continuity is that $f(x, y)$ is continuous at a point (x_0, y_0) if (i) it is defined at this point and (ii) if for every $\epsilon > 0$ there exists a $\delta > 0$ such that for all points (x, y) in a δ-neighborhood of (x_0, y_0) the relation $|f(x, y) - f(x_0, y_0)| < \epsilon$ is satisfied.

The function $z = z(x, y)$ represents a surface and points on this surface can be described by the position vector $\vec{r} = \vec{r}(x, y) = x\,\widehat{\mathbf{e}}_1 + y\,\widehat{\mathbf{e}}_2 + z(x, y)\,\widehat{\mathbf{e}}_3$. The curves $\vec{r} = \vec{r}(x_0, y)$ and $\vec{r} = \vec{r}(x, y_0)$ represent curves on this surface called coordinate curves. These curves can be viewed as the intersection of the planes $y = y_0$ and $x = x_0$ with the surface $z = z(x, y)$. At the point (x_0, y_0, z_0) of the surface, where $z_0 = z(x_0, y_0)$, the vectors $\left.\frac{\partial \vec{r}}{\partial x}\right|_{(x_0, y_0, z_0)}$ and $\left.\frac{\partial \vec{r}}{\partial y}\right|_{(x_0, y_0, z_0)}$ represent tangent vectors to the surface curves $\vec{r} = \vec{r}(x, y_0)$ and $\vec{r} = \vec{r}(x_0, y)$ respectively. A normal vector $\vec{\mathcal{N}}$ to the surface $z = z(x, y)$, at the point (x_0, y_0), can be calculated from the cross product $\vec{\mathcal{N}} = \frac{\partial \vec{r}}{\partial x} \times \frac{\partial \vec{r}}{\partial y}$ evaluated at the point (x_0, y_0, z_0). Note that the vector $-\vec{\mathcal{N}}$ is also normal to the surface.

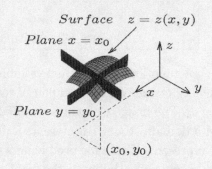

A surface is called a smooth surface if $\frac{\partial z}{\partial x}$ and $\frac{\partial z}{\partial y}$ are well defined continuous functions over the surface. This implies that a smooth surface has a well defined normal at each point on the surface. A sketch of the planes $x = x_0$ and $y = y_0$ intersecting the surface $z = z(x, y)$ shows the coordinate surface curves. The partial derivatives $\frac{\partial z}{\partial x}$ and $\frac{\partial z}{\partial y}$, evaluated at the point of intersection of the coordinate curves, represent the slopes of these coordinate curves at the point of intersection.

The locus of points (x, y) which satisfy $z(x, y) = constant$ produces a curve on the surface where z has a constant level and so such curves are called level curves. By selecting various constants $c_1, c_2, c_3, \ldots$ one can construct level curves $z(x, y) = c_i$ for $i = 1, 2, \ldots, n$. These level curves represent an intersection of the surface $z = z(x, y)$ with the planes $z = c_i = constant$, for $i = 1, 2, \ldots$. There exists many graphical packages for computer usage which will graph these level curves. Some of the more advanced graphical packages make it possible to sketch both the given surface and the level curves associated with the surface. These packages can be extremely helpful in illustrating the surface represented by a given function. A representative sketch of a set of level curves or contour plot associated with a specific surface is illustrated in the figure 1-3.

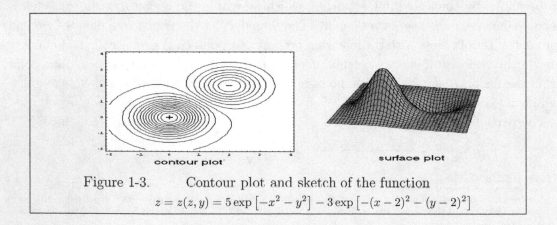

Figure 1-3.   Contour plot and sketch of the function
$$z = z(z, y) = 5 \exp\left[-x^2 - y^2\right] - 3\exp\left[-(x-2)^2 - (y-2)^2\right]$$

Higher ordered derivatives

Higher ordered partial derivatives are defined as derivatives of derivatives. For example, if the first ordered partial derivatives $\frac{\partial z}{\partial x}$ and $\frac{\partial z}{\partial y}$ are differentiated with respect to x and y there results four possible second ordered derivatives. These four possibilities are

$$\frac{\partial}{\partial x}\left(\frac{\partial z}{\partial x}\right) = \frac{\partial^2 z}{\partial x^2}, \quad \frac{\partial}{\partial x}\left(\frac{\partial z}{\partial y}\right) = \frac{\partial^2 z}{\partial x \partial y}, \quad \frac{\partial}{\partial y}\left(\frac{\partial z}{\partial x}\right) = \frac{\partial^2 z}{\partial y \partial x}, \quad \frac{\partial}{\partial y}\left(\frac{\partial z}{\partial y}\right) = \frac{\partial^2 z}{\partial y^2}$$

Here the notation $\frac{\partial^2 z}{\partial x \partial y}$ denotes first a differentiation with respect to x which is to be followed by a differentiation with respect to y. In the case that all the derivatives are continuous over a common domain of definition, then the mixed partial derivatives are equal so that

$$\frac{\partial^2 z}{\partial x \partial y} = \frac{\partial^2 z}{\partial y \partial x}. \tag{1.6}$$

This result is known as Clairaut's theorem. Higher ordered mixed derivatives have a similar property. For example, the third partial derivatives of the function z can be denoted

$$\frac{\partial^3 z}{\partial x^3} = \frac{\partial}{\partial x}\left(\frac{\partial^2 z}{\partial x^2}\right), \quad \frac{\partial^3 z}{\partial y^2 \partial x} = \frac{\partial}{\partial y}\left(\frac{\partial^2 z}{\partial y \partial x}\right), \quad \text{etc.} \tag{1.7}$$

Using Clairaut's theorem one can say that if all the mixed third partial derivatives are defined and continuous over a common domain of definition then it can be shown that

$$\frac{\partial^3 z}{\partial x \partial y^2} = \frac{\partial^3 z}{\partial y \partial x \partial y} = \frac{\partial^3 z}{\partial y^2 \partial x} \quad \text{and} \quad \frac{\partial^3 z}{\partial y \partial x^2} = \frac{\partial^3 z}{\partial x \partial y \partial x} = \frac{\partial^3 z}{\partial x^2 \partial y}$$

with similar results applying to mixed higher derivatives. The above concepts can be generalized to functions of n-real variables.

Partial derivatives are often times represented using a subscript notation. For example, if $z = z(x, y)$, then the first and second partial derivatives can be represented by the following subscript notation

$$z_x = \frac{\partial z}{\partial x}, \quad z_y = \frac{\partial z}{\partial y}, \quad z_{xx} = \frac{\partial^2 z}{\partial x^2}, \quad z_{xy} = \frac{\partial^2 z}{\partial x \partial y}, \quad z_{yy} = \frac{\partial^2 z}{\partial y^2} \tag{1.8}$$

As another example, if $F = F(x, y, w, \frac{\partial w}{\partial x}, \frac{\partial w}{\partial y}) = F(x, y, w, w_x, w_y)$, and we treat each variable as being independent, then the first partial derivatives of F can be represented

$$F_x = \frac{\partial F}{\partial x}, \quad F_y = \frac{\partial F}{\partial y}, \quad F_w = \frac{\partial F}{\partial w}, \quad F_{w_x} = \frac{\partial F}{\partial w_x}, \quad F_{w_y} = \frac{\partial F}{\partial w_y} \tag{1.9}$$

Curves in space

A set of parametric equations of the form

$$x = x(t), \quad y = y(t), \quad z = z(t) \quad \text{with parameter } t \tag{1.10}$$

defines a space curve C. The position vector to a general point on the curve C is written as

$$\vec{r} = \vec{r}(t) = x(t)\,\widehat{\mathbf{e}}_1 + y(t)\,\widehat{\mathbf{e}}_2 + z(t)\,\widehat{\mathbf{e}}_3. \tag{1.11}$$

where $\widehat{e}_1$, $\widehat{e}_2$, $\widehat{e}_3$ are unit basis vectors in the directions of the x, y and z axes respectively. These basis vectors are sometimes written as the $\hat{i}, \hat{j}, \hat{k}$ unit vectors in the directions of the x, y and z axes. Consider a point P_1 on the curve C which is defined by the parametric value $t = t_1$ and having the position vector defined by $\vec{r}_1 = \vec{r}(t_1)$. The tangent vector to the curve C at the point P_1 is given by

$$\frac{d\vec{r}}{dt} = \frac{dx}{dt}\widehat{e}_1 + \frac{dy}{dt}\widehat{e}_2 + \frac{dz}{dt}\widehat{e}_3 \bigg|_{t=t_1} = x'(t_1)\widehat{e}_1 + y'(t_1)\widehat{e}_2 + z'(t_1)\widehat{e}_3 \qquad (1.12)$$

Let ds denote an element of arc length along the curve C. An element of arc length squared is written

$$ds^2 = d\vec{r} \cdot d\vec{r}, \quad \text{or} \quad |\frac{d\vec{r}}{dt}| = \frac{ds}{dt} = \sqrt{\frac{d\vec{r}}{dt} \cdot \frac{d\vec{r}}{dt}} \qquad (1.13)$$

Hence, one can write a unit tangent vector to a point on the curve C from the relation

$$\vec{T} = \frac{\frac{d\vec{r}}{dt}}{|\frac{d\vec{r}}{dt}|} = \frac{\frac{d\vec{r}}{dt}}{\frac{ds}{dt}} = \frac{d\vec{r}}{ds} \qquad (1.14)$$

where s denotes arc length along the curve. When equation (1.14) is evaluated at the parametric value $t = t_1$ one obtains the unit tangent vector $\vec{T}$ to the point P_1.

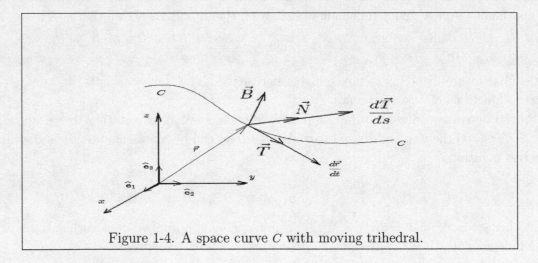

Figure 1-4. A space curve C with moving trihedral.

The rate of change of the unit tangent vector $\vec{T}$ with respect to arc length s along the curve C gives a vector which is normal to the unit tangent vector. The reason for this is that at any point on the curve C the unit tangent vector $\vec{T}$ satisfies $\vec{T} \cdot \vec{T} = 1$ and, consequently, if one differentiates this relation with respect to arc length s, there results

$$\frac{d}{ds}\left(\vec{T} \cdot \vec{T}\right) = \frac{d}{ds}(1) \Rightarrow \vec{T} \cdot \frac{d\vec{T}}{dt} + \frac{d\vec{T}}{ds} \cdot \vec{T} = 2\vec{T} \cdot \frac{d\vec{T}}{ds} = 0, \quad \text{or} \quad \vec{T} \cdot \frac{d\vec{T}}{ds} = 0$$

This last equation tells us that $\frac{d\vec{T}}{ds}$ is perpendicular to $\vec{T}$ since their dot product is zero. From the infinite number of vectors perpendicular to $\vec{T}$, the unit vector in the direction $\frac{d\vec{T}}{ds}$ is given the special name of principal unit normal to the curve C at the point P_1, and this principal unit normal is denoted by the symbol $\vec{N}$. At a point on the curve C, where the unit tangent is constructed, one can write

$$\frac{d\vec{T}}{ds} = \kappa\vec{N}, \qquad \vec{N} \cdot \vec{N} = 1, \qquad \frac{d\vec{T}}{ds} \cdot \frac{d\vec{T}}{ds} = \kappa^2 \qquad (1.15)$$

where κ is a scalar called the curvature of the curve C at a specific point where $\vec{T}$ is constructed. The quantity $\rho = 1/\kappa$ is called the radius of curvature of the curve C at this point. The unit vector $\vec{B}$ defined by $\vec{B} = \vec{T} \times \vec{N}$ is called the binormal vector to the curve C and is perpendicular to both $\vec{T}$ and $\vec{N}$. The three unit vectors $\vec{T}$, $\vec{N}$, $\vec{B}$ form a right-handed localized coordinate system at each point along the curve C and is often called a moving trihedral as the arc length s changes. A nominal situation is illustrated in the figure 1-4.

The vectors $\vec{T}$, $\vec{N}$, $\vec{B}$ satisfy the Frenet-Serret formulas

$$\frac{d\vec{T}}{ds} = \kappa\vec{N}, \qquad \frac{d\vec{N}}{ds} = \tau\vec{B} - \kappa\vec{T}, \qquad \frac{d\vec{B}}{ds} = -\tau\vec{N} \qquad (1.16)$$

where τ is a scalar called the torsion and the quantity $\sigma = 1/\tau$ is called the radius of torsion. The torsion τ is a measure of the twisting of a space curve out of a plane. If $\tau = 0$, then the curve C is a plane curve. The curvature κ and radius of curvature ρ measure the turning of the curve C in relation to a localized circle which just touches the curve C at a point. If the curvature $\kappa = 0$, then the curve C is a straight line.

The plane containing the unit tangent vector $\vec{T}$ and the principal normal vector $\vec{N}$ is called the osculating plane. The plane containing the unit vectors $\vec{B}$ and $\vec{N}$ is perpendicular to the unit tangent vector and is called the normal plane to the curve. The plane containing the unit vectors $\vec{B}$ and $\vec{T}$ which is perpendicular to the unit principal normal vector is called the rectifying plane.

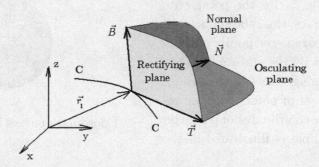

It is an easy exercise to show that the equation of the tangent line to the space curve C at the point P_1 is given by $(\vec{r} - \vec{r}_1) \times \vec{T} = \vec{0}$. Another easy exercise is to show the equations of the osculating, normal and rectifying planes constructed at the point P_1 are given by

Osculating plane	Normal plane	Rectifying plane	
$(\vec{r} - \vec{r}_1) \cdot \vec{B} = 0$	$(\vec{r} - \vec{r}_1) \cdot \vec{T} = 0$	$(\vec{r} - \vec{r}_1) \cdot \vec{N} = 0$	(1.17)

where $\vec{r} = x\,\widehat{e}_1 + y\,\widehat{e}_2 + z\,\widehat{e}_3$ is the position vector to a variable point on the line or plane being constructed.

Curvilinear coordinates

Parametric equations of the form

$$x = x(u,v), \qquad y = y(u,v), \qquad z = z(u,v), \tag{1.18}$$

which involve two independent parameters u and v, are used to define a surface. If one can solve for u and v from two of the equations, then the results can be substituted into the third equation to obtain the surface in the form $F(x,y,z) = 0$. The special case where equations (1.18) have the form

$$x = u, \qquad y = v, \qquad z = z(u,v) \tag{1.19}$$

produces the surface in the form $z = z(x,y)$.

The position vector

$$\vec{r} = \vec{r}(u,v) = x(u,v)\,\widehat{\mathbf{e}}_1 + y(u,v)\,\widehat{\mathbf{e}}_2 + z(u,v)\,\widehat{\mathbf{e}}_3 \tag{1.20}$$

defines a general point on the surface. By setting $u = u_1 = constant$, one obtains a curve on the surface given by

$$\vec{r}(u_1,v) = x(u_1,v)\,\widehat{\mathbf{e}}_1 + y(u_1,v)\,\widehat{\mathbf{e}}_2 + z(u_1,v)\,\widehat{\mathbf{e}}_3, \quad v \text{ is parameter} \tag{1.21}$$

Similarly, if one sets the parameter $v = v_1 = constant$ one obtains the curves

$$\vec{r}(u,v_1) = x(u,v_1)\,\widehat{\mathbf{e}}_1 + y(u,v_1)\,\widehat{\mathbf{e}}_2 + z(u,v_1)\,\widehat{\mathbf{e}}_3, \quad u \text{ is parameter} \tag{1.22}$$

These curves are called coordinate curves.

If one selects equally spaced constant values $u_1, u_2, u_3, \ldots$ and $v_1, v_2, v_3, \ldots$ for the parametric values in the above curves, then the surface will be covered by a two parameter family of coordinate curves. A point on the surface can then characterized by assigning values to the parameters u and v and the set of points (u,v) are referred to as curvilinear coordinates of points on the surface. An example is illustrated in the accompanying figure.

Coordinate curves for unit sphere
$x = \sin u \cos v, \;\; y = \sin u \sin v, \;\; z = \cos u$

The partial derivatives $\frac{\partial \vec{r}}{\partial u}$ and $\frac{\partial \vec{r}}{\partial v}$ represent tangent vectors to the coordinate curves at a point (u,v) on the surface and consequently a normal vector $\vec{N}$ to the surface is given by

$$\vec{N} = \frac{\partial \vec{r}}{\partial u} \times \frac{\partial \vec{r}}{\partial v}$$

so that a unit normal vector $\widehat{n}$ to the surface is given by

$$\widehat{n} = \pm \frac{\vec{N}}{|\vec{N}|} = \pm \frac{\frac{\partial \vec{r}}{\partial u} \times \frac{\partial \vec{r}}{\partial v}}{|\,\frac{\partial \vec{r}}{\partial u} \times \frac{\partial \vec{r}}{\partial v}\,|} \tag{1.23}$$

Note that there are always two unit normals at any point on a surface. These unit normals are given by $\hat{n}$ and $-\hat{n}$. If the surface is a closed surface then there is an outward pointing and inward pointing unit normal at each point on the surface. Therefore, you must select which unit normal you want.

The differential $d\vec{r}$ of the position vector $\vec{r} = \vec{r}(u, v)$ is written

$$d\vec{r} = \frac{\partial \vec{r}}{\partial u}\, du + \frac{\partial \vec{r}}{\partial v}\, dv \tag{1.24}$$

so that the square of an element of arc length on the surface is given by

$$ds^2 = d\vec{r} \cdot d\vec{r} = \frac{\partial \vec{r}}{\partial u} \cdot \frac{\partial \vec{r}}{\partial u}\, du^2 + 2\frac{\partial \vec{r}}{\partial u} \cdot \frac{\partial \vec{r}}{\partial v}\, du\, dv + \frac{\partial \vec{r}}{\partial v} \cdot \frac{\partial \vec{r}}{\partial v}\, dv^2$$

This is often written in the form

$$ds^2 = E\, du^2 + 2F\, du\, dv + G\, dv^2 \tag{1.25}$$

where

$$E = \frac{\partial \vec{r}}{\partial u} \cdot \frac{\partial \vec{r}}{\partial u}, \qquad F = \frac{\partial \vec{r}}{\partial u} \cdot \frac{\partial \vec{r}}{\partial v}, \qquad G = \frac{\partial \vec{r}}{\partial v} \cdot \frac{\partial \vec{r}}{\partial v} \tag{1.26}$$

The differential $d\vec{r}$ defines a vector lying in the tangent plane to the surface and can be thought of as the diagonal of an elemental area parallelogram on the surface having sides $\frac{\partial \vec{r}}{\partial u}\, du$ and $\frac{\partial \vec{r}}{\partial v}\, dv$. The area of this elemental parallelogram is given by

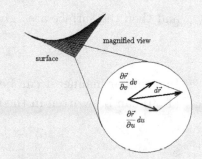

$$d\sigma = \left| \frac{\partial \vec{r}}{\partial u} \times \frac{\partial \vec{r}}{\partial v} \right|\, du\, dv = \sqrt{\left(\frac{\partial \vec{r}}{\partial u} \times \frac{\partial \vec{r}}{\partial v} \right) \cdot \left(\frac{\partial \vec{r}}{\partial u} \times \frac{\partial \vec{r}}{\partial v} \right)}\, du\, dv$$

One can employ the vector identity $(\vec{A} \times \vec{B}) \cdot (\vec{C} \times \vec{D}) = (\vec{A} \cdot \vec{C})(\vec{B} \cdot \vec{D}) - (\vec{A} \cdot \vec{D})(\vec{B} \cdot \vec{C})$ to represent the element of surface area in the form

$$d\sigma = \sqrt{EG - F^2}\, du\, dv \tag{1.27}$$

In the special case the surface is defined by the parametric equations having the form $x = u$, $y = v$, $z = z(u, v) = z(x, y)$, then the element of surface area reduces to

$$d\sigma = \sqrt{1 + \left(\frac{\partial z}{\partial x} \right)^2 + \left(\frac{\partial z}{\partial y} \right)^2}\, dx\, dy \tag{1.28}$$

Using this same idea but changing symbols around, a surface defined by the parametric equations $x = u$, $y = y(u, v) = y(x, z)$, $z = v$, has an element of surface area in the form

$$d\sigma = \sqrt{1 + \left(\frac{\partial y}{\partial x} \right)^2 + \left(\frac{\partial y}{\partial z} \right)^2}\, dz\, dx \tag{1.29}$$

Similarly, a surface defined by $x = x(u,v) = x(y,z)$, $y = u$, $z = v$, has the element of surface area

$$d\sigma = \sqrt{1 + \left(\frac{\partial x}{\partial y}\right)^2 + \left(\frac{\partial x}{\partial z}\right)^2} \, dy \, dz \tag{1.30}$$

If the surface is given in the implicit form $F(x,y,z) = 0$, then one can construct a normal vector to the surface from the equation $\vec{N} = \text{grad}\, F = \nabla F$, with unit normal vector $\hat{n} = \pm\frac{\nabla F}{|\nabla F|}$, because the gradient vector evaluated at a surface point is perpendicular to the surface defined by $F(x,y,z) = 0$. To show this, write the differential of F as follows

$$dF = \frac{\partial F}{\partial x}\, dx + \frac{\partial F}{\partial y}\, dy + \frac{\partial F}{\partial z}\, dz = \left(\frac{\partial F}{\partial x}\, \hat{e}_1 + \frac{\partial F}{\partial y}\, \hat{e}_2 + \frac{\partial F}{\partial z}\, \hat{e}_3\right) \cdot (dx\, \hat{e}_1 + dy\, \hat{e}_2 + dz\, \hat{e}_3) = \text{grad}\, F \cdot d\vec{r} = 0.$$

This shows that the vector $\text{grad}\, F$ is perpendicular to the vector $d\vec{r}$ which lies in the tangent plane to the surface, and so, must be perpendicular to the surface at the surface point of evaluation. In the special case $F = z(x,y) - z = 0$ defines the surface, then a unit normal vector to the surface is given by

$$\hat{n} = \frac{\frac{\partial z}{\partial x}\, \hat{e}_1 + \frac{\partial z}{\partial y}\, \hat{e}_2 - \hat{e}_3}{\sqrt{1 + \left(\frac{\partial z}{\partial x}\right)^2 + \left(\frac{\partial z}{\partial y}\right)^2}} \tag{1.31}$$

and then the surface area given by equation (1.28) can be written in the form

$$d\sigma = \frac{dx\, dy}{|\,\hat{n} \cdot \hat{e}_3\,|} \tag{1.32}$$

In a similar manner it can be shown that the surface elements given by equations (1.29) and (1.30) can be written in the alternative forms

$$d\sigma = \frac{dx\, dz}{|\,\hat{n} \cdot \hat{e}_2\,|} \qquad \text{and} \qquad d\sigma = \frac{dy\, dz}{|\,\hat{n} \cdot \hat{e}_1\,|} \tag{1.33}$$

where $\hat{n}$ is a unit normal to the surface. These equations have the physical interpretation of representing the projections of the surface element $d\sigma$ onto the x-y, x-z or y-z planes.

Integration

Integration is sometimes referred to as an anti-derivative. If you know a differentiation formula, then you can immediately obtain an integration formula. That is, if

$$\frac{dF(x)}{dx} = f(x), \quad \text{then} \quad \int f(x)\, dx = F(x) + C, \tag{1.34}$$

where C is a constant of integration. In the use of definite integrals the above relations are written

$$\text{if} \quad \frac{dF(x)}{dx} = f(x), \quad \text{then} \quad \int_a^b f(x)\, dx = F(x)\Big|_a^b = F(b) - F(a). \tag{1.35}$$

If we replace the upper limit of integration by a variable quantity x, then equation (1.35) can be written as

$$F(x) = F(a) + \int_a^x f(x)\, dx, \quad \text{with} \quad \frac{dF(x)}{dx} = f(x).$$

There are many instances in the study of calculus of variations where integration by parts is recommended. The integration by parts in one dimension can be written using either of the notations

$$\int_a^b U \, dV = UV \Big|_a^b - \int_a^b V \, dU$$

or $$\int_{x_1}^{x_2} g(x) \frac{df(x)}{dx} \, dx = g(x)f(x) \Big|_{x_1}^{x_2} - \int_{x_1}^{x_2} f(x) \frac{dg(x)}{dx} \, dx. \tag{1.36}$$

First mean value theorem for integrals

If (i) $f(x)$ is a continuous function over an interval $a \le x \le b$ and (ii) there exists constants m and M such that $m \le f(x) \le M$ for all $x \in [a, b]$ and (iii) $g(x)$ is an integrable function over the interval (a, b), then there exists at least one point ξ satisfying $a \le \xi \le b$ such that

$$\int_a^b f(x)g(x) \, dx = f(\xi) \int_a^b g(x) \, dx \tag{1.37}$$

Line integrals

Let $\vec{F}(x, y) = P(x, y) \, \hat{e}_1 + Q(x, y) \, \hat{e}_2$ denote a two-dimensional vector field where $\hat{e}_1$ and $\hat{e}_2$ are unit vectors in the directions of the x and y axis respectively. The work done (force times distance) in moving through this vector field from point (a_1, b_1) to point (a_2, b_2), along a curve C is given by a line integral. If the curve C is defined by the position vector $\vec{r} = x \, \hat{e}_1 + y \, \hat{e}_2$, then the work done is denoted by the line integral

$$W = \int_C \vec{F} \cdot d\vec{r} = \int_C \left(P(x, y) \, \hat{e}_1 + Q(x, y) \, \hat{e}_2 \right) \cdot \left(dx \, \hat{e}_1 + dy \, \hat{e}_2 \right) = \int_{(a_1, b_1)}^{(a_2, b_2)} P(x, y) \, dx + Q(x, y) \, dy. \tag{1.38}$$

If the curve C is defined using the notation $y = f(x)$, then the line integral given by equation (1.38) reduces to the ordinary integral

$$W = \int_{a_1}^{a_2} P(x, f(x)) \, dx + Q(x, f(x)) \, f'(x) \, dx. \tag{1.39}$$

If the curve C is defined by the parametric equations $x = x(t)$, $y = y(t)$ for $t_a \le t \le t_b$, then the line integral given by equation (1.38) becomes the ordinary integral

$$W = \int_{t_a}^{t_b} P(x(t), y(t)) \, x'(t) \, dt + Q(x(t), y(t)) \, y'(t) \, dt. \tag{1.40}$$

If the curve C is defined by an equation having the form $x = x(y)$, then the line integral given by equation (1.38) can be expressed as the ordinary integral.

$$W = \int_{b_1}^{b_2} P(x(y), y) \, x'(y) \, dy + Q(x(y), y) \, dy. \tag{1.41}$$

In each case above, you must substitute values of x and y from the curve C over which the integration is performed.

The above ideas carry over to higher dimensions. Define the three-dimensional vector field $\vec{F}(x, y, z) = F_1(x, y, z)\,\hat{e}_1 + F_2(x, y, z)\,\hat{e}_2 + F_3(x, y, z)\,\hat{e}_3$ and let C denote a curve defined by the position vector $\vec{r} = x\,\hat{e}_1 + y\,\hat{e}_2 + z\,\hat{e}_3$. The work done in moving from point (x_1, y_1, z_1) to the point (x_2, y_2, z_2), along a curve C joining these points, is denoted by the line integral

$$W = \int_C \vec{F} \cdot d\vec{r} = \int_{(x_1,y_1,z_1)}^{(x_2,y_2,z_2)} F_1(x, y, z)\,dx + F_2(x, y, z)\,dy + F_3(x, y, z)\,dz \qquad (1.42)$$

Whenever the space curve C is defined by a set of parametric equations having the form $C: \; x = x(t), \; y = y(t), \; z = z(t)$ for $t_1 \leq t \leq t_2$, then the line integral given by equation (1.42) reduces to the ordinary integral

$$W = \int_{t_1}^{t_2} F_1(x(t), y(t), z(t))\,x'(t)\,dt + F_2(x(t), y(t), z(t))\,y'(t)\,dt + F_3(x(t), y(t), z(t))\,z'(t)\,dt.$$

Consider a function $f = f(x, y, z)$ which is defined and continuous along a curve C. The curve C is assumed to be a continuous curve so that it is possible to partition the curve C into N elements of length Δs_i between two distinct points on the curve. One can then select arbitrary points (x_i, y_i, z_i) lying on the curve and within the arc length segment Δs_i. The line integral of a function $f = f(x, y, z)$ with respect to arc length s along a curve C between two points on the curve C is then defined

$$\int_C f(x, y, z)\,ds = \lim_{\substack{N \to \infty \\ \Delta s_i \to 0}} \sum_{i=1}^{N} f(x_i, y_i, z_i)\,\Delta s_i \qquad (1.43)$$

if this limit exists. The limits in equation (1.43) must be independent of how the subdivisions Δs_i are construct. The limits must also be independent of where the points (x_i, y_i, z_i) are selected within a subdivision.

In general, integrals of the form of equations (1.38), (1.42), or (1.43) where all quantities are evaluated along a curve C are called line integrals. For example, to evaluate the integral (1.43) along a given curve C, we must know the parametric representation of the curve C. Assume that you can construct a parametric representation $x = x(t), \; y = y(t), \; z = z(t)$ of the curve C, then from this representation one can calculate the element of arc length ds from the relation

$$\left(\frac{ds}{dt}\right)^2 = \frac{d\vec{r}}{dt} \cdot \frac{d\vec{r}}{dt} = \left(\frac{dx}{dt}\right)^2 + \left(\frac{dy}{dt}\right)^2 + \left(\frac{dz}{dt}\right)^2$$

and express the line integral (1.43) in the form of an ordinary integral

$$\int_C f(x, y, z)\,ds = \int_{t_1}^{t_2} f(x(t), y(t), z(t)) \sqrt{\left(\frac{dx}{dt}\right)^2 + \left(\frac{dy}{dt}\right)^2 + \left(\frac{dz}{dt}\right)^2}\,dt \qquad (1.44)$$

If the direction of integration along a curve C is reversed, then there is a change in sign of the line integral. For example, the line integral $\int_{-C} \vec{F} \cdot d\vec{r} = -\int_C \vec{F} \cdot d\vec{r}$ if there is a direction change in the integration. Whenever the curve C can be broken up into a union of a finite set of curves $C_1, C_2, \ldots, C_n$, then the line integral can be written

$$\int_C \vec{F} \cdot d\vec{r} = \int_{C_1} \vec{F} \cdot d\vec{r} + \int_{C_2} \vec{F} \cdot d\vec{r} + \cdots + \int_{C_n} \vec{F} \cdot d\vec{r}.$$

Simple closed curves

A plane curve C in the x, y-plane can be defined by a set of parametric equations $\{x(t), y(t)\}$ for $t_1 \leq t \leq t_2$. The curve C is called a closed curve if the end points coincide. In such a case the end conditions satisfy $(x(t_1), y(t_1)) = (x(t_2), y(t_2))$. If (x_0, y_0) is a point on the curve C, which is not an end point, and there exists more than one value of the parameter t such that $(x(t), y(t)) = (x_0, y_0)$, then the point (x_0, y_0) is called a multiple point or point where the curve C crosses itself. A curve C is called a simple closed curve if the end points meet and it has no multiple points. A similar definition applies to curves in higher dimensional spaces. If C is a simple closed curve, then the line integral representing the work done around a simple closed curve is denoted using the notation

$$W = \oint_C \vec{F} \cdot d\vec{r}. \tag{1.45}$$

where the arrow indicates the direction taken around the closed curve.

In two-dimensions, curves are referred to as plane curves. A region R in the plane is called a simply connected region if it has the property that any simple closed plane curve within the region can be continuously shrunk to a point without leaving the region. Regions which are not simply-connected are called multiply-connected.

Green's theorem in the plane

Consider the two-dimensional vector field defined by

$$\vec{F}(x, y) = P(x, y)\, \widehat{e}_1 + Q(x, y)\, \widehat{e}_2$$

where the components P, Q and their derivatives $\frac{\partial P}{\partial y}$, $\frac{\partial Q}{\partial x}$ are single-valued and continuous in a simply-connected region R which is bounded by a simple closed plane curve C. The Green's theorem in the plane can be written

$$\oint_C \vec{F} \cdot d\vec{r} = \oint_C P(x, y)\, dx + Q(x, y)\, dy = \iint_R \left(\frac{\partial Q}{\partial x} - \frac{\partial P}{\partial y} \right) dxdy \tag{1.46}$$

where the line integral around the curve C is taken in the positive sense.[†] The Green's theorem in the plane states that a line integral around a simple closed plane curve C can be replaced by an area integral over the region R which is enclosed by the boundary curve C. The notation ∂R is often used to denote the boundary curve C which encloses the region R.

Example 1-1.   Area integral

Consider Green's theorem in the plane in the special case where the vector field is defined $\vec{F} = -y\, \widehat{e}_1 + x\, \widehat{e}_2$. For this special vector field verify that the Green's theorem integral relation becomes

$$\oint_C x\, dy - y\, dx = \iint_R \left(\frac{\partial(x)}{\partial x} - \frac{\partial(-y)}{\partial y} \right) dxdy = 2 \iint_R dxdy = 2A.$$

[†] A line integral taken in the positive sense can be described as follows. Imagine walking around the boundary curve C with your head in the direction of the unit normal to the plane or surface in which the curve lies. If the area enclosed by the curve C is always on your left, then the direction of your path is referred to as being taken in the positive sense.

This special result tells us that the area A inside the simple closed curve $C = \partial R$ is given by

$$\iint_R dx\,dy = A = \frac{1}{2}\oint_C x\,dy - y\,dx. \tag{1.47}$$

For example, by knowing values of (x, y) on the boundary of the region R, it is possible to calculate the area inside the region R. This is the theory behind the mechanical engineering device known as a planimeter. You calibrate a planimeter by moving a pointer around the perimeter of a known area A_0, say a known area on a map, and then recording the distance traveled which is obtained from a measuring device on the planimeter. After resetting the planimeter, the pointer is moved around an enclosed area A, on the same map, and then recording the new distance traveled. One can then set up the ratio

$$\frac{\text{distance around known area}}{A_0} = \frac{\text{distance around enclosed area}}{A}$$

and from this ratio solve for the area A.

As an example, let $\{C: \ x = r\cos\theta, \ y = r\sin\theta, \ \text{for} \ 0 \le \theta \le 2\pi\}$, denote the values of (x, y) on the boundary of a circle with radius r. For r constant, we calculate $dx = -r\sin\theta\,d\theta$ and $dy = r\cos\theta\,d\theta$ and substitute x, y, dx, dy into the line integral (1.47) to obtain

$$A = \frac{1}{2}\int_0^{2\pi}(r\cos\theta)(r\cos\theta\,d\theta) - (r\sin\theta)(-r\sin\theta\,d\theta) = \frac{1}{2}r^2\int_0^{2\pi}d\theta = \pi r^2.$$

∎

Stokes' theorem

Let C denote a simple closed curve on a surface which encloses a surface area S and let

$$\vec{F} = \vec{F}(x, y, z) = F_1(x, y, z)\,\widehat{\mathbf{e}}_1 + F_2(x, y, z)\,\widehat{\mathbf{e}}_2 + F_3(x, y, z)\,\widehat{\mathbf{e}}_3$$

denote a vector field which is continuous and differentiable. The Stokes' theorem can be written

$$\iint_S (\nabla \times \vec{F}) \cdot \widehat{n}\,d\sigma = \oint_C \vec{F} \cdot d\vec{r} \tag{1.48}$$

where $d\sigma$ is an element of surface area on the surface, $\widehat{n}$ is a unit normal to the surface, and $\vec{r}$ is a position vector describing the curve C on the surface and the line integral around C is traversed in the positive sense. The element of surface area $d\sigma$ is to be summed over the surface area S inside the closed curve C lying on the surface. Note that there are always two unit normals $\widehat{n}$ and $-\widehat{n}$ to a smooth open surface and you must select one of them. The Stokes' theorem states that a summation of the normal component of the curl of $\vec{F}$ over the surface S is equivalent to a line integral of the tangential component of $\vec{F}$ summed around the simple closed curve C which is the boundary of the surface S. Observe that the Green's theorem in the plane is a special case of the Stokes' theorem.

Gauss divergence theorem

Let S denote a given surface area on an open surface in the vicinity of a vector field

$$\vec{F}(x,y,z) = F_1(x,y,z)\,\widehat{\mathbf{e}}_1 + F_2(x,y,z)\,\widehat{\mathbf{e}}_2 + F_3(x,y,z)\,\widehat{\mathbf{e}}_3$$

which is continuous with continuous derivatives. Denote by $\widehat{n}$ a unit normal to the surface, with $d\sigma$ denoting an element of surface area. The surface integral $\iint_S \vec{F} \cdot \widehat{n}\,d\sigma$ is called the net flux crossing the surface area S. This surface integral represents the normal components of the vector field $\vec{F}$ summed over the portion S of the surface placed in the vicinity of the vector field. In the special case the surface S encloses a volume V, then the total flux across the closed surface is written $\oiint_S \vec{F} \cdot \widehat{n}\,d\sigma$ where $\widehat{n}$ is an exterior unit normal to the surface. As an example, if $I = \iint_S \vec{F} \cdot \widehat{n}\,d\sigma$ where $\vec{F}$ denotes a velocity field with units $[cm/sec]$, $d\sigma$ is an element of area with units $[cm^2]$ ($\widehat{n}$ is dimensionless), then the surface integral I has the flux units of $[cm^3/sec]$ and represents the volume of fluid crossing the surface per second. If the above surface integral is multiplied by the density of the fluid, then the flux integral represents the mass flow rate across the surface.

Let S denote a surface which encloses a volume V. Let $\widehat{n}$ denote the outward unit normal to the given closed surface and let $\vec{F}$ denote a continuous vector field which also has continuous derivatives, then the Gauss divergence theorem can be written in the form

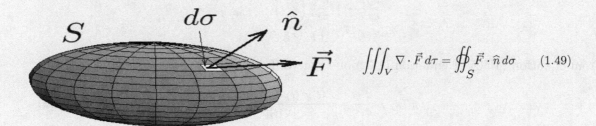

$$\iiint_V \nabla \cdot \vec{F}\,d\tau = \oiint_S \vec{F} \cdot \widehat{n}\,d\sigma \qquad (1.49)$$

where $d\tau$ is an element of volume and $d\sigma$ is an element of surface area. The Gauss divergence theorem states that the divergence of $\vec{F}$ summed over the volume V, which is enclosed by a surface S, equals the summation of the normal component of $\vec{F}$ taken over the surface S. The surface integral on the right-hand side of equation (1.49) is called the flux of the vector field summed over the surface S. The above result is sometimes used to show that the divergence represents a flux per unit volume at a point P. Such a flux is defined by the limiting process

$$\operatorname{div}\vec{F} = \lim_{\substack{\Delta V \to 0 \\ \Delta S \to 0}} \frac{\oiint_{\Delta S} \vec{F} \cdot \widehat{n}\,d\sigma}{\Delta V}.$$

Here ΔS denotes a small surface, about the point P, which encloses the volume ΔV. In the limit as ΔS and ΔV approach zero, the above ratio represents the flux per unit volume of the vector field at the point P. If the divergence is positive, then a source is said to exist at the point P. If the divergence is negative, then a sink is said to exist at the point P. If the divergence equals zero, then the vector field $\vec{F}$ is called solenoidal at the point P.

Differentiation of composite functions

If y is a function of x and x is a function of t, this can be denoted using the notation $y = y(x)$, where $x = x(t)$. The derivative of y with respect to t is obtained using the chain rule for differentiation which can be expressed[‡]

$$y'(t) = \frac{dy}{dt} = \frac{dy}{dx}\frac{dx}{dt} = y'(x)x'(t) \qquad \text{or} \qquad y'(x) = \frac{y'(t)}{x'(t)} \tag{1.50}$$

The second derivative of y with respect to t is obtained by differentiating the first derivative to obtain

$$\frac{d^2y}{dt^2} = \frac{d}{dt}\left(\frac{dy}{dt}\right) = \frac{d}{dt}\left(\frac{dy}{dx}\frac{dx}{dt}\right). \tag{1.51}$$

Using the product rule for differentiating and remembering that $y'(x)$ is a function of x which must be differentiated using the chain rule, one finds

$$\begin{aligned}
\frac{d^2y}{dt^2} &= \frac{dy}{dx}\frac{d}{dt}\left(\frac{dx}{dt}\right) + \left[\frac{d}{dt}\left(\frac{dy}{dx}\right)\right]\frac{dx}{dt} \\
&= \frac{dy}{dx}\frac{d^2x}{dt^2} + \left[\frac{d}{dx}\left(\frac{dy}{dx}\right)\frac{dx}{dt}\right]\frac{dx}{dt} \\
&= \frac{dy}{dx}\frac{d^2x}{dt^2} + \frac{d^2y}{dx^2}\left(\frac{dx}{dt}\right)^2 \\
&= y'(x)x''(t) + y''(x)(x'(t))^2
\end{aligned} \tag{1.52}$$

From equation (1.50) or equation (1.52) one can verify that

$$y''(x) = \frac{d^2y}{dx^2} = \frac{x'(t)y''(t) - y'(t)x''(t)}{[x'(t)]^3} \tag{1.53}$$

Higher ordered derivatives are obtained in a similar manner. In calculating higher order derivatives remember to always employ the chain rule for differentiating a function of x. For example, if x is a function of t, then

$$\frac{d}{dt}\mathcal{A}(x) = \frac{d\mathcal{A}}{dx}\frac{dx}{dt} \tag{1.54}$$

where $\mathcal{A}(x)$ denotes any continuous function of x. In equation (1.52), where the second derivative was calculated, note that we have used the above rule with $\mathcal{A}(x) = y'(x)$.

[‡] The prime $'$ always denotes differentiation with respect to the argument of the function. Thus, $x'(t) = \frac{dx}{dt}$ and $y'(x) = \frac{dy}{dx}$.

Parametric equations

Portions of a curve $y = f(x)$ are sometimes represented by a set of parametric equations $x = x(t)$ and $y = y(t)$ where t ranges over some domain $a \leq t \leq b$. The parametric representation of a curve $y = f(x)$ is not unique. For example, consider the parabola $y = x^2$ for all values of the real variable x. There are many parametric representations for portions of the parabolic curve or all of the parabolic curve. Some example parametric representations of the curve $y = x^2$ are the following.

$$x = t, \qquad y = t^2, \qquad 0 \leq t \leq 1 \qquad\qquad x = -\cosh t, \qquad y = \cosh^2 t, \qquad t \geq 0$$

$$x = \sin t, \qquad y = \sin^2 t, \qquad 0 \leq t \leq 2\pi \qquad\qquad x = t, \qquad y = t^2, \qquad -\infty < t < \infty$$

$$x = \sinh t, \qquad y = \sinh^2 t, \qquad t \geq 0 \qquad\qquad x = e^t, \qquad y = e^{2t}, \qquad -\infty < t < \infty$$

It is left as an exercise to plot the above curves to determine which portions of the parabola are represented.

Given a set of parametric equations $x = x(t)$ and $y = y(t)$, one can obtain y as a function of x by eliminating the parameter t from the given equations. If we assume that $\frac{dx}{dt} \neq 0$, then the equation (1.50) can be expressed in the form

$$\frac{dy}{dx} = \frac{\frac{dy}{dt}}{\frac{dx}{dt}} = \frac{y'(t)}{x'(t)}. \tag{1.55}$$

which represents the derivative of y with respect to x. The equation for $x = x(t)$ together with the equation (1.55) is a parametric representation of the derivative $\frac{dy}{dx}$ as t varies over some domain $a \leq t \leq b$. Higher derivatives can be obtained using the chain rule for differentiation. If we differentiate both sides of equation (1.55) with respect to t and use the rule for differentiating a quotient, we obtain

$$\frac{d}{dt}\left(\frac{dy}{dx}\right) = \frac{d}{dt}\left(\frac{\frac{dy}{dt}}{\frac{dx}{dt}}\right) = \frac{\frac{dx}{dt}\frac{d^2y}{dt^2} - \frac{dy}{dt}\frac{d^2x}{dt^2}}{\left(\frac{dx}{dt}\right)^2} = \frac{x'(t)y''(t) - y'(t)x''(t)}{[x'(t)]^2} \tag{1.56}$$

The chain rule applied to the left-hand side of equation (1.56) gives

$$\frac{d}{dt}\left(\frac{dy}{dx}\right) = \frac{d}{dx}\left(\frac{dy}{dx}\right)\frac{dx}{dt} = \frac{d^2y}{dx^2}\frac{dx}{dt}. \tag{1.57}$$

Substituting this result into equation (1.56) produces the parametric representation for the second derivative as the set of parametric equations

$$x = x(t), \qquad \frac{d^2y}{dx^2} = \frac{\frac{dx}{dt}\frac{d^2y}{dt^2} - \frac{dy}{dt}\frac{d^2x}{dt^2}}{\left(\frac{dx}{dt}\right)^3} = \frac{x'(t)y''(t) - y'(t)x''(t)}{[x'(t)]^3}.$$

This is the same result as equation (1.53). Higher ordered derivatives are obtained in a similar manner.

Function defined by an integral

Consider the function $I = I(x)$ defined by the integral

$$I = I(x) = \int_{g(x)}^{h(x)} f(t, x)\, dt \tag{1.58}$$

where the limits of integration are functions of x and t is the variable of integration. It is assumed that the integrand f is both continuous and differentiable with respect x. The differentiation of a function defined by an integral containing a parameter x is given by the Leibnitz rule

$$\frac{dI}{dx} = I'(x) = \int_{g(x)}^{h(x)} \frac{\partial f(t, x)}{\partial x}\, dt + f(h(x), x)\frac{dh}{dx} - f(g(x), x)\frac{dg}{dx}. \tag{1.59}$$

The above result follows from the definition of a derivative. Consider first the special case of the integral $I(x) = \int_g^h f(t, x)\, dt$ where g and h are constants. Calculate the difference

$$I(x + \Delta x) - I(x) = \int_g^h [f(t, x + \Delta x) - f(t, x)]\, dt$$

and then employ the mean value theorem to write

$$I(x + \Delta x) - I(x) = \int_g^h \frac{\partial f(t, x + \theta \Delta x)}{\partial x} \Delta x\, dt$$

where $0 < \theta < 1$. Letting $\Delta x \to 0$ gives the derivative

$$\frac{\partial I}{\partial x} = \lim_{\Delta x \to 0} \frac{I(x + \Delta x) - I(x)}{\Delta x} = \lim_{\Delta x \to 0} \int_g^h \frac{\partial f(t, x + \theta \Delta x)}{\partial x}\, dt = \int_g^h \frac{\partial f(t, x)}{\partial x}\, dt. \tag{1.60}$$

In the special case that both the upper and lower limits of integration are functions of x, one can employ chain rule differentiation and express the derivative

$$\frac{dI}{dx} = \frac{\partial I}{\partial x} + \frac{\partial I}{\partial g}\frac{dg}{dx} + \frac{\partial I}{\partial h}\frac{dh}{dx}. \tag{1.61}$$

The equation (1.61) simplifies to the result given by equation (1.59).

Example 1-2.   Jacobi elliptic functions

In 1793, Adrien Marie Legendre defined the three functions

$$u = F(\theta, k) = \int_0^\theta \frac{d\theta}{\sqrt{1 - k^2 \sin^2 \theta}} \tag{1.62}$$

$$v = E(\theta, k) = \int_0^\theta \sqrt{1 - k^2 \sin^2 \theta}\, d\theta \tag{1.63}$$

$$w = \Pi(\theta, k, n) = \int_0^\theta \frac{d\theta}{(1 + n \sin^2 \theta)\sqrt{1 - k^2 \sin^2 \theta}} \tag{1.64}$$

where k, $0 \le k < 1$, and n are constants. These functions are called respectively the elliptic integrals of the first, second and third kind. In the first integral (1.62) the upper limit θ is called the amplitude of u and is written $\theta = \operatorname{am} u$ with inverse function denoted $u = \operatorname{am}^{-1}\theta$. Let $x = \sin\theta$, then one can write $x = \sin \operatorname{am} u$. This is abbreviated using the notation

$$x = \operatorname{sn}(u,k) \quad \text{with inverse function} \quad u = \operatorname{sn}^{-1}(x,k) \tag{1.65}$$

In a similar manner the following abbreviations can be defined

$$\sqrt{1-x^2} = \cos\theta = \cos \operatorname{am} u = \operatorname{cn}(u,k) \tag{1.66}$$

$$\sqrt{1-k^2\sin^2\theta} = \sqrt{1-k^2\sin^2 \operatorname{am} u} = \operatorname{dn}(u,k). \tag{1.67}$$

The functions $\operatorname{sn}, \operatorname{cn}, \operatorname{dn}$ are the three basic Jacobi elliptic functions obtained by inverting the elliptic integral of the first kind. These function names are pronounced by saying each letter. Thus, the $\operatorname{sn} u$ function is called "s", "n" of u. Similarly, $\operatorname{cn} u$ is called the "c", "n" of u function, where each letter in the name is read. In addition to the three basic functions one can define the additional Jacobi elliptic functions

$$
\begin{aligned}
\operatorname{ns}(u,k) &= \frac{1}{\operatorname{sn}(u,k)} & \operatorname{cs}(u,k) &= \frac{\operatorname{cn}(u,k)}{\operatorname{sn}(u,k)} & \operatorname{ds}(u,k) &= \frac{\operatorname{dn}(u,k)}{\operatorname{sn}(u,k)} \\
\operatorname{nc}(u,k) &= \frac{1}{\operatorname{cn}(u,k)} & \operatorname{sc}(u,k) &= \frac{\operatorname{sn}(u,k)}{\operatorname{cn}(u,k)} & \operatorname{dc}(u,k) &= \frac{\operatorname{dn}(u,k)}{\operatorname{cn}(u,k)} \\
\operatorname{nd}(u,k) &= \frac{1}{\operatorname{dn}(u,k)} & \operatorname{sd}(u,k) &= \frac{\operatorname{sn}(u,k)}{\operatorname{dn}(u,k)} & \operatorname{cd}(u,k) &= \frac{\operatorname{cn}(u,k)}{\operatorname{dn}(u,k)}
\end{aligned}
\tag{1.68}
$$

Differentiating the integral (1.62) with respect to θ gives

$$\frac{du}{d\theta} = \frac{1}{\sqrt{1-k^2\sin^2\theta}} = \frac{1}{\operatorname{dn}(u,k)} \quad \text{or} \quad \frac{d\theta}{du} = \operatorname{dn}(u,k) \tag{1.69}$$

Consequently, one can develop the differentiation formula

$$\frac{d}{du}\operatorname{sn}(u,k) = \frac{d}{du}\sin\theta = \frac{d}{d\theta}(\sin\theta)\frac{d\theta}{du} = \cos\theta\,\operatorname{dn}(u,k) = \operatorname{cn}(u,k)\operatorname{dn}(u,k) \tag{1.70}$$

$$\frac{d}{du}\operatorname{cn}(u,k) = \frac{d}{du}\cos\theta = \frac{d}{d\theta}(\cos\theta)\frac{d\theta}{du} = -\sin\theta\,\operatorname{dn}(u,k) = -\operatorname{sn}(u,k)\operatorname{dn}(u,k) \tag{1.71}$$

$$\frac{d}{du}\operatorname{dn}(u,k) = \frac{d}{du}\sqrt{1-k^2\sin^2\theta} = -\frac{k^2\sin\theta\cos\theta}{\sqrt{1-k^2\sin^2\theta}}\frac{d\theta}{du} = -k^2\operatorname{sn}(u,k)\operatorname{cn}(u,k) \tag{1.72}$$

From the differentiation formula one can immediately obtain the integration formula

$$\int \operatorname{cn}(u,k)\operatorname{dn}(u,k)\,du = \operatorname{sn}(u,k) + C \tag{1.73}$$

$$\int \operatorname{sn}(u,k)\operatorname{dn}(u,k)\,du = -\operatorname{cn}(u,k) + C \tag{1.74}$$

$$\int \operatorname{sn}(u,k)\operatorname{cn}(u,k)\,du = -\frac{1}{k^2}\operatorname{dn}(u,k) + C \tag{1.75}$$

where C is a constant of integration.

20

The Jacobi elliptic functions sn, cn, sc have properties closely resembling the trigonometric functions sin, cos, tan. Graphs of the functions $sn(u, 1/2), cn(u, 1/2)$ and $sc(u, 1/2)$ are illustrated in the figure 1-5. The elliptic integrals of the first, second and third kind and their derivatives as well as the Jacobi elliptic functions are used in many areas of mathematics, science and engineering.

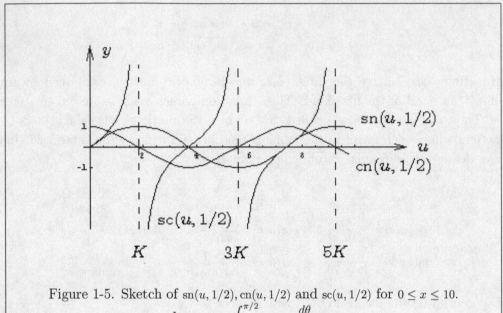

Figure 1-5. Sketch of $sn(u, 1/2), cn(u, 1/2)$ and $sc(u, 1/2)$ for $0 \le x \le 10$. where $K = \int_0^{\pi/2} \dfrac{d\theta}{\sqrt{1 + (\frac{1}{2})^2 \sin^2 \theta}}$

Implicit functions

Recall from calculus that if there exists a point (x_0, y_0) which satisfies an equation of the form $F(x, y) = 0$, and in addition $\frac{\partial F}{\partial y} \ne 0$, then the equation $F(x, y) = 0$ is said to implicitly define y as some function of x in the neighborhood of the point (x_0, y_0). It is important for at least one point (x_0, y_0) to exist and satisfy the given equation $F(x, y) = 0$ in order that an implicit function be defined. For example, given the equation

$$F = F(x, y) = y^6 - y^2 - x = 0 \tag{1.76}$$

it is not immediately obvious how one would solve explicitly for y when given a value for x. A sketch of the function $f(y) = y^6 - y^2$ vs y shows what values of x can be selected such that a value of y will exist satisfying $f(y) = x$.

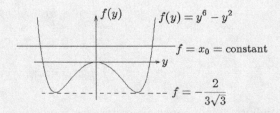

Here the given equation defines y as a function of x implicitly only for $x \geq -2/3\sqrt{3}$ because if this condition is satisfied, then one can solve numerically for at least one real value for y given a value for $x = x_0$. Observe that for an implicit function to exist one must be able to solve numerically for one or more real values of y corresponding to any given real x_0 value substituted into the equation $F(x, y) = 0$. One can then select specific values of y to create a 1-to-1 mapping defining a single-valued function. This is illustrated in the following example.

Example 1-3. **Implicit functions** Using the function $F(x, y) = y^3 - y^2 - x = 0$ one can define three single-valued implicit functions

$$y = f_1(x), \quad x \geq -4/27 \qquad \triangle \cdots \triangle \cdots \triangle$$

$$y = f_2(x), \quad -4/27 \leq x \leq 0 \qquad \cdots \cdots \cdots$$

$$y = f_3(x), \quad x \leq 0 \qquad \circ \cdots \circ \cdots \circ$$

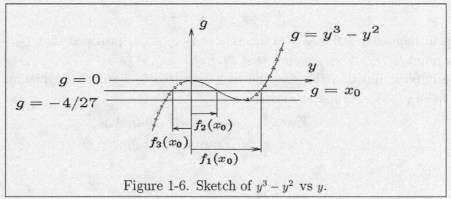

Figure 1-6. Sketch of $y^3 - y^2$ vs y.

These single-valued functions can be inferred from a sketch of $g = y^3 - y^2$ vs y illustrated in the figure 1-6. Note that there is always one or more real values of y where the line $g = x_0$ intersects the $g = y^3 - y^2$ curve. These one or more intersections represent the real values $y_1 = f_1(x_0)$, $y_2 = f_2(x_0)$ and $y_3 = f_3(x_0)$ corresponding to the fixed x_0 value. ∎

To understand the importance of having at least one point (x_0, y_0), which satisfies $F(x_0, y_0) = 0$, consider the following physical interpretation associated with the equation $F(x, y) = 0$. Imagine the intersection of the two surfaces $z = F(x, y)$ and $z = 0$. The plane $z = 0$ is the x, y-plane and the equation $F(x, y) = 0$ represents an intersection of the surface $z = F(x, y)$ with this plane. If the surface does not intersect the plane, then an implicit function will not exist. Hence, the requirement that at least one point (x_0, y_0) exists such that $F(x_0, y_0) = 0$.

Implicit function theorem

The implicit function theorem from calculus states that if there exists a point (x_0, y_0) where the given function $F(x, y)$ satisfies $F(x_0, y_0) = 0$, and the function $F(x, y)$ possesses continuous partial derivatives $\frac{\partial F}{\partial x}$ and $\frac{\partial F}{\partial y}$ in a neighborhood of the point (x_0, y_0) where $\frac{\partial F(x_0, y_0)}{\partial y} \neq 0$, then there exists a unique continuous function $y = y(x)$ which satisfies the conditions $y_0 = y(x_0)$ and $F(x, y(x)) = 0$, for all x in some neighborhood of x_0. This unique function has a continuous derivative $\frac{dy}{dx} = y'(x)$ which exists in some neighborhood of the point x_0.

The implicit function theorem can be applied to equations with more than two variables. Consider the equation

$$F = F(x, y, z) = 0, \tag{1.77}$$

involving three variables x, y and z. One can select two of the variables as independent variables and the remaining variable as the dependent variable. If z is selected as the dependent variable, and if $\frac{\partial F}{\partial z} \neq 0$, and there exists a point (x_0, y_0, z_0) such that $F(x_0, y_0, z_0) = 0$, then the equation (1.77) is said to define z implicitly as a function of x and y in the neighborhood of a point (x_0, y_0, z_0).

In general, equations with $n + 1$ variables of the form

$$F = F(x_1, x_2, \ldots, x_n, w) = 0 \tag{1.78}$$

are said to implicitly define w as a function of $x_1, x_2, \ldots, x_n$ provided that $\frac{\partial F}{\partial w} \neq 0$ and there exists a point $(x_1^0, x_2^0, \ldots, x_n^0, w^0)$ such that $F(x_1^0, x_2^0, \ldots, x_n^0, w^0) = 0$.

The implicit function theorem can be generalized to systems of equations. Given m-equations in $(n + m)$-unknowns, such as

$$
\begin{aligned}
F_1(x_1, x_2, \ldots, x_n, w_1, w_2, \ldots, w_m) &= 0 \\
F_2(x_1, x_2, \ldots, x_n, w_1, w_2, \ldots, w_m) &= 0 \\
&\vdots \qquad\qquad\vdots \\
F_m(x_1, x_2, \ldots, x_n, w_1, w_2, \ldots, w_m) &= 0,
\end{aligned}
\tag{1.79}
$$

where each equation is a function of $(n + m)$-variables, we assume that all the equations are identically satisfied at a point $P^0 = (x_1^0, x_2^0, \ldots, x_n^0, w_1^0, \ldots, w_m^0)$. Each of the functions F_i, $i = 1, \ldots, m$ are assumed to possess first order partial derivatives with respect to all the variables in some neighborhood of the point P^0. In addition, it is assumed that the Jacobian determinant defined by

$$
\frac{\partial(F_1, F_2, \ldots, F_m)}{\partial(w_1, w_2, \ldots, w_m)} =
\begin{vmatrix}
\frac{\partial F_1}{\partial w_1} & \frac{\partial F_1}{\partial w_2} & \cdots & \frac{\partial F_1}{\partial w_m} \\
\frac{\partial F_2}{\partial w_1} & \frac{\partial F_2}{\partial w_2} & \cdots & \frac{\partial F_2}{\partial w_m} \\
\vdots & \vdots & \ddots & \vdots \\
\frac{\partial F_m}{\partial w_1} & \frac{\partial F_m}{\partial w_2} & \cdots & \frac{\partial F_m}{\partial w_m}
\end{vmatrix}
\tag{1.80}
$$

is different from zero. Theoretically, one can then solve for each of the quantities $w_1, w_2, \ldots, w_m$ as functions of the n variables $x_1, x_2, \ldots, x_n$. The implicit function theorem states that in some neighborhood of the point P^0 there exists a unique system of continuous functions

$$
\begin{aligned}
w_1 &= w_1(x_1, x_2, \ldots, x_n) \\
w_2 &= w_2(x_1, x_2, \ldots, x_n) \\
&\vdots \\
w_m &= w_m(x_1, x_2, \ldots, x_n)
\end{aligned}
\tag{1.81}
$$

where each function possesses continuous first order partial derivatives with respect to the variables $x_1, \ldots, x_n$. Here the number of dependent variables is equal to the number of equations and the number of independent variables is equal to n.

Given a system of equations of the form (1.79), the quantities $w_1, w_2, \ldots, w_m$ are said to be defined implicitly. Let us investigate ways to obtain the various derivatives of the implicitly defined functions represented by the system of equations (1.81).

System of two equations with four unknowns

Given two equations

$$F_1(x, y, w_1, w_2) = 0$$
$$F_2(x, y, w_1, w_2) = 0 \tag{1.82}$$

in the four unknowns x, y, w_1, w_2 which satisfy the conditions of the implicit function theorem, then theoretically one can solve the equations (1.82) for w_1 and w_2 by elimination to obtain w_1 and w_2 as functions of x and y. Here it is implied that if the Jacobian determinant

$$\frac{\partial(F_1, F_2)}{\partial(w_1, w_2)}$$

is different from zero, then one can obtain from the equations (1.82) a set of functions

$$w_1 = w_1(x, y) \qquad w_2 = w_2(x, y). \tag{1.83}$$

The equations (1.82) are said to define the equations (1.83) implicitly whenever these equations cannot be obtained explicitly.

System of three equations with five unknowns

As another example, consider a system of equations of the form

$$F_1(x, y, z, u, v) = 0$$
$$F_2(x, y, z, u, v) = 0 \tag{1.84}$$
$$F_3(x, y, z, u, v) = 0$$

in the five unknowns x, y, z, u, v, which satisfy the conditions of the implicit function theorem. Here we have three equations and so any three variables can be selected as the dependent variables. If we select x, y, z as the dependent variables, then theoretically one can solve for x, y and z as functions of u and v provided the Jacobian determinant

$$\frac{\partial(F_1, F_2, F_3)}{\partial(x, y, z)}$$

is different from zero. The system of equations (1.84) are said to define implicitly a set of parametric equations of the form

$$x = x(u, v), \qquad y = y(u, v), \qquad z = z(u, v). \tag{1.85}$$

Chain rule differentiation for functions of more than one variable

Assume that the functions given are well defined and possess continuous partial derivatives over their domain of definition. We use the notation $z = z(x, y)$ to denote z as a function of x and y, where both x and y are functions of another variable t, say $x = x(t)$ and $y = y(t)$, then the chain rule for differentiating a function of two variables is given by

$$\frac{dz}{dt} = \frac{\partial z}{\partial x}\frac{dx}{dt} + \frac{\partial z}{\partial y}\frac{dy}{dt} \tag{1.86}$$

This can be generalized to handle more than one independent variable. For example, if $z = z(x, y)$ and $x = x(u, v)$ and $y = y(u, v)$ the chain rule is written

$$\frac{\partial z}{\partial u} = \frac{\partial z}{\partial x}\frac{\partial x}{\partial u} + \frac{\partial z}{\partial y}\frac{\partial y}{\partial u}$$
$$\frac{\partial z}{\partial v} = \frac{\partial z}{\partial x}\frac{\partial x}{\partial v} + \frac{\partial z}{\partial y}\frac{\partial y}{\partial v} \tag{1.87}$$

A still more general application of the chain rule is when z is a function of n independent variables $z = z(x_1, x_2, \ldots, x_n)$ where each variable is a function of m other variables
$$x_1 = x_1(u_1, u_2, \ldots, u_m),\ x_2 = x_2(u_1, u_2, \ldots, u_m), \ldots, x_n = x_n(u_1, u_2, \ldots, u_m).$$
In this case the derivatives of z with respect to $u_1, \ldots, u_m$ are expressed using the chain rule for differentiation

$$\frac{\partial z}{\partial u_1} = \frac{\partial z}{\partial x_1}\frac{\partial x_1}{\partial u_1} + \frac{\partial z}{\partial x_2}\frac{\partial x_2}{\partial u_1} + \cdots + \frac{\partial z}{\partial x_n}\frac{\partial x_n}{\partial u_1}$$
$$\frac{\partial z}{\partial u_2} = \frac{\partial z}{\partial x_1}\frac{\partial x_1}{\partial u_2} + \frac{\partial z}{\partial x_2}\frac{\partial x_2}{\partial u_2} + \cdots + \frac{\partial z}{\partial x_n}\frac{\partial x_n}{\partial u_2}$$
$$\vdots$$
$$\frac{\partial z}{\partial u_m} = \frac{\partial z}{\partial x_1}\frac{\partial x_1}{\partial u_m} + \frac{\partial z}{\partial x_2}\frac{\partial x_2}{\partial u_m} + \cdots + \frac{\partial z}{\partial x_n}\frac{\partial x_n}{\partial u_m} \tag{1.88}$$

The differentials of the above functions are written as follows. If $z = z(x, y)$, then

$$dz = \frac{\partial z}{\partial x}dx + \frac{\partial z}{\partial y}dy \tag{1.89}$$

If $z = z(x_1, x_2, \ldots, x_n)$, then the differential of z is given by

$$dz = \frac{\partial z}{\partial x_1}dx_1 + \frac{\partial z}{\partial x_2}dx_2 + \cdots + \frac{\partial z}{\partial x_n}dx_n \tag{1.90}$$

Note that if one finds a differential relation of the form

$$df = A_1 dx_1 + A_2 dx_2 + \cdots + A_n dx_n,$$

then one can conclude that $A_i = \dfrac{\partial f}{\partial x_i}$. This follows from the assumption that if all the variables x_j, $j \neq i$ are held constant, then $dx_j = 0$ for $j \neq i$, and consequently $df = A_i dx_i$. This implies that $A_i = \frac{df}{dx_i} = \frac{\partial f}{\partial x_i}$. One can employ this result to find derivatives of functions which are defined implicitly.

Derivatives of implicit functions

Given an equation of the form $f(x, y) = 0$, let us assume there exists a point (x_0, y_0) such that $f(x_0, y_0) = 0$, then one can assume that y is a implicit function of x with continuous derivatives in the neighborhood of this point. One can then differentiate the given equation with respect to x to obtain

$$\frac{\partial f}{\partial x} + \frac{\partial f}{\partial y} y' = 0 \tag{1.91}$$

or one can write

$$y' = \frac{dy}{dx} = -\frac{\frac{\partial f}{\partial x}}{\frac{\partial f}{\partial y}} \qquad \text{provided that} \qquad \frac{\partial f}{\partial y} \neq 0. \tag{1.92}$$

Higher derivatives are obtained by differentiation. For example, differentiating the equation (1.91) with respect to x gives the equation

$$\frac{\partial^2 f}{\partial x^2} + \frac{\partial^2 f}{\partial y \partial x} y' + \left(\frac{\partial^2 f}{\partial x \partial y} + \frac{\partial^2 f}{\partial y^2} y' \right) y' + \frac{\partial f}{\partial y} y'' = 0 \tag{1.93}$$

from which the second derivative can be determined.

Given an equation of the form $F(x, y, z) = 0$, which satisfies the conditions of the implicit function theorem, then one can assume that z is a function of x and y which has continuous partial derivatives in some region R. One can differentiate the given equation with respect to x and y to obtain

$$\frac{\partial F}{\partial x} + \frac{\partial F}{\partial z} \frac{\partial z}{\partial x} = 0 \tag{1.94}$$

$$\frac{\partial F}{\partial y} + \frac{\partial F}{\partial z} \frac{\partial z}{\partial y} = 0 \tag{1.95}$$

Differentiating the equation (1.94) with respect to x gives

$$\frac{\partial^2 F}{\partial x^2} + \frac{\partial^2 F}{\partial x \partial z} \frac{\partial z}{\partial x} + \left(\frac{\partial^2 F}{\partial z \partial x} + \frac{\partial^2 F}{\partial z^2} \frac{\partial z}{\partial x} \right) \frac{\partial z}{\partial x} + \frac{\partial F}{\partial z} \frac{\partial^2 z}{\partial x^2} = 0 \tag{1.96}$$

which is an equation for determining $\frac{\partial^2 z}{\partial x^2}$. In a similar manner, the equations (1.94) and (1.95) can be differentiated with respect to y to obtain expressions for determining the derivatives $\frac{\partial^2 z}{\partial x \partial y}$ and $\frac{\partial^2 z}{\partial y^2}$.

The above ideas can be extended to higher dimensions. Given an equation of the form $F(x_1, x_2, \ldots, x_n, w) = 0$, which satisfies the conditions of the implicit function theorem, then one can assume that w is implicitly defined as a function of $x_1, x_2, \ldots, x_n$ with continuous derivatives in some n-dimensional region R_n. One can then differentiate the given equation to obtain the derivatives

$$\frac{\partial F}{\partial x_1} + \frac{\partial F}{\partial w} \frac{\partial w}{\partial x_1} = 0$$

$$\frac{\partial F}{\partial x_2} + \frac{\partial F}{\partial w} \frac{\partial w}{\partial x_2} = 0$$

$$\vdots \tag{1.97}$$

$$\frac{\partial F}{\partial x_n} + \frac{\partial F}{\partial w} \frac{\partial w}{\partial x_n} = 0$$

Again, higher ordered derivatives of w are obtained by differentiating the expressions containing the lower ordered derivatives.

Example 1-4.   Differentiation of composite function

If $w = xy^2$ where x and y are implicit functions defined by the equations

$$x + y^3 = t, \quad \text{and} \quad x^2 + y^4 = t^3,$$

then find $\frac{dw}{dt}$.

Solution: Differentiate the equations defining x and y implicitly to obtain

$$\frac{dx}{dt} + 3y^2 \frac{dy}{dt} = 1 \quad \text{and} \quad 2x\frac{dx}{dt} + 4y^3\frac{dy}{dt} = 3t^2.$$

These equations represent two equations in the two unknowns $\frac{dx}{dt}$ and $\frac{dy}{dt}$. Solving for these derivatives one obtains

$$\frac{dx}{dt} = \frac{\begin{vmatrix} 1 & 3y^2 \\ 3t^2 & 4y^3 \end{vmatrix}}{\begin{vmatrix} 1 & 3y^2 \\ 2x & 4y^3 \end{vmatrix}} = \frac{4y^3 - 9t^2y^2}{4y^3 - 6xy^2}, \quad \text{and} \quad \frac{dy}{dt} = \frac{\begin{vmatrix} 1 & 1 \\ 2x & 3t^2 \end{vmatrix}}{\begin{vmatrix} 1 & 3y^2 \\ 3t^2 & 4y^3 \end{vmatrix}} = \frac{3t^2 - 2x}{4y^3 - 6xy^2}$$

Hence,

$$\frac{dw}{dt} = \frac{\partial w}{\partial x}\frac{dx}{dt} + \frac{\partial w}{\partial y}\frac{dy}{dt} = y^2\left(\frac{4y^3 - 9t^2y^2}{4y^3 - 6xy^2}\right) + 2xy\left(\frac{3t^2 - 2x}{4y^3 - 6xy^2}\right).$$

One equation in many variables

An equation of the form

$$F(x, y, z, u, v, w, \ldots) = 0 \tag{1.98}$$

implicitly defines any one of the variables in terms of the other variables, provided the partial derivative of F with respect to the variable chosen is not zero. The total differential of F in equation (1.98) is written

$$dF = \frac{\partial F}{\partial x}dx + \frac{\partial F}{\partial y}dy + \frac{\partial F}{\partial z}dz + \frac{\partial F}{\partial u}du + \frac{\partial F}{\partial v}dv + \frac{\partial F}{\partial w}dw + \cdots = 0 \tag{1.99}$$

If we set all differentials equal to zero except two, say du and dy, then equation (1.99) reduces to

$$\frac{\partial F}{\partial y}dy + \frac{\partial F}{\partial u}du = 0$$

from which one can obtain

$$\frac{dy}{du} = -\frac{\frac{\partial F}{\partial u}}{\frac{\partial F}{\partial y}}$$

where all variables, except for u and y, are held constant. To indicate which variables are being held constant it is customary to write these variables as subscripts. For example, the above equation would be written

$$\left(\frac{dy}{du}\right)_{xzvw\ldots} = -\frac{\frac{\partial F}{\partial u}}{\frac{\partial F}{\partial y}} \tag{1.100}$$

The reason for this notation is discussed in the following example.

Example 1-5. Derivative notation

Consider an equation of the form $F(x, y, z) = 0$ which defines implicitly any one of the variables x, y, z in terms of the other variables. The total differential is given by

$$dF = \frac{\partial F}{\partial x} \, dx + \frac{\partial F}{\partial y} \, dy + \frac{\partial F}{\partial z} \, dz = 0. \tag{1.101}$$

If we hold x constant in equation (1.101), then one can write

$$\left(\frac{dy}{dz}\right)_x = -\frac{\frac{\partial F}{\partial z}}{\frac{\partial F}{\partial y}} \tag{1.102}$$

If we hold y constant in equation (1.101), then one can write

$$\left(\frac{dz}{dx}\right)_y = -\frac{\frac{\partial F}{\partial x}}{\frac{\partial F}{\partial z}} \tag{1.103}$$

If we hold z constant in equation (1.101), then one can write

$$\left(\frac{dx}{dy}\right)_z = -\frac{\frac{\partial F}{\partial y}}{\frac{\partial F}{\partial x}} \tag{1.104}$$

Multiplying the equations (1.102), (1.103), and (1.104) produces the result

$$\left(\frac{dy}{dz}\right)_x \left(\frac{dz}{dx}\right)_y \left(\frac{dx}{dy}\right)_z = -1 \tag{1.105}$$

If in equations (1.102), (1.103), and (1.104) we had used the notations $\frac{\partial y}{\partial z}$, $\frac{\partial z}{\partial x}$ and $\frac{\partial x}{\partial y}$ to denote the derivatives, then equation (1.105) would have the form

$$\frac{\partial y}{\partial z} \frac{\partial z}{\partial x} \frac{\partial x}{\partial y} = -1 \tag{1.106}$$

This notation is frowned upon because many students treat the partial differential symbols as quantities that can be canceled. If one does this cancelation in equation (1.106) one obtains a ridiculous result that 1=-1. To prevent this idea of cancelation of differentials, the subscript notation was adopted.

■

System of equations

Consider the system of equations

$$\begin{aligned} F_1(x, y, z, u, v) &= 0 \\ F_2(x, y, z, u, v) &= 0 \\ F_3(x, y, z, u, v) &= 0 \end{aligned} \tag{1.107}$$

which satisfy the conditions of the implicit function theorem. These equations can be thought of as defining x, y, z implicitly in terms of the variables u and v. We desired to find representations for the derivatives

$$\frac{\partial x}{\partial u}, \quad \frac{\partial x}{\partial v}, \quad \frac{\partial y}{\partial u}, \quad \frac{\partial y}{\partial v}, \quad \frac{\partial z}{\partial u}, \quad \frac{\partial z}{\partial v}$$

directly from the equations (1.107). To accomplish this one can proceed as follows. Observe that the differentials of the equations (1.107) must equal zero and so one can write

$$dF_1 = \frac{\partial F_1}{\partial x}dx + \frac{\partial F_1}{\partial y}dy + \frac{\partial F_1}{\partial z}dz + \frac{\partial F_1}{\partial u}du + \frac{\partial F_1}{\partial v}dv = 0$$

$$dF_2 = \frac{\partial F_2}{\partial x}dx + \frac{\partial F_2}{\partial y}dy + \frac{\partial F_2}{\partial z}dz + \frac{\partial F_2}{\partial u}du + \frac{\partial F_2}{\partial v}dv = 0 \qquad (1.108)$$

$$dF_3 = \frac{\partial F_3}{\partial x}dx + \frac{\partial F_3}{\partial y}dy + \frac{\partial F_3}{\partial z}dz + \frac{\partial F_3}{\partial u}du + \frac{\partial F_3}{\partial v}dv = 0$$

Write the equations (1.108) in the form of a linear system of equations from which we can solve for the differentials dx, dy and dz. One can write the system of equations (1.108) in the form

$$\frac{\partial F_1}{\partial x}dx + \frac{\partial F_1}{\partial y}dy + \frac{\partial F_1}{\partial z}dz = -\frac{\partial F_1}{\partial u}du - \frac{\partial F_1}{\partial v}dv$$

$$\frac{\partial F_2}{\partial x}dx + \frac{\partial F_2}{\partial y}dy + \frac{\partial F_2}{\partial z}dz = -\frac{\partial F_2}{\partial u}du - \frac{\partial F_2}{\partial v}dv \qquad (1.109)$$

$$\frac{\partial F_3}{\partial x}dx + \frac{\partial F_3}{\partial y}dy + \frac{\partial F_3}{\partial z}dz = -\frac{\partial F_3}{\partial u}du - \frac{\partial F_3}{\partial v}dv$$

One can use Cramer's rule and the properties of determinants to solve the system of equations (1.109) for the quantities dx, dy and dz. The determinants can be expressed in the Jacobian determinant notation to produce the results

$$dx = -\frac{\begin{vmatrix} \frac{\partial F_1}{\partial u} & \frac{\partial F_1}{\partial y} & \frac{\partial F_1}{\partial z} \\ \frac{\partial F_2}{\partial u} & \frac{\partial F_2}{\partial y} & \frac{\partial F_2}{\partial z} \\ \frac{\partial F_3}{\partial u} & \frac{\partial F_3}{\partial y} & \frac{\partial F_3}{\partial z} \end{vmatrix}}{\begin{vmatrix} \frac{\partial F_1}{\partial x} & \frac{\partial F_1}{\partial y} & \frac{\partial F_1}{\partial z} \\ \frac{\partial F_2}{\partial x} & \frac{\partial F_2}{\partial y} & \frac{\partial F_2}{\partial z} \\ \frac{\partial F_3}{\partial x} & \frac{\partial F_3}{\partial y} & \frac{\partial F_3}{\partial z} \end{vmatrix}} du - \frac{\begin{vmatrix} \frac{\partial F_1}{\partial v} & \frac{\partial F_1}{\partial y} & \frac{\partial F_1}{\partial z} \\ \frac{\partial F_2}{\partial v} & \frac{\partial F_2}{\partial y} & \frac{\partial F_2}{\partial z} \\ \frac{\partial F_3}{\partial v} & \frac{\partial F_3}{\partial y} & \frac{\partial F_3}{\partial z} \end{vmatrix}}{\begin{vmatrix} \frac{\partial F_1}{\partial x} & \frac{\partial F_1}{\partial y} & \frac{\partial F_1}{\partial z} \\ \frac{\partial F_2}{\partial x} & \frac{\partial F_2}{\partial y} & \frac{\partial F_2}{\partial z} \\ \frac{\partial F_3}{\partial x} & \frac{\partial F_3}{\partial y} & \frac{\partial F_3}{\partial z} \end{vmatrix}} dv = -\frac{\frac{\partial(F_1,F_2,F_3)}{\partial(u,\,y,\,z)}}{\frac{\partial(F_1,F_2,F_3)}{\partial(x,\,y,\,z)}} du - \frac{\frac{\partial(F_1,F_2,F_3)}{\partial(v,\,y,\,z)}}{\frac{\partial(F_1,F_2,F_3)}{\partial(x,\,y,\,z)}} dv$$

$$dy = -\frac{\begin{vmatrix} \frac{\partial F_1}{\partial x} & \frac{\partial F_1}{\partial u} & \frac{\partial F_1}{\partial z} \\ \frac{\partial F_2}{\partial x} & \frac{\partial F_2}{\partial u} & \frac{\partial F_2}{\partial z} \\ \frac{\partial F_3}{\partial x} & \frac{\partial F_3}{\partial u} & \frac{\partial F_3}{\partial z} \end{vmatrix}}{\begin{vmatrix} \frac{\partial F_1}{\partial x} & \frac{\partial F_1}{\partial y} & \frac{\partial F_1}{\partial z} \\ \frac{\partial F_2}{\partial x} & \frac{\partial F_2}{\partial y} & \frac{\partial F_2}{\partial z} \\ \frac{\partial F_3}{\partial x} & \frac{\partial F_3}{\partial y} & \frac{\partial F_3}{\partial z} \end{vmatrix}} du - \frac{\begin{vmatrix} \frac{\partial F_1}{\partial x} & \frac{\partial F_1}{\partial v} & \frac{\partial F_1}{\partial z} \\ \frac{\partial F_2}{\partial x} & \frac{\partial F_2}{\partial v} & \frac{\partial F_2}{\partial z} \\ \frac{\partial F_3}{\partial x} & \frac{\partial F_3}{\partial v} & \frac{\partial F_3}{\partial z} \end{vmatrix}}{\begin{vmatrix} \frac{\partial F_1}{\partial x} & \frac{\partial F_1}{\partial y} & \frac{\partial F_1}{\partial z} \\ \frac{\partial F_2}{\partial x} & \frac{\partial F_2}{\partial y} & \frac{\partial F_2}{\partial z} \\ \frac{\partial F_3}{\partial x} & \frac{\partial F_3}{\partial y} & \frac{\partial F_3}{\partial z} \end{vmatrix}} dv = -\frac{\frac{\partial(F_1,F_2,F_3)}{\partial(x,\,u,\,z)}}{\frac{\partial(F_1,F_2,F_3)}{\partial(x,\,y,\,z)}} du - \frac{\frac{\partial(F_1,F_2,F_3)}{\partial(x,\,v,\,z)}}{\frac{\partial(F_1,F_2,F_3)}{\partial(x,\,y,\,z)}} dv$$

$$dz = -\frac{\begin{vmatrix} \frac{\partial F_1}{\partial x} & \frac{\partial F_1}{\partial y} & \frac{\partial F_1}{\partial u} \\ \frac{\partial F_2}{\partial x} & \frac{\partial F_2}{\partial y} & \frac{\partial F_2}{\partial u} \\ \frac{\partial F_3}{\partial x} & \frac{\partial F_3}{\partial y} & \frac{\partial F_3}{\partial u} \end{vmatrix}}{\begin{vmatrix} \frac{\partial F_1}{\partial x} & \frac{\partial F_1}{\partial y} & \frac{\partial F_1}{\partial z} \\ \frac{\partial F_2}{\partial x} & \frac{\partial F_2}{\partial y} & \frac{\partial F_2}{\partial z} \\ \frac{\partial F_3}{\partial x} & \frac{\partial F_3}{\partial y} & \frac{\partial F_3}{\partial z} \end{vmatrix}} du - \frac{\begin{vmatrix} \frac{\partial F_1}{\partial x} & \frac{\partial F_1}{\partial y} & \frac{\partial F_1}{\partial v} \\ \frac{\partial F_2}{\partial x} & \frac{\partial F_2}{\partial y} & \frac{\partial F_2}{\partial v} \\ \frac{\partial F_3}{\partial x} & \frac{\partial F_3}{\partial y} & \frac{\partial F_3}{\partial v} \end{vmatrix}}{\begin{vmatrix} \frac{\partial F_1}{\partial x} & \frac{\partial F_1}{\partial y} & \frac{\partial F_1}{\partial z} \\ \frac{\partial F_2}{\partial x} & \frac{\partial F_2}{\partial y} & \frac{\partial F_2}{\partial z} \\ \frac{\partial F_3}{\partial x} & \frac{\partial F_3}{\partial y} & \frac{\partial F_3}{\partial z} \end{vmatrix}} dv = -\frac{\frac{\partial(F_1,F_2,F_3)}{\partial(x,\,y,\,u)}}{\frac{\partial(F_1,F_2,F_3)}{\partial(x,\,y,\,z)}} du - \frac{\frac{\partial(F_1,F_2,F_3)}{\partial(x,\,y,\,v)}}{\frac{\partial(F_1,F_2,F_3)}{\partial(x,\,y,\,z)}} dv$$

We know that $dx = \frac{\partial x}{\partial u}du + \frac{\partial x}{\partial v}dv$ and so comparing this result with the differential dx just obtained, we find the partial derivatives of x with respect to u and v can be written.

$$\frac{\partial x}{\partial u} = -\frac{\frac{\partial(F_1,F_2,F_3)}{\partial(u,\,y,\,z)}}{\frac{\partial(F_1,F_2,F_3)}{\partial(x,\,y,\,z)}}, \quad \text{and} \quad \frac{\partial x}{\partial v} = -\frac{\frac{\partial(F_1,F_2,F_3)}{\partial(v,\,y,\,z)}}{\frac{\partial(F_1,F_2,F_3)}{\partial(x,\,y,\,z)}} \qquad (1.110)$$

Similarly, from the differentials

$$dy = \frac{\partial y}{\partial u}\,du + \frac{\partial y}{\partial v}\,dv, \quad \text{and} \quad dz = \frac{\partial z}{\partial u}\,du + \frac{\partial z}{\partial v}\,dv$$

compared with the results for dy and dz previously calculated, one finds the partial derivative representations

$$\frac{\partial y}{\partial u} = -\frac{\frac{\partial(F_1,F_2,F_3)}{\partial(x,\,u,\,z)}}{\frac{\partial(F_1,F_2,F_3)}{\partial(x,\,y,\,z)}}, \quad \text{and} \quad \frac{\partial y}{\partial v} = -\frac{\frac{\partial(F_1,F_2,F_3)}{\partial(x,\,v,\,z)}}{\frac{\partial(F_1,F_2,F_3)}{\partial(x,\,y,\,z)}}$$

$$\frac{\partial z}{\partial u} = -\frac{\frac{\partial(F_1,F_2,F_3)}{\partial(x,\,y,\,u)}}{\frac{\partial(F_1,F_2,F_3)}{\partial(x,\,y,\,z)}}, \quad \text{and} \quad \frac{\partial z}{\partial v} = -\frac{\frac{\partial(F_1,F_2,F_3)}{\partial(x,\,y,\,v)}}{\frac{\partial(F_1,F_2,F_3)}{\partial(x,\,y,\,z)}} \tag{1.111}$$

Example 1-6.   Two equations with three unknowns

Given the system of equations

$$F(x,u,v) = 0, \qquad G(x,u,v) = 0,$$

which satisfy the conditions of the implicit function theorem, then one can select two of the variables as dependent. For example, if u,v are selected as the dependent variables, then one can verify that the derivatives are given by the relations

$$\frac{du}{dx} = -\frac{\frac{\partial(F,G)}{\partial(x,v)}}{\frac{\partial(F,G)}{\partial(u,v)}}, \qquad \frac{dv}{dx} = -\frac{\frac{\partial(F,G)}{\partial(u,x)}}{\frac{\partial(F,G)}{\partial(u,v)}}$$

provided that the Jacobian determinant $\dfrac{\partial(F,G)}{\partial(u,v)} \neq 0$.

∎

Example 1-7.   System of two equations with four unknowns

Given the system of equations

$$F(x,y,u,v) = 0 \qquad G(x,y,u,v) = 0 \tag{1.112}$$

which satisfy the conditions of the implicit function theorem, then one can select two of the variables as dependent. If u,v are selected as the dependent variables, then one can verify the following derivatives

$$\frac{\partial u}{\partial x} = -\frac{\frac{\partial(F,G)}{\partial(x,v)}}{\frac{\partial(F,G)}{\partial(u,v)}}, \qquad \frac{\partial v}{\partial x} = -\frac{\frac{\partial(F,G)}{\partial(u,x)}}{\frac{\partial(F,G)}{\partial(u,v)}},$$

$$\frac{\partial u}{\partial y} = -\frac{\frac{\partial(F,G)}{\partial(y,v)}}{\frac{\partial(F,G)}{\partial(u,v)}}, \qquad \frac{\partial v}{\partial y} = -\frac{\frac{\partial(F,G)}{\partial(u,y)}}{\frac{\partial(F,G)}{\partial(u,v)}},$$

provided $\dfrac{\partial(F,G)}{\partial(u,v)} \neq 0$.

∎

Example 1-8. Parametric equations

Consider an example where you have one function $F = F(x, y)$ and two equations

$$u = u(x, y, t) = 0, \quad v = v(x, y, t) = 0. \tag{1.113}$$

which satisfy the conditions of the implicit function theorem. The two given equations $u = u(x, y, t) = 0$ and $v = v(x, y, t) = 0$ can be interpreted as defining implicitly the variables x and y as functions of t. Thus, if we treat x and y as functions of t we can obtain the derivatives

$$\frac{du}{dt} = \frac{\partial u}{\partial x}\frac{dx}{dt} + \frac{\partial u}{\partial y}\frac{dy}{dt} + \frac{\partial u}{\partial t} = 0$$
$$\frac{dv}{dt} = \frac{\partial v}{\partial x}\frac{dx}{dt} + \frac{\partial v}{\partial y}\frac{dy}{dt} + \frac{\partial v}{\partial t} = 0. \tag{1.114}$$

One can now solve the system of equations (1.114) and obtain expressions for the derivatives $\frac{dx}{dt}$ and $\frac{dy}{dt}$. We employ Cramer's rule and solve the system of equations (1.114) to obtain

$$\frac{dx}{dt} = \frac{\begin{vmatrix} -\frac{\partial u}{\partial t} & \frac{\partial u}{\partial y} \\ -\frac{\partial v}{\partial t} & \frac{\partial v}{\partial y} \end{vmatrix}}{\begin{vmatrix} \frac{\partial u}{\partial x} & \frac{\partial u}{\partial y} \\ \frac{\partial v}{\partial x} & \frac{\partial v}{\partial y} \end{vmatrix}} = -\frac{\frac{\partial(u,v)}{\partial(t,y)}}{\frac{\partial(u,v)}{\partial(x,y)}}$$

$$\frac{dy}{dt} = \frac{\begin{vmatrix} \frac{\partial u}{\partial x} & -\frac{\partial u}{\partial t} \\ \frac{\partial v}{\partial x} & -\frac{\partial v}{\partial t} \end{vmatrix}}{\begin{vmatrix} \frac{\partial u}{\partial x} & \frac{\partial u}{\partial y} \\ \frac{\partial v}{\partial x} & \frac{\partial v}{\partial y} \end{vmatrix}} = -\frac{\frac{\partial(u,v)}{\partial(x,t)}}{\frac{\partial(u,v)}{\partial(x,y)}} \tag{1.115}$$

provided the Jacobian $\dfrac{\partial(u,v)}{\partial(x,y)}$ is different from zero. The derivatives $\frac{dx}{dt}$ and $\frac{dy}{dt}$ can now be used to calculate the derivative of F with respect to t using the relation

$$\frac{dF}{dt} = \frac{\partial F}{\partial x}\frac{dx}{dt} + \frac{\partial F}{\partial y}\frac{dy}{dt}. \tag{1.116}$$

∎

System of three equations with six unknowns

Consider the three equations

$$F(x, y, z, u, v, w) = 0$$
$$G(x, y, z, u, v, w) = 0 \tag{1.117}$$
$$H(x, y, z, u, v, w) = 0$$

in the six unknowns x, y, z, u, v, w, which satisfy the conditions of the implicit function theorem. Here any three variables can be selected as the dependent variables. If one assumes that u, v, w are functions of x, y and z, then the derivative of u, v and w with respect to x, y and z can be determined provided the Jacobian determinant

$$\frac{\partial(F, G, H)}{\partial(x, y, z)} \neq 0.$$

If we assume that u,v and w are functions of x,y and z, then one can differentiate the equations (1.117) with respect to x to obtain the equations

$$\frac{\partial F}{\partial x} + \frac{\partial F}{\partial u}\frac{\partial u}{\partial x} + \frac{\partial F}{\partial v}\frac{\partial v}{\partial x} + \frac{\partial F}{\partial w}\frac{\partial w}{\partial x} = 0$$

$$\frac{\partial G}{\partial x} + \frac{\partial G}{\partial u}\frac{\partial u}{\partial x} + \frac{\partial G}{\partial v}\frac{\partial v}{\partial x} + \frac{\partial G}{\partial w}\frac{\partial w}{\partial x} = 0 \qquad (1.118)$$

$$\frac{\partial H}{\partial x} + \frac{\partial H}{\partial u}\frac{\partial u}{\partial x} + \frac{\partial H}{\partial v}\frac{\partial v}{\partial x} + \frac{\partial H}{\partial w}\frac{\partial w}{\partial x} = 0$$

By hypothesis the Jacobian $\dfrac{\partial(F,G,H)}{\partial(x,y,z)} \neq 0$ and so the system of equations (1.118) can be solved to determine the derivatives $\frac{\partial u}{\partial x}$, $\frac{\partial v}{\partial x}$ and $\frac{\partial w}{\partial x}$. One finds

$$\frac{\partial u}{\partial x} = -\frac{\frac{\partial(F,G,H)}{\partial(x,v,w)}}{\frac{\partial(F,G,H)}{\partial(u,v,w)}}, \qquad \frac{\partial v}{\partial x} = -\frac{\frac{\partial(F,G,H)}{\partial(u,x,w)}}{\frac{\partial(F,G,H)}{\partial(u,v,w)}}, \qquad \frac{\partial w}{\partial x} = -\frac{\frac{\partial(F,G,H)}{\partial(u,v,x)}}{\frac{\partial(F,G,H)}{\partial(u,v,w)}} \qquad (1.119)$$

Differentiating the equations (1.117) with respect to y produces the equations

$$\frac{\partial F}{\partial y} + \frac{\partial F}{\partial u}\frac{\partial u}{\partial y} + \frac{\partial F}{\partial v}\frac{\partial v}{\partial y} + \frac{\partial F}{\partial w}\frac{\partial w}{\partial y} = 0$$

$$\frac{\partial G}{\partial y} + \frac{\partial G}{\partial u}\frac{\partial u}{\partial y} + \frac{\partial G}{\partial v}\frac{\partial v}{\partial y} + \frac{\partial G}{\partial w}\frac{\partial w}{\partial y} = 0 \qquad (1.120)$$

$$\frac{\partial H}{\partial y} + \frac{\partial H}{\partial u}\frac{\partial u}{\partial y} + \frac{\partial H}{\partial v}\frac{\partial v}{\partial y} + \frac{\partial H}{\partial w}\frac{\partial w}{\partial y} = 0$$

One can now solve this system of equations for the derivatives $\frac{\partial u}{\partial y}, \frac{\partial v}{\partial y}$ and $\frac{\partial w}{\partial y}$. It is left as an exercise to show these derivatives are determined from the equations

$$\frac{\partial u}{\partial y} = -\frac{\frac{\partial(F,G,H)}{\partial(y,v,w)}}{\frac{\partial(F,G,H)}{\partial(u,v,w)}}, \qquad \frac{\partial v}{\partial y} = -\frac{\frac{\partial(F,G,H)}{\partial(u,y,w)}}{\frac{\partial(F,G,H)}{\partial(u,v,w)}}, \qquad \frac{\partial w}{\partial y} = -\frac{\frac{\partial(F,G,H)}{\partial(u,v,y)}}{\frac{\partial(F,G,H)}{\partial(u,v,w)}} \qquad (1.121)$$

Differentiating the equations (1.117) with respect to z produces the equations

$$\frac{\partial F}{\partial z} + \frac{\partial F}{\partial u}\frac{\partial u}{\partial z} + \frac{\partial F}{\partial v}\frac{\partial v}{\partial z} + \frac{\partial F}{\partial w}\frac{\partial w}{\partial z} = 0$$

$$\frac{\partial G}{\partial z} + \frac{\partial G}{\partial u}\frac{\partial u}{\partial z} + \frac{\partial G}{\partial v}\frac{\partial v}{\partial z} + \frac{\partial G}{\partial w}\frac{\partial w}{\partial z} = 0 \qquad (1.122)$$

$$\frac{\partial H}{\partial z} + \frac{\partial H}{\partial u}\frac{\partial u}{\partial z} + \frac{\partial H}{\partial v}\frac{\partial v}{\partial z} + \frac{\partial H}{\partial w}\frac{\partial w}{\partial z} = 0$$

One can now solve this system of equations for the derivatives $\frac{\partial u}{\partial z}, \frac{\partial v}{\partial z}$ and $\frac{\partial w}{\partial z}$. It is left as an exercise to show these derivatives are determined from the equations

$$\frac{\partial u}{\partial z} = -\frac{\frac{\partial(F,G,H)}{\partial(z,v,w)}}{\frac{\partial(F,G,H)}{\partial(u,v,w)}}, \qquad \frac{\partial v}{\partial z} = -\frac{\frac{\partial(F,G,H)}{\partial(u,z,w)}}{\frac{\partial(F,G,H)}{\partial(u,v,w)}}, \qquad \frac{\partial w}{\partial z} = -\frac{\frac{\partial(F,G,H)}{\partial(u,v,z)}}{\frac{\partial(F,G,H)}{\partial(u,v,w)}} \qquad (1.123)$$

Transformations

A transformation from a region A to a region B is a rule which produces a one-to-one mapping where each point in the region A has a unique image point in the region B. For example, in two dimensions one can map points from a two-dimensional (u, v)-plane to a two-dimensional (x, y)-plane by defining functions $x = x(u, v)$ and $y = y(u, v)$ which maps all points $(u, v) \in A$ to unique image points $(x, y) \in B$. A typical situation is illustrated in the figure 1-7.

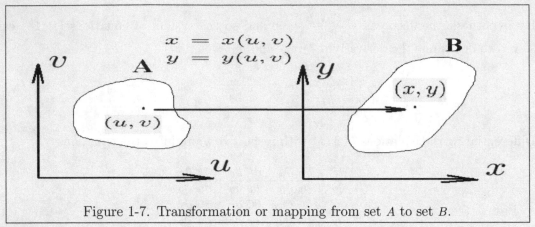

Figure 1-7. Transformation or mapping from set A to set B.

The idea of a transformation can be generalized to n-dimensions. The system of functions

$$y_1 = y_1(x_1, x_2, \ldots, x_n)$$

$$y_2 = y_2(x_1, x_2, \ldots, x_n)$$

$$\vdots$$

$$y_n = y_n(x_1, x_2, \ldots, x_n),$$

where the number of functions and the number of variables are equal, defines a mapping or transformation from a set A containing points $x^0 = (x_1^0, x_2^0, \ldots, x_n^0)$ to a set B containing points $y^0 = (y_1^0, y_2^0, \ldots, y_n^0)$. The mapping or transformation is called regular if

(i) The mapping is one-to-one with each image point unique.

(ii) All the functions y_i, $i = 1, \ldots, n$ posses continuous partial derivatives.

(iii) The Jacobian determinant $\dfrac{\partial(y_1, y_2, \ldots, y_n)}{\partial(x_1, x_2, \ldots, x_n)}$ is different from zero.

If the mapping is regular in the neighborhood of a point x^0, then there exists an inverse mapping where $x_1, x_2, \ldots, x_n$ can be represented as functions of $y_1, y_2, \ldots, y_n$. The inverse mapping is regular in the neighborhood of the image point y^0. Further, the Jacobian determinants of the mapping and inverse mapping satisfy the reciprocal relation

$$\frac{\partial(y_1, y_2, \ldots, y_n)}{\partial(x_1, x_2, \ldots, x_n)} = \frac{1}{\dfrac{\partial(x_1, x_2, \ldots, x_n)}{\partial(y_1, y_2, \ldots, y_n)}}. \tag{1.124}$$

Example 1-9. Coordinate transformation

A special case of the equation (1.112) are equations having the form

$$x = x(u,v), \quad y = y(u,v) \qquad \text{or} \qquad F = x(u,v) - x = 0, \quad \text{and} \quad G = y(u,v) - y = 0 \qquad (1.125)$$

which represents a change of variable from an (x,y) coordinate system to a (u,v) coordinate system. Theoretically one can obtain the inverse transformation from the equation (1.125) by solving for u and v in terms of x and y. This can be done provided the Jacobian determinant $\frac{\partial(x,y)}{\partial(u,v)} \neq 0$. Note that one can obtain from the equations (1.125) the differentials

$$dx = \frac{\partial x}{\partial u}\,du + \frac{\partial x}{\partial v}\,dv, \qquad dy = \frac{\partial y}{\partial u}\,du + \frac{\partial y}{\partial v}\,dv.$$

One can now solve for du and dv to obtain

$$du = \frac{\frac{\partial y}{\partial v}\,dx - \frac{\partial x}{\partial v}\,dy}{\frac{\partial x}{\partial u}\frac{\partial y}{\partial v} - \frac{\partial x}{\partial v}\frac{\partial y}{\partial u}}, \qquad dv = \frac{-\frac{\partial y}{\partial u}\,dx + \frac{\partial x}{\partial u}\,dy}{\frac{\partial x}{\partial u}\frac{\partial y}{\partial v} - \frac{\partial x}{\partial v}\frac{\partial y}{\partial u}}. \qquad (1.126)$$

The coefficients of dx and dy in equation (1.126) give the partial derivatives

$$\frac{\partial u}{\partial x} = -\frac{\frac{\partial(F,G)}{\partial(x,v)}}{\frac{\partial(F,G)}{\partial(u,v)}} = \frac{\frac{\partial y}{\partial v}}{\frac{\partial x}{\partial u}\frac{\partial y}{\partial v} - \frac{\partial x}{\partial v}\frac{\partial y}{\partial u}} \qquad \frac{\partial v}{\partial x} = -\frac{\frac{\partial(F,G)}{\partial(u,x)}}{\frac{\partial(F,G)}{\partial(u,v)}} = \frac{-\frac{\partial y}{\partial u}}{\frac{\partial x}{\partial u}\frac{\partial y}{\partial v} - \frac{\partial x}{\partial v}\frac{\partial y}{\partial u}}$$

$$\frac{\partial u}{\partial y} = -\frac{\frac{\partial(F,G)}{\partial(y,v)}}{\frac{\partial(F,G)}{\partial(u,v)}} = \frac{-\frac{\partial x}{\partial v}}{\frac{\partial x}{\partial u}\frac{\partial y}{\partial v} - \frac{\partial x}{\partial v}\frac{\partial y}{\partial u}} \qquad \frac{\partial v}{\partial y} = -\frac{\frac{\partial(F,G)}{\partial(u,y)}}{\frac{\partial(F,G)}{\partial(u,v)}} = \frac{\frac{\partial x}{\partial u}}{\frac{\partial x}{\partial u}\frac{\partial y}{\partial v} - \frac{\partial x}{\partial v}\frac{\partial y}{\partial u}}$$

provided the Jacobian $\frac{\partial(x,y)}{\partial(u,v)} \neq 0$. ∎

It is left as an exercise to examine the special case of equations (1.117) given by the system of equations

$$\begin{aligned} x &= x(u,v,w) \\ y &= y(u,v,w) \\ z &= z(u,v,w). \end{aligned} \qquad (1.127)$$

These system of equations represent a coordinate transformation from (x,y,z) coordinates to (u,v,w) coordinates. From advanced calculus it is well known that if the Jacobian determinant satisfies the relation $\frac{\partial(x,y,z)}{\partial(u,v,w)} \neq 0$, then an inverse transformation exists.

Whenever the Jacobian determinant is zero $\frac{\partial(x,y,z)}{\partial(u,v,w)} = 0$ this is an indication that there is a functional relationship between the variables x, y and z and, consequently, the given equations do not represent a coordinate transformation.

Example 1-10. Coordinate transformation

Transform the Laplace equation

$$\nabla^2 \psi = \frac{\partial^2 \psi}{\partial x^2} + \frac{\partial^2 \psi}{\partial y^2} = 0, \qquad \psi = \psi(x,y), \qquad \begin{aligned} -\infty &< x < \infty \\ -\infty &< y < \infty \end{aligned} \qquad (1.128)$$

to polar coordinates.

Solution: We use the transformation equations

$$x = r\cos\theta, \quad y = r\sin\theta, \qquad r > 0, \quad 0 \le \theta < 2\pi \tag{1.129}$$

to convert to polar coordinates. These equations have the inverse transform

$$r = \sqrt{x^2 + y^2}, \qquad \theta = \arctan\frac{y}{x}. \tag{1.130}$$

If ψ is treated as a function of r and θ, then one can use the chain rule to calculate the derivatives

$$\begin{aligned}
\frac{\partial\psi}{\partial x} &= \frac{\partial\psi}{\partial r}\frac{\partial r}{\partial x} + \frac{\partial\psi}{\partial\theta}\frac{\partial\theta}{\partial x} \\
\frac{\partial\psi}{\partial y} &= \frac{\partial\psi}{\partial r}\frac{\partial r}{\partial y} + \frac{\partial\psi}{\partial\theta}\frac{\partial\theta}{\partial y}.
\end{aligned} \tag{1.131}$$

Differentiate the inverse transformation equations (1.130) and show that

$$\begin{aligned}
\frac{\partial r}{\partial x} &= \frac{x}{r} = \cos\theta, & \frac{\partial\theta}{\partial x} &= -\frac{y}{r^2} = -\frac{\sin\theta}{r} \\
\frac{\partial r}{\partial y} &= \frac{y}{r} = \sin\theta, & \frac{\partial\theta}{\partial y} &= \frac{x}{r^2} = \frac{\cos\theta}{r}
\end{aligned} \tag{1.132}$$

The derivatives given by equation (1.131) can then be written as

$$\begin{aligned}
\frac{\partial\psi}{\partial x} &= \frac{\partial\psi}{\partial r}\cos\theta - \frac{\partial\psi}{\partial\theta}\frac{\sin\theta}{r} \\
\frac{\partial\psi}{\partial y} &= \frac{\partial\psi}{\partial r}\sin\theta + \frac{\partial\psi}{\partial\theta}\frac{\cos\theta}{r}
\end{aligned} \tag{1.133}$$

The second derivatives are derivatives of first derivatives and so one can write

$$\frac{\partial^2\psi}{\partial x^2} = \frac{\partial}{\partial x}\left(\frac{\partial\psi}{\partial x}\right) = \frac{\partial}{\partial r}\left(\frac{\partial\psi}{\partial x}\right)\frac{\partial r}{\partial x} + \frac{\partial}{\partial\theta}\left(\frac{\partial\psi}{\partial x}\right)\frac{\partial\theta}{\partial x} \tag{1.134}$$

Here one must remember that $\frac{\partial\psi}{\partial x}$ is being treated as a function of r and θ and so one must use the chain rule for differentiating this function. In equation (1.134) one can replace x by y to obtain the second derivative

$$\frac{\partial^2\psi}{\partial y^2} = \frac{\partial}{\partial y}\left(\frac{\partial\psi}{\partial y}\right) = \frac{\partial}{\partial r}\left(\frac{\partial\psi}{\partial y}\right)\frac{\partial r}{\partial y} + \frac{\partial}{\partial\theta}\left(\frac{\partial\psi}{\partial y}\right)\frac{\partial\theta}{\partial y} \tag{1.135}$$

which is another application of the chain rule for differentiation. Now substitute the appropriate derivative from equations (1.133) into the equations (1.134) and (1.135) to obtain the second derivatives

$$\begin{aligned}
\frac{\partial^2\psi}{\partial x^2} &= \frac{\partial}{\partial r}\left(\frac{\partial\psi}{\partial r}\cos\theta - \frac{\partial\psi}{\partial\theta}\frac{\sin\theta}{r}\right)\frac{\partial r}{\partial x} + \frac{\partial}{\partial\theta}\left(\frac{\partial\psi}{\partial r}\cos\theta - \frac{\partial\psi}{\partial\theta}\frac{\sin\theta}{r}\right)\frac{\partial\theta}{\partial x} \\
\frac{\partial^2\psi}{\partial y^2} &= \frac{\partial}{\partial r}\left(\frac{\partial\psi}{\partial r}\sin\theta + \frac{\partial\psi}{\partial\theta}\frac{\cos\theta}{r}\right)\frac{\partial r}{\partial y} + \frac{\partial}{\partial\theta}\left(\frac{\partial\psi}{\partial r}\sin\theta + \frac{\partial\psi}{\partial\theta}\frac{\cos\theta}{r}\right)\frac{\partial\theta}{\partial y}
\end{aligned} \tag{1.136}$$

One can now expand the equations (1.136) to obtain the second derivatives

$$\begin{aligned}
\frac{\partial^2\psi}{\partial x^2} &= \frac{\partial^2\psi}{\partial r^2}\cos^2\theta - 2\frac{\partial^2\psi}{\partial r\partial\theta}\frac{\sin\theta\cos\theta}{r} + 2\frac{\partial\psi}{\partial\theta}\frac{\sin\theta\cos\theta}{r^2} + \frac{\partial\psi}{\partial r}\frac{\sin^2\theta}{r} + \frac{\partial^2\psi}{\partial\theta^2}\frac{\sin^2\theta}{r^2} \\
\frac{\partial^2\psi}{\partial y^2} &= \frac{\partial^2\psi}{\partial r^2}\sin^2\theta + 2\frac{\partial^2\psi}{\partial r\partial\theta}\frac{\sin\theta\cos\theta}{r} - 2\frac{\partial\psi}{\partial\theta}\frac{\sin\theta\cos\theta}{r^2} + \frac{\partial\psi}{\partial r}\frac{\cos^2\theta}{r} + \frac{\partial^2\psi}{\partial\theta^2}\frac{\cos^2\theta}{r^2}
\end{aligned} \tag{1.137}$$

Adding the equations (1.137) one obtains the representation of the Laplace equation in polar coordinates

$$\nabla^2\psi = \frac{\partial^2\psi}{\partial x^2} + \frac{\partial^2\psi}{\partial y^2} = \frac{\partial^2\psi}{\partial r^2} + \frac{1}{r}\frac{\partial\psi}{\partial r} = \frac{1}{r^2}\frac{\partial^2\psi}{\partial\theta^2} = 0 \tag{1.138}$$

∎

Example 1-11.   Second law of thermodynamics

In thermodynamics one is concerned with the following variables

$$p = \text{pressure [N/m}^2\text{]}, \qquad U = \text{internal energy [J]},$$

$$T = \text{temperature [}^\circ\text{K]}, \qquad V = \text{volume [m}^3\text{]}.$$

For a given gas or substance the above variables are related by two equations. Consequently, any two of these variables can be selected as the dependent variables and so the remaining two variables constitute the independent variables. When U and p are selected as the dependent variables, the second law of thermodynamics is written

$$\frac{\partial U}{\partial V} - T\frac{\partial p}{\partial T} + p = 0 \tag{1.139}$$

where each term has units of [N/m^2]. Find the representation of the second law of thermodynamics if U and V are selected the dependent variables.

Solution: By hypothesis, if $U = U(V, T)$ and $p = p(V, T)$, then

$$dU = \frac{\partial U}{\partial V}\,dV + \frac{\partial U}{\partial T}\,dT, \qquad dp = \frac{\partial p}{\partial V}\,dV + \frac{\partial p}{\partial T}\,dT \tag{1.140}$$

implies the second law of thermodynamics has the form

$$\frac{\partial U}{\partial V} - T\frac{\partial p}{\partial T} + p = 0. \tag{1.141}$$

If we assume that $U = U(p, T)$ and $V = V(p, T)$, then

$$dU = \frac{\partial U}{\partial p}\,dp + \frac{\partial U}{\partial T}\,dT, \qquad dV = \frac{\partial V}{\partial p}\,dp + \frac{\partial V}{\partial T}\,dT \tag{1.142}$$

In order to relate the equations (1.140) and (1.142) we solve the equations (1.142) for the quantities dU and dp in terms of dV and dT. For example, take the second of the equations (1.142) and solve for dp to obtain

$$dp = \frac{1}{\frac{\partial V}{\partial p}}\,dV - \frac{\frac{\partial V}{\partial T}}{\frac{\partial V}{\partial p}}\,dT \tag{1.143}$$

and substitute this result into the first of the equations (1.142) to obtain

$$dU = \frac{\frac{\partial U}{\partial p}}{\frac{\partial V}{\partial p}}\,dV - \left(\frac{\frac{\partial U}{\partial p}\frac{\partial V}{\partial T} - \frac{\partial U}{\partial T}\frac{\partial V}{\partial p}}{\frac{\partial V}{\partial p}}\right)\,dT. \tag{1.144}$$

From the equations (1.143) and (1.144) one can obtain the derivatives

$$\frac{\partial U}{\partial V} = \frac{\frac{\partial U}{\partial p}}{\frac{\partial V}{\partial p}}, \qquad \text{and} \qquad \frac{\partial p}{\partial T} = -\frac{\frac{\partial V}{\partial T}}{\frac{\partial V}{\partial p}} \tag{1.145}$$

in terms of the new dependent and independent variables. Substituting the results from equation (1.145) into the second law of thermodynamics, given by equation (1.141), one obtains

$$\frac{\frac{\partial U}{\partial p}}{\frac{\partial V}{\partial p}} + T\frac{\frac{\partial V}{\partial T}}{\frac{\partial V}{\partial p}} + p = 0$$

36

which simplifies to

$$\frac{\partial U}{\partial p} + T\frac{\partial V}{\partial T} + p\frac{\partial V}{\partial p} = 0, \qquad U, V \text{ are dependent variables.} \tag{1.146}$$

This is the form of the second law of thermodynamics using the dependent variables U, V and independent variables p, T.

There are other ways to represent the second law of thermodynamics by selecting a different set of dependent variables. It is left as an exercise to verify the following alternative forms for the second law of thermodynamics

$$\begin{aligned}
\frac{\partial T}{\partial p} - T\frac{\partial V}{\partial U} + p\frac{\partial(V,T)}{\partial(U,p)} &= 0 \quad T, V \text{ are dependent variables} \\
\frac{\partial T}{\partial V} + T\frac{\partial p}{\partial U} - p\frac{\partial T}{\partial U} &= 0 \quad T, p \text{ are dependent variables} \\
T\frac{\partial(p,V)}{\partial(T,U)} - p\frac{\partial V}{\partial U} - 1 &= 0 \quad p, V \text{ are dependent variables} \\
T - p\frac{\partial T}{\partial p} + \frac{\partial(T,U)}{\partial(V,p)} &= 0 \quad T, U \text{ are dependent variables.}
\end{aligned} \tag{1.147}$$

■

Directional derivatives

Given a smooth scalar function of position, defined by $F = F(x, y)$, together with a unit vector $\hat{\mathbf{e}}_\alpha = \cos\alpha\,\hat{\mathbf{e}}_1 + \sin\alpha\,\hat{\mathbf{e}}_2$ determined by an angle α, the directional derivative of F at a point (x_0, y_0) in the direction $\hat{\mathbf{e}}_\alpha$ is defined as follows. Construct a line through the given point (x_0, y_0) in the direction $\hat{\mathbf{e}}_\alpha$. A point on this line can be described by the parametric equations

$$x = x_0 + s\cos\alpha, \qquad y = y_0 + s\sin\alpha \tag{1.148}$$

where s represents a distance along the line measured from the point (x_0, y_0). The situation is illustrated in the figure 1-8.

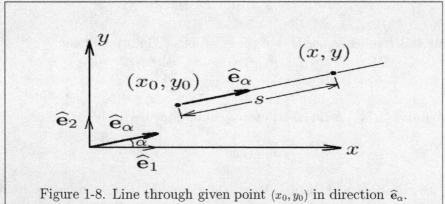

Figure 1-8. Line through given point (x_0, y_0) in direction $\hat{\mathbf{e}}_\alpha$.

Substituting the equations (1.148) into the scalar function of position gives

$$F = F(x, y) = F(x_0 + s\cos\alpha, y_0 + s\sin\alpha) \tag{1.149}$$

which gives a representation of F as a function of distance s along the line in the direction of the unit vector $\widehat{\mathbf{e}}_\alpha$. One can use the chain rule to differentiate the equation (1.149). When this derivative is evaluated at the point where $s = 0$, the derivative is called the directional derivative of F at the point (x_0, y_0) in the direction $\widehat{\mathbf{e}}_\alpha$. The directional derivative is represented using one of the notations

$$\frac{dF}{ds} = \frac{\partial F}{\partial x}\frac{dx}{ds} + \frac{\partial F}{\partial y}\frac{dy}{ds}\bigg|_{s=0} = \frac{\partial F}{\partial x}\cos\alpha + \frac{\partial F}{\partial y}\sin\alpha\bigg|_{s=0}$$

$$\text{or} \qquad \nabla_\alpha F = \frac{\partial F}{\partial x}\cos\alpha + \frac{\partial F}{\partial y}\sin\alpha\bigg|_{s=0} \tag{1.150}$$

Consider the family of level curves $F(x,y) = c$, where c is a constant. The gradient vector, $\operatorname{grad} F$ evaluated at (x_0, y_0), is normal to the special level curve $F(x,y) = c_0$ which passes through the given point (x_0, y_0). The equation (1.150) indicates that the directional derivative can be written in the vector form

$$\nabla_\alpha F = \frac{dF}{ds} = \operatorname{grad} F \cdot \widehat{\mathbf{e}}_\alpha\bigg|_{s=0} = \left(\frac{\partial F}{\partial x}\widehat{\mathbf{e}}_1 + \frac{\partial F}{\partial y}\widehat{\mathbf{e}}_2\right)\cdot(\cos\alpha\,\widehat{\mathbf{e}}_1 + \sin\alpha\,\widehat{\mathbf{e}}_2)\bigg|_{s=0}. \tag{1.151}$$

A physical interpretation of the equation (1.151) is as follows. If the unit vector $\widehat{\mathbf{e}}_\alpha$ is moved to the point where the gradient is represented, then the directional derivative is just a projection of the gradient vector upon the unit vector $\widehat{\mathbf{e}}_\alpha$. A representation of the vectors involved in the projections is illustrated in the figure 1-9.

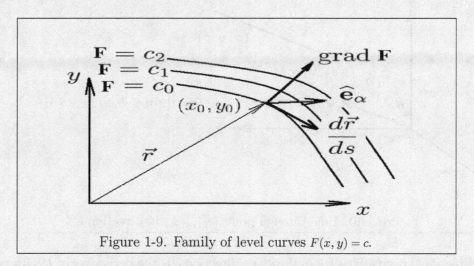

Figure 1-9. Family of level curves $F(x,y) = c$.

Note that the projection of the gradient vector on $\widehat{\mathbf{e}}_\alpha$ is a maximum when $\widehat{\mathbf{e}}_\alpha$ has the same direction as the vector $\operatorname{grad} F$. Thus, the gradient vector of a scalar field, evaluated at a point, always points in the direction of maximum rate of change of the scalar field. The projection of the gradient vector on $\widehat{\mathbf{e}}_\alpha$ is zero when the direction $\widehat{\mathbf{e}}_\alpha$ is perpendicular to the vector $\operatorname{grad} F$. An alternative viewpoint one can take is to treat equation (1.150) as a function of the angle α. Then the directional derivative has a maximum or minimum value when

$$\frac{d}{d\alpha}\left[\frac{dF}{ds}\right] = \frac{d}{d\alpha}\left[\frac{\partial F}{\partial x}\cos\alpha + \frac{\partial F}{\partial y}\sin\alpha\right] = 0$$

$$\text{or} \qquad -\frac{\partial F}{\partial x}\sin\alpha + \frac{\partial F}{\partial y}\cos\alpha = 0$$

(1.152)

Define the angle α_1 as the angle corresponding to the direction of the gradient vector and define the angle α_2 as the angle corresponding to the direction of the unit tangent vector $\frac{d\vec{r}}{ds}$ to the curve $F(x,y) = c_0$ at the point (x_0, y_0) as illustrated in the figure 1-9. Here we have the situation where the directional derivative is zero when the tangent of the angle α_2 satisfies $\tan\alpha_2 = -\frac{\frac{\partial F}{\partial x}}{\frac{\partial F}{\partial y}} = m_2$ and the directional derivative is a maximum when the angle α_1 satisfies $\tan\alpha_1 = \frac{\frac{\partial F}{\partial y}}{\frac{\partial F}{\partial x}} = m_1$. Note that the slopes m_1 and m_2 satisfy $m_1 m_2 = -1$ which shows these directions are perpendicular.

Directional derivatives associated with functions of more than two variables are treated similar to the two dimensional case. For example, if $F = F(x, y, z)$ is a function of three variables, then the directional derivative of F in a direction $\hat{e}$, evaluated at a point (x_0, y_0, z_0), is obtained by first constructing a line through the point (x_0, y_0, z_0) in the direction $\hat{e}$ as illustrated in the figure 1-10, and then calculating the derivative with respect to distance s along this line.

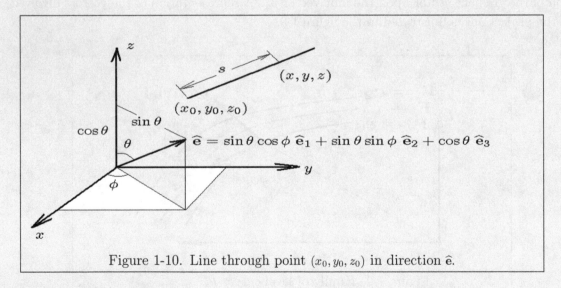

Figure 1-10. Line through point (x_0, y_0, z_0) in direction $\hat{e}$.

The directional derivative of F in the given direction $\hat{e}$ is the derivative of F with respect to the distance s in the direction $\hat{e}$. For example, let the direction $\hat{e}$ be defined by the unit vector $\hat{e} = \sin\theta\cos\phi\,\hat{e}_1 + \sin\theta\sin\phi\,\hat{e}_2 + \cos\theta\,\hat{e}_3$. This represents a unit vector defining a direction in terms of the angles θ and ϕ illustrated in the figure 1-10. Let (x_0, y_0, z_0) denote a given point, then a general point (x, y, z) on the line through (x_0, y_0, z_0) in the direction $\hat{e}$ is given by

$$x = x_0 + s\sin\theta\cos\phi, \quad y = y_0 + s\sin\theta\sin\phi, \quad z = z_0 + s\cos\theta$$

(1.153)

where s is a parameter representing distance along the line measured from the point (x_0, y_0, z_0) in the direction of the unit vector $\widehat{\mathbf{e}}$. Along this line the function F has the values

$$F = F(x, y, z) = F(x_0 + s \sin\theta \cos\phi, y_0 + s \sin\theta \sin\phi, z_0 + s \cos\theta) \tag{1.154}$$

where now F is a function of the distance s along the line as the angles θ and ϕ are treated as constants. The directional derivative is calculated by taking the derivative of F with respect to the distance s along the line. This directional derivative can be written

$$\frac{dF}{ds} = \frac{\partial F}{\partial x}\frac{dx}{ds} + \frac{\partial F}{\partial y}\frac{dy}{ds} + \frac{\partial F}{\partial z}\frac{dz}{ds}$$

and it simplifies to

$$\frac{dF}{ds} = \frac{\partial F}{\partial x}\sin\theta\cos\phi + \frac{\partial F}{\partial y}\sin\theta\sin\phi + \frac{\partial F}{\partial z}\cos\theta$$

which can be written in the vector form

$$\frac{dF}{ds} = \left(\frac{\partial F}{\partial x}\widehat{\mathbf{e}}_1 + \frac{\partial F}{\partial y}\widehat{\mathbf{e}}_2 + \frac{\partial F}{\partial z}\widehat{\mathbf{e}}_3\right) \cdot (\sin\theta\cos\phi\,\widehat{\mathbf{e}}_1 + \sin\theta\sin\phi\,\widehat{\mathbf{e}}_2 + \cos\theta\,\widehat{\mathbf{e}}_3)$$

$$\frac{dF}{ds} = \operatorname{grad} F \cdot \widehat{\mathbf{e}}.$$

Alternatively, one can define the direction cosines of the line through the point (x_0, y_0, z_0) by $\cos\alpha, \cos\beta, \cos\gamma$ where

$$\cos\alpha = \sin\theta\cos\phi, \quad \cos\beta = \sin\theta\sin\phi, \quad \cos\gamma = \cos\theta. \tag{1.155}$$

The directional derivative can then be represented in the form

$$\frac{dF}{ds} = \frac{\partial F}{\partial x}\cos\alpha + \frac{\partial F}{\partial y}\cos\beta + \frac{\partial F}{\partial z}\cos\gamma \tag{1.156}$$

The directional derivative represents the rate of change of F with respect to distance s as s moves in the direction of a unit vector $\widehat{\mathbf{e}}$. Note the following special cases for the directional derivative.

$$\text{If } \widehat{\mathbf{e}} = \widehat{\mathbf{e}}_1, \text{ then } \frac{dF}{ds} = \frac{\partial F}{\partial x}$$
$$\text{If } \widehat{\mathbf{e}} = \widehat{\mathbf{e}}_2, \text{ then } \frac{dF}{ds} = \frac{\partial F}{\partial y} \tag{1.157}$$
$$\text{If } \widehat{\mathbf{e}} = \widehat{\mathbf{e}}_3, \text{ then } \frac{dF}{ds} = \frac{\partial F}{\partial z}.$$

The second directional derivative is defined as a directional derivative of a directional derivative and can be written

$$\frac{d^2 F}{ds^2} = \frac{d}{ds}\left(\frac{dF}{ds}\right) = \operatorname{grad}\left(\frac{dF}{ds}\right) \cdot \widehat{\mathbf{e}}. \tag{1.158}$$

For example, in two dimensions, if $F = F(x, y)$ denotes a scalar field which changes with position, then the first and second directional derivative of F, in a direction $\widehat{\mathbf{e}}_\alpha$ which is to be evaluated at a point (x_0, y_0), are given by the equations

$$\frac{dF}{ds} = \text{grad}\, F \cdot \widehat{\mathbf{e}}_\alpha = \frac{\partial F}{\partial x}\cos\alpha + \frac{\partial F}{\partial y}\sin\alpha$$

$$\frac{d^2F}{ds^2} = \text{grad}\left(\frac{dF}{ds}\right)\cdot\widehat{\mathbf{e}}_\alpha$$

$$\frac{d^2F}{ds^2} = \frac{\partial}{\partial x}\left(\frac{\partial F}{\partial x}\cos\alpha + \frac{\partial F}{\partial y}\sin\alpha\right)\cos\alpha + \frac{\partial}{\partial y}\left(\frac{\partial F}{\partial x}\cos\alpha + \frac{\partial F}{\partial y}\sin\alpha\right)\sin\alpha$$

$$\frac{d^2F}{ds^2} = \frac{\partial^2 F}{\partial x^2}\cos^2\alpha + 2\frac{\partial^2 F}{\partial x \partial y}\sin\alpha\cos\alpha + \frac{\partial^2 F}{\partial y^2}\sin^2\alpha$$

where it is understood that all the derivatives are to be evaluated at the point (x_0, y_0).

Euler's theorem for homogeneous functions

A function $f = f(x, y)$ is said to be homogeneous of degree m in a region R if for all $x, y \in R$ and m having a fixed value, the function f satisfies the relation

$$f(tx, ty) = t^m f(x, y) \tag{1.159}$$

Functions which are homogeneous of degree m satisfy the Euler relation

$$x\frac{\partial f}{\partial x} + y\frac{\partial f}{\partial y} = mf(x, y). \tag{1.160}$$

This result is obtained by differentiating both sides of the equation (1.159) with respect to t and then setting the variable t to have the value $t = 1$.

The Euler's theorem can be generalized to functions of n-variables. If for a fixed value of m a function $f = f(x_1, x_2, \ldots, x_n)$ satisfies the relation

$$f(tx_1, tx_2, \ldots, tx_n) = t^m f(x_1, x_2, \ldots, x_n), \tag{1.161}$$

then the function f satisfies the partial differential equation

$$x_1\frac{\partial f}{\partial x_1} + x_2\frac{\partial f}{\partial x_2} + \cdots + x_n\frac{\partial f}{\partial x_n} = mf(x_1, x_2, \ldots, x_n). \tag{1.162}$$

This result is derived by differentiating the equation (1.161) with respect to t and then setting the variable t to the value $t = 1$.

Taylor series

The Taylor series expansion of a function $f(x)$ of a single real variable x about a point x_0 is given by the power series expansion

$$f(x_0 + h) = f(x_0) + f'(x_0)h + f''(x_0)\frac{h^2}{2!} + f'''(x_0)\frac{h^3}{3!} + \cdots + f^{(n)}(x_0)\frac{h^n}{n!} + R_n \tag{1.163}$$

where R_n is called a remainder term. The remainder term can be written in one of the following forms

$$R_n = f^{(n+1)}(\xi)\frac{h^{n+1}}{(n+1)!}, \qquad x_0 < \xi < x_0 + h \tag{1.164}$$

called the Lagrange remainder term or

$$R_n = \int_{x_0}^{x} \frac{(x-t)^n}{n!} f^{(n+1)}(t)\,dt \tag{1.165}$$

called the integral form of the remainder. Replacing h by $x - x_0$ the Taylor series expansion can also be written in the form

$$f(x) = \sum_{j=0}^{n} \frac{1}{j!} f^{(j)}(x_0)(x-x_0)^j + \frac{1}{(n+1)!} f^{(n+1)}(\xi)(x-x_0)^{(n+1)} \tag{1.166}$$

for $x_0 < \xi < x$.

Example 1-12.   Taylor series

A Taylor series expansion of the function $\sin x$ about the point x_0 which is truncated after the second derivative term is represented

$$\sin x = \sin x_0 + \cos x_0 \frac{(x-x_0)}{1!} - \sin x_0 \frac{(x-x_0)^2}{2!} - \cos\xi \frac{(x-x_0)^3}{3!}$$

where $x_0 < \xi < x$.

■

The Taylor series expansion of a function $f(x,y)$ of two variables x and y about a point (x_0, y_0) is given by the power series

$$f(x_0 + h, y_0 + k) = f(x_0, y_0) + \left(h\frac{\partial}{\partial x} + k\frac{\partial}{\partial y}\right)_0 f + \frac{1}{2!}\left(h\frac{\partial}{\partial x} + k\frac{\partial}{\partial y}\right)_0^2 f$$
$$+ \frac{1}{3!}\left(h\frac{\partial}{\partial x} + k\frac{\partial}{\partial y}\right)_0^3 f + \cdots + \frac{1}{n!}\left(h\frac{\partial}{\partial x} + k\frac{\partial}{\partial y}\right)_0^n f + R_n \tag{1.167}$$

where

$$R_n = \frac{1}{(n+1)!}\left(h\frac{\partial}{\partial x} + k\frac{\partial}{\partial y}\right)_{(\xi,\eta)}^{n+1} f, \qquad x_0 < \xi < x_0 + h, \quad y_0 < \eta < y_0 + k. \tag{1.168}$$

is the Lagrange remainder term. The subscript (ξ, η) in equation (1.168) denotes that the partial derivatives are to be evaluated at an interior point (ξ, η) of the domain given by $R = \{(x,y)\,|\,x_0 \le x \le x_0 + h,\ y_0 \le y \le y_0 + k\}$. The terms $\left(h\frac{\partial}{\partial x} + k\frac{\partial}{\partial y}\right)^m$, with m an integer, are differential operators and the subscript 0 in equation (1.167) denotes that the partial derivatives are to be evaluated at the point (x_0, y_0). Some examples using these operators are

$$\left(h\frac{\partial}{\partial x} + k\frac{\partial}{\partial y}\right) f = h\frac{\partial f}{\partial x} + k\frac{\partial f}{\partial y}$$
$$\left(h\frac{\partial}{\partial x} + k\frac{\partial}{\partial y}\right)^2 f = h^2\frac{\partial^2 f}{\partial x^2} + 2hk\frac{\partial^2 f}{\partial x\partial y} + k^2\frac{\partial^2 f}{\partial y^2}$$
$$\left(h\frac{\partial}{\partial x} + k\frac{\partial}{\partial y}\right)^3 f = h^3\frac{\partial^3 f}{\partial x^3} + 3h^2 k\frac{\partial^3 f}{\partial x^2\partial y} + 3hk^2\frac{\partial^3 f}{\partial x\partial y^2} + k^3\frac{\partial^3 f}{\partial y^3}$$

with higher order terms $\left(h\dfrac{\partial}{\partial x} + k\dfrac{\partial}{\partial y} \right)^n$ being expanded using the binomial expansion.

Replacing h by $(x - x_0)$ and k by $(y - y_0)$, one finds that the Taylor series given by equation (1.167) can also be represented in the form

$$f(x,y) = \sum_{i=0}^{n} \sum_{j=0}^{i} \frac{1}{j!(i-j)!}(x-x_0)^j(y-y_0)^{i-j}\frac{\partial^i f(x_0,y_0)}{\partial x^j \partial y^{i-j}} + R_n \tag{1.169}$$

where

$$R_n = \sum_{j=0}^{n+1} \frac{1}{j!(n+1-j)!}(x-x_0)^j(y-y_0)^{n+1-j}\frac{\partial^{n+1} f(\xi,\eta)}{\partial x^j \partial y^{n+1-j}} \tag{1.170}$$

is the remainder term.

As an example, the Taylor series expansion of a function $f = f(x,y)$ about the point (x_0, y_0) through second order terms can be expressed in terms of a 2×2 matrix in the form

$$f(x,y) = f(x_0,y_0) + \frac{\partial f}{\partial x}\bigg|_{(x_0,y_0)}(x-x_0) + \frac{\partial f}{\partial y}\bigg|_{(x_0,y_0)}(y-y_0) + \frac{1}{2!}\left[(x-x_0)\ (y-y_0)\right]\begin{bmatrix} \frac{\partial^2 f}{\partial x^2} & \frac{\partial^2 f}{\partial x \partial y} \\ \frac{\partial^2 f}{\partial x \partial y} & \frac{\partial^2 f}{\partial y^2} \end{bmatrix}_{(\xi,\eta)}\begin{bmatrix} (x-x_0) \\ (y-y_0) \end{bmatrix}$$

where $x_0 < \xi < x$ and $y_0 < \eta < y$. The 2×2 square matrix with second derivative elements $\frac{\partial^2 f}{\partial x^2}$, $\frac{\partial^2 f}{\partial x \partial y}$ and $\frac{\partial^2 f}{\partial y^2}$ evaluated at (ξ, η) is called the Hessian matrix associated with the function f.

For functions $f = f(x,y,z)$ of three variables the Taylor series expansion about a point (x_0, y_0, z_0) is represented

$$f(x_0 + h, y_0 + k, z_0 + \ell) = f(x_0,y_0,z_0) + \left(h\frac{\partial}{\partial x} + k\frac{\partial}{\partial y} + \ell\frac{\partial}{\partial z} \right)_0 f + \frac{1}{2!}\left(h\frac{\partial}{\partial x} + k\frac{\partial}{\partial y} + \ell\frac{\partial}{\partial z} \right)_0^2 f$$
$$+ \frac{1}{3!}\left(h\frac{\partial}{\partial x} + k\frac{\partial}{\partial y} + \ell\frac{\partial}{\partial z} \right)_0^3 f + \cdots + \frac{1}{n!}\left(h\frac{\partial}{\partial x} + k\frac{\partial}{\partial y} + \ell\frac{\partial}{\partial z} \right)_0^n f + R_{n+1} \tag{1.171}$$

where the subscript 0 denotes that the derivatives are to be evaluated at the point (x_0, y_0, z_0).

Functions with more than three variables have similar Taylor series expansions. Consider a function $f = f(x_1, x_2, x_3, \ldots, x_n)$ of n-variables. Let $\bar{x}_0 = (x_1^0, x_2^0, x_3^0, \ldots, x_n^0)$ denote a fixed point, then a Taylor series expansion of the function f about the point $\bar{x}_0$ through second order terms can be written

$$f(\bar{x}) = f(\bar{x}_0) + \frac{\partial f}{\partial x_1}\bigg|_{\bar{x}_0}(x_1 - x_1^0) + \frac{\partial f}{\partial x_2}\bigg|_{\bar{x}_0}(x_2 - x_2^0) + \cdots + \frac{\partial f}{\partial x_n}\bigg|_{\bar{x}_0}(x_n - x_n^0) +$$

$$\frac{1}{2!}[(x_1 - x_1^0), (x_2 - x_2^0), \ldots, (x_n - x_n^0)]\begin{bmatrix} \frac{\partial^2 f}{\partial x_1^2} & \frac{\partial^2 f}{\partial x_1 \partial x_2} & \cdots & \frac{\partial^2 f}{\partial x_1 \partial x_n} \\ \frac{\partial^2 f}{\partial x_2 \partial x_1} & \frac{\partial^2 f}{\partial x_2^2} & \cdots & \frac{\partial^2 f}{\partial x_2 \partial x_n} \\ \vdots & \vdots & \ddots & \vdots \\ \frac{\partial^2 f}{\partial x_n \partial x_1} & \frac{\partial^2 f}{\partial x_n \partial x_2} & \cdots & \frac{\partial^2 f}{\partial x_n^2} \end{bmatrix}_{\bar{\xi}}\begin{bmatrix} (x_1 - x_1^0) \\ (x_2 - x_2^0) \\ \vdots \\ (x_n - x_n^0) \end{bmatrix}$$

where the last term contains the quadratic form $(\bar{x} - \bar{x}_0)H(\bar{\xi})(\bar{x} - \bar{x}_0)^T$ with H a $n \times n$ square matrix having $\frac{\partial^2 f}{\partial x_i \partial x_j}$, evaluated at $\bar{\xi}$, for the element in the ith row and jth column of H. The matrix H is called the Hessian matrix of f at the point $\bar{\xi}$.

Linear differential equations

Define the n-th order linear differential operator

$$L(y) = a_0(x)\frac{d^n y}{dx^n} + a_1(x)\frac{d^{n-1}y}{dx^{n-1}} + \cdots + a_{n-2}(x)\frac{d^2 y}{dx^2} + a_{n-1}(x)\frac{dy}{dx} + a_n(x)y \qquad (1.172)$$

with real coefficients $a_0, a_1, \ldots, a_n$ with $a_0(x)$ not identically zero. The differential equation $L(y) = 0$ is called a n-th order linear homogeneous differential equation while the equation $L(y) = f(x)$ is called a n-th order linear nonhomogeneous differential equation. The general procedure to solve the n-th order linear nonhomogeneous differential equation $L(y) = f(x)$ is as follows.

(i) Determine a particular solution y_p which satisfies $L(y_p) = f(x)$.

(ii) Find n-independent solutions or fundamental set of solutions $\{y_1, y_2, \ldots, y_n\}$ to the n-th order linear homogeneous equation $L(y) = 0$.

(iii) The general solution y_h to the homogeneous differential equation $L(y) = 0$ is then any linear combination of the functions from the fundamental set of solutions and one can write this general solution in the form

$$y_h = c_1 y_1 + c_2 y_2 + \cdots + c_n y_n \qquad (1.173)$$

where $c_1, c_2, \ldots, c_n$ are arbitrary constants.

(iv) The general solution to the nonhomogeneous differential equation $L(y) = f(x)$ is then given by

$$y = y_h + y_p \qquad \text{where} \quad L(y_h) = 0 \quad \text{and} \quad L(y_p) = f(x) \qquad (1.174)$$

Constant coefficients

Whenever the coefficients in the linear operator (1.172) are real constants, then one usually assumes an exponential solution $y = e^{mx}$ to the linear homogeneous differential equation

$$L(y) = a_0 y^{(n)} + a_1 y^{(n-1)} + \cdots + a_{n-1}y'' + a_{n-1}y' + a_n y = 0 \qquad (1.175)$$

Substituting $y = e^{mx}$ into the homogeneous differential equation (1.175) produces the characteristic or auxiliary equation

$$a_0 m^n + a_1 m^{n-1} + \cdots + a_{n-1}m + a_n = 0 \qquad (1.176)$$

This is an n-th order algebraic equation with n-characteristic roots $\{m_1, m_2, \ldots, m_n\}$.

Case 1: (Distinct roots)

If $\{m_1, m_2, \ldots, m_n\}$ are distinct roots of the characteristic equation, then $\{e^{m_1 x}, e^{m_2 x}, \ldots, e^{m_n x}\}$ is a fundamental set of solutions and so the general solution can be written

$$y_h = c_1 e^{m_1 x} + c_2 e^{m_2 x} + \cdots + c_n e^{m_n x} \qquad (1.177)$$

where $c_1, c_2, \ldots, c_n$ are arbitrary constants.

Case 2: (Repeated real roots)

If the characteristic equation (1.176) has a repeated root m, occuring k times, it is called a root of multiplicity k. That part of the fundamental set of solutions which corresponds to the k repeated roots $\underbrace{\{m, m, \ldots, m\}}_{k \ values}$ is given by

$$\{e^{mx}, xe^{mx}, x^2 e^{mx}, \ldots, x^{k-1} e^{mx}\}$$

If the remaining roots are distinct, then the fundamental set corresponding to these roots is given by

$$\{e^{m_{k+1}x}, e^{m_{k+2}x}, \ldots, e^{m_n x}\}$$

The general solution to the homogeneous equation with constant coefficients can then be written

$$y_h = \left(c_1 + c_2 x + c_3 x^2 + \cdots + c_k x^{k-1}\right) e^{mx} + c_{k+1} e^{m_{k+1}x} + c_{k+2} e^{m_{k+2}x} + \cdots + c_n e^{m_n x}$$

where $c_1, c_2, \ldots, c_n$ are arbitrary constants.

Case 3: (Conjugate roots)

Whenever the characteristic equation (1.176) has a root which is a complex number $m_1 = a + ib$, the complex conjugate $m_2 = a - ib$ must also be a root. This is because equation (1.176) is assumed to have real coefficients. That is, complex roots of the characteristic equation (1.176) occur in conjugate pairs $a \pm ib$. The corresponding part of the fundamental set of solutions corresponding to these conjugate roots is given by

$$\{y_1, y_2\} = \{e^{m_1 x}, e^{m_2 x}\} = \{e^{(a+ib)x}, e^{(a-ib)x}\}$$

One can then employ the Euler identity

$$e^{i\theta} = \cos\theta + i\sin\theta \tag{1.178}$$

and write

$$y_1 = e^{ax}(\cos bx + i \sin bx) \qquad y_2 = e^{ax}(\cos bx - i \sin bx)$$

as solutions to the homogeneous equation $L(y) = 0$. One can then form the linear combinations

$$Y_1 = \frac{1}{2} y_1 + \frac{1}{2} y_2 = e^{ax} \cos bx \qquad \text{and} \qquad Y_2 = \frac{1}{2i} y_1 - \frac{1}{2i} y_2 = e^{ax} \sin bx$$

then $\{Y_1, Y_2\}$ are real solutions to the homogeneous differential equation $L(y) = 0$. That part of the fundamental set of solutions which corresponds to the conjugate root $a \pm ib$ can then be written in the form

$$\{e^{ax} \cos bx, \ e^{ax} \sin bx\}$$

If the remaining roots are distinct, then the fundamental set corresponding to these roots is given by

$$\{e^{m_3 x}, e^{m_4 x}, \ldots, e^{m_n x}\}$$

The general solution to the homogeneous equation with constant coefficients can then be written

$$y_h = c_1 e^{ax} \cos bx + c_2 e^{ax} \sin bx + c_3 e^{m_3 x} + c_4 e^{m_4 x} + \cdots + c_n e^{m_n x}$$

where $c_1, c_2, \ldots, c_n$ are arbitrary constants.

Case 4: (Repeated conjugate roots)

If $a \pm ib$ are conjugate roots of the characteristic equation (1.176) which are multiple roots of multiplicity k, then the corresponding solutions in the fundamental set of solutions can be written

$$\{e^{ax}\sin bx, xe^{ax}\sin bx, x^2 e^{ax}\sin bx, \ldots, x^{k-1}e^{ax}\sin bx, e^{ax}\cos bx, xe^{ax}\cos bx, x^2 e^{ax}\cos bx, \ldots, x^{k-1}e^{ax}\cos bx\}$$

If the remaining roots are distinct roots $\{m_{2k+1}, m_{2k+2}, \ldots, m_n\}$, then the general solution to the homogeneous differential equation can be written

$$y_h = (c_1 + c_2 x + \ldots + c_k x^{k-1})e^{ax}\cos bx + (c_{k+1} + c_{k+2}x + \cdots + c_{2k}x^{k-1})e^{ax}\sin bx + c_{2k+1}e^{m_{2k+1}x} + \cdots + c_n e^{m_n x}$$

where $c_1, c_2, \ldots, c_n$ are arbitrary constants.

Example 1-13.

Find the general solution to the differential equation $\dfrac{d^2 y}{dx^2} - y = 0$

Solution:

Assume an exponential solution $y = e^{mx}$ and substitute y and its derivatives into the given differential equation to obtain the characteristic equation $m^2 - 1 = 0$ which can also be written in the factored form $(m-1)(m+1) = 0$. The characteristic roots are therefore $\{1, -1\}$ and the fundamental set of solutions can be expressed $\{e^x, e^{-x}\}$. The general solution is any linear combination of functions from the fundamental set and can be written in the form $y_h = c_1 e^x + c_2 e^{-x}$.

Note that one can form any linear combination of functions from the fundamental set to construct other solutions to the differential equation. For example the linear combinations

$$\frac{1}{2}e^x + \frac{1}{2}e^{-x} = \cosh x \qquad \text{and} \qquad \frac{1}{2}e^x - \frac{1}{2}e^{-x} = \sinh x$$

produce the hyperbolic functions $\sinh x$ and $\cosh x$ which are also solutions to the given differential equations as one can readily verify by substitution of these functions into the given differential equation. One could then use the set $\{\cosh x, \sinh x\}$ as the fundamental set and write the general solution in the form $y_h = C_1 \cosh x + C_2 \sinh x$ where C_1 and C_2 are arbitrary constants.

∎

Example 1-14.

Find the general solution to the differential equation $\dfrac{d^2 y}{dx^2} + 2\dfrac{dy}{dx} + y = 0$

Solution:

Assume an exponential solution $y = e^{mx}$ and verify that this produces the characteristic equation $m^2 + 2m + 1 = 0$ which can be written in the factored form $(m+1)(m+1) = 0$. The characteristic equation has the characteristic roots $\{-1, -1\}$ which are repeated roots. The characteristic roots are used to produce the fundamental set of solutions $\{e^{-x}, xe^{-x}\}$. The general solution is then any linear combination of functions from the fundamental set and can be written in the form $y_h = (c_1 + c_2 x)e^{-x}$.

∎

46

Example 1-15.

Find the general solution to the differential equation $\frac{d^2y}{dx^2} + 2\frac{dy}{dx} + 2y = 0$

Solution:

Assume an exponential solution $y = e^{mx}$ and verify that the characteristic equation can be written $m^2 + 2m + 2 = (m+1)^2 + 1 = 0$. One obtains the characteristic roots $\{-1+i, -1-i\}$ which are conjugate roots. The characteristic roots produce the fundamental set of solutions $\{e^{-x}\cos x, e^{-x}\sin x\}$. The general solution is written as any linear combination of the functions in the fundamental set of solutions. This gives the general solution $y_h = e^{-x}\left(c_1\cos x + c_2\sin x\right)$.

∎

Example 1-16.

Find the general solution to the differential equation

$$\frac{d^9y}{dx^9} + 20\frac{d^8y}{dx^8} + 193\frac{d^7y}{dx^7} + 1168\frac{d^6y}{dx^6} + 4811\frac{d^5y}{dx^5} + 13828\frac{d^4y}{dx^4} + 27367\frac{d^3y}{dx^3} + 35192\frac{d^2y}{dx^2} + 25948\frac{dy}{dx} + 8112y = 0$$

Solution:

Assume a solution to the linear homogeneous differential equation of the form $y = e^{mx}$ with derivatives of the form $\frac{d^n y}{dx^n} = m^n e^{mx}$ for $n = 1, 2, 3, \ldots, 9$. Substitute y and its derivatives into the differential equation to obtain the characteristic equation

$$m^9 + 20\,m^8 + 193\,m^7 + 1168\,m^6 + 4811\,m^5 + 13828\,m^4 + 27367\,m^3 + 35192\,m^2 + 25948\,m + 8112 = 0$$

One can use synthetic division and show that the characteristic equation can be written in the factored form

$$(m + 2 \mp 3i)^2(m+2)^2(m+1)(m+3)(m+4) = 0$$

The characteristic roots are given by

$$\{-2 \pm 3i, -2 \pm 3i, -2, -2, -1, -3, -4\}$$

The roots $-2 \pm 3i$ are repeated roots and the root -2 is a repeated root. The remaining roots are distinct. The fundamental set of solutions corresponding to the characteristic roots is given by

$$\{e^{-2x}\cos 3x, e^{-2x}\sin 3x, xe^{-2x}\cos 3x, xe^{-2x}\sin 3x, e^{-2x}, xe^{-2x}, e^{-x}, e^{-3x}, e^{-4x}\}$$

The general solution to the given differential equation is then any linear combination of functions from the fundamental set of solutions and can be written in the form

$$y_h = (c_1 + c_2 x)\,e^{-2x}\cos 3x + (c_3 + c_4 x)\,e^{-2x}\sin 3x + (c_5 + c_6 x)\,e^{-2x} + c_7 e^{-x} + c_8 e^{-3x} + c_9 e^{-4x}$$

where $c_1, c_2, \ldots, c_9$ are arbitrary constants.

∎

Exercises Chapter 1

▶ **1.**

Find the first derivatives $\dfrac{dy}{dx}$ and define the region where the given functions are well defined.

(a) $y = \sin^{-1} x$ (b) $y = \sinh^{-1} x$ (c) $x = y/(y^2 - x^2)$

▶ **2.**   Given that $x = t^2$ and $y = t^3$, find $\frac{dy}{dx}$ and $\frac{d^2y}{dx^2}$.

▶ **3.**   Given that $x = 1 - t^2$ and $y = t - t^2$, find $\frac{dy}{dx}$ and $\frac{d^2y}{dx^2}$.

▶ **4.**   Find the first and second partial derivatives $\dfrac{\partial w}{\partial x}$, $\dfrac{\partial w}{\partial y}$, $\dfrac{\partial^2 w}{\partial x^2}$, $\dfrac{\partial^2 w}{\partial x \partial y}$ and $\dfrac{\partial^2 w}{\partial y^2}$ associated with the given functions.

(a) $w = e^x \cos(y^2)$ (b) $w = \ln(x^2 + y^2)$ (c) $w = \tan^{-1} \dfrac{y}{x}$

▶ **5.**   For a, b, c, h constants, use the Green's theorem in the plane to find the area of the given figures.

(a)
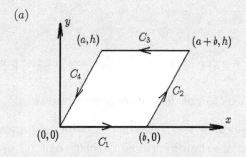

(b)
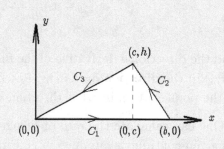

▶ **6.**

(a) Find the directional derivative of the function $f = f(x, y) = xy^2 + x^2y$ at the point $(1, 2)$ in the direction $\widehat{e} = \frac{1}{\sqrt{2}} \widehat{e}_1 + \frac{1}{\sqrt{2}} \widehat{e}_2$.

(b) At the point $(1, 2)$, in what direction is the directional derivative a maximum?

[†] BCE ("Before Common Era") replaces B.C. ("Before Christ") usage.

► **7.**

(a) Use the Green's theorem to show that the area A of the triangle formed by the three points (x_1, y_1), (x_2, y_2) and (x_3, y_3) can be represented by the determinant

$$A = \pm \frac{1}{2} \begin{vmatrix} 1 & x_1 & y_1 \\ 1 & x_2 & y_2 \\ 1 & x_3 & y_3 \end{vmatrix}$$

Note that the sign is selected to make the area positive.

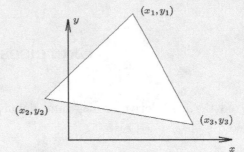

(b) Show that if the area is zero then the three points lie on a straight line.

► **8.** Given an implicit function $f(x,y) = 0$. Show that if $\frac{\partial f}{\partial y} \neq 0$, then

(a) $\quad \dfrac{dy}{dx} = -\dfrac{\dfrac{\partial f}{\partial x}}{\dfrac{\partial f}{\partial y}},$ $\qquad$ (b) $\quad \dfrac{d^2 y}{dx^2} = -\dfrac{\dfrac{\partial^2 f}{\partial x^2}\left(\dfrac{\partial f}{\partial y}\right)^2 - 2\dfrac{\partial^2 f}{\partial x \partial y}\dfrac{\partial f}{\partial x}\dfrac{\partial f}{\partial y} + \dfrac{\partial^2 f}{\partial y^2}\left(\dfrac{\partial f}{\partial x}\right)^2}{\left(\dfrac{\partial f}{\partial y}\right)^3}$

► **9.** Differentiate the given equations to find the first derivatives $\dfrac{dy}{dx}$.

$\quad$ (a) $\quad x^3 + y^3 = 27$ $\qquad$ (c) $\quad \arctan\dfrac{y}{x} + \cos(x+y) - x^2 = 0$

$\quad$ (b) $\quad e^x + e^y = e^x e^y$ $\qquad$ (d) $\quad \ln(x^2 + y^2) - x^2 - y^2 = 0$

► **10.** If $f(u,v) = 0$ with $u = \alpha x + \beta y + \gamma z$ and $v = x^2 + y^2 + z^2$, with α, β, γ constants, then show that z satisfies the partial differential equation

$$(\gamma y - \beta z)\frac{\partial z}{\partial x} + (\alpha z - \gamma x)\frac{\partial z}{\partial y} = (\beta x - \alpha y).$$

► **11.**

(a) Find the directional derivative of the function $f = f(x,y,z) = xyz + 3x^2 z - 4xy^2$ at the point $(1,1,1)$ in the direction $\hat{\mathbf{e}} = \frac{2}{7}\hat{\mathbf{e}}_1 + \frac{6}{7}\hat{\mathbf{e}}_2 + \frac{3}{7}\hat{\mathbf{e}}_3$.

(b) At the point $(1,1,1)$, in what direction is the directional derivative a maximum?

► **12.** One can calculate the rate of change of a function $f = f(x,y)$ as the point (x,y) moves along a given continuous curve $y = y(x)$. First calculate the directional derivative of $f(x,y)$ in the direction of the tangent to the given curve and evaluate this directional derivative at the point (x,y) on the given curve where the tangent is constructed. Show this derivative is

$$\frac{df}{ds} = \frac{\frac{\partial f}{\partial x} + \frac{\partial f}{\partial y}y'(x)}{\sqrt{1 + [y'(x)]^2}} \quad \text{evaluated at point } (x,y) \text{ on the given curve } y = y(x).$$

This defines a directional derivative of $f = f(x,y)$ as the point (x,y) moves along the given continuous curve $y = y(x)$.

▶ **13.** Show that the function $f = f(x, y + \epsilon\eta, y' + \epsilon\eta')$, where $y' = \frac{dy}{dx}$, $\eta' = \frac{d\eta}{dx}$ and ϵ is a small constant, can be expanded in a Taylor series in powers of ϵ having the form

$$f(x, y + \epsilon\eta, y' + \epsilon\eta') = f(x,y,y') + \frac{\epsilon}{1!}\frac{\partial f(x, y + \epsilon\eta, y' + \epsilon\eta')}{\partial \epsilon}\bigg|_{\epsilon=0} + \frac{\epsilon^2}{2!}\frac{\partial^2 f(x, y + \epsilon\eta, y' + \epsilon\eta')}{\partial \epsilon^2}\bigg|_{\epsilon=0}$$

$$+ \cdots + \frac{\epsilon^{(n-1)}}{(n-1)!}\frac{\partial^{(n-1)} f(x, y + \epsilon\eta, y' + \epsilon\eta')}{\partial \epsilon^{(n-1)}}\bigg|_{\epsilon=0} + R_n$$

where R_n is the Lagrange remainder term. Calculate $\dfrac{\partial f}{\partial \epsilon}\bigg|_{\epsilon=0}$ and $\dfrac{\partial^2 f}{\partial \epsilon^2}\bigg|_{\epsilon=0}$.

▶ **14.** Show that with the change of variables $x = u - v$ and $y = u + v$ the quantity $\dfrac{\partial^2 \phi}{\partial x^2} - \dfrac{\partial^2 \phi}{\partial y^2}$ becomes $-\dfrac{\partial^2 \phi}{\partial u \partial v}$

▶ **15.** Show that with the change of variable $x = u\cosh v$ and $y = u\sinh v$ the quantity $\dfrac{\partial^2 \phi}{\partial x^2} - \dfrac{\partial^2 \phi}{\partial y^2}$ becomes $\dfrac{\partial^2 \phi}{\partial u^2} - \dfrac{1}{u^2}\dfrac{\partial^2 \phi}{\partial v^2} + \dfrac{1}{u}\dfrac{\partial \phi}{\partial u}$

▶ **16.** Find the first and second partial derivatives of the given functions.

$$(a) \quad f = xy^2 + yx^2, \qquad (b) \quad f = \sqrt{x^2 - y^2}, \qquad (c) \quad f = e^x \cos y$$

▶ **17.** Determine $\dfrac{\partial z}{\partial x}$ and $\dfrac{\partial z}{\partial y}$ from the given equations

$$(a) \quad x + y + z = xyz, \qquad (b) \quad x^2 + y^2 + z^2 = 1, \qquad (c) \quad xy + xz + \sin^{-1}\frac{y}{z} = 0$$

▶ **18.**

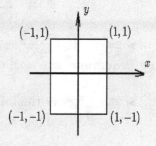

(a) Find all points (x, y) and directions α where the directional derivative of $F = F(x, y) = x^2 + xy$ is a maximum if the points (x, y) are restricted to lie on the perimeter of the square illustrated.

(b) Evaluate the line integral $I = \oint_C F(x, y)\, ds$ where C is the perimeter of the square and ds denotes an element of arc length.

▶ **19.** Examine the system of linear equations $x = \alpha_0 u + \beta_0 v + \gamma_0$ and $y = \alpha_1 u + \beta_1 v + \gamma_1$ where $\alpha_0, \alpha_1, \beta_0, \beta_1, \gamma_0, \gamma_1$ are constants.

(a) Show that the Jacobian of the transformation is given by $\frac{\partial(x,y)}{\partial(u,v)} = \alpha_0\beta_1 - \beta_0\alpha_1$

(b) Show that if the Jacobian of the transformation is different from zero, then one can solve the given equations for the variables u and v.

(c) Show that if the Jacobian of the transformation is zero, then the inverse transformation does not exist.

(d) Show that if the Jacobian of the transformation is zero, then there exists a relationship between x and y given by $\beta_1 x - \beta_0 y = \gamma_0\beta_1 - \beta_0\gamma_1$

▶ **20.** Consider a general fixed point P_0 having coordinates (x_0, y_0, z_0) on the surface of the paraboloid $z = \alpha x^2 + \beta y^2$ where α and β are constants. Find the equation of the tangent plane to the surface at the point P_0.

▶ **21.** Show that if $u = u(x, y)$ and $v = v(x, y)$ are well defined functions which are everywhere differentiable and have derivatives which satisfy the equations $\dfrac{\partial u}{\partial x} = \dfrac{\partial v}{\partial y}$ and $\dfrac{\partial u}{\partial y} = -\dfrac{\partial v}{\partial x}$, then the directional derivatives satisfy $\nabla_\alpha u = \nabla_{\alpha + \frac{\pi}{2}} v$ for all angles α.

▶ **22.** Given a function $U = U(x, y)$ which is everywhere continuous and differentiable together with a transformation from rectangular (x, y)-coordinates to polar (r, θ) coordinates given by the equations $x = x(r, \theta) = r \cos \theta$ and $y = y(r, \theta) = r \sin \theta$. The coordinate curves in polar coordinates are given by the lines $\theta = a$ *constant* and the circles $r = a$ *constant*.

(a) Show the directional derivative of $U = U(x, y)$ in the direction of a polar coordinate line $\theta = a$ *constant* is given by $\nabla_\theta U = \dfrac{\partial U}{\partial r}$

(b) Show the directional derivative of $U = U(x, y)$ in the direction of the tangent to a coordinate circle $r = a$ *constant* is given by $\nabla_{\theta + \frac{\pi}{2}} U = \dfrac{1}{r} \dfrac{\partial U}{\partial \theta}$

▶ **23.** **Chain rule differentiation**

(a) If $w = w(x, y)$ where $x = x(t)$ and $y = y(t)$, then find $\frac{dw}{dt}$

(b) If $w = w(x, y)$ where $x = x(u, v)$ and $y = y(u, v)$, then find   (i) $\dfrac{\partial w}{\partial u}$ (ii) $\dfrac{\partial w}{\partial v}$

(c) If $w = w(x_1, x_2, \ldots, x_n)$ where

$x_1 = x_1(u_1, u_2, \ldots, u_m)$, $x_2 = x_2(u_1, u_2, \ldots, u_m), \ldots, x_n = x_n(u_1, u_2, \ldots, u_m)$, then find

(i) $\dfrac{\partial w}{\partial u_1}$ (ii) $\dfrac{\partial w}{\partial u_2}$ $\ldots$ (iii) $\dfrac{\partial w}{\partial u_m}$

▶ **24.** If $z = x^2 \sin y$ where x and y are implicit functions of t which are defined by the equations $x^2 + \sin x + t^2 + 2t - 5 = 0$, $ty + y + y^2 - t^2 = 0$, then find $\dfrac{dz}{dt}$.

▶ **25.** Verify the following alternative forms for the second law of thermodynamics
(See pages 35-36)

(a) $\dfrac{\partial T}{\partial p} - T \dfrac{\partial V}{\partial U} + p \dfrac{\partial(V, T)}{\partial(U, p)} = 0$ T, V are dependent variables

(b) $\dfrac{\partial T}{\partial V} + T \dfrac{\partial p}{\partial U} - p \dfrac{\partial T}{\partial U} = 0$ T, p are dependent variables

(c) $T \dfrac{\partial(p, V)}{\partial(T, U)} - p \dfrac{\partial V}{\partial U} - 1 = 0$ p, V are dependent variables

(d) $T - p \dfrac{\partial T}{\partial p} + \dfrac{\partial(T, U)}{\partial(V, p)} = 0$ T, U are dependent variables.

▶ **26.** Use the Green's theorem in the plane to find the area of the ellipse defined by the parametric equations $x = a \cos \theta$, $y = b \sin \theta$ where a and b are nonzero positive constants.

▶ **27.** Given an equation of the form $F(P, V, T) = 0$ relating pressure (P), volume (V) and temperature (T) of a gas. If the coefficient of expansion of the gas is given by $\alpha = \dfrac{1}{V}\left(\dfrac{dV}{dT}\right)_P$ with the pressure held constant and the gas has a modulus of elasticity given by $E = -V\left(\dfrac{dP}{dV}\right)_T$ when the temperature is held constant. Show that the rate of change of pressure with respect to temperature is given by $\left(\dfrac{dP}{dT}\right)_V = \alpha E$ when the volume is held constant.

▶ **28.**

(a) Show that under a change of variables $x = x(u, v)$ and $y = y(u, v)$ an element of area $dA = dxdy$ is transformed to the form $dA = \dfrac{\partial(x, y)}{\partial(u, v)}\, du\, dv$

(b) Show that under a change of variables $x = x(u, v, w)$, $y = y(u, v, w)$, $z = z(u, v, w)$ an element of volume $d\tau = dxdydz$ is transformed to the form $d\tau = \dfrac{\partial(x, y, z)}{\partial(u, v, w)}\, du dv dw$

   Hint: Sketch some coordinate curves.

▶ **29.** For $\vec{F} = xy\,\widehat{e}_1 + yz\,\widehat{e}_2 + x^2\,\widehat{e}_3$ calculate the flux integral $I = \displaystyle\iint_S \vec{F}\cdot\widehat{n}\, d\sigma$ where S is that portion of the paraboloid $z = 1 - x^2 - y^2$ above the plane $z = 0$ and $\widehat{n}$ is a unit outward normal to the surface.

▶ **30.**

(a) Given a transformation of coordinates, say $x = x(u, v)$ and $y = y(u, v)$ having the Jacobian $J = \frac{\partial(x,y)}{\partial(u,v)}$ different from zero. Show that for the inverse transformation the following derivatives exist

$$(i)\quad \frac{\partial u}{\partial x} = \frac{1}{J}\frac{\partial y}{\partial v}, \qquad (ii)\quad \frac{\partial u}{\partial y} = \frac{-1}{J}\frac{\partial x}{\partial v}, \qquad (iii)\quad \frac{\partial v}{\partial x} = \frac{-1}{J}\frac{\partial y}{\partial u}, \qquad (iv)\quad \frac{\partial v}{\partial y} = \frac{1}{J}\frac{\partial x}{\partial u}$$

(b) Given the coordinate transformation $x = r\cos\theta$, $y = r\sin\theta$, find the following derivatives

$$(i)\quad \frac{\partial r}{\partial x}, \qquad (ii)\quad \frac{\partial r}{\partial y}, \qquad (iii)\quad \frac{\partial\theta}{\partial x}, \qquad (iv)\quad \frac{\partial\theta}{\partial y}$$

▶ **31.** Given a transformation of coordinates, say $x = x(u, v, w)$, $y = y(u, v, w)$ and $z = z(u, v, w)$ with nonzero Jacobian $J = \frac{\partial(x,y,z)}{\partial(u,v,w)}$, show that for the inverse transformation the following derivatives exist.

$$(a)\quad \frac{\partial u}{\partial x} = \frac{1}{J}\frac{\partial(y, z)}{\partial(v, w)}, \qquad (b)\quad \frac{\partial u}{\partial y} = \frac{1}{J}\frac{\partial(z, x)}{\partial(v, w)}, \qquad (c)\quad \frac{\partial u}{\partial z} = \frac{1}{J}\frac{\partial(x, y)}{\partial(v, w)}$$

$$(d)\quad \frac{\partial v}{\partial x} = \frac{1}{J}\frac{\partial(y, z)}{\partial(w, u)}, \qquad (e)\quad \frac{\partial v}{\partial y} = \frac{1}{J}\frac{\partial(z, x)}{\partial(w, u)}, \qquad (f)\quad \frac{\partial v}{\partial z} = \frac{1}{J}\frac{\partial(x, y)}{\partial(w, u)}$$

$$(g)\quad \frac{\partial w}{\partial x} = \frac{1}{J}\frac{\partial(y, z)}{\partial(u, v)}, \qquad (h)\quad \frac{\partial w}{\partial y} = \frac{1}{J}\frac{\partial(z, x)}{\partial(u, v)}, \qquad (i)\quad \frac{\partial w}{\partial z} = \frac{1}{J}\frac{\partial(x, y)}{\partial(u, v)}$$

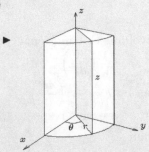

The transformation from rectangular (x, y, z) coordinates to cylindrical (r, θ, z) coordinates is given by the equations
$$x = r\cos\theta, \qquad y = r\sin\theta, \qquad z = z$$

(a) Calculate the Jacobian $\frac{\partial(x,y,z)}{\partial(r,\theta,z)}$
(b) Calculate the inverse transformation
(c) Calculate the derivatives (*i*) $\frac{\partial r}{\partial y}$, (*ii*) $\frac{\partial\theta}{\partial x}$, (*iii*) $\frac{\partial\theta}{\partial y}$

▶ **33.**

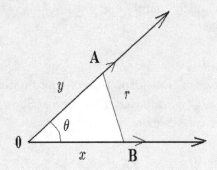

The transformation from rectangular (x, y, z) coordinates to spherical (ρ, θ, ϕ) coordinates is given by the equations
$$x = \rho\sin\theta\cos\phi, \qquad y = \rho\sin\theta\sin\phi, \qquad z = \rho\cos\theta$$

(a) Calculate the Jacobian $\frac{\partial(x,y,z)}{\partial(\rho,\theta,\phi)}$
(b) Calculate the inverse transformation
(c) Calculate the derivatives (*i*) $\frac{\partial\rho}{\partial y}$, (*ii*) $\frac{\partial\theta}{\partial z}$, (*iii*) $\frac{\partial\phi}{\partial x}$

▶ **34.**
Two highways intersect at a constant angle of θ as illustrated. Assume $0 < \theta < \frac{\pi}{2}$ with car A is moving away from the intersection on one highway with constant velocity v_A while simultaneously car B is moving away from the intersection on the other highway with constant velocity v_B. Find the rate $\frac{dr}{dt}$ at which the distance $r = \overline{AB}$ is increasing at the instant when $\overline{OA} = y$ and $\overline{OB} = x$.

▶ **35.** Evaluate the given line integrals around the square illustrated

(a) $I_a = \oint_C x\,dy - y\,dx$

(b) $I_b = \oint_C (x^2 + y^2)\,ds$

(c) $I_c = \oint_C xy\,dx + (x + y)\,dy$

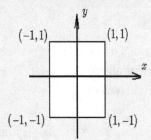

▶ **36.** Evaluate the surface integral $I = \iint_S (\nabla \times \vec{F}) \cdot \hat{n}\,d\sigma$ where $\vec{F}$ represents the vector field $\vec{F} = x\,\hat{e}_1 + (y - 2xz)\,\hat{e}_2 - yz\,\hat{e}_3$ and S is the surface of the sphere $x^2 + y^2 + z^2 = 1$ above the xy-plane with $\hat{n}$ a unit outward normal to the surface.

▶ **37.**

Evaluate the given line integrals along the curve specified.

(a) $I_a = \int_C y\,dx + z\,dy + x\,dz$ along the straight line from $(0,0,0)$
    to $(1,1,1)$

(b) $I_b = \int_C y\,dx + z\,dy + x\,dz$
    along the curve $\vec{r} = \vec{r}(t) = t\,\widehat{e}_1 + t^2\,\widehat{e}_2 + t^3\,\widehat{e}_3$ where $0 \le t \le 1$.

(c) $I_c = \int_C yz\,dx + xz\,dy + xy\,dz$

from $(0,0,0)$ to $(1,1,1)$ along the path segments $C_1 + C_2 + C_3 = C$
illustrated in the figure.

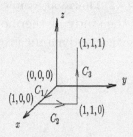

▶ **38.** Evaluate the surface integral $I = \oiint_S \vec{F} \cdot \widehat{n}\,d\sigma$ where $\vec{F} = x\,\widehat{e}_1 + y\,\widehat{e}_2 + z\,\widehat{e}_3$, S is the surface
of the sphere $x^2 + y^2 + z^2 = 1$ and $\widehat{n}$ is a unit outward normal to the surface.

▶ **39.** Show that for $\phi = \phi(x,y,z)$ and $\psi = \psi(x,y,z)$ continuous functions

$$\iiint_V (\phi\nabla^2\psi - \psi\nabla^2\phi)\,d\tau = \iint_S (\phi\nabla\psi - \psi\nabla\phi) \cdot \widehat{n}\,d\sigma$$

Hint: Let $\vec{F} = \phi\nabla\psi$ in the divergence theorem.

▶ **40. Special directional derivatives**

(a) If $u = u(x,y)$ is continuous with continuous derivatives over
a closed region R with boundary given by a simple closed curve
$C = \partial R$, defined by a set of parametric equations $x = x(s)$ and
$y = y(s)$, with s denoting arc length, then the normal derivative of
u evaluated on the boundary of the region R is defined $\dfrac{\partial u}{\partial n} = \nabla u \cdot \widehat{n}$
where $\widehat{n}$ is a unit outward normal vector to the boundary. Show
that

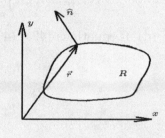

$$\frac{\partial u}{\partial n} - \nabla u \cdot \widehat{n} - \left(\frac{\partial u}{\partial x}\,\widehat{e}_1 + \frac{\partial u}{\partial y}\,\widehat{e}_2 \right) \cdot \left(\frac{dy}{ds}\,\widehat{e}_1 - \frac{dx}{ds}\,\widehat{e}_2 \right)$$

where s is an arc length parameter along the boundary curve.

(b) If $u = u(x,y,z)$, then the normal derivative to a surface $z = z(x,y)$ in space is defined
$\dfrac{\partial u}{\partial n} = \nabla u \cdot \widehat{n}$ where $\widehat{n}$ is a unit normal to the surface. Show that

$$\frac{\partial u}{\partial n} = \nabla u \cdot \widehat{n} = \left(\frac{\partial u}{\partial x}\,\widehat{e}_1 + \frac{\partial u}{\partial y}\,\widehat{e}_2 + \frac{\partial u}{\partial z}\,\widehat{e}_3 \right) \cdot \widehat{n}, \quad \text{where} \quad \widehat{n} = \frac{1}{\sqrt{\left(\frac{\partial z}{\partial x}\right)^2 + \left(\frac{\partial z}{\partial y}\right)^2 + 1}} \left(-\frac{\partial z}{\partial x}\,\widehat{e}_1 - \frac{\partial z}{\partial y}\,\widehat{e}_2 + \widehat{e}_3 \right).$$

▶ **41. Volume integral**

Show that the three-dimensional analog of equation (1.47) is the following. If the value of
(x,y,z) is known everywhere on a piecewise smooth closed surface S which encloses a volume
V, then the volume is given by $V = \dfrac{1}{3} \oiint_S \vec{r} \cdot \widehat{n}\,d\sigma$, where $\vec{r}$ is a position vector to a point on the
surface S and $\widehat{n}$ is a unit exterior normal to the surface at that point and $d\sigma$ is an element
of surface area.

Hint: Use the divergence theorem.

▶ **42.   Thermodynamics**

In the study of thermodynamics one will encounter the following relations associated with the thermodynamic properties of an ideal gas having an equation of state of the form $Pv = RT$

$$du = T\,ds - P\,dv \qquad\qquad (42a)$$

$$dh = T\,ds + v\,dP \qquad\qquad (42b)$$

$$a = u - T\,s \qquad\qquad (42c)$$

$$g = h - T\,s \qquad\qquad (42d)$$

where R, $[JK^{-1}mol^{-1}]$ is the gas constant, a, $[J/kg]$, is the Helmholtz function, g $[J/kg]$ is the Gibbs function, u, $[J/kg]$ is the specific internal energy, h, $[J/kg]$ is the specific enthalpy, s, $[J/kg\,K]$ is the specific entropy, T, $[K]$, is the temperature, P, $[N/m^2]$ is the gas pressure and v, $[m^3/kg]$ is the specific volume. The specific enthalpy of an ideal gas is a function of temperature with $h = u + Pv$.

(a) Show that $da = -P\,dv - s\,dT$ $\qquad\qquad$ (42e)

(b) Show that $dg = v\,dP - s\,dT$ $\qquad\qquad$ (42f)

(c) If equations (42a),(42b),(42e) and (42f) are exact differentials, then show

(i) $\quad \left(\dfrac{\partial u}{\partial s}\right)_v = T, \quad \left(\dfrac{\partial u}{\partial v}\right)_s = -P \qquad (iii) \quad \left(\dfrac{\partial a}{\partial v}\right)_T = -P, \quad \left(\dfrac{\partial a}{\partial T}\right)_v = -s$

(ii) $\quad \left(\dfrac{\partial h}{\partial s}\right)_P = T, \quad \left(\dfrac{\partial h}{\partial P}\right)_s = v \qquad (iv) \quad \left(\dfrac{\partial g}{\partial P}\right)_T = v, \quad \left(\dfrac{\partial g}{\partial T}\right)_P = -s$

(d) If equations (42a),(42b),(42e) and (42f) are exact differentials, then show

(i) $\quad \left(\dfrac{\partial T}{\partial v}\right)_s = -\left(\dfrac{\partial P}{\partial s}\right)_v \qquad (iii) \quad \left(\dfrac{\partial P}{\partial T}\right)_v = \left(\dfrac{\partial s}{\partial v}\right)_T$

(ii) $\quad \left(\dfrac{\partial T}{\partial P}\right)_s = \left(\dfrac{\partial v}{\partial s}\right)_P \qquad (iv) \quad \left(\dfrac{\partial v}{\partial T}\right)_P = -\left(\dfrac{\partial s}{\partial P}\right)_T$

(e) Show that $\quad (i) \quad h = g - T\left(\dfrac{\partial g}{\partial T}\right)_P, \qquad (ii) \quad a = g - P\left(\dfrac{\partial g}{\partial P}\right)_T$

▶ **43.   Thermodynamics** For an ideal gas $Pv = RT$ where R is the gas constant. Also the following relations exist for an ideal gas. $u = C_v T, \quad h = C_p T = u + RT, \quad \gamma = C_p/C_v$ where $C_P = \left(\dfrac{\partial h}{\partial T}\right)_p$ is the specific heat at constant pressure and $\left(\dfrac{\partial u}{\partial T}\right)_v = C_v$ is the specific heat at constant volume. Here u denotes internal energy and h denotes enthalpy.

(a) Show that the above equations imply that $C_p = C_v + R$.

(b) The first law of thermodynamics states that the change in thermal energy dq is related to change in internal energy du and work done dw by the relation $dq = du + dw$. If the work done by the system is given by $dw = Pdv$, then the first law of thermodynamics takes on the form $dq = C_v dT + Pdv$. Show that this implies $dq = C_v\,d\left(\dfrac{Pv}{R}\right) + Pdv$.

(c) Show for an adiabatic process $(dq = 0)$ the first law of thermodynamics implies $-\gamma\dfrac{dv}{v} = \dfrac{dP}{P}$.

(d) Show that for an adiabatic process the part (b) implies that $Pv^\gamma = constant$ where γ is the ratio of the specific heats.

Chapter 2
Maxima and Minima

In this chapter we investigate methods for determining points where the maximum and minimum values of a given function occur. We begin by studying extreme values associated with real functions $y = y(x)$ of a single independent real variable x. These functions are assumed to be well defined and continuous over a given interval $R = \{x \mid a \le x \le b\}$. Concepts introduced in the study of extreme values associated with functions of a single variable are then generalized to investigate extreme values associated with real functions $f = f(x, y)$ of two independent real variables, x and y. The functions $f = f(x, y)$ investigated are assumed to be well defined and continuous over a given region R of the x, y-plane. In addition to the functions of one and two real variables being well defined, we assume these functions have derivatives through the second order that are also well defined and continuous. The concepts introduced in the study of maximum and minimum values for real functions of one and two real variables are then generalize to investigate maximum and minimum values associated with real functions $f = f(x_1, x_2, \ldots, x_n)$ of several real variables. These functions are assumed to be well defined and have derivatives everywhere in a given region R of n-dimensional space. In the investigation of local maximum and minimum values of functions we develop several new concepts such as directional derivative, Hessian matrices and Lagrange multipliers to aid in the investigation of extreme values of a function. We conclude this chapter with an introduction to mathematical programming.

Functions of a single real variable

Let $y = f(x)$ denote a function that is well defined and continuous over an interval $a \le x \le b$. A δ-neighborhood of a point x_0 within this interval is defined as the set of points x satisfying $|x - x_0| < \delta$, where δ is a small positive number. Functions of a real single variable $y = f(x)$ are said to have a relative or local maximum or minimum value at a point x_0 if for all values of x in a δ-neighborhood of x_0 certain inequalities are satisfied. In particular, at points x_0 where a local maximum value occurs, one can say

$$f(x_0) \ge f(x) \text{ for all values of } x \text{ satisfying } |x - x_0| < \delta.$$

At points x_0 where a local minimum occurs, one can say

$$f(x_0) \le f(x) \text{ for all values of } x \text{ satisfying } |x - x_0| < \delta,$$

where δ is a small positive number.

Associated with a function $y = f(x)$, that is well defined over a closed interval $[a, b]$, there are certain points called critical points. Critical points are defined as either

(i) Those points x where the derivative of the function is zero, or

(ii) Those points x where the derivative $f'(x)$ fails to exist.

The point or points where the derivative is zero satisfies the equation $f'(x) = 0$. These points are called stationary points.

56

A curve $y = f(x)$ is called concave upward over an interval if the graph of f lies above all of the tangent lines to the curve on the interval. A curve $y = f(x)$ is called concave downward over an interval if the graph of f lies below all of its tangent lines on the interval. A point on a curve $y = f(x)$ where the concavity changes is called an inflection point.

A smooth curve is found to be concave up in regions where $f''(x) > 0$ and concave downward in regions where $f''(x) < 0$. Those points x that satisfy $f''(x) = 0$ are the points of inflection. These are the x values where the concavity of the curve changes. Note that relative maximum and minimum values of a function $y = f(x)$, within the interval (a, b) of definition, can occur at critical points and at stationary points. The end points $x = a$ and $x = b$ must be tested separately to determine if a local maximum or minimum value exists. The figure 2-1 illustrates several functions defined over an interval $a \leq x \leq b$ that have critical points. These critical points illustrate local maximum values or local minimum values that occur at points where either the slope is zero or not defined. Observe in figure 2-1 that at points where a sharp corner occurs the left and right-handed derivatives are not the same at these points and so the derivative fails to exist at these points. Also note that at the end points of a given interval a function can have relative or local maximum or minimum values. When testing for relative or local maximum and minimum values over an interval $a \leq x \leq b$, the end points at $x = a$ and $x = b$ should be tested separately.

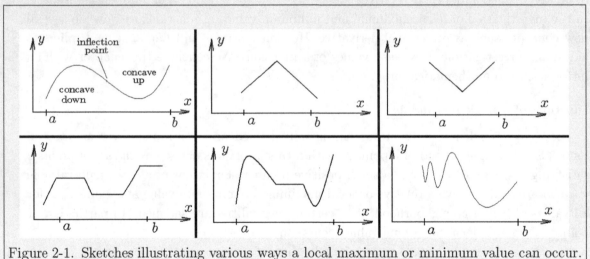

Figure 2-1. Sketches illustrating various ways a local maximum or minimum value can occur.

The terminology extrema, extremum or relative extreme values is a way of referring to both the relative maximum and minimum values associated with a real function. Relative maxima or minima values are referred to as local extrema of the function $y = f(x)$ over the interval $[a, b]$ where the function is defined. Points where $f'(x) = 0$ are referred to as stationary points since these are points where the slope is zero.

A function $y = f(x)$ is said to have an absolute maximum at a point x_0 in an interval (a, b) if $f(x_0) \geq f(x)$ for all $x \in (a, b)$. A function $y = f(x)$ is said to have an absolute minimum at a point x_0 in an interval (a, b) if $f(x_0) \leq f(x)$ for all $x \in (a, b)$. The absolute maximum or minimum value of a function $y = f(x)$ can occur either at a critical point of $f(x)$ within the interval or at one of the end points of a closed interval. One should make it a habit to always test separately the end points of a closed interval for absolute maximum or minimum values of the function.

On an open interval $(a < x < b)$, where the end points are not included, some functions do not possess a maximum or minimum value. For example, the function $y = y(x) = x$ on the interval $(0 < x < 1)$. On closed intervals $(a \leq x \leq b)$ the extremum values may occur at the end points or boundaries. The Weierstrass theorem states that a continuous function on a closed interval attains its maximum or minimum values on a boundary or at an interior point.

Tests for maximum and minimum values

In every calculus course one learns to test the critical points associated with a continuous functions $y = f(x)$ defined over an interval $a \leq x \leq b$. The critical points can then be tested to see if a relative maximum or minimum value occurs. The first derivative test and second derivative test are familiar tests for examining the behavior of smooth functions at a critical point.

First derivative test

Recall that the first derivative test states that if $x = x_0$ is a critical point associated with a continuous differentiable function $y = f(x)$, then $f(x_0)$ is called a relative minimum value of the function $f(x)$ if the following conditions hold true

$$(i) \qquad f'(x) < 0 \text{ for } x < x_0 \quad \text{and} \qquad (ii) \qquad f'(x) > 0 \text{ for } x > x_0 \qquad (2.1)$$

That is, the derivative changes sign from negative to zero to positive as x moves across the critical point in the positive direction.

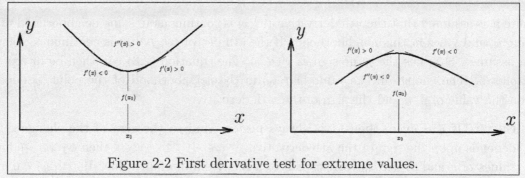

Figure 2-2 First derivative test for extreme values.

If $x = x_0$ is a critical point associated with the a continuous differentiable function $y = f(x)$, then $f(x_0)$ is called a relative maximum value of the function $f(x)$ if the following conditions hold

$$(i) \qquad f'(x) > 0 \text{ for } x < x_0 \quad \text{and} \quad (ii) \qquad f'(x) < 0 \text{ for } x > x_0 \qquad (2.2)$$

That is, the derivative changes sign from positive to zero to negative as x moves across the critical point in the positive direction. Conditions for a first derivative test for an extreme value are illustrated in the figure 2-2.

Second derivative test

The second derivative test for relative maximum and minimum values is based upon the concavity of the curve in the vicinity of a critical point. Assume the second derivative $f''(x)$ is continuous in the vicinity of a critical point x_0. If the second derivative at the critical point is positive, then the given curve will be concave upward. If the second derivative at the critical point is negative, then the given curve will be concave downward. Consequently, one has the following second derivative test for local extreme values

(i) If $f'(x_0) = 0$ and $f''(x_0) > 0$, then $f(x_0)$ represents a local minimum value.

(ii) If $f'(x_0) = 0$ and $f''(x_0) < 0$, then $f(x_0)$ represents a local maximum value.

(iii) If $f'(x_0) = 0$ and $f''(x_0) = 0$ or $f''(x_0)$ does not exist,

then the second derivative test fails.

If the second derivative test fails, then the first derivative test can be applied.

Further investigation of critical points

Assume a given function $y = f(x)$ is defined over a closed interval $[a, b]$ and is such that it has a Taylor series expansion about an interior point x_0 so that one can write

$$f(x_0 + h) = f(x_0) + f'(x_0)h + f''(x_0)\frac{h^2}{2!} + \cdots + f^{(n-1)}(x_0)\frac{h^{n-1}}{(n-1)!} + f^{(n)}(\xi)\frac{h^n}{n!}$$

where the last term represents the Lagrange remainder term with $x_0 < \xi < x_0 + h$.

Now if x_0 is a critical point that satisfies $f'(x_0) = 0$ and if in addition one has

$$f''(x_0) = f'''(x_0) = \cdots = f^{(n-1)}(x_0) = 0$$

and $f^{(n)}(x_0) \neq 0$, then one can obtain from the Taylor series with Lagrange remainder the relation

$$\Delta f(x_0) = f(x_0 + h) - f(x_0) = f^{(n)}(\xi)\frac{h^n}{n!} \tag{2.3}$$

where it is assumed that the nth derivative $f^{(n)}(x)$ is continuous in some neighborhood of the point x_0 and ξ lies in this neighborhood. If the nth derivative $f^{(n)}(x)$ is continuous, then we can assume $f^{(n)}(\xi)$ has the same sign as $f^{(n)}(x_0)$. The equation (2.3) can then be interpreted as follows. The change in the value of $f(x_0)$ in the neighborhood of the point x_0 depends upon the value of n, h and the sign of the n-th derivative.

Case 1: If n is even, then h^n is always positive and so the sign of the change $\Delta f(x_0)$ depends upon the sign of the nth derivative $f^{(n)}(x_0)$. If $f^{(n)}(x_0) > 0$, then $\Delta f(x_0) > 0$ for all values of h and so $x = x_0$ corresponds to a relative minimum value. If $f^{(n)}(x_0) < 0$, then $\Delta f(x_0) < 0$ for all values of h and so $x = x_0$ corresponds to a relative maximum value.

Case 2: If n is odd, then h^n takes on different signs depending upon whether $h > 0$ or $h < 0$. Hence $\Delta f(x_0)$ changes signs in the neighborhood of the critical point x_0. That is, for $x = x_0 + h$ there will be values of h for which $f(x) > f(x_0)$ and values for which $f(x) < f(x_0)$ and hence the point $x = x_0$ corresponds to a point of inflection.

One can use arguments similar to those given above to analyze critical points associated with functions of several variables.

Example 2-1. Law of reflection

We examine the situation where light travels in a straight line and is reflected from a smooth polished surface, say a mirror. Assume ℓ, h_1 and h_2 are positive quantities and we are given the two points $(0, h_1)$ and (ℓ, h_2) together with a general point x on the x-axis satisfying $0 < x < \ell$. Let L_1 denote the distance from the point $(0, h_1)$ to the point x and let L_2 denote the distance from the point x to the point (ℓ, h_2). Find the position of the point x such that the sum of the distances L_1 and L_2 is a minimum.

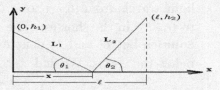

Solution: The sum of the distances L_1 and L_2 can be written as a function of x. We have $L_1 = \sqrt{x^2 + h_1^2}$ and $L_2 = \sqrt{(\ell - x)^2 + h_2^2}$ so that the sum can be written

$$S = L_1 + L_2 = \sqrt{x^2 + h_1^2} + \sqrt{(\ell - x)^2 + h_2^2}, \quad \text{with} \quad \frac{dS}{dx} = \frac{x}{\sqrt{x^2 + h_1^2}} + \frac{x - \ell}{\sqrt{(\ell - x)^2 + h_2^2}}.$$

At an extremum we require $\dfrac{dS}{dx} = 0$ which requires that

$$\frac{x}{\sqrt{x^2 + h_1^2}} = \cos\theta_1 = \frac{\ell - x}{\sqrt{(\ell - x)^2 + h_2^2}} = \cos\theta_2 \qquad (2.4)$$

The equation (2.4) can be interpreted as finding the point x^* where two graphs intersect. We find that the equation (2.4) has a unique solution x^* where the curves $y = x/\sqrt{x^2 + h_1^2}$ and $y = (\ell - x)/\sqrt{(\ell - x)^2 + h_2^2}$ intersect as illustrated in the accompanying figure.

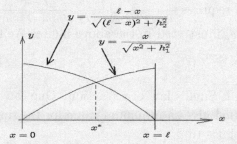

The equation (2.4) implies that $\theta_1 = \theta_2$ at this critical value for x. When the x-axis is a mirror this requires that the angle of incidence equal the angle of reflection. The second derivative test can be used to show that S has a minimum value under these conditions. One can verify that the second derivative reduces to

$$\frac{d^2 S}{dx^2} = \frac{h_1^2}{\sqrt{x^2 + h_1^2}} + \frac{h_2^2}{\sqrt{(\ell - x)^2 + h_2^2}}.$$

This is a quantity that is always positive and hence, by the second derivative test, the distance S has a minimum value at the critical point x^* where $\theta_1 = \theta_2$.

$\blacksquare$

Example 2-2. Snell's law of refraction

We examine the situation where light changes direction as it travels from one medium to another. Assume ℓ, h_1 and h_2 are positive quantities and we are given the two points $(0, h_1)$ and $(\ell, -h_2)$ together with a general point x on the x-axis.

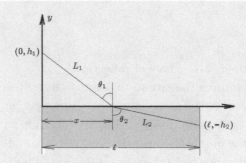

The region $y > 0$ represents a medium where the speed of light is c_1 and the region $y < 0$ represents a medium where the speed of light is c_2. Find the point $(x, 0)$ such that light travels from $(0, h_1)$ to $(x, 0)$ and then to $(\ell, -h_2)$ in the shortest time.

Solution Let L_1 and L_2 denote the distances that the light travels in the media above and below the x-axis. We find

$$L_1 = \sqrt{h_1^2 + x^2} \qquad \text{and} \qquad L_2 = \sqrt{h_2^2 + (\ell - x)^2}.$$

Using the formula $distance = (velocity)(time)$, one can verify that the time it takes for light to travel from $(0, h_1)$ to $(\ell, -h_2)$ is given by

$$T = T(x) = \frac{1}{c_1}\sqrt{h_1^2 + x^2} + \frac{1}{c_2}\sqrt{h_2^2 + (\ell - x)^2}.$$

The time is an extremum for the value of x that satisfies

$$\frac{dT}{dx} = \frac{1}{c_1}\frac{x}{\sqrt{h_1^2 + x^2}} - \frac{1}{c_2}\frac{\ell - x}{\sqrt{h_2^2 + (\ell - x)^2}} = 0.$$

This requires $\dfrac{1}{c_1}\dfrac{x}{\sqrt{h_1^2 + x^2}} = \dfrac{1}{c_2}\dfrac{\ell - x}{\sqrt{h_2^2 + (\ell - x)^2}}$ which can be written in terms of the angles θ_1 and θ_2 illustrated in the above figure. We find

$$\frac{1}{c_1}\sin\theta_1 = \frac{1}{c_2}\sin\theta_2.$$

This is Snell's law of refraction. One can verify, using the second derivative test, that this solution gives a minimum value for the time of travel. One can verify the second derivative simplifies to

$$\frac{d^2T}{dx^2} = \frac{1}{c_1}\frac{h_1^2}{\sqrt{h_1^2 + x^2}} + \frac{1}{c_2}\frac{h_2^2}{\sqrt{h_2^2 + (\ell - x)^2}}.$$

The second derivative is always positive and so the critical point corresponds to a minimum time.

■

Functions of two variables

Let $z = z(x, y)$ denote a function of x and y that is defined everywhere in a domain R of the x, y-plane. Further we assume that this function is continuous with derivatives that are also continuous. Here z can be thought of as the height of a continuous surface above the x, y-plane. The function $z = z(x, y)$ is said to have a relative or local maximum value at a point (x_0, y_0) in the interior of the region R if

$$z(x_0, y_0) \geq z(x, y) \tag{2.5}$$

for all points (x, y) in the delta neighborhood $N_\delta = \{(x, y) \mid (x - x_0)^2 + (y - y_0)^2 \leq \delta^2\}$, of the point (x_0, y_0), where δ is a small positive number. If the inequality in equation (2.5) holds for all points (x, y) interior to the region R and for points (x, y) on the boundary ∂R of the region R, then $z = z(x_0, y_0)$ is called an absolute maximum value over the region R.

Similarly, the function $z = z(x, y)$ is said to have a relative or local minimum at a point (x_0, y_0) interior to the region R if

$$z(x_0, y_0) \leq z(x, y) \tag{2.6}$$

for all points (x, y) in the delta neighborhood N_δ. Again, if the inequality in equation (2.6) holds for all points (x, y) interior to the region R and for points (x, y) on the boundary ∂R of the region R, then $z(x_0, y_0)$ is called an absolute minimum value over the region R. Boundary points of the region R must be tested separately for maximum and minimum values.

A function $z = z(x, y)$ can also be thought of as representing a scalar field associated with the points (x, y) within a region R. The directional derivative of z in the direction of a unit vector $\widehat{\mathbf{e}}_\alpha = \cos \alpha \, \widehat{\mathbf{e}}_1 + \sin \alpha \, \widehat{\mathbf{e}}_2$ and evaluated at a point (x_0, y_0) is given by

$$\frac{dz}{ds} = \operatorname{grad} z \cdot \widehat{\mathbf{e}}_\alpha = \left(\frac{\partial z}{\partial x} \widehat{\mathbf{e}}_1 + \frac{\partial z}{\partial y} \widehat{\mathbf{e}}_2 \right) \cdot \left(\cos \alpha \, \widehat{\mathbf{e}}_1 + \sin \alpha \, \widehat{\mathbf{e}}_2 \right) = \frac{\partial z}{\partial x} \cos \alpha + \frac{\partial z}{\partial y} \sin \alpha \tag{2.7}$$

where all derivatives are evaluated at the point (x_0, y_0). Points on the surface $z = z(x, y)$ that correspond to stationary points are those points where the directional derivative is zero for all directions α. Therefore, at a stationary point we will have

$$\frac{\partial z}{\partial x} = 0, \quad \text{and} \quad \frac{\partial z}{\partial y} = 0. \tag{2.8}$$

Stationary points are those points where the tangent plane to the surface is parallel to the x, y-plane. Here we assume that points (x, y) used to define $z = z(x, y)$ are restricted to a region R of the x, y-plane. We convert the problem of analyzing stationary points, for determining maximum and minimum values for functions of two variables, to a familiar one dimensional problem as follows. If (x_0, y_0) is a stationary point to be tested, then slide the free vector $\widehat{\mathbf{e}}_\alpha$ to the point (x_0, y_0) and construct a plane normal to the plane $z = 0$, such that this plane contains the vector $\widehat{\mathbf{e}}_\alpha$. The constructed plane intersects the given surface in a curve that can be represented by the equation

$$z = z(s) = z(x_0 + s \cos \alpha, y_0 + s \sin \alpha) \tag{2.9}$$

where s represents distance in the direction $\widehat{\mathbf{e}}_\alpha$. We can now analyze the change in z along the curve of intersection for all directions α. Methods from calculus can now be applied to analyze maximum and minimum values associated with the curve of intersection formed by the plane and surface. The situation is illustrated in the figure 2-3.

At a stationary point we must have $\frac{dz}{ds} = 0$ for all directions α. In addition the second directional derivative

$$\frac{d^2 z}{ds^2} = \operatorname{grad} \left(\frac{dz}{ds} \right) \cdot \widehat{\mathbf{e}}_\alpha$$

$$\frac{d^2 z}{ds^2} = \left[\left(\frac{\partial^2 z}{\partial x^2} \cos \alpha + \frac{\partial^2 z}{\partial y \partial x} \sin \alpha \right) \widehat{\mathbf{e}}_1 + \left(\frac{\partial^2 z}{\partial x \partial y} \cos \alpha + \frac{\partial^2 z}{\partial y^2} \sin \alpha \right) \widehat{\mathbf{e}}_2 \right] \cdot [\cos \alpha \, \widehat{\mathbf{e}}_1 + \sin \alpha \, \widehat{\mathbf{e}}_2] \tag{2.10}$$

$$\frac{d^2 z}{ds^2} = \frac{\partial^2 z}{\partial x^2} \cos^2 \alpha + 2 \frac{\partial^2 z}{\partial x \partial y} \sin \alpha \cos \alpha + \frac{\partial^2 z}{\partial y^2} \sin^2 \alpha$$

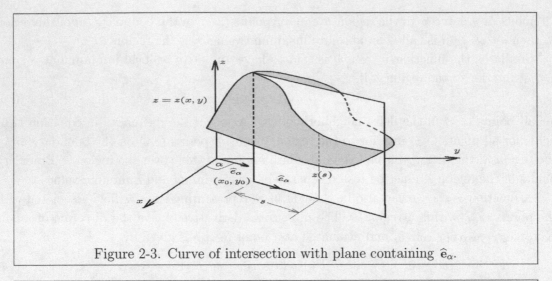

Figure 2-3. Curve of intersection with plane containing $\widehat{e}_\alpha$.

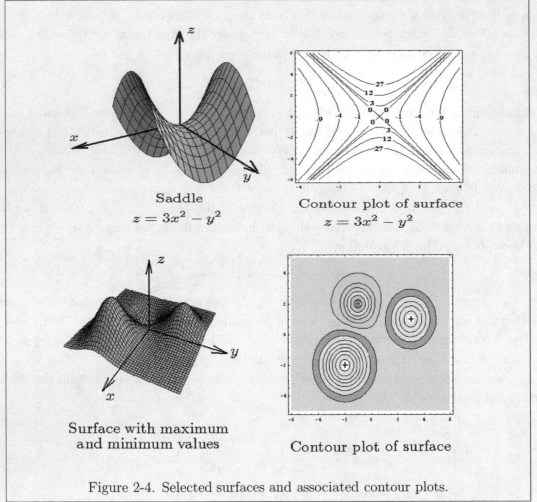

Saddle
$z = 3x^2 - y^2$

Contour plot of surface
$z = 3x^2 - y^2$

Surface with maximum
and minimum values

Contour plot of surface

Figure 2-4. Selected surfaces and associated contour plots.

can be evaluated at the stationary point. If this second directional derivative is positive for all directions α, then the stationary point corresponds to a relative minimum. If the second directional derivative is negative for all directions α, then the stationary point corresponds to a relative maximum of the function $z = z(x, y)$ at the stationary point.

A function $z = z(x, y)$ is said to have a saddle point at a stationary point (x_0, y_0) if there exists a delta neighborhood N_δ of the point (x_0, y_0) such that for some points $(x, y) \in N_\delta$ we have $z(x, y) > z(x_0, y_0)$ and for other points $(x, y) \in N_\delta$ we have $z(x, y) < z(x_0, y_0)$.

If graphical software is available it is sometimes advantages to plot graphs of the surfaces being tested in order to display where maximum and minimum values occur. One can also examine numerous level curves called contour plots that represent the intersection of the surface $z = z(x, y)$ with the plane $z = c = $ constant for selected values of the constant c. The figure 2-4 illustrates some sketches of surfaces and the corresponding level curves associated with the surfaces.

Analysis of second directional derivative

To analyze a stationary value associated with a function $z = z(x, y)$ one must be able to analyze the second directional derivative given by equation (2.10). Let

$$A = \frac{\partial^2 z}{\partial x^2}, \qquad B = \frac{\partial^2 z}{\partial x \partial y}, \qquad C = \frac{\partial^2 z}{\partial y^2} \tag{2.11}$$

denote the second derivatives in equation (2.10) evaluated at a stationary point (x_0, y_0). The second directional derivative given by equation (2.10) can then be written in a more tractable form for analysis purposes. We write equation (2.10) in the form

$$\frac{d^2 z}{ds^2}\bigg|_0 = A \cos^2 \alpha + 2B \cos \alpha \sin \alpha + C \sin^2 \alpha \tag{2.12}$$

and then factor out the leading term followed by a completing the square operation on the first two terms to obtain

$$\begin{aligned}
\frac{d^2 z}{ds^2}\bigg|_0 &= A \left[\cos^2 \alpha + 2\frac{B}{A} \cos \alpha \sin \alpha + \frac{C}{A} \sin^2 \alpha \right] \\
&= A \left[\left(\cos \alpha + \frac{B}{A} \sin \alpha \right)^2 + \frac{(AC - B^2)}{A^2} \sin^2 \alpha \right]
\end{aligned} \tag{2.13}$$

Case 1: Assume that $(AC - B^2) = \frac{\partial^2 z}{\partial x^2}\frac{\partial^2 z}{\partial y^2} - \left(\frac{\partial^2 z}{\partial x \partial y} \right)^2 = 0$, then in those directions α that satisfy $\cos \alpha + \frac{B}{A} \sin \alpha = 0$, the second directional derivative vanishes. For all other values of α the second directional derivative is of constant sign that is the same sign as A. If the above condition is satisfied, then the second directional derivative test fails. Note that if $AC - B^2 = 0$ there may be a local minimum, a local maximum or neither maximum or minimum at the point being tested, hence the test is inconclusive.

Case 2: Assume that $(AC - B^2) = \frac{\partial^2 z}{\partial x^2}\frac{\partial^2 z}{\partial y^2} - \left(\frac{\partial^2 z}{\partial x \partial y} \right)^2 < 0$, then the second directional derivative is not of constant sign. It assumes different signs in different directions α. In particular, if $\alpha = 0$ we have $\frac{d^2 z}{ds^2} = A$ and for α satisfying $\cos \alpha + \frac{B}{A} \sin \alpha = 0$, we have $\frac{d^2 z}{ds^2} = \frac{A(AC - B^2)}{A^2} \sin^2 \alpha$. Hence if

$A > 0$, then $A(AC - B^2)$ is negative and if $A < 0$, then $A(AC - B^2)$ is positive. This shows the second directional derivative is of nonconstant sign. In this situation the stationary point (x_0, y_0) is said to correspond to a saddle point.

Case 3: Assume that $(AC - B^2) = \frac{\partial^2 z}{\partial x^2} \frac{\partial^2 z}{\partial y^2} - \left(\frac{\partial^2 z}{\partial x \partial y}\right)^2 > 0$, then the second directional derivative is of constant sign, which is the sign of A.

    (i) If $A > 0$, then $\frac{d^2 z}{ds^2} > 0$ so that the curve $z = z(s)$ is concave upward for all directions α and consequently the stationary point corresponds to a relative minimum.

    (ii) If $A < 0$, then $\frac{d^2 z}{ds^2} < 0$ so that the curve $z = z(s)$ is concave downward for all directions α and consequently the stationary point corresponds to a relative maximum.

Generalization

The analysis of a function having maximum and minimum values can be approached by way of a Taylor series expansion and quadratic forms. For a function of one-variable a Taylor series expansion can be written

$$f(x_0 + \Delta x) = f(x_0) + f'(x_0)\Delta x + \frac{1}{2!}f''(\xi)(\Delta x)^2 \tag{2.14}$$

where the last term is the Lagrange remainder term. Now if $f'(x_0) = 0$, then the difference $f(x_0 + \Delta x) - f(x_0) > 0$ if $f''(\xi) > 0$ and $f(x_0 + \Delta x) - f(x_0) < 0$ if $f''(\xi) < 0$. These inequalities give the definition of a maximum and minimum value at x_0. Observe that if $f''(x)$ is continuous in a δ-neighborhood of x_0, then δ can be selected small enough so that the conditions $f''(\xi) > 0$ and $f''(\xi) < 0$ can be replaced by the conditions $f''(x_0) > 0$ and $f''(x_0) < 0$ as the tests for minimum and maximum values respectively. That is, if $f''(x)$ is continuous, then there exists a δ-neighborhood of x_0 where $f''(x_0)$ and $f''(\xi)$ have the same sign everywhere in the neighborhood.

For functions of two-variables we have the Taylor series expansion

$$\begin{aligned} f(x_0 + \Delta x, y_0 + \Delta y) =& f(x_0, y_0) + \frac{\partial f}{\partial x}\bigg|_0 \Delta x + \frac{\partial f}{\partial y}\bigg|_0 \Delta y \\ &+ \frac{1}{2!}\left(\frac{\partial^2 f}{\partial x^2}(\Delta x)^2 + 2\frac{\partial^2 f}{\partial x \partial y}\Delta x \Delta y + \frac{\partial^2 f}{\partial y^2}(\Delta y)^2\right)_{(\xi, \eta)} \end{aligned} \tag{2.15}$$

where the subscript 0 denotes that the derivatives are to be evaluated at the point $P_0 = (x_0, y_0)$. Now if $\frac{\partial f}{\partial x}\big|_0 = 0$ and $\frac{\partial f}{\partial y}\big|_0 = 0$, then the equation (2.15) can be written in matrix notation

$$f(x_0 + \Delta x, y_0 + \Delta y) - f(x_0, y_0) = \frac{1}{2!}[\Delta x, \Delta y]\begin{bmatrix} \frac{\partial^2 f}{\partial x^2} & \frac{\partial^2 f}{\partial x \partial y} \\ \frac{\partial^2 f}{\partial y \partial x} & \frac{\partial^2 f}{\partial y^2} \end{bmatrix}_{(\xi, \eta)}\begin{bmatrix} \Delta x \\ \Delta y \end{bmatrix} \tag{2.16}$$

where the right-hand side of equation (2.16) contains a quadratic form evaluated at the point (ξ, η). The matrix of the quadratic form is called the Hessian matrix. If the Hessian matrix of the quadratic form is positive definite, then

$$f(x_0 + \Delta x, y_0 + \Delta y) - f(x_0, y_0) > 0,$$

so that $f(x_0, y_0)$ corresponds to a local minimum value. Similarly, if the Hessian matrix of the quadratic form is negative definite, then $f(x_0 + \Delta x, y_0 + \Delta y) - f(x_0, y_0) < 0$, so that $f(x_0, y_0)$ corresponds to a local maximum value. If the second order partial derivatives of f are continuous in a δ-neighborhood of the point (x_0, y_0), then the Hessian matrix of the quadratic form can be tested at the point (x_0, y_0) instead of the point (ξ, η). Recall from linear algebra that $\bar{x} H \bar{x}^T$, where H is a symmetric matrix, is said to be positive definite if $\bar{x} H \bar{x}^T > 0$ for all $\bar{x} \neq \bar{0}$ and it is called negative definite if $\bar{x} H \bar{x}^T < 0$ for all $\bar{x} \neq \bar{0}$. If $\bar{x} H \bar{x}^T$ takes on different signs, then it is called indefinite. By convention the matrix H associated with the quadratic form is given the same labeling as the quadratic form associated with the matrix. That is, the matrix H is called positive definite, negative definite or indefinite depending upon the sign assigned to the quadratic form $\bar{x} H \bar{x}^T$.

Generalizing the above concepts to functions of n-variables is accomplished as follows. Denote a scalar function of position by $f = f(x_1, x_2, \ldots, x_n)$ and let $\bar{x}_0 = (x_1^0, x_2^0, \ldots, x_n^0)$ denote a fixed point and $\bar{x} = (x_1, x_2, \ldots, x_n)$ denote a variable point in a region R of n-dimensional space $\mathbb{R}^n$. Then a shorthand notation for representing the function f is given by $f = f(\bar{x})$. One can then define a δ-neighborhood $N_\delta(\bar{x}_0)$ of the point $\bar{x}_0$ as consisting of the set of points $\bar{x}$ that satisfy

$$N_\delta(\bar{x}_0) = \{(x_1, x_2, \ldots, x_n) \mid (x_1 - x_1^0)^2 + (x_2 - x_2^0)^2 + \cdots + (x_n - x_n^0)^2 < \delta^2\} \qquad (2.17)$$

This represents all points within an n-dimensional sphere about the point $\bar{x}_0$. By definition, one can say that $f(\bar{x})$ has a local maximum value at $\bar{x}_0$ if there exists a δ-neighborhood of $\bar{x}_0$ such that $f(\bar{x}_0) \geq f(\bar{x})$ for $\bar{x} \in N_\delta(\bar{x}_0)$. Similarly, $f(\bar{x})$ has a local minimum at a point $\bar{x}_0$ if there exists a δ-neighborhood of $\bar{x}_0$ such that $f(\bar{x}) \geq f(\bar{x}_0)$ for $\bar{x} \in N_\delta(\bar{x}_0)$. We assume $f = f(\bar{x})$ has derivatives through the second order that are continuous. A point $\bar{x}_0$ is called a critical or stationary point if

$$\left.\frac{\partial f}{\partial x_1}\right|_{\bar{x}_0} = 0, \quad \left.\frac{\partial f}{\partial x_2}\right|_{\bar{x}_0} = 0, \cdots, \left.\frac{\partial f}{\partial x_n}\right|_{\bar{x}_0} = 0. \qquad (2.18)$$

Thus, to find critical or stationary points we require that $df = 0$ at the point $\bar{x}_0$. A Taylor series expansion about the point $\bar{x}_0$ can be written

$$f(\bar{x}) = f(\bar{x}_0) + \left.\frac{\partial f}{\partial x_1}\right|_{\bar{x}_0}(x_1 - x_1^0) + \left.\frac{\partial f}{\partial x_2}\right|_{\bar{x}_0}(x_2 - x_2^0) + \cdots + \left.\frac{\partial f}{\partial x_n}\right|_{\bar{x}_0}(x_n - x_n^0) +$$

$$\frac{1}{2!}[(x_1 - x_1^0), (x_2 - x_2^0), \ldots, (x_n - x_n^0)] \begin{bmatrix} \frac{\partial^2 f}{\partial x_1^2} & \frac{\partial^2 f}{\partial x_1 \partial x_2} & \cdots & \frac{\partial^2 f}{\partial x_1 \partial x_n} \\ \frac{\partial^2 f}{\partial x_2 \partial x_1} & \frac{\partial^2 f}{\partial x_2^2} & \cdots & \frac{\partial^2 f}{\partial x_2 \partial x_n} \\ \vdots & \vdots & \ddots & \vdots \\ \frac{\partial^2 f}{\partial x_n \partial x_1} & \frac{\partial^2 f}{\partial x_n \partial x_2} & \cdots & \frac{\partial^2 f}{\partial x_n^2} \end{bmatrix}_{\bar{\xi}} \begin{bmatrix} (x_1 - x_1^0) \\ (x_2 - x_2^0) \\ \vdots \\ (x_n - x_n^0) \end{bmatrix}$$

where the last term contains the quadratic form $(\bar{x} - \bar{x}_0) H(\bar{\xi})(\bar{x} - \bar{x}_0)^T$ with H a $n \times n$ square matrix having $\frac{\partial^2 f}{\partial x_i \partial x_j}$, evaluated at $\bar{\xi}$, for the element in the ith row and jth column of H. The matrix

$$H = \begin{bmatrix} \frac{\partial^2 f}{\partial x_1^2} & \frac{\partial^2 f}{\partial x_1 \partial x_2} & \cdots & \frac{\partial^2 f}{\partial x_1 \partial x_n} \\ \frac{\partial^2 f}{\partial x_2 \partial x_1} & \frac{\partial^2 f}{\partial x_2^2} & \cdots & \frac{\partial^2 f}{\partial x_2 \partial x_n} \\ \vdots & \vdots & \ddots & \vdots \\ \frac{\partial^2 f}{\partial x_n \partial x_1} & \frac{\partial^2 f}{\partial x_n \partial x_2} & \cdots & \frac{\partial^2 f}{\partial x_n^2} \end{bmatrix}$$

is called the Hessian matrix of the function f. A submatrix of order m is formed from H by deleting the last $n-m$ rows and columns of the matrix H. By definition the Hessian function H_m is the determinant of this submatrix. For example

$$H_1 = \frac{\partial^2 f}{\partial x_1{}^2}, \quad H_2 = \begin{vmatrix} \frac{\partial^2 f}{\partial x_1{}^2} & \frac{\partial^2 f}{\partial x_1 \partial x_2} \\ \frac{\partial^2 f}{\partial x_2 \partial x_1} & \frac{\partial^2 f}{\partial x_2{}^2} \end{vmatrix}, \quad H_3 = \begin{vmatrix} \frac{\partial^2 f}{\partial x_1{}^2} & \frac{\partial^2 f}{\partial x_1 \partial x_2} & \frac{\partial^2 f}{\partial x_1 \partial x_3} \\ \frac{\partial^2 f}{\partial x_2 \partial x_1} & \frac{\partial^2 f}{\partial x_2{}^2} & \frac{\partial^2 f}{\partial x_2 \partial x_3} \\ \frac{\partial^2 f}{\partial x_3 \partial x_1} & \frac{\partial^2 f}{\partial x_3 \partial x_2} & \frac{\partial^2 f}{\partial x_3{}^2} \end{vmatrix}, \quad \dots$$

are the first three Hessian functions associated with the matrix H.

Observe that if H is positive definite, then f has a local minimum at $\bar{x}_0$. and if H is negative definite, then f has a local maximum at $\bar{x}_0$. Whenever the second order derivatives of f are continuous functions, then there will exist a δ-neighborhood of the point $\bar{x}_0$, such that the quadratic form with Hessian matrix H evaluated at the point $\bar{x}_0$ will have the same sign as the quadratic form with Hessian matrix H evaluated at the point $\bar{\xi}$. The Hessian matrix evaluated at $\bar{x}_0$ can then be used to test for maximum and minimum values. If H is $n \times n$ and symmetric, then the following can be used as tests for local maximum and minimum values.

If $\bar{x}_0$ is a critical point of $f(\bar{x})$ and one of the following is true, then $f(\bar{x}_0)$ has a local minimum at $\bar{x}_0$

    (i) H is positive definite.

    (ii) The determinants of all the leading principal submatrics produce Hessian functions $H_1, H_2, \dots, H_n$ with positive values.

    (iii) Each eigenvalue of H is positive.

If $\bar{x}_0$ is a critical point of $f(\bar{x})$ and one of the following is true, then $f(\bar{x}_0)$ has a local maximum at $\bar{x}_0$.

    (i) H is negative definite.

    (ii) The determinants of all the leading principal submatrics produce the Hessian functions $H_1, H_2, \dots, H_n$. A local maximum occurs whenever H_{2m-1} have negative values and H_{2m} have positive values for $m = 1, 2, \dots$. (i.e. The Hessian functions oscillate in sign.)

    (iii) Each eigenvalue of H is negative.

The critical point $\bar{x}_0$ corresponds to a saddle point whenever H is indefinite.

Derivative test for functions of two-variables

Our previous analysis of the second directional derivative is equivalent to the following. The Hessian functions H_1 and H_2 associated with the function $z = z(x, y)$ are defined

$$H_1 = \frac{\partial^2 z}{\partial x^2}, \qquad H_2 = \begin{vmatrix} A & B \\ B & C \end{vmatrix} = \begin{vmatrix} \frac{\partial^2 z}{\partial x^2} & \frac{\partial^2 z}{\partial x \partial y} \\ \frac{\partial^2 z}{\partial y \partial x} & \frac{\partial^2 z}{\partial y^2} \end{vmatrix} = \frac{\partial^2 z}{\partial x^2} \frac{\partial^2 z}{\partial y^2} - \left(\frac{\partial^2 z}{\partial x \partial y} \right)^2 \tag{2.19}$$

and can be evaluated at all points in the domain of $z = z(x, y)$. In terms of Hessian functions one can write a derivative test for analysis of stationary values for functions of two-variables as follows.

Assume $z = z(x, y)$ is continuous everywhere in a region R of the x, y-plane. Further assume that z possess derivatives through the second order that are also continuous everywhere in the region R. If $P_0 = (x_0, y_0)$ is a stationary point of $z = z(x, y)$ interior to the region R, then $\frac{\partial z}{\partial x}\Big|_{P_0} = 0$ and $\frac{\partial z}{\partial y}\Big|_{P_0} = 0$.

(a) If $H_1(P_0) > 0$ and $H_2(P_0) > 0$, then $z(x_0, y_0)$ corresponds to a local minimum value for the function $z = z(x, y)$.

(b) If $H_1(P_0) < 0$ and $H_2(P_0) > 0$, then $z(x_0, y_0)$ corresponds to a local maximum value for the function $z = z(x, y)$

(c) If $H_2(P_0) < 0$, then P_0 gives a saddle point for the function $z = z(x, y)$.

(d) The boundary points of the region R, if a boundary exists,

   must be tested separately for maximum and minimum values.

For functions of three or more variables one can rely upon graphical methods only in special cases. The derivative test for determining maximum and minimum values of a function of three variables over a region R is similar to the derivative test for functions of two variables.

Derivative test for function of three-variables

Assume $f = f(x, y, z)$ is continuous everywhere in a region R of x, y, z-space. Further assume that f possess derivatives through the second order that are also continuous everywhere in the region R. If $P_0 = (x_0, y_0, z_0)$ is a stationary point of $f = f(x, y, z)$ interior to the region R, then $\frac{\partial f}{\partial x}\Big|_{P_0} = 0$, $\frac{\partial f}{\partial y}\Big|_{P_0} = 0$, and $\frac{\partial f}{\partial z}\Big|_{P_0} = 0$. Let $f(P_0) = f(x_0, y_0, z_0)$ and calculate the Hessian matrix

$$H = H(P_0) = \begin{bmatrix} \frac{\partial^2 f}{\partial x^2} & \frac{\partial^2 f}{\partial x \partial y} & \frac{\partial^2 f}{\partial x \partial z} \\ \frac{\partial^2 f}{\partial y \partial x} & \frac{\partial^2 f}{\partial y^2} & \frac{\partial^2 f}{\partial y \partial z} \\ \frac{\partial^2 f}{\partial z \partial x} & \frac{\partial^2 f}{\partial z \partial y} & \frac{\partial^2 f}{\partial z^2} \end{bmatrix}_0 = \begin{bmatrix} H_{1,1} & H_{1,2} & H_{1,3} \\ H_{2,1} & H_{2,2} & H_{2,3} \\ H_{3,1} & H_{3,2} & H_{3,3} \end{bmatrix} \tag{2.20}$$

from which one can calculate the Hessian functions H_1 as the element in the $H_{1,1}$ position of the Hessian matrix, H_2 as the determinant of the submatrix $\begin{bmatrix} H_{1,1} & H_{1,2} \\ H_{2,1} & H_{2,2} \end{bmatrix}$ and H_3 as the determinant of H. This gives the Hessian functions

$$H_1(P_0) = \frac{\partial^2 f}{\partial x^2}, \quad H_2(P_0) = \begin{vmatrix} \frac{\partial^2 f}{\partial x^2} & \frac{\partial^2 f}{\partial x \partial y} \\ \frac{\partial^2 f}{\partial y \partial x} & \frac{\partial^2 f}{\partial y^2} \end{vmatrix}, \quad H_3(P_0) = \begin{vmatrix} \frac{\partial^2 f}{\partial x^2} & \frac{\partial^2 f}{\partial x \partial y} & \frac{\partial^2 f}{\partial x \partial z} \\ \frac{\partial^2 f}{\partial y \partial x} & \frac{\partial^2 f}{\partial y^2} & \frac{\partial^2 f}{\partial y \partial z} \\ \frac{\partial^2 f}{\partial z \partial x} & \frac{\partial^2 f}{\partial z \partial y} & \frac{\partial^2 f}{\partial z^2} \end{vmatrix}$$

where all the derivatives in the determinants are to be evaluated at the stationary point P_0. One can then state that

(a) If $H_1(P_0) > 0$, $H_2(P_0) > 0$ and $H_3(P_0) > 0$, then $f(P_0)$ corresponds to a local minimum value for $f = f(x, y, z)$.

(b) If $H_1(P_0) < 0$, $H_2(P_0) > 0$ and $H_3(P_0) < 0$, then $f(P_0)$ corresponds to a local maximum value for $f = f(x, y, z)$

(c) If $H_2(P_0) < 0$, then $f(P_0)$ corresponds to a saddle point for $f = f(x, y, z)$.

(d) If $H_1(P_0) \leq 0$, and $H_3(P_0) > 0$, then $f(P_0)$ corresponds to a saddle point.

(e) If $H_1(P) \geq 0$, and $H_3(P_0) < 0$, then $f(P_0)$ corresponds to a saddle point.

Example 2-3. Find extremum values of $z(x, y) = 3x^2 + 4y^2 - 6x - 8y + 8$ for x, y restricted to the triangular region R having vertices $(0, 0), (0, 3)$ and $(3, 0)$.

Solution:

We must test $z = z(x, y)$ at points (x, y) interior to R and then we must test the given function for points (x, y) on the boundary of the region R. We calculate the first derivatives

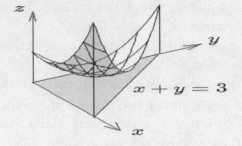

$$\frac{\partial z}{\partial x} = 6x - 6, \quad \frac{\partial z}{\partial y} = 8y - 8.$$

At a stationary value it is required that these derivatives be zero simultaneously. This gives the critical point $(1, 1)$. The second derivatives are calculated

$$\frac{\partial^2 z}{\partial x^2} = 6, \quad \frac{\partial^2 z}{\partial x \partial y} = 0, \quad \frac{\partial^2 z}{\partial y^2} = 8$$

and consequently one finds the Hessian functions

$$H_1 = 6 > 0, \quad \text{and} \quad H_2 = \begin{vmatrix} 6 & 0 \\ 0 & 8 \end{vmatrix} = 48 > 0.$$

Therefore, at the critical point $(1, 1)$ we find $z(1, 1) = 1$ that prior to further investigation we label a local minimum. On the boundary $y = 0$ one can verify $z(x, 0) = 3x^2 - 6x + 8$ for $0 \le x \le 3$ with end conditions $z(0, 0) = 8$ and $z(3, 0) = 17$ and minimum value $z(1, 0) = 5$. On the boundary $x = 0$ we have $z(0, y) = 4y^2 - 8y + 8$ for $0 \le y \le 3$ which gives the end conditions $z(0, 0) = 8$ and $z(0, 3) = 20$ and minimum value $z(0, 1) = 4$. On the line $y = 3 - x$ we have $z(x, 3 - x) = 7x^2 - 22x + 20$ for $0 \le x \le 3$ which gives the end conditions $z(0, 3) = 20$ and $z(3, 0) = 17$ and minimum value $z(11/7, 10/7) = 19/7$. Consequently, $z(0, 3) = 20$ corresponds to an absolute maximum over R and $z(1, 1) = 1$ corresponds to an absolute minimum over the region R. The situation is illustrated in the accompanying figure. ∎

Example 2-4.

A rectangular box, without a top, is to be constructed such that it has a volume of four cubic feet. Find the dimensions of the box with the smallest surface area.

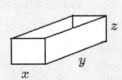

Solution: Let S denote the surface area of the box without a top that has the dimensions x, y and z illustrated and let V denote the volume of the box. One can verify that

$$S = xy + 2xz + 2yz \quad \text{and} \quad V = xyz = 4.$$

Since the volume is a constant, one can substitute $z = 4/xy$ into the surface area relation to obtain

$$S = xy + \frac{8}{y} + \frac{8}{x}.$$

At a stationary value we require that the partial derivatives satisfy the equations

$$\frac{\partial S}{\partial x} = y - \frac{8}{x^2} = 0, \qquad \text{and} \qquad \frac{\partial S}{\partial y} = x - \frac{8}{y^2} = 0.$$

In order for these equations to be satisfied simultaneously we require $y = 8/x^2$ and $y^2 = 8/x$ or $64/x^4 = 8/x$ which simplifies to $x^3 = 8$. The only realistic critical value is $x = 2$ with $y = 8/x^2 = 2$. We find at $x = 2$ and $y = 2$ that $S = 12$ and $z = 1$ so that the height of the box is one-half the size of an edge. We test the second derivatives evaluate at the critical point $(2, 2)$ and find

$$\frac{\partial^2 S}{\partial x^2} = \frac{16}{x^3}\Big|_{x=2} = 2, \qquad \frac{\partial^2 S}{\partial y^2} = \frac{16}{y^3}\Big|_{y=2} = 2, \qquad \frac{\partial^2 S}{\partial x \partial y} = 1.$$

This gives

$$H_2 = \begin{vmatrix} \frac{\partial^2 S}{\partial x^2} & \frac{\partial^2 S}{\partial x \partial y} \\ \frac{\partial^2 S}{\partial x \partial y} & \frac{\partial^2 S}{\partial y^2} \end{vmatrix} = \begin{vmatrix} 2 & 1 \\ 1 & 2 \end{vmatrix} = 3 > 0$$

which together with $H_1 = \frac{\partial^2 S}{\partial x^2} > 0$ shows that $S = 12$ square feet corresponds to a minimum surface area.

■

Example 2-5.

Find the local maximum and minimum values on the surface $z = z(x, y) = -x^3 - y^3 + 3x^2 + 12y$ over the region R consisting of all x, y values.

Solution: One can verify that

$$\frac{\partial z}{\partial x} = -3x^2 + 6x, \quad \frac{\partial z}{\partial y} = -3y^2 + 12, \quad \frac{\partial^2 z}{\partial x^2} = -6x + 6, \quad \frac{\partial^2 z}{\partial x \partial y} = 0, \quad \frac{\partial^2 z}{\partial y^2} = -6y.$$

We find that when $\frac{\partial z}{\partial x} = 0$ and $\frac{\partial z}{\partial y} = 0$ there results the following critical points $(0, -2)$, $(0, 2)$, $(2, -2)$ and $(2, 2)$.

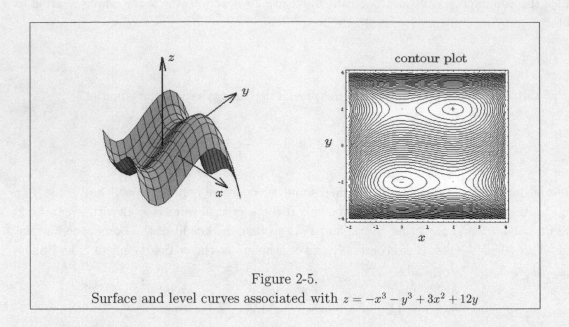

Figure 2-5.
Surface and level curves associated with $z = -x^3 - y^3 + 3x^2 + 12y$

The Hessian matrix is given by

$$H(x,y) = \begin{bmatrix} \frac{\partial^2 z}{\partial x^2} & \frac{\partial^2 z}{\partial x \partial y} \\ \frac{\partial^2 z}{\partial x \partial y} & \frac{\partial^2 z}{\partial y^2} \end{bmatrix} = \begin{bmatrix} -6x + 6 & 0 \\ 0 & -6y \end{bmatrix}.$$

The Hessian matrix gives us the Hessian functions that can be used to analyze the critical points.

At $(0, 2)$ we have $H_1 > 0$, and $H_2 < 0$ which corresponds to a saddle point.

At $(0, -2)$ we have $H_1 > 0$, and $H_2 > 0$ which corresponds to a local minimum value.

At $(2, 2)$ we have $H_1 < 0$ and $H_2 > 0$ which corresponds to a local maximum value.

At $(2, -2)$ we have $H_1 < 0$ and $H_2 < 0$ which corresponds to a saddle point.

The local minimum value is $z(0, -2) = -16$ and the local maximum value is $z(2, 2) = 20$.

■

Example 2-6. A circular plate of radius one meter has a temperature distribution given by $T = T(x,y) = x^2 + y^2 - y - x + 1$. Find the points of local maximum and minimum temperature.
Solution: The stationary points occur where

$$\frac{\partial T}{\partial x} = 2x - 1 = 0, \qquad \text{and} \qquad \frac{\partial T}{\partial y} = 2y - 1 = 0.$$

This gives the critical point $P_0 = (\frac{1}{2}, \frac{1}{2})$. The second partial derivatives give $\frac{\partial^2 T}{\partial x^2} = 2$, $\frac{\partial^2 T}{\partial x \partial y} = 0$, $\frac{\partial^2 T}{\partial y^2} = 2$, with $H_1 > 0$ and $H_2 > 0$ so that P_0 corresponds to a local minimum value.

The boundary of the region must be tested separately. This can be accomplished by representing the boundary using a set of parametric equations to represent points on the boundary. Let $x = \cos\theta$ and $y = \sin\theta$, for $0 \le \theta \le 2\pi$, denote the boundary curve, then the temperature along the boundary curve is given by

$$T = T(\theta) = \cos^2\theta + \sin^2\theta - \sin\theta - \cos\theta + 1 = 2 - \sin\theta - \cos\theta.$$

We test for stationary points by calculating the derivative and setting it equal to zero. We find

$$\frac{\partial T}{\partial \theta} = -\cos\theta + \sin\theta = 0 \quad \text{for} \quad \theta = \frac{\pi}{4} \quad \text{and} \quad \theta = \frac{5\pi}{4}.$$

We use the second derivative test and find that $\theta = \frac{\pi}{4}$ corresponds to a local minimum value and $\theta = \frac{5\pi}{4}$ corresponds to a local maximum value. A sketch of the temperature distribution over the boundary of the circular plate is illustrated.

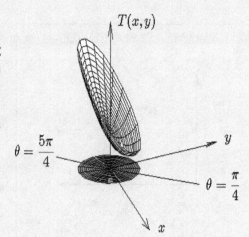

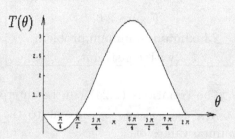

Lagrange multipliers

Consider the problem of finding stationary values associated with a function $f = f(x, y)$ subject to the constraint condition that $g = g(x, y) = 0$. Recall that a necessary condition for $f = f(x, y)$ to have an extremum value at a point (a, b) requires that the differential $df = 0$ or

$$df = \frac{\partial f}{\partial x}\, dx + \frac{\partial f}{\partial y}\, dy = 0. \tag{2.21}$$

Whenever the small changes dx and dy are independent, one obtains the necessary conditions that

$$\frac{\partial f}{\partial x} = 0 \quad \text{and} \quad \frac{\partial f}{\partial y} = 0$$

at a critical point. Whenever a constraint condition is required to be satisfied, then the small changes dx and dy are no longer independent and one must find the relationship between the small changes dx and dy as the point (x, y) moves along the constraint curve. From the differential relation $dg = 0$ one finds that

$$dg = \frac{\partial g}{\partial x}\, dx + \frac{\partial g}{\partial y}\, dy = 0$$

must be satisfied. Assume that $\frac{\partial g}{\partial y} \ne 0$, then one can obtain

$$dy = \frac{-\frac{\partial g}{\partial x}}{\frac{\partial g}{\partial y}}\, dx \tag{2.22}$$

as the dependent relationship between the small changes dx and dy.

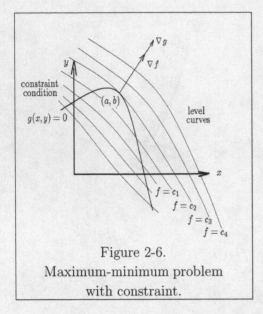

Figure 2-6.
Maximum-minimum problem
with constraint.

Substitute the dy from equation (2.22) into the equation (2.21) to produce the result

$$df = \frac{1}{\frac{\partial g}{\partial y}} \left(\frac{\partial f}{\partial x} \frac{\partial g}{\partial y} - \frac{\partial f}{\partial y} \frac{\partial g}{\partial x} \right) dx = 0 \qquad (2.23)$$

that must hold for an arbitrary change dx. This gives the following necessary condition. The critical points (x, y) of the function f, subject to the constraint equation $g(x, y) = 0$, must satisfy the equations

$$\begin{aligned} \frac{\partial f}{\partial x} \frac{\partial g}{\partial y} - \frac{\partial f}{\partial y} \frac{\partial g}{\partial x} &= 0 \\ g(x, y) &= 0 \end{aligned} \qquad (2.24)$$

simultaneously.

The equations (2.24) can be interpreted that when a member of the family of curves $f(x, y) = c = constant$ is tangent to the constraint curve $g(x, y) = 0$, then we will have the common values of

$$\frac{dy}{dx} = \frac{-\frac{\partial f}{\partial x}}{\frac{\partial f}{\partial y}} = \frac{-\frac{\partial g}{\partial x}}{\frac{\partial g}{\partial y}} \qquad \Rightarrow \qquad \frac{\partial f}{\partial x} \frac{\partial g}{\partial y} - \frac{\partial f}{\partial y} \frac{\partial g}{\partial x} = 0.$$

One can give a physical picture of the problem. Think of the constraint condition given by $g = g(x, y) = 0$ as defining a curve in the x, y-plane and then consider the family of level curves $f = f(x, y) = c$, where c is some constant. A representative sketch of the curve $g(x, y) = 0$, together with several level curves from the family, $f = c$ are illustrated in the figure 2-6. Among all the level curves that intersect the constraint condition curve $g(x, y) = 0$ we desire that curve for which c has the largest or smallest value. We shall assume that the constraint curve $g(x, y) = 0$ is a smooth curve without singular points.

If (a, b) denotes a point of tangency between a curve of the family $f = c$ and the constraint curve $g(x, y) = 0$, then at this point both curves will have gradient vectors that are collinear and so one can write $\nabla f + \lambda \nabla g = \vec{0}$ for some constant λ called a Lagrange multiplier. This relationship together with the constraint equation produces the three scalar equations

$$\begin{aligned} \frac{\partial f}{\partial x} + \lambda \frac{\partial g}{\partial x} &= 0 \\ \frac{\partial f}{\partial y} + \lambda \frac{\partial g}{\partial y} &= 0 \\ g(x, y) &= 0. \end{aligned} \qquad \Rightarrow \qquad \frac{\partial f}{\partial x} \frac{\partial g}{\partial y} - \frac{\partial f}{\partial y} \frac{\partial g}{\partial x} = 0 \qquad (2.25)$$

Lagrange viewed the above problem in the following way. Define the function

$$F(x, y, \lambda) = f(x, y) + \lambda g(x, y) \qquad (2.26)$$

where $f(x, y)$ is called an objective function and represents the function to be maximized or minimized. The parameter λ is called a Lagrange multiplier and the function $g(x, y)$ is obtained from the constraint condition. Lagrange observed that a stationary value of the function F, without constraints, is equivalent to the problem of stationary values of f with a constraint condition because one would have at a stationary value of F the conditions

$$\frac{\partial F}{\partial x} = \frac{\partial f}{\partial x} + \lambda \frac{\partial g}{\partial x} = 0$$
$$\frac{\partial F}{\partial y} = \frac{\partial f}{\partial y} + \lambda \frac{\partial g}{\partial y} = 0 \qquad\qquad (2.27)$$
$$\frac{\partial F}{\partial \lambda} = g(x, y) = 0 \qquad \text{The constraint condition.}$$

These represent three equations in the three unknowns x, y, λ that must be solved. The equations (2.26) and (2.27) are known as the Lagrange rule for the method of Lagrange multipliers.

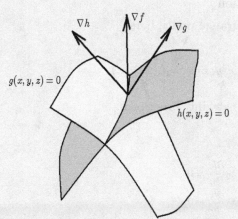

The method of Lagrange multipliers can be applied in higher dimensions. For example, consider the problem of finding maximum and minimum values associated with a function $f = f(x, y, z)$ subject to the constraint conditions $g(x, y, z) = 0$ and $h(x, y, z) = 0$. Here the equations $g(x, y, z) = 0$ and $h(x, y, z) = 0$ describe two surfaces that may or may not intersect. Assume the surfaces intersect to give a space curve.

The problem is to find an extremal value of $f = f(x, y, z)$ as (x, y, z) varies along the curve of intersection of surfaces $g = 0$ and $h = 0$. At a critical point where a stationary value exists, the directional derivative of f along this curve must be zero. Here the directional derivative is given by $\frac{df}{ds} = \nabla f \cdot \hat{\mathbf{e}}_t$, where $\hat{\mathbf{e}}_t$ is a unit tangent vector to the space curve and $\nabla f = \text{grad}\, f$. Note that if the directional derivative is zero, then ∇f must lie in a plane normal to the curve of intersection.

Another way to view the problem, and also suggest that the concepts can be extended to higher dimensional spaces, is to introduce the notation $\bar{x} = (x_1, x_2, x_3) = (x, y, z)$ to denote a vector to a point on the curve of intersection of the two surfaces $g(x_1, x_2, x_3) = 0$ and $h(x_1, x_2, x_3) = 0$. At a stationary value of f one must have

$$df = \frac{\partial f}{\partial x_1} dx_1 + \frac{\partial f}{\partial x_2} dx_2 + \frac{\partial f}{\partial x_3} dx_3 = \text{grad}\, f \cdot d\bar{x} = 0$$

This implies that $\text{grad}\, f$ is normal to the curve of intersection since it is perpendicular to the tangent vector $d\bar{x}$ to the curve of intersection. At a stationary point, the normal plane containing the vector $\text{grad}\, f$ also contains the vectors ∇g and ∇h since $dg = \text{grad}\, g \cdot d\bar{x} = 0$ and

$dh = \operatorname{grad} h \cdot d\vec{x} = 0$ at the stationary point. Hence, if these three vectors are noncollinear, then there will exist scalars λ_1 and λ_2 such that

$$\nabla f + \lambda_1 \nabla g + \lambda_2 \nabla h = 0 \tag{2.28}$$

at a stationary point. The equation (2.28) is a vector equation and is equivalent to the three scalar equations

$$\frac{\partial f}{\partial x} + \lambda_1 \frac{\partial g}{\partial x} + \lambda_2 \frac{\partial h}{\partial x} = 0 \qquad \frac{\partial f}{\partial x_1} + \lambda_1 \frac{\partial g}{\partial x_1} + \lambda_2 \frac{\partial h}{\partial x_1} = 0$$

$$\frac{\partial f}{\partial y} + \lambda_1 \frac{\partial g}{\partial y} + \lambda_2 \frac{\partial h}{\partial y} = 0 \quad \text{or} \quad \frac{\partial f}{\partial x_2} + \lambda_1 \frac{\partial g}{\partial x_2} + \lambda_2 \frac{\partial h}{\partial x_2} = 0$$

$$\frac{\partial f}{\partial z} + \lambda_1 \frac{\partial g}{\partial z} + \lambda_2 \frac{\partial h}{\partial z} = 0. \qquad \frac{\partial f}{\partial x_3} + \lambda_1 \frac{\partial g}{\partial x_3} + \lambda_2 \frac{\partial h}{\partial x_3} = 0.$$

depending upon the notation you are using. These three equations together with the constraint equations $g = 0$ and $h = 0$ gives us five equations in the five unknowns $x, y, z, \lambda_1, \lambda_2$ that must be satisfied at a stationary point.

By the Lagrangian rule one can form the function

Lagrange multipliers

$$F = F(x, y, z, \lambda_1, \lambda_2) = f(x, y, z) + \lambda_1 g(x, y, z) + \lambda_2 h(x, y, z)$$

Objective
function

constraint functions

Observe that F has a stationary value where

$$\frac{\partial F}{\partial x} = \frac{\partial f}{\partial x} + \lambda_1 \frac{\partial g}{\partial x} + \lambda_2 \frac{\partial h}{\partial x} = 0$$

$$\frac{\partial F}{\partial y} = \frac{\partial f}{\partial y} + \lambda_1 \frac{\partial g}{\partial y} + \lambda_2 \frac{\partial h}{\partial y} = 0$$

$$\frac{\partial F}{\partial z} = \frac{\partial f}{\partial z} + \lambda_1 \frac{\partial g}{\partial z} + \lambda_2 \frac{\partial h}{\partial z} = 0 \tag{2.29}$$

$$\frac{\partial F}{\partial \lambda_1} = g(x, y, z) = 0$$

$$\frac{\partial F}{\partial \lambda_2} = h(x, y, z) = 0$$

These are the same five equations, with unknowns $x, y, z, \lambda_1, \lambda_2$, for determining the stationary points as previously noted.

Generalization of Lagrange multipliers

In general, to find an extremal value associated with a n-dimensional function given by $f = f(\bar{x}) = f(x_1, x_2, \ldots, x_n)$ subject to k constraint conditions that can be written in the form $g_i(\bar{x}) = g_i(x_1, x_2, \ldots, x_n) = 0$, for $i = 1, 2, \ldots, k$, where k is less than n. We require that the gradient vectors $\nabla g_1, \nabla g_2, \ldots, \nabla g_k$ be linearly independent vectors, then one can employ the method of Lagrange multipliers as follows. The Lagrangian rule requires that we form the function $F = f + \sum_{i=1}^{k} \lambda_i g_i$ that can be written in the expanded form

Lagrange multipliers

$$F(\bar{x}; \bar{\lambda}) = f + \lambda_1 g_1 + \lambda_2 g_2 + \cdots + \lambda_k g_k \tag{2.30}$$

objective function

constraint functions

which contains the objective function f, summed with each of the constraint functions g_i, multiplied by a Lagrange multiplier λ_i, for the index i having the values $i = 1, \ldots, k$. Here the function F and consequently the function f has stationary values at those points where the following equations are satisfied

$$\frac{\partial F}{\partial x_i} = 0, \quad \text{for } i = 1, \ldots, n$$
$$\frac{\partial F}{\partial \lambda_j} = 0, \quad \text{for } j = 1, \ldots, k \tag{2.31}$$

The equations (2.31) represent a system of $(n + k)$ equations in the $(n + k)$ unknowns $x_1, x_2, \ldots, x_n, \lambda_1, \lambda_2, \ldots, \lambda_k$ for determining the stationary points. In general, the stationary points will be found in terms of the λ_i values. The vector $(\bar{x}_0, \bar{\lambda}_0)$ where $\bar{x}_0$ and $\bar{\lambda}_0$ are solutions of the system of equations (2.31) can be thought of as critical points associated with the Lagrangian function $F(\bar{x}, \bar{\lambda})$ given by equation (2.30). The resulting stationary points must then be tested to determine whether they correspond to a relative maximum value, minimum value or saddle point. One can form the Hessian matrix associated with the function $F(\bar{x}; \bar{\lambda})$ and analyze this matrix at the critical points. Whenever the determinant of the Hessian matrix is zero at a critical point, then the critical point $(\bar{x}_0, \bar{\lambda}_0)$ is said to be degenerate and one must seek an alternative method to test for an extremum.

Note: There can arise situations where F of equation (2.30) is defined using minus signs and written in the form

$$F(\bar{x}; \bar{\lambda}) = f - \lambda_1 g_1 - \lambda_2 g_2 - \cdots - \lambda_k g_k \tag{2.32}$$

In the case of equality constraints it doesn't matter what sign the Lagrange multipliers are assigned. However, in the case of linear programming problems the constraint conditions can be written to define regions by using inequality signs. In such a case the constraints are written as $g_i(\bar{x}) \geq 0$ or $g_i(\bar{x}) \leq 0$. Whenever a constraint involves an inequality of the form $g_i(\bar{x}) \leq 0$ it is usually the custom to write it in the form $-g_i(\bar{x}) \geq 0$ and inequalities of the form $g_i(\bar{x}) \geq 0$ are left in this form with a positive sign. The signs assigned to the Lagrange multipliers λ_i are such that the quantity F becomes a convex function. An n-dimensional region R_n is called convex if $\bar{x} = (x_1, x_2, \ldots, x_n)$ and $\bar{y} = (y_1, y_2, \ldots, y_n)$ are two distinct points within the region R_n such that for all θ satisfying $0 \leq \theta \leq 1$ their convex combination $\theta \bar{x} + (1 - \theta)\bar{y}$ also belongs to the region R_n. The condition that a region R_n is convex can be given the geometric interpretation that a straight line joining any two points within the region also remains within the region. Also required in the study of optimization problems are knowledge of monotone functions and convex functions.

Definition: Monotone function

A function $f(x)$ of a single variable x is said to be monotone increasing over an interval if for any two points x_1 and x_2 selected from the interval, with $x_1 < x_2$, the relation $f(x_1) \leq f(x_2)$ is always satisfied. If the condition $f(x_1) \geq f(x_2)$ always holds when $x_1 < x_2$, the function f is called monotone decreasing over the interval.

Definition: Convex function

A function $f(x)$ of a single variable x is said to be concave upward or convex over an interval if for any two points x_1 and x_2 on the interval, where $x_1 < x_2$, the inequality

$$f\left(\frac{x_1 + x_2}{2}\right) \leq \frac{1}{2}\left(f(x_1) + f(x_2)\right) \tag{2.33}$$

is satisfied. This has the geometric interpretation that the function evaluated at the midpoint of the interval (x_1, x_2) is always less than or equal to the average value of the end point ordinates. An alternative way of writing this definition is to say that f is a convex function if for all values of θ satisfying $0 \leq \theta \leq 1$, the following inequality holds

$$f\left((1-\theta)x_1 + \theta x_2\right) \leq (1-\theta)f(x_1) + \theta f(x_2) \tag{2.34}$$

The definition of a convex function can be generalized to functions of n-variables $f(\bar{x})$ which are real valued functions defined in a convex region R_n. The more general definition states that if $\bar{x}$ and $\bar{y}$ are two points in a convex region R_n, then a real-valued function f is said to be convex in R_n if the following inequality is satisfied

$$f\left((1-\theta)\bar{x} + \theta\bar{y}\right) \leq (1-\theta)f(\bar{x}) + \theta f(\bar{y}), \qquad \text{for all } \theta \text{ satisfying } 0 \leq \theta \leq 1 \tag{2.35}$$

It can be shown, using an extended mean value theorem, that the local minimum of a convex function is a global minimum.

Example 2-7.

Determine the extreme values of the function $f = x^2 + y^2$ as the point (x, y) moves along the curve $xy = 1$.

Solution 1: Substitute the value $y = 1/x$ into the objective function f and write the objective function in the form of a function of one variable $f = f(x) = x^2 + \dfrac{1}{x^2}$. Stationary values occur where $\dfrac{df}{dx} = 2x - \dfrac{2}{x^3} = 0$ or $x^4 - 1 = (x^2 - 1)(x^2 + 1) = 0$. The only real critical values occur at $x = 1$ and $x = -1$. At these critical values one finds that $\dfrac{d^2 f}{dx^2} = 2 + \dfrac{6}{x^4} > 0$ and consequently the critical values correspond to a relative minimum of the objective function f.

Solution 2: We use Lagrange multipliers and write

$$F = f + \lambda g = x^2 + y^2 + \lambda(xy - 1) \tag{2.36}$$

At a stationary value of F and f we have

$$\frac{\partial F}{\partial x} = 2x + \lambda y = 0$$
$$\frac{\partial F}{\partial y} = 2y + \lambda x = 0$$
$$\frac{\partial F}{\partial \lambda} = xy - 1 = 0$$

The first two equations above imply the relation $\lambda = -2x/y = -2y/x$ must be satisfied. This requires $x^2 - y^2 = (x - y)(x + y) = 0$, which gives two cases to examine. The case where $x = y$ and the case where $x = -y$. The first case $x = y$ gives $xy = x^2 = 1$ with corresponding critical values $x = 1$ and $x = -1$. Both critical values produce the objective function value $f = 2$. For $f = x^2 + y^2$ we have the Hessian function

$$H_2 = \begin{vmatrix} 2 & 0 \\ 0 & 2 \end{vmatrix} > 0$$

which demonstrates that the value $f = 2$ corresponds to a relative minimum value at the critical points $x = 1$ and $x = -1$.

The case where $x = -y$ gives $xy = -x^2 = 1$ which has no real solution and so this case is discarded.

∎

Example 2-8.

Find the shortest distance from the hyperbola $15x^2 + 34xy + 15y^2 + 32 = 0$ to the origin.

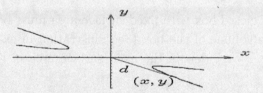

Solution: If (x, y) is a point on the hyperbola, then the distance from the origin to this point squared is $d^2 = x^2 + y^2$. Let $f = f(x, y) = d^2 = x^2 + y^2$ denote the objective function to be minimized and introduce a Lagrange multiplier λ to form the function

$$F = x^2 + y^2 + \lambda(15x^2 + 34xy + 15y^2 + 32)$$

containing the objective function and constraint condition. A critical point occurs where $\frac{\partial F}{\partial x} = 0$, $\frac{\partial F}{\partial y} = 0$ and $\frac{\partial F}{\partial \lambda} = 0$. We form this system of equations and obtain

$$\frac{\partial F}{\partial x} = 2x + \lambda(30x + 34y) = 0 \tag{2.37}$$

$$\frac{\partial F}{\partial y} = 2y + \lambda(34x + 30y) = 0 \tag{2.38}$$

$$\frac{\partial F}{\partial \lambda} = 15x^2 + 34xy + 15y^2 + 32 = 0 \tag{2.39}$$

The equations (2.37) and (2.38) can be written in the form

$$(1 + 15\lambda)x + 17\lambda y = 0$$
$$17\lambda x + (1 + 15\lambda)y = 0$$

(2.40)

The point (x, y) where a minimum occurs cannot be $(0, 0)$ and therefore in order for the system of equations (2.40) to have a nonzero solution we require that the determinant of the coefficient matrix be zero. This requires λ satisfy the condition

$$\begin{vmatrix} (1 + 15\lambda) & 17\lambda \\ 17\lambda & (1 + 15\lambda) \end{vmatrix} = 1 + 30\lambda - 64\lambda^2 = 0.$$

(2.41)

The quadratic equation (2.41) has the roots $\lambda = 1/2$ and $\lambda = -1/32$. We now examine the critical points associated with each of these cases.

Case 1:

For the case $\lambda = 1/2$ observe that both the equations (2.37) and (2.38) reduce to the equation $y = -x$. We substitute this result into the equation (2.39) to obtain $x^2 = 8$. Consequently, $y^2 = 8$ and the objective function becomes $f = x^2 + y^2 = 16$. The shortest distance is therefore given by $d = \sqrt{f} = 4$.

Case 2:

For the case $\lambda = -1/32$, the equations (2.37) and (2.38) reduce to the equation $y = x$. When this result is substituted into the equation (2.39) one obtains the result $64x^2 = -32$ which has no real solutions and so this case is discarded.

■

Example 2-9.

Show the largest rectangular closed box with fixed surface area must be a cube.

Solution: Let $V = xyz$ denote the volume of the box (See illustration in example 2-4, given on page 68), and let $A = 2yz + 2xz + 2xy$ denote the fixed surface area of the box. We use Lagrange multipliers and write

$$f = xyz + \lambda(2yz + 2xz + 2xy - A)$$

as the function to be maximized. At a stationary value we require that the following equations be satisfied

$$\frac{\partial f}{\partial x} = yz + \lambda(2z + 2y) = 0$$
$$\frac{\partial f}{\partial y} = xz + \lambda(2z + 2x) = 0$$
$$\frac{\partial f}{\partial z} = xy + \lambda(2y + 2x) = 0$$
$$\frac{\partial f}{\partial \lambda} = 2yz + 2xz + 2xy - A = 0$$

One can solve the above system of equations and verify that $x = y = z = \sqrt{A/6}$

■

Example 2-10.

Find the maximum and minimum value of $f = x^2 + y^2 + z^2$ as the point (x, y, z) moves along the curve defined by the intersection of the following surfaces.

The ellipsoid $\quad \dfrac{x^2}{4} + \dfrac{y^2}{9} + \dfrac{z^2}{36} = 1 \quad$ and the plane $\quad z = x + y.$

Solution: We introduce two Lagrange multipliers λ_1 and λ_2 and define the function

$$F = x^2 + y^2 + z^2 + \lambda_1 \left(\frac{x^2}{4} + \frac{y^2}{9} + \frac{z^2}{36} - 1 \right) + \lambda_2 (x + y - z) \tag{2.42}$$

composed of the objective function and the constraint conditions multiplied by the Lagrange multipliers. The functions F and f have stationary values where $\frac{\partial F}{\partial x} = \frac{\partial F}{\partial y} = \frac{\partial F}{\partial z} = \frac{\partial F}{\partial \lambda_1} = \frac{\partial F}{\partial \lambda_2} = 0$. The required equations to solve are given by

$$\frac{\partial F}{\partial x} = 2x + \lambda_1 \left(\frac{2x}{4} \right) + \lambda_2 (1) = 0 \tag{2.43}$$

$$\frac{\partial F}{\partial y} = 2y + \lambda_1 \left(\frac{2y}{9} \right) + \lambda_2 (1) = 0 \tag{2.44}$$

$$\frac{\partial F}{\partial z} = 2z + \lambda_1 \left(\frac{2z}{36} \right) + \lambda_2 (-1) = 0 \tag{2.45}$$

$$\frac{\partial F}{\partial \lambda_1} = \frac{x^2}{4} + \frac{y^2}{9} + \frac{z^2}{36} - 1 = 0 \tag{2.46}$$

$$\frac{\partial F}{\partial \lambda_2} = x + y - z = 0 \tag{2.47}$$

The equations (2.43), (2.44), (2.45) give the results

$$x = \frac{-\lambda_2}{2 + \frac{2\lambda_1}{4}}, \qquad y = \frac{-\lambda_2}{2 + \frac{2\lambda_1}{9}}, \qquad z = \frac{\lambda_2}{2 + \frac{2\lambda_1}{36}}. \tag{2.48}$$

If we substitute these results into the equation (2.47) and simplify the resulting equation one obtains the quadratic equation

$$3 + \frac{7}{9}\lambda_1 + \frac{49}{1296}\lambda_1^2 = 0 \tag{2.49}$$

with roots $\lambda_1 = -108/7$ and $\lambda_1 = -36/7$. In the special case $\lambda_1 = -108/7$ one substitutes λ_1 into the equations (2.48) to obtain

$$x = \frac{7\lambda_2}{40}, \qquad y = \frac{7\lambda_2}{10}, \qquad z = \frac{7\lambda_2}{8} \tag{2.50}$$

If we substitute these values into the constraint equation (2.46) and solve for λ_2 one obtains $\lambda_2 = +\frac{120\sqrt{2}}{49}$ This gives the critical points

$$P_1 = \left(\frac{3\sqrt{2}}{7}, \frac{12\sqrt{2}}{7}, \frac{15\sqrt{2}}{7} \right), \qquad \text{and} \qquad P_2 = \left(\frac{-3\sqrt{2}}{7}, \frac{-12\sqrt{2}}{7}, \frac{-15\sqrt{2}}{7} \right)$$

At these critical points the objective function has the value $f = x^2 + y^2 + z^2 = \frac{108}{7}$.

Using the value $\lambda_1 = -\frac{36}{7}$, the equations (2.48) give the values

$$x = \frac{7\lambda_2}{4}, \qquad y = \frac{-7\lambda_2}{6}, \qquad z = \frac{7\lambda_2}{12}. \tag{2.51}$$

Substituting these values into the constraint equation (2.46) and solving for λ_2 gives the value $\lambda_2 = \pm\frac{36\sqrt{2}}{49}$. This gives the critical points

$$P_3 = (\frac{9\sqrt{2}}{7}, \frac{-6\sqrt{2}}{7}, \frac{3\sqrt{2}}{7}), \qquad \text{and} \qquad P_4 = (\frac{-9\sqrt{2}}{7}, \frac{6\sqrt{2}}{7}, \frac{-3\sqrt{2}}{7}) \tag{2.52}$$

At these critical points the value of the objective function is $f = x^2 + y^2 + z^2 = \frac{36}{7}$.

Observe that the directional derivative of f along the curve of intersection of the ellipsoid and plane can be calculated using the relations

$$\vec{N} = \text{grad}\, g_1 \times \text{grad}\, g_2, \qquad \hat{e} = \frac{1}{|\vec{N}|}\vec{N}, \qquad \frac{df}{ds} = \text{grad}\, f \cdot \hat{e} \tag{2.53}$$

where $g_1 = \frac{x^2}{4} + \frac{y^2}{9} + \frac{z^2}{36} - 1 = 0$ and $g_2 = x + y - z = 0$ are the given surfaces. One finds that the directional derivative is zero at the critical points. The second directional derivative is given by $\frac{d^2 f}{ds^2} = \text{grad}\left(\frac{df}{ds}\right) \cdot \hat{e}$. One finds that the second directional derivative is negative at the critical points P_1 and P_2 so that the value of f corresponds to a local maximum. The second directional derivative is positive at the critical points P_3 and P_4 so that the value of f corresponds to a local minimum value.

■

Example 2-11.

Find the extreme values of z on the ellipse formed by the intersection of the ellipsoid $x^2 + \frac{y^2}{4} + \frac{z^2}{16} = 1$ and the plane $x + y + z = 1$.

Solution: Using Lagrange multipliers one can form the function

$$F = z + \lambda_1(x + y + z - 1) + \lambda_2\left(x^2 + \frac{y^2}{4} + \frac{z^2}{16} - 1\right), \tag{2.54}$$

then the critical points of F and $f = z$ occur at those points where the derivatives $\frac{\partial F}{\partial x}, \frac{\partial F}{\partial y}, \frac{\partial F}{\partial z}, \frac{\partial F}{\partial \lambda_1}, \frac{\partial F}{\partial \lambda_2}$ are zero simultaneously. We calculate these derivatives and set them equal to zero to obtain.

$$\frac{\partial F}{\partial x} = \lambda_1 + 2\lambda_2 x = 0 \tag{2.55}$$

$$\frac{\partial F}{\partial y} = \lambda_1 + 2\lambda_2 y/4 = 0 \tag{2.56}$$

$$\frac{\partial F}{\partial z} = 1 + \lambda_1 + 2\lambda_2 z/16 = 0 \tag{2.57}$$

$$\frac{\partial F}{\partial \lambda_1} = x + y + z - 1 = 0 \tag{2.58}$$

$$\frac{\partial F}{\partial \lambda_2} = x^2 + \frac{y^2}{4} + \frac{z^2}{16} - 1 = 0 \tag{2.59}$$

x

Subtract the equations (2.55) and (2.56) to obtain

$$4\lambda_2 x - 4\lambda_2 y/4 = 0 \qquad \text{or} \qquad \lambda_2(4x - y) = 0.$$

Hence, $\lambda_2 = 0$ or $y = 4x$. If $\lambda_2 = 0$, then the equation (2.55) requires that $\lambda_1 = 0$ and as a consequence the equation (2.57) is not satisfied. Consequently we assume that $\lambda_2 \neq 0$ and examine the case where $y = 4x$. By substituting $y = 4x$ into the equation (2.58) and then solving for x one finds $x = (1-z)/5$. The values $x = (1-z)/5$ and $y = 4x = 4(1-z)/5$ substituted into the equation (2.59) gives

$$\frac{(1-z)^2}{25} + \frac{16(1-z)^2}{100} + \frac{z^2}{16} = 1, \quad \text{that simplifies to} \quad 21z^2 - 32z - 64 = (3z-8)(7z+8) = 0.$$

This gives the critical values $z = z_{max} = 8/3$ and $z = z_{min} = -8/7$. Here we are only interested in the z-value and so it is not necessary to solve for the other $x, y, \lambda_1, \lambda_2$ values.

Mathematical Programming

A mathematical programming problem is defined as an optimization problem consisting of an objective function $f = f(x_1, x_2, \ldots, x_n)$ to be optimized subject to certain constraint conditions. A general nonlinear mathematical programming problem has the general form

$$\text{Optimize:} \quad f = f(x_1, x_2, \ldots, x_n)$$
$$\text{Subject to constraints:} \quad g_1(x_1, x_2, \ldots, x_n) = 0$$
$$g_2(x_1, x_2, \ldots, x_n) = 0$$
$$\vdots$$
$$g_m(x_1, x_2, \ldots, x_n) = 0 \tag{2.60}$$

This type of problem is referred to as an unconstrained mathematical programming problem. In contrast, a mathematical program having the form

$$\text{Optimize:} \quad f = f(x_1, x_2, \ldots, x_n)$$
$$\text{Subject to constraints:} \quad g_1(x_1, x_2, \ldots, x_n) \leq c_1$$
$$g_2(x_1, x_2, \ldots, x_n) \leq c_2$$
$$\vdots$$
$$g_m(x_1, x_2, \ldots, x_n) \leq c_m \tag{2.61}$$

where $c_1, c_2, \ldots, c_m$ are constants, is referred to as a constrained mathematical programming problem. The less than or equal signs appearing in the constraint equations of the system (2.61) can be replaced by $\leq$, $=$ or $\geq$ signs to form other types of constrained mathematical programming problems.

Whenever both the constraint conditions and objective function f are linear functions having the forms

$$f = f(x_1, x_2, \ldots, x_n) = \alpha_1 x_1 + \alpha_2 x_2 + \cdots + \alpha_n x_n$$
$$\text{and} \quad g_i = g_i(x_1, x_2, \ldots, x_n) = \beta_{i1} x_1 + \beta_{i2} x_2 + \cdots + \beta_{in} x_n \tag{2.62}$$

where α_i and β_{ij} for $i = 1, \ldots, m$ and $j = 1, \ldots, n$ are given constants, then the mathematical programming problem is referred to as a linear programming problem. In the special case where the constraints g_i are linear and the objective function has the form

$$f = f(x_1, x_2, \ldots, x_n) = \sum_{i=1}^{n} \sum_{j=1}^{n} \gamma_{ij} x_i x_j + \sum_{i=1}^{n} \alpha_i x_i, \tag{2.63}$$

where γ_{ij} and α_i are given constants, then the mathematical programming problem is referred to as a quadratic programming problem.

Linear Programming

Consider the linear programming problem to maximize or minimize

$$f(\bar{x}) = f(x_1, x_2, \ldots, x_n) = \sum_{i=1}^{n} \alpha_k x_k = \alpha_1 x_1 + \alpha_2 x_2 + \cdots + \alpha_n x_n \tag{2.64}$$

subject to the constraint conditions having one of the forms

$$\sum_{j=1}^{n} \beta_{ij} x_j > c, \quad \sum_{j=1}^{n} \beta_{ij} x_j \geq c_i, \quad \sum_{j=1}^{n} \beta_{ij} x_j = c_i, \quad \sum_{j=1}^{n} \beta_{ij} x_j \leq c_i, \quad \sum_{j=1}^{n} \beta_{ij} x_j < c_i, \tag{2.65}$$

for $i = 1, \ldots, m$. For illustrative purposes we consider the special case where $n = 2$.

$$\text{Maximize or minimize} \quad f = f(x_1, x_2) = \alpha_1 x_1 + \alpha_2 x_2, \tag{2.66}$$

with α_1 and α_2 constants, where f is subject to constraint conditions having one of the forms

$$\beta_{i1} x_1 + \beta_{i2} x_2 \begin{Bmatrix} > \\ \geq \\ = \\ \leq \\ < \end{Bmatrix} c_i, \qquad i = 1, 2, \ldots, m \tag{2.67}$$

where only one of the conditions $>, \geq, =, \leq, <$ is selected for each value of i. Let $x_1 = x$ and $x_2 = y$ with constants $\beta_{i1} = a_i$ and $\beta_{i2} = b_i$, then each inequality given by equation (2.67) is satisfied by all points (x, y) in some half-plane. The half-plane is constructed by replacing the inequality with an equality sign to form the equation of a straight line. That is, the line $a_i x + b_i y - c_i = 0$ is constructed and then a point is selected in one of the half-planes to determine the side of the line where the inequality is valid. An example situation is illustrated in the figure 2-7. If there are constraints involving the equality sign, then this restricts the (x, y) values to lie upon the given line.

A half-plane is called closed if the points on the boundary line are part of the half-plane, otherwise the half-plane is called open. If a solution exists, then one can plot all half-planes determined by the given inequalities and determine a region R containing points (x, y) that satisfies all the inequalities.

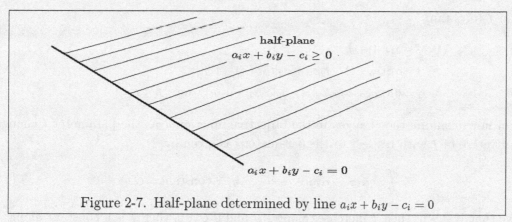

Figure 2-7. Half-plane determined by line $a_i x + b_i y - c_i = 0$

The set of points (x, y) that satisfies all the inequalities of the type given by equation (2.67) is called the graph of the system of linear inequalities. The graph represents a region formed by the intersection of the m-half-planes. Of course, one must assume that the region formed by the intersection of the half-planes is nonempty. The region R formed by the intersection of the half-planes will, in most cases, be a polygonal region since the region is bounded by straight lines. The corners of the polygonal region are called vertices. The coordinates of the vertices are determined by solving an appropriate pair of linear equations. However, if the region determined by the constraint conditions extends indefinitely in a direction, then the region is said to be unbounded. Polygonal regions defined by specific line segments are called bounded regions. Recall our earlier definition of a convex region. We can take any two points A and B in a region R and let $\overline{AB}$ denote the line segment connecting these points. If $\overline{AB}$ also belongs to R for all points A and B in the region R, then the region R is called a convex region. Note that each half-plane is a convex region. Hence, the region of intersection formed by all half-planes is also a convex region.

Consider two arbitrary points (x_0, y_0) and (x_1, y_1) in the region R formed by the intersection of the m-half-planes. We now examine the values of $f = \alpha_1 x + \alpha_2 y$ on the line segment joining the above two points. We shall demonstrate that either

(i) f has a constant value-or

(ii) f is smallest at one end and largest (2.68)
at the other end of the line segment.

To show this we construct a set of parametric equations to describe the line segment and then calculate the value of f at a general point (x, y) on the line segment. One can verify that one set of parametric equations describing the line segment connecting the points (x_0, y_0) and (x_1, y_1) can be written

$$x = tx_0 + (1 - t)x_1, \quad \text{and} \quad y = ty_0 + (1 - t)y_1 \tag{2.69}$$

where the parameter t satisfies the inequality $0 \leq t \leq 1$. The value of f on this line segment can be calculated by substituting the parametric equations for x and y into the objective

84

function f to obtain

$$f = \alpha_1 x + \alpha_2 y$$
$$f = \alpha_1[tx_0 + (1-t)x_1] + \alpha_2[ty_0 + (1-t)y_1] \tag{2.70}$$
$$f = t(\alpha_1 x_0 + \alpha_2 y_0) + (1-t)[\alpha_1 x_1 + \alpha_2 y_1] \quad \text{for} \quad 0 \le t \le 1$$

One can now examine the changes in the objective function f as the parameter t changes. The derivative of f with respect to the parameter t is given by

$$\frac{df}{dt} = \alpha_1 x_0 + \alpha_2 y_0 - \alpha_1 x_1 - \alpha_2 y_1 = \text{constant} = C \tag{2.71}$$

Observe that if $C > 0$, then f increases along $\overline{AB}$ and if $C = 0$, then f is a constant along $\overline{AB}$ and if $C < 0$, then f decreases along $\overline{AB}$. These results are used to form the conclusions given by (2.68).

Let us examine the maximum and minimum values of the objective function f at a general point P inside the convex region R determined by the constraint conditions. Assume that one can construct a straight line through the point P which has its end points on the boundary of the convex region R. Using the results (2.68) the value of f, at one of the end points, is at least as great as the value at the point P. Hence, the maximum and minimum values of f, if they exist, must occur on the boundary of the region R. If one of the sides of the region R is finite in length, the maximum value of f along this line must occur at one end. If all sides of the region R are finite in length, then the maximum value of f must occur at a vertex. If one side of the region R is infinite, then there may not be a maximum value for f. If a maximum value occurs, it must occur at a finite end point of a boundary line. If the region defined by the constraint conditions is a polygonal region with a finite number of vertices, then the maximum and minimum values of the objective function must occur at a vertex of the region. Hence, if one lists the coordinates of all the vertices and calculates the value of the objective function at each vertex, then one can easily determine where the maximum and minimum values occur for the region R.

Example 2-12. Find the maximum and minimum values of $f = 6x + 8y$ subject to the constraint conditions

$$x + y \le 1$$
$$4x + 2y \ge 1$$
$$x \ge 0$$
$$y \ge 0$$

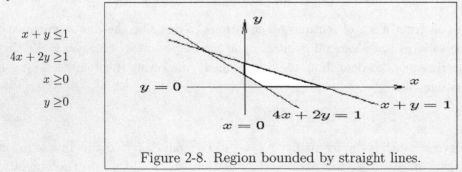

Figure 2-8. Region bounded by straight lines.

Solution: Draw the straight lines $x + y = 1$, $4x + 2y = 1$, $x = 0$ and $y = 0$ and verify that points (x, y) inside the region bounded by the straight lines satisfy all of the inequalities simultaneously. The situation is illustrated in the figure 2-8.

One can now calculate the vertices associated with the above region and then evaluate the objective function at these points. One can produce a table of values, such as the one given, in order to find the maximum and minimum values for the objective function.

Vertex	Value of $f = 6x + 8y$	
$(0, 1)$	8	maximum
$(1, 0)$	6	
$(0, 1/2)$	4	
$(1/4, 0)$	3/2	minimum

Maxwell-Boltzmann distribution

Consider the number N of gas particles in a container. This number N can be very large, say on the order of $(10)^{23}$ particles or larger. It is not possible to keep track of each individual particle and so statistics must be used. We shall assume that the system under study consists of a large collection of identical but distinguishable particles. Each particle will be considered as a point mass with momentum components p_x, p_y, p_z and position coordinates (x, y, z). We consider a collection of these particles in an element of phase space $d\tau = dp_x dp_y dp_z dx dy dz$. We assume that all particles within an element of phase space are in the same energy state since they all have the same phase space coordinates. The basic assumption used in Boltzmann statistics is that there is a discrete distribution of particles in the phase space. Let n_i denote the number of particles in an element $d\tau$ of phase space that have energy E_i. Imagine the phase space to be divided up into a collection of volume elements $d\tau$. The system can then be viewed as representing a macrostate or set of distribution numbers n_i representing the number of particles in the ith energy state E_i. The set of distribution numbers which produces the largest number of macrostates is called the most probable distribution, sometimes referred to as the highest thermodynamic probability. By using distinguishable particles one can interchange particles having the same energy states to give a new microstate or configuration of particles. That is, there exists many different distributions of particles which lead to the same energy state. The number of permutations of N distinguishable particles between energy states is $N!$. One must consider the fact that permutations of identical particles with a given fixed energy state does not produce a new macrostate and so these type of permutations must be excluded from our consideration. In an energy state E_i there are n_i particles and so $n_i!$ permutations of these particles does not really change the macrostate of the system. This reduces the number of permutations $N!$ by a factor of $n_i!$. Consequently, one can state that the probability P of any distribution is given by

$$P = \frac{N!}{n_1! n_2! n_3! \cdots}$$

(2.72)

where N denotes the total number of particles and n_i represents the number of particles with energy state E_i. We want to maximize P subject to the constraint conditions that we have

(i) Conservation of particle number $N = n_1 + n_2 + n_3 + \cdots = \sum_m n_m$ and

(ii) Conservation of energy $E = n_1 E_1 + n_2 E_2 + n_3 E_3 + \cdots = \sum_m n_m E_m$ is the total energy.

The probability function P given by equation (2.72) is difficult to work with because of all the factorial terms which can become quite large. It has been found that it is more convenient to work with $\ln P$ instead of P because the function $\ln P$ is a monotone increasing function so that if we can maximize $\ln p$ then we obtain a maximum for P also. In working with $\ln n!$ one can employ the Stirling's approximation

$$\ln n! \approx n \ln n - n = n(\ln n - 1). \tag{2.73}$$

Our problem then becomes to maximize $\ln \dfrac{N!}{n_1! n_2! n_3! \cdots}$ subject to the constraint conditions that $N = \sum_m n_m$ and $E = \sum_m n_m E_m$. To solve this problem we introduce the Lagrange multipliers λ_1 and λ_2 and form the function

$$F = \ln \frac{N!}{n_1! n_2! n_3! \cdots} - \lambda_1 \left(\sum_m n_m - N \right) - \lambda_2 \left(\sum_m n_m E_m - E \right) \tag{2.74}$$

then at a maximum value for F we require that $\frac{\partial F}{\partial n_i} = 0$ for all values of $i = 1, 2, 3, \ldots$ together with the constraint conditions obtained from the partial derivatives $\frac{\partial F}{\partial \lambda_1} = 0$ and $\frac{\partial F}{\partial \lambda_2} = 0$. Using the Stirling's approximation we find

$$\frac{\partial F}{\partial n_i} = -\ln n_i - \lambda_1 - \lambda_2 E_i = 0 \tag{2.75}$$

which implies

$$n_i = e^{-\lambda_1} e^{-\lambda_2 E_i} \tag{2.76}$$

which is known as the Boltzmann distribution law. The conservation of particle number requires that

$$N = \sum_m n_m = \sum_m e^{-\lambda_1} e^{-\lambda_2 E_m} = e^{-\lambda_1} \sum_m e^{-\lambda_2 E_m} \tag{2.77}$$

or

$$e^{-\lambda_1} = \frac{N}{\sum\limits_m e^{-\lambda_2 E_m}} \tag{2.78}$$

Substituting equation (2.78) into equation (2.76) one can verify that

$$n_i = \left[\frac{N}{\sum\limits_m e^{-\lambda_2 E_m}} \right] e^{-\lambda_2 E_i} \tag{2.79}$$

which is known as a Maxwell-Boltzmann distribution and consequently the conservation of energy can be written

$$E = \sum_j n_j E_j = \sum_j \left[\frac{N}{\sum\limits_m e^{-\lambda_2 E_m}} \right] E_j e^{-\lambda_2 E_j} \tag{2.80}$$

"Mathematics is the tool specially suited for dealing with abstract concepts of any kind and there is no limit to its power in this field.

Paul Adrien Maurice Dirac (1902-1984)

Exercises Chapter 2

▶ **1.** Locate the critical points, classify them and sketch a graph of the functions.

(a) $y = x + \frac{4}{x}$, (b) $y = 2\sin x + \sin 2x$, $0 \le x \le 2\pi$, (c) $y = (x-1)^2(x+2)^3$

▶ **2.** Find the critical points and test for maxima and minima.

(a) $z = 2x^2 - xy - 3y^2 - 3x + 7y$ (b) $z = \sin x \sin y$, $-1 \le x \le 4$, $-1 \le y \le 4$

▶ **3.** Consider all rectangles having a given area A_0.

(a) Find the rectangle with the smallest perimeter.

(b) Find the rectangle with the shortest diagonal.

▶ **4.** Find the critical points of the given functions and classify them. Sketch some level curves $z = C$ where C is a constant. (a) $z = xy$, (b) $z = x^2 - 2x + y^2$

▶ **5.** Water waves in deep water have the approximate velocity $v = \sqrt{\left(\frac{g\lambda}{2\pi} + \frac{2\pi\tau}{\rho\lambda}\right)}$ where g is the acceleration of gravity, τ is the surface tension, ρ is the density of the fluid and λ is the wave-length. Assume τ, ρ and g are constants and determine λ such that v is a minimum. What is the minimum wave velocity?

▶ **6.** Find the minimum distance between the lines L_1 and L_2 given by the following parametric equations L_1: $x = 10 + 2t$, $y = 8 + 2t$, $z = 6 - 2t$ and L_2: $x = 1 + t$, $y = 6 - 2t$, $z = t$.

▶ **7.** Given n points in space $p_1, p_2, \ldots, p_n$ where $p_i = (x_i, y_i, z_i)$ for $i = 1, 2, \ldots, n$. Find the point $P = (x, y, z)$ for which $f(P) = \sum_{i=1}^n |P - p_i|^2$ is a minimum, where $|P - p_i|$ denotes the distance between the points p_i and P. Give a physical interpretation for the point P in relation to the other points.

▶ **8.** Find the point (x, y) on the line $x + y - 1 = 0$ such that the distance squared from the origin to the point (x, y) is minimized.

▶ **9.** Find the critical points of $z = x^2 + 24xy + 8y^2$ subject to the constraint $x^2 + y^2 = 25$.

▶ **10.** Find the stationary points of $w = xyz$ as the point (x, y, z) moves along the curve defined by the intersection of the surfaces $x^2 + y^2 = 1$ and $z - x = 0$. Classify the critical points.

▶ **11.** Find the maximum and minimum values of $z = xy$ as x and y move around the circle $x^2 + y^2 = 1$. Solve this problem by two different methods.

▶ **12.** Find the points on the surface $xyz = 8$ which are nearest the origin. Solve this problem by two different methods.

▶ **13.** Let x and y denote the lengths of the sides of a rectangular base of a tank and let z denote the height of the tank. Find the dimensions of the tank with volume of 32 cubic meters if the tank is to be designed such that it has a minimal surface area associated with its base and side walls. Solve this problem by two different methods.

▶ **14.** For a, b, c, d given nonzero constants:
(a) Find the minimum distance from the line $ax + by - c = 0$ to the origin.
(b) Find the point on the line $ax + by - c = 0$ which is nearest the origin.
(c) Find the minimum distance from the plane $ax + by + cz - d = 0$ to the origin.
(d) Find the point on the plane $ax + by + cz - d = 0$ which is nearest the origin.

▶ **15.** The frustum of a right circular cone has a fixed height of h (meters) and a fixed lower base of radius R (meters). The upper base of the frustum has a variable radius of x (meters) as illustrated. The variable x is restricted to lie in the interval $0 \leq x \leq R$ so that when $x = 0$ a right circular cone results and when $x = R$ a cylinder results.

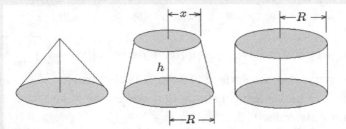

The lateral surface area of the above frustum is given by $A_\ell = \pi(x + R)\sqrt{h^2 + (R - x)^2}$. Find the values of x such that the lateral surface area is a maximum or a minimum. Be sure to consider the cases where $2h^2 > R^2$ and $2h^2 < R^2$.
Hint: See the reference, R.A. Johnson, "A Problem in Maxima and Minima", American Mathematical Monthly, Vol. 35, (1928), pp.187-188.

▶ **16.** Find the minimum distance between the curves $y = x^2$ and $y = x - b$, where b is a constant having the value $b = (1 + 20\sqrt{2})/4$.

▶ **17.** Given the triangle with vertices $B(0, b), C(0, 0)$ and $A(a, 0)$ and pick an arbitrary point (x, y) inside the triangle and construct from this point the distances $d_1 =$ distance to vertex $(0, 0)$, $d_2 =$ distance to vertex $(0, b)$ and $d_3 =$ distance to vertex $(a, 0)$. Find the point (x, y) inside the triangle such that $f = d_1^2 + d_2^2 + d_3^2$ is a minimum.

▶ **18.** Find the maximum and minimum distance from the point (x_0, y_0, z_0)
to a point on the sphere $x^2 + y^2 + z^2 = 1$.
      (a) Assume that $x_0^2 + y_0^2 + z_0^2 > 1$ (b) Assume that $x_0^2 + y_0^2 + z_0^2 < 1$

▶ **19.** Analyze the critical points associated with the function $y = \dfrac{1 - x}{1 + x^2}$

▶ **20.** A light L is to be placed on a wall at a height h to illuminate a point P on the floor. The brightness at the point P is known to vary inversely as the square of the distance from the light source and directly as the cosine of the angle which the light ray makes with the normal to the surface at the point P. Find the height h on the wall to place the light L such that the point P, a distance a from the wall, receives the maximum brightness. Consider the cases where h is the height of the ceiling and (i) $h > a/\sqrt{2}$ and (ii) $h < a/\sqrt{2}$.

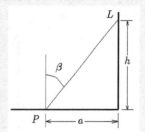

▶ **21.** The intensity of a light source at point A is I_A and the intensity of a light source at point B is I_B and the distance between the two light sources A and B is ℓ as illustrated in the figure below. Assume that the illumination at a point P on the line joining the light sources is proportional to the intensity of the sources and inversely proportional to the square of the distance from the light source.

(a) Let x denote the distance of the point P from the light source A with $(\ell - x)$ the distance of the point P from the light source B. Show that the intensity of the light at point P is given by

$$I = k\frac{I_A}{x^2} + k\frac{I_B}{(\ell - x)^2}$$

where $k > 0$ is a proportionality constant.

(b) Find the position of the point P on the line joining the light sources such that the illumination is a minimum.

(c) Show that at minimum intensity $\dfrac{x}{\ell - x} = \left(\dfrac{I_A}{I_B}\right)^{1/3}$

(d) Show that $\frac{d^2I}{dx^2} > 0$ at the point P

▶ **22.** Find the maximum and minimum values for y if $y^3 = 6xy - x^3 - 1$.

▶ **23.** Find the maximum and minimum values for the function $y = \cos t \cos 2t$ for $0 \le t \le \pi$.

▶ **24.** Find all stationary points and extreme values for the functions

$$(a) \quad y = \frac{x(x-1)}{(x+1)(x-5)} \qquad (b) \quad y = \frac{x(x-1)}{x-2}$$

▶ **25.** Find and classify the stationary values associated with the function
$$z = z(x,y) = x^3 - y^3 + 3x^2 + 3y^2 - 9x$$

▶ **26.** In billiards let $P_0(x_0, y_0)$ denote the position of the cue ball and let $P_1(x_1, y_1)$ denote the position of another ball. Let $y = c$ (c constant), denote a side wall of the billiard table with point $P(x, c)$ a point on this wall. Let d_0 denote the distance $\overline{P_0P}$ and let d_1 denote the distance $\overline{PP_1}$ as illustrated below.

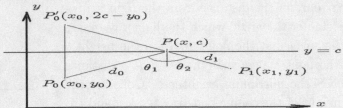

(a) Find the point $P(x, c)$ such that $d_0 + d_1$ is minimized.

(b) Show that the point P which minimizes $d_0 + d_1$ is such that $\theta_1 = \theta_2$.

(c) Let P_0' denote the mirror image of the point P_0 with respect to the side wall. Show that the points P_0', P, P_1 all lie on a straight line.

▶ **27.**

(a) Given a function $F = F(x, y, z)$, the directional derivative of F, at a point P, in the direction of a unit vector $\hat{e} = n_1 \hat{e}_1 + n_2 \hat{e}_2 + n_3 \hat{e}_3$, where $|\hat{e}| = n_1^2 + n_2^2 + n_3^2 = 1$, is given by

$$\frac{dF}{ds} = \text{grad} F \cdot \hat{e} = \frac{\partial F}{\partial x} n_1 + \frac{\partial F}{\partial y} n_2 + \frac{\partial F}{\partial z} n_3.$$

Examine a physical interpretation of the dot product $\text{grad} F \cdot \hat{e}$ and show the maximum rate of change of F at the point P, (i.e. the maximum directional derivative) takes place in the direction specified by $\text{grad} F$ evaluated at the point P.

(b) The potential function V due to a charge distribution is given in spherical coordinates (ρ, θ, ϕ) by the equation $V = \dfrac{\cos \theta}{\rho^2}$. Show the maximum directional derivative at any point is given by $\dfrac{\sqrt{\sin^2 \theta + 4 \cos^2 \theta}}{\rho^3}$. Hint: Express $\text{grad} V$ in spherical coordinates.

▶ **28.** Find the maximum and minimum value of the function $f = xy + yz$ subject to the constraint conditions $x^2 + y^2 = 2$ and $yz = 2$.

▶ **29.** **(Method of least squares)**

(a) Given the five points $(-2, y_1)$, $(-1, y_2)$, $(0, y_3)$, $(1, y_4)$, $(2, y_5)$, where y_1, y_2, y_3, y_4, y_5 denote 5 given numbers, it is impossible to find a quadratic curve $y = y(x) = ax^2 + bx + c$ with a, b, c constant, which passes through all five of the given points. However, one can select the constants a, b, c such that the total square error E given by

$$E = E(a, b, c) = [y(-2) - y_1]^2 + [y(-1) - y_2]^2 + [y(0) - y_3]^2 + [y(1) - y_4]^2 + [y(2) - y_5]^2$$

is a minimum. Find how to determine the constants a, b, c such that E is a minimum. The resulting curve is called the best fit parabola in the least squares sense. Sketch a diagram of what you are trying to do.

(b) Test your result from part (a) and find the best fit parabola in the least squares sense associated with the following data. $(-2, 4)$, $(-1, 2)$, $(0, 1/2)$, $(1, 3)$, $(2, 3)$.

▶ **30.**

(a) Consider all rectangular plots of land with a given perimeter of 20 kilometers. Find the plot of land with maximum area.

(b) Consider all right circular cylinders with a given volume V. Find the cylinder which has the least surface area on its sides, top and bottom.

▶ **31.** Find the maximum and minimum values associated with the function
$$f = f(x,y) = x^3 + y^3 - 12x - 27y + 60$$

▶ **32.** Construct a triangle having the vertices $C(0,0), B(0,b)$ and $A(a,0)$ and then select an arbitrary point (x,y) inside the triangle. Construct the distances $d_1 =$ perpendicular distance from (x,y) to the line $\overline{AB}$, $d_2 =$ perpendicular distance from (x,y) to the line $\overline{BC}$, and the distance $d_3 =$ perpendicular distance from (x,y) to the line $\overline{AC}$. Find the point (x,y) inside the triangle such that $f = d_1^2 + d_2^2 + d_3^2$ is a minimum.

▶ **33.** A rectangular tank is to be constructed to hold 32 cubic meters. The tank is to have an open top with surface area of the base and sides a minimum. Find the dimensions of the tank with minimal area of base and sides.

▶ **34.** The work done in an air compressor operating at a pressure p is given by the formula $W = k \left\{ \left(\dfrac{p_1}{p} \right)^{\frac{n-1}{n}} + \left(\dfrac{p}{p_2} \right)^{\frac{n-1}{n}} - 2 \right\}$ where the quantities k, n, p_1, p_2 are given constants. Determine the value of p such that the work done is a minimum.

▶ **35.** Find the dimensions of a rectangular box, without a top, which has maximum volume subject to the constraint that the surface area is 108 square inches.

▶ **36.**

(a) Show that the arithmetic mean of n given positive numbers $x_1, x_2, \ldots, x_n$ is a number $\bar{x}$ such that $S = \displaystyle\sum_{i=1}^{n}(x_i - \bar{x})^2$ is a minimum.

(b) Show that the harmonic mean of n given positive numbers $x_1, x_2, \ldots, x_n$ is a number $h = \frac{1}{\bar{y}}$ such that $S = \displaystyle\sum_{i=1}^{n} \left(\frac{1}{x_i} - \bar{y} \right)^2$ is a minimum.

(c) The geometric mean of n given positive numbers is $g = \sqrt[n]{x_1 x_2 x_3 \ldots x_n}$.
Construct, for $n = 2, 3, 4$, a set of numbers $(x_1, x_2, \ldots, x_n)$ to test the inequality $h \le g \le \bar{x}$.

▶ **37.** Find the minimum of the function $f = f(x,y) = x^2 + y^2$ subject to the constraint condition that $xy = C^2$ where C is a given constant.

▶ **38.** A set of functions $\{\phi_n(x)\}$, for $n = 1, 2, 3, \ldots$, is said to be an orthogonal set over the interval (a, b), with respect to a weight function $w(x)$, if the inner product integral $(\phi_n, \phi_m) = \int_a^b w(x)\phi_n(x)\phi_m(x)\,dx = \parallel \phi_n \parallel^2 \delta_{nm}$, where n and m are integers with δ_{nm} the Kronecker delta, define by $\delta_{nm} = \begin{cases} 1 & \text{if } n = m \\ 0 & \text{if } n \neq m \end{cases}$ and the quantity $\parallel \phi_n \parallel^2$ is called a norm squared and is defined by the integral $(\phi_n, \phi_n) = \parallel \phi_n \parallel^2 = \int_a^b w(x)\phi_n^2(x)\,dx$. We desired to represent a function $f(x)$ as a series of orthogonal functions in the form of a series given by $f(x) = \sum_{n=1}^{\infty} c_n \phi_n(x)$ where c_n are constants. The constants c_n are to be selected such that the error $E = \int_a^b w(x)\left(f(x) - \sum_{n=1}^{\infty} c_n \phi_n(x)\right)^2 dx$ is minimized. Show that E is minimized if the constants c_n are selected to satisfy the ratio

$$c_n = \frac{(f(x), \phi_n(x))}{\parallel \phi_n \parallel^2} = \frac{\int_a^b w(x)f(x)\phi_n(x)\,dx}{\int_a^b w(x)\phi_n^2(x)\,dx}, \quad \text{for } n = 1, 2, 3, \ldots$$

which represents an inner product divided by a norm squared.

▶ **39.** Show the minimum lateral surface area of a right circular cone with a fixed volume V is given by $A_\ell = 3\left(\dfrac{\sqrt{3}}{2}\pi V^2\right)^{1/3}$
Hint: The lateral surface area of a cone is given by
$$A_\ell = \pi r \sqrt{r^2 + h^2} = \pi r \ell$$

$Volume = \frac{1}{3}\pi r^2 h$

▶ **40.** Show that if $f = f(x, y, z)$ is to have a maximum or minimum value where x, y and z are restricted to lie on the surface $g(x, y, z) = 0$, then a necessary condition to be satisfied by x, y and z is that the following equations must be satisfied.

$$g(x, y, z) = 0, \qquad \frac{\partial f}{\partial x}\frac{\partial g}{\partial z} - \frac{\partial f}{\partial z}\frac{\partial g}{\partial x} = 0, \qquad \frac{\partial f}{\partial y}\frac{\partial g}{\partial z} - \frac{\partial f}{\partial z}\frac{\partial g}{\partial y} = 0$$

▶ **41.** Show that for $f = f(x, y)$ to have a maximum or minimum value when x and y are restricted to lie on the curve $g(x, y) = 0$, a necessary condition to be satisfied by x and y is that the following equations must be satisfied.

$$g(x, y) = 0, \qquad \frac{\partial f}{\partial x}\frac{\partial g}{\partial y} - \frac{\partial f}{\partial y}\frac{\partial g}{\partial x} = 0$$

▶ **42.**
(a) Graph the function $y(x) = x^n e^{-x}$ over the interval $0 \leq x \leq 3$ for the following cases:

$$(i) \quad n = 0, \qquad (ii) \quad n = 1/2, \qquad (iii) \quad n = 1, \qquad (iv) \quad n = 2$$

(b) For each graph above find the local maximum and minimum values over the interval $0 \leq x \leq 3$.

▶ **43.**   **Maxwell-Boltzmann distribution**

In thermodynamics the Gibbs free energy is given by $G = E - TS$ where

$$S = k_B \ln \frac{N!}{n_1! n_2! n_3! \cdots} \qquad \text{denotes a quantity called entropy}$$

$$E = n_1 E_1 + n_2 E_2 + n_3 E_3 + \cdots \qquad \text{denotes the internal energy of the system}$$

where T is the absolute temperature in degree Kelvin, k_B is the Boltzmann constant given by $k_B = 1.38054\,(10)^{-23}\ J/K.$ and n_i denotes the number of particles which have energy E_i where the n_i are subject to the constraint condition that

$$N = n_1 + n_2 + n_3 + n_4 + \cdots \tag{43 - 1}$$

denotes the total number of identical but distinguishable particles. Assume that each n_i is large such that the Stirling approximation

$$\ln n_i! = n_i \ln n_i - n_i = n_i(\ln n_i - 1)$$

is valid. The problem is to minimize the free energy G subject to the constraint condition that equation (43−1) is satisfied. Here one can assume that T, E_i, k_B and N are given constants.

(a) Use the method of Lagrange multipliers and show that a necessary condition to be satisfied for the free energy to be a minimum, is that

$$E_i + k_B T \ln n_i - \lambda = 0$$

where λ is a Lagrange multiplier.

(b) Show that at a minimum one has $n_i = \exp\left(\dfrac{\lambda - E_i}{k_B T}\right)$

(c) Show that $N = \exp\left(\dfrac{\lambda}{k_B T}\right) \sum_m \exp\left(-\dfrac{E_m}{k_B T}\right)$

(d) Show that at a minimum one obtains $n_i = N\left[\dfrac{\exp\left(-\dfrac{E_i}{k_B T}\right)}{\sum_m \exp\left(-\dfrac{E_m}{k_B T}\right)}\right]$ which is called a Maxwell-Boltzmann distribution. The denominator in this equation is known as the canonical partition function.

▶ **44.**   Find the minimum distance between the curves $y = 1 + \frac{x^2}{4}$ and $y = \frac{x}{2} + 3/4 - \sqrt{5}/2$.

▶ **45.**   Find the largest rectangle that can be inscribed inside the ellipse $\dfrac{x^2}{a^2} + \dfrac{y^2}{b^2} = 1$.

▶ **46.**

Given a circle of radius R as illustrated. Construct an upper segment of the circle by drawing a line $y = h$ where $0 < h < R$. Determine the largest rectangle that can be inscribed in this upper segment.

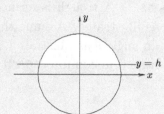

▶ **47.** Consider the curve $y = y(x) = \dfrac{1 - \alpha^2}{1 - 2\alpha \cos(x - \xi) + \alpha^2}$ where $0 < \alpha < 1$ is a constant.

(a) Holding ξ constant, show that the given function can be represented in the form

$$y = y(x) = \left(\frac{1-\alpha}{1+\alpha}\right) \frac{\sec^2\left(\frac{x-\xi}{2}\right)}{\left(\frac{1-\alpha}{1+\alpha}\right)^2 + \tan^2\left(\frac{x-\xi}{2}\right)}$$

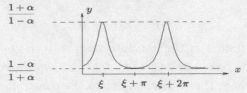

(b) Show the maximum and minimum values for y occur at the points $x = \xi$, $\pi + \xi$, $2\pi + \xi, \ldots$ where $\sin(x - \xi) = 0$.

(c) Show the maximum values are given by $\frac{1+\alpha}{1-\alpha}$ and the minimum values are given by $\frac{1-\alpha}{1+\alpha}$.

▶ **48.** Find the shortest distance from the hyperbola $12x^2 - 26xy + 12y^2 + 25 = 0$ to the origin.

▶ **49.** Find the shortest distance from the hyperbola $2x^2 + 5xy + 2y^2 + 18 = 0$ to the origin.

▶ **50.** Find the local maximum and minimum values associated with the function
$y = (x - 1)^2(x - 6)^3$ and then sketch the function.

▶ **51.**

(a) If $y = u + v$ is subject to the constraint condition $uv = C$ where C is a constant, then find the local maximum and minimum values for y if they exist.

(b) If $y = uv$ is subject to the constraint condition $u + v = C$ where C is a constant, then find the local maximum and minimum values for y if they exist.

▶ **52.** (**Nuclear reactor theory**) It is desired to find an extremum value for the volume of a nuclear reactor having the shape of a right circular cylinder of radius R and height H. Neutron diffusion theory requires that the constraint $\left(\frac{\alpha}{R}\right)^2 + \left(\frac{\pi}{H}\right)^2 = $ a constant be satisfied, where α is a constant, found from the smallest root of the Bessel Function $J_0(x) = 0$.

(a) Find the ratio H/R that produces an extremum for the volume.

(b) Is the extremum a maximum or a minimum?

▶ **53.** A trough is formed from a strip of metal of width ℓ by bending up the sides as illustrated. Assume the bent sides are of equal length x and that they are bent to an angle θ as illustrated. Find the width x and the angle θ such that the trough will have a maximum capacity.

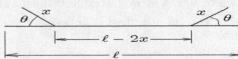

Chapter 3
Introduction to the Calculus of Variations

We continue our investigation of finding maximum and minimum values associated with various quantities. Instead of finding points where functions have a relative maximum or minimum value over some domain $x_1 \leq x \leq x_2$, we examine situations where certain curves have the property of producing a maximum or minimum value. For example, the problem of determining the curve or shape which minimizes drag force and maximizes lift force on an airplane wing moving at a given speed requires that one find a special function $y = y(x)$ defining the shape of the curve which "best" achieves the desired objective.

We begin by studying a function and possibly some of its derivatives that determine the value of an integral. We vary the function and determine its effect on the value of the integral. We try to find the function which makes the integral have a maximum or minimum value. This is a basic calculus of variations problem. The methods developed in the study of the variational calculus introduce new concepts and principles. These new variational principles can then be employed to view certain problems in physics and mechanics from a new and different viewpoint. We begin with the simplest calculus of variations problems which have fixed end points.

Functionals

A functional is a mapping which assigns a real number to each function or curve associated with some class of functions. Some examples of functionals are the following.

1. In Cartesian coordinates consider all plane curves $y = y(x)$ which pass through two given points (x_1, y_1) and (x_2, y_2). When one of these curves is rotated about the x-axis a surface of revolution is produced. The surface area S which is generated is given by

$$S = 2\pi \int_{x_1}^{x_2} y \sqrt{1 + \left(\frac{dy}{dx}\right)^2}\, dx \qquad (3.1)$$

and the scalar value obtained depends upon the function $y(x)$ selected. Finding the particular function $y = y(x)$ which produces the minimum surface area is an example of a calculus of variation problem.

2. In Cartesian coordinates consider all plane curves $y = y(x)$ which pass through two given points (x_1, y_1) and (x_2, y_2). The length ℓ along one particular curve between the given points is obtained by integrating the element of arc length $ds = \sqrt{1 + \left(\frac{dy}{dx}\right)^2}\, dx$ between the limits x_1 and x_2 to obtain

96

$$\ell = \int_{x_1}^{x_2} \sqrt{1 + \left(\frac{dy}{dx}\right)^2}\, dx \qquad (3.2)$$

The scalar value representing the length depends upon the curve $y(x)$ selected. The problem of finding the curve $y = y(x)$, which produces the minimum length, is a calculus of variations problem.

3. In Cartesian coordinates consider all plane curves $x = x(y)$ which pass through two given points (x_1, y_1) and (x_2, y_2). The length ℓ along one particular curve between the given end points is determined from the integral

$$\ell = \int_{y_1}^{y_2} \sqrt{1 + \left(\frac{dx}{dy}\right)^2}\, dy \qquad (3.3)$$

and this length depends upon the curve $x = x(y)$ selected. Finding the curve which produces the minimum length is a calculus of variations problem. This is the same problem as the previous example but formulated with x as the dependent variable and y as the independent variable.

4. In polar coordinates consider all plane curves $r = r(\theta)$ which pass through the points (r_1, θ_1) and (r_2, θ_2). The length ℓ along one particular curve between the given points is obtained by integrating the element of arc length $ds = \sqrt{r^2 + \left(\frac{dr}{d\theta}\right)^2}\, d\theta$

$$\ell = \int_{\theta_1}^{\theta_2} \sqrt{r^2 + \left(\frac{dr}{d\theta}\right)^2}\, d\theta \qquad (3.4)$$

and the value of this integral depends upon the curve selected. Here the problem of calculating length along a curve is posed in polar coordinates. Note that the same problem can be formulated in many different ways by making a change of variables.

5. In Cartesian coordinates consider all parametric equations $x = x(t)$, $y = y(t)$ for $a \le t \le b$ which satisfy the end point conditions $x(a) = x_1$, $y(a) = y_1$ and $x(b) = x_2$, $y(b) = y_2$ where (x_1, y_1) and (x_2, y_2) are given fixed points. The length ℓ of one of these curves between the given points is determined by integrating the element of arc length to obtain the relation

$$\ell = \ell(x, y) = \int_a^b \sqrt{\left(\frac{dx}{dt}\right)^2 + \left(\frac{dy}{dt}\right)^2}\, dt \qquad (3.5)$$

and the calculated length depends upon the curve selected through the end points. This is the same type of problem as the previous two problems. However, it is formulated in the form of finding parametric equations $x = x(t)$ and $y = y(t)$ which define the curve producing a minimum value for the arc length.

6. In three dimensional space consider all curves which pass through two given points (x_1, y_1, z_1) and (x_2, y_2, z_2). This family of curves can be represented by the position vector

$$\vec{r} = \vec{r}(t) = t\,\hat{e}_1 + y(t)\,\hat{e}_2 + z(t)\,\hat{e}_3 \qquad x_1 \le t \le x_2$$

where $\hat{e}_1$, $\hat{e}_2$, $\hat{e}_3$ are unit base vectors in the directions of the x, y and z axes and the functions $y = y(t)$ and $z = z(t)$ are assumed to have continuous derivatives and satisfy the conditions $y(x_1) = y_1$, $y(x_2) = y_2$ and $z(x_1) = z_1$, $z(x_2) = z_2$. The arc length ℓ of one of these curves is given by the integral

$$\ell = \ell(y, z) = \int_{x_1}^{x_2} \sqrt{1 + \left(\frac{dy}{dt}\right)^2 + \left(\frac{dz}{dt}\right)^2}\, dt$$

If we desire to find the minimum value for ℓ, then one must find the parametric functions $y = y(t)$ and $z = z(t)$ which produce this minimum value. This is the three-dimensional version of the previous problem.

7. Consider a family of curves $y = y(x)$ which are continuous and twice differentiable and pass through the points (x_1, y_1) and (x_2, y_2). Let each curve $y = y(x)$ in the family satisfy the conditions $y(x_1) = y_1$ and $y(x_2) = y_2$. Consider an integral of the form

$$I = I(y) = \int_{x_1}^{x_2} f(x, y(x), y'(x))\, dx \tag{3.6}$$

where the integrand $f = f(x, y(x), y'(x))$ is a given continuous function of x, y, y'. This is an example of a general functional where the value of the integral I depends upon the smooth function $y = y(x)$ through the given points. The problem of finding a smooth function $y = y(x)$ from the family of curves which makes the functional have a maximum or minimum value is a typical calculus of variations problem. In this introductory development we restrict our study to finding smooth curves which produce an extreme value. Later we shall admit discontinuous curves into our class of functions that can be substituted into the given integrals.

Basic lemma used in the calculus of variations

Consider the integral

$$\delta I = \int_a^b \eta(x)\beta(x)\, dx \tag{3.7}$$

where $\eta(x)$ is an arbitrary function which is defined and continuous over the interval $[a, b]$ and satisfies the end conditions $\eta(a) = 0$ and $\eta(b) = 0$. If $\delta I = 0$ for all arbitrary functions $\eta(x)$, satisfying the given conditions, then what can be said about the function $\beta(x)$? The answer to this question is given by the following lemma.

Basic Lemma

If $\beta = \beta(x)$ is continuous over the interval $a \le x \le b$ and if the integral

$$\delta I = \delta I(\eta) = \int_a^b \eta(x)\beta(x)\, dx = 0$$

for every continuous function $\eta(x)$ which satisfies $\eta(a) = \eta(b) = 0$, then necessarily $\beta(x) = 0$ for all values of $x \in [a, b]$.

The proof of the above lemma is a proof by contradiction. Assume that $\beta(x)$ is nonzero, say positive, for some point in the interval $[a, b]$. By hypothesis, the function $\beta(x)$ is continuous and so it must be positive for all values of x in some subinterval $[x_1, x_2]$ contained in the interval $[a, b]$. If this is true, then the integral $I = I(\eta)$ cannot be identically zero for every function $\eta(x)$. Consider the special function

$$\eta(x) = \begin{cases} 0, & a \leq x \leq x_1 \\ (x - x_1)^2(x - x_2)^2, & x_1 \leq x \leq x_2 \\ 0, & x_2 \leq x \leq b \end{cases} \tag{3.8}$$

which is positive over the subinterval where $\beta(x)$ is positive. Substituting this special function into the integral (3.7) and using the mean value theorem for integrals, there results

$$\delta I = \delta I(\eta) = \int_a^b \eta(x)\beta(x)\,dx = \int_{x_1}^{x_2} \eta(x)\beta(x)\,dx = (x_2 - x_1)\eta(x^*)\beta(x^*) > 0$$

for some value x^* satisfying $x_1 < x^* < x_2$. This is a contradiction to our original assumption that $I = I(\eta) = 0$ for all functions $\eta(x)$.

If we assume that the given integral (3.7) is zero for all possible functions $\eta = \eta(x)$ then we need only select $\eta = \beta(x)$ in order to establish the lemma. In this text we are only concerned with those $\eta = \eta(x)$ which have derivatives which are continuous functions.

The above lemma can be generalized to double, triple and multiple integrals of product functions $\eta\beta$. For example, consider an integral of the form

$$\delta I = \delta I(\eta) = \iint_R \eta(x, y)\beta(x, y)\,dxdy, \tag{3.9}$$

where the integration is over a region R of the x, y-plane. We use the notation ∂R to denote the curve representing the boundary of the region R. It is assumed that the region R is bounded by a simple closed curve ∂R. If the integral given by equation (3.9) is zero for all continuous functions η and in addition the function η is zero when evaluated at a boundary point (x, y) on the boundary curve ∂R of the region R, then it follows that $\beta(x, y) = 0$ for $(x, y) \in R$. The proof is very similar to the proof given for the single integral.

Notation

Let $f = f(x_1, x_2, \ldots, x_n)$ denote a real function of n-real variables defined and continuous over a region R. The notation, "f belongs to the class $\mathcal{C}^{(n)}$ in a region R", is employed to denote the condition that f and all its derivatives up to and including the nth order, exist and are continuous in the region R. This notation is sometimes shortened to the form, $f \in \mathcal{C}^{(n)}$ in a region R. For example, if $f = f(x, y, y')$, for $x_1 \leq x \leq x_2$, is such that $f \in \mathcal{C}^{(2)}$, then

$$f, \quad \frac{\partial f}{\partial x}, \quad \frac{\partial f}{\partial y}, \quad \frac{\partial f}{\partial y'}, \quad \frac{\partial^2 f}{\partial x^2}, \quad \frac{\partial^2 f}{\partial y^2}, \quad \frac{\partial^2 f}{\partial x \partial y}, \quad \frac{\partial^2 f}{\partial x \partial y'}, \quad \frac{\partial^2 f}{\partial y \partial y'} \quad \text{and} \quad \frac{\partial^2 f}{\partial y'^2}$$

all exist and are continuous over the interval $x_1 \leq x \leq x_2$.

General approach

We present an overview of the basic ideas that will be employed to analyze functionals such as

$$I = I(y) = \int_{x_1}^{x_2} f(x, y(x), y'(x))\, dx \tag{3.10}$$

We consider changes in $I(y)$ as we vary the functions $y = y(x)$ that are used in the evaluation of the functional. Our general approach is as follows.

(i) If we can find a curve $y = y(x)$ for $x \in [x_1, x_2]$ such that for all other curves $Y = Y(x)$, belonging to some class, we have $I(y) \geq I(Y)$, then the curve y produces a maximum value for the functional $I(y)$.

(ii) If we can find a function $y = y(x)$ for $x \in [x_1, x_2]$ such that for all other curves $Y = Y(x)$, the inequality $I(y) \leq I(Y)$ is satisfied, then the curve y produces a minimum value for the functional $I(y)$.

(iii) Curves $Y = Y(x)$, which differ slightly from the curve y which produces an extreme value, can be denoted by $Y = Y(x) = y(x) + \epsilon\eta(x)$ where ϵ is a small quantity and the function η can be any arbitrary curve through the end points of the integral. Note that by varying the parameter ϵ we are varying the function which occurs in the functional. Upon substituting $Y = Y(x)$ into the functional (3.10) one obtains

$$I = I(\epsilon) = \int_{x_1}^{x_2} f(x, Y, Y')\, dx = \int_{x_1}^{x_2} f(x, y + \epsilon\eta, y' + \epsilon\eta')\, dx$$

which can then be viewed as a function of ϵ which has an extreme value at $\epsilon = 0$. If $I(\epsilon)$ is a continuous function of ϵ, then $\frac{dI}{d\epsilon}\Big|_{\epsilon=0} = 0$ corresponds to a stationary value for the functional.

We will use this general approach to find a way of determining all curves y which produce an extreme value for the functional I. We develop necessary conditions to be satisfied by each member of a family of curves which make a given functional have a extreme value.

We use the terminology "find an extremum for the functional" to mean– find the necessary conditions to be satisfied by the functions which produce a stationary value associated with the functional I above. To determine if a given curve produces an extreme value of a maximum or minimum value associated with a given functional requires further testing. Recall that the first derivative $\frac{dI}{d\epsilon} = 0$ is only a necessary condition for an extreme value to exist. The condition $\frac{dI}{d\epsilon} = 0$ gives us stationary values. An examination of the second derivative is required to analyze the stationary values to determine if the stationary values correspond to a maximum, minimum or saddle point.

In this chapter we study various types of functionals and develop necessary conditions to be satisfied such that these functionals take on a stationary value. We begin by considering only variations of functions which have fixed values at the end points of an interval or functions which have specified values on the boundary of a region in two or three-dimensions. These special functions are easier to handle. We examine other types of boundary conditions in the next chapter.

In the following discussions we examine functionals represented by various types of integrals over some region R. We label these different functionals by the type of integrand they have and we use the notation $[f1]$, $[f2]$, $[f3]$, ... to denote these integrands for future reference.

[f1]: Integrand $f(x, y, y')$

Consider the functional

$$I = \int_{x_1}^{x_2} f(x, y(x), y'(x)) \, dx \tag{3.11}$$

where $f = f(x, y, y')$ is a given integrand and $f \in C^{(2)}$ over the interval (x_1, x_2). Let us examine how the integral given by equation (3.11) changes as the function $y(x)$ changes. The variation of a function $y(x)$ and its derivative $y'(x)$ within an integral (3.11) is called a variational problem and is the most fundamental problem in the development of the variational calculus.

We require that the functions available for substitution into the functional (3.11) are such that (i) the integral exists and (ii) $y = y(x) \in C^{(2)}$ over the region $R = \{x \,|\, x_1 \le x \le x_2\}$ and (iii) the functions $y = y(x)$ considered are required to satisfy the end point conditions

$$y_1 = y(x_1) \qquad \text{and} \qquad y_2 = y(x_2). \tag{3.12}$$

We now illustrate a procedure that can be used to construct a differential equation whose solution family makes the integral I, given by equation (3.11), have a stationary value. We begin by assuming that we know the function $y(x)$ which defines a curve C that makes the given functional have a stationary value and then consider a comparison function

$$Y = Y(x) = y(x) + \epsilon \eta(x) \tag{3.13}$$

which defines a curve C^*, where ϵ is a small parameter and $\eta(x)$ is an arbitrary function which is defined and continuous over the interval $[x_1, x_2]$ and satisfies the end conditions $\eta(x_1) = 0$ and $\eta(x_2) = 0$.

Here the end conditions on $\eta(x)$ have been selected such that the comparison function satisfies the same end conditions as the curve which produces a stationary value. The comparison function therefore satisfies the end point conditions $Y(x_1) = y_1$ and $Y(x_2) = y_2$ for all values of the parameter ϵ. The situation is illustrated in the figure 3-1.

A weak variation is said to exist if the function η is independent of ϵ and in the limit as ϵ tends to zero the comparison curve C^* approaches the optimal curve C and simultaneously the slopes along C^* approach the slopes of the curve C for all values of x satisfying $x_1 \le x \le x_2$. That is, for a weak variation we assume that

$$|Y(x) - y(x)| \qquad \text{and} \qquad |Y'(x) - y'(x)|,$$

are both small and approach zero as epsilon tends toward zero. If these conditions are not satisfied, then a strong variation is said to exist. Note that comparison functions of the form $Y(x) = y(x) + \epsilon \eta(x, \epsilon)$, where η is a function of both x and ϵ, produces the functional

$$I(\epsilon) = \int_{x_1}^{x_2} f(x, y(x) + \epsilon \eta(x, \epsilon), y'(x) + \epsilon \eta'(x, \epsilon)) \, dx \tag{3.14}$$

which may or may not approach the functional of equation (3.11) as ϵ tends toward zero. Weierstrass gave the following example of a strong variation, $\eta = \eta(x, \epsilon) = \sin\left[\frac{(x-x_1)\pi}{\epsilon^n}\right]$, where n is a positive integer. In this case we have the limits $\lim_{\epsilon \to 0} |Y(x) - y(x)| = \lim_{\epsilon \to 0} \epsilon \eta(x, \epsilon) = 0$ but $\lim_{\epsilon \to 0} |Y'(x) - y'(x)| \neq 0$. We begin by considering only weak variations because weak variations can lead to maximum and minimum values of the functional given by equation (3.11). In contrast, maximum and minimum values for the functional (3.11) may or may not occur if strong variations are considered.

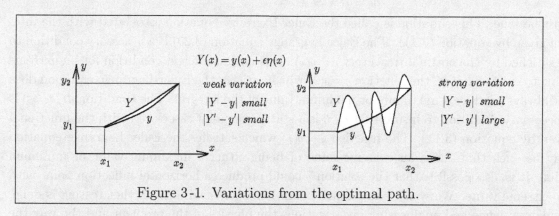

Figure 3-1. Variations from the optimal path.

Substituting the comparison function given by equation (3.13) into the integral given by equation (3.11) gives

$$I = I(\epsilon) = \int_{x_1}^{x_2} f(x, Y, Y') \, dx = \int_{x_1}^{x_2} f(x, y(x) + \epsilon \eta(x), y'(x) + \epsilon \eta'(x)) \, dx \qquad (3.15)$$

which by assumption has a stationary value when $\epsilon = 0$. Treating $I = I(\epsilon)$ as a continuous function of ϵ which has a stationary value at $\epsilon = 0$ requires that the condition $\frac{dI(\epsilon)}{d\epsilon} = I'(\epsilon)$ equal zero at $\epsilon = 0$. This is a necessary condition which must be satisfied in order for a stationary value to exist at $\epsilon = 0$. We calculate the derivative $I'(\epsilon)$ and find

$$\frac{dI}{d\epsilon} = I'(\epsilon) = \int_{x_1}^{x_2} \left(\frac{\partial f}{\partial Y}\frac{\partial Y}{\partial \epsilon} + \frac{\partial f}{\partial Y'}\frac{\partial Y'}{\partial \epsilon}\right) dx = \int_{x_1}^{x_2} \left(\frac{\partial f}{\partial Y}\eta + \frac{\partial f}{\partial Y'}\eta'\right) dx. \qquad (3.16)$$

The necessary condition for a stationary value is obtained by setting $\epsilon = 0$ in this derivative. This is equivalent to letting $Y = y$ and $Y' = y'$ in equation (3.16) so that at the stationary value we have

$$\frac{dI}{d\epsilon}\bigg|_{\epsilon=0} = I'(0) = \int_{x_1}^{x_2} \frac{\partial f}{\partial y}\eta \, dx + \int_{x_1}^{x_2} \frac{\partial f}{\partial y'}\eta' \, dx = 0. \qquad (3.17)$$

In equation (3.17) we integrate the second term by parts to obtain

$$\frac{dI}{d\epsilon}\bigg|_{\epsilon=0} = I'(0) = \int_{x_1}^{x_2} \frac{\partial f}{\partial y}\eta \, dx + \frac{\partial f}{\partial y'}\eta(x)\bigg|_{x_1}^{x_2} - \int_{x_1}^{x_2} \frac{d}{dx}\left(\frac{\partial f}{\partial y'}\right)\eta(x) \, dx = 0$$

which can also be written in the form

$$\frac{dI}{d\epsilon}\bigg|_{\epsilon=0} = I'(0) = \frac{\partial f}{\partial y'}\eta(x)\bigg|_{x_1}^{x_2} + \int_{x_1}^{x_2}\left[\frac{\partial f}{\partial y} - \frac{d}{dx}\left(\frac{\partial f}{\partial y'}\right)\right]\eta(x) \, dx = 0. \qquad (3.18)$$

Observe that the boundary conditions $\eta(x_1) = \eta(x_2) = 0$ insures that the first term of equation (3.18) is zero. Consequently, the necessary condition for a stationary value can be written

$$I'(0) = \int_{x_1}^{x_2} \left[\frac{\partial f}{\partial y} - \frac{d}{dx} \left(\frac{\partial f}{\partial y'} \right) \right] \eta(x) \, dx = 0. \tag{3.19}$$

For arbitrary functions $\eta(x)$ the basic lemma previously considered requires that the condition

$$\frac{\partial f}{\partial y} - \frac{d}{dx} \left(\frac{\partial f}{\partial y'} \right) = 0, \quad x_1 \leq x \leq x_2 \tag{3.20}$$

must be true. This equation is called the Euler-Lagrange equation associated with the integral given by equation (3.11). The Euler-Lagrange equation (3.20) is a necessary condition to be satisfied by the optimal trajectory $y = y(x)$. It is not a sufficient condition for an extreme value to exist. That is, the function $y = y(x)$ which satisfies the Euler-Lagrange equation does not always produce a maximum or minimum value of the integral. The condition $dI/d\epsilon = 0$ is a necessary condition to insure that $\epsilon = 0$ is a stationary point associated with the functional given by equation (3.11). The function $y = y(x)$ which satisfies the Euler-Lagrange equation may be such that I has an extreme value of being either a maximum value or minimum value. It is also possible that the solution y could produce a horizontal inflection point with no extreme value. We won't know which condition is satisfied until further testing is done. In many science and engineering investigations the physics of the problem and the way the problem is formulated, sometimes suggests that a maximum or minimum value for the functional exists. Under such circumstances the solution $y = y(x)$ of the Euler-Lagrange equation can be said to produce an extreme value for the functional.

Note that the variables x, y, y' occurring in the integrand of equation (3.11) have been treated as independent variables. Consequently, the derivative term in the Euler-Lagrange equation (3.20) is evaluated

$$\frac{d}{dx} \left(\frac{\partial f}{\partial y'} \right) = \frac{\partial^2 f}{\partial y' \partial x} + \frac{\partial^2 f}{\partial y' \partial y} y' + \frac{\partial^2 f}{\partial y'^2} y'' \tag{3.21}$$

and so the expanded form of the Euler-Lagrange equation (3.20) can be written

$$\left(\frac{\partial^2 f}{\partial y'^2} \right) \frac{d^2 y}{dx^2} + \left(\frac{\partial^2 f}{\partial y' \partial y} \right) \frac{dy}{dx} + \left(\frac{\partial^2 f}{\partial y' \partial x} - \frac{\partial f}{\partial y} \right) = 0. \tag{3.22}$$

Observe that if the second partial derivative $\frac{\partial^2 f}{\partial y'^2}$ is different from zero, then the equation (3.22) is a second order ordinary differential equation which is to be solved. This second order differential equation may be linear or nonlinear depending upon the functional considered. In many instances when a difficult nonlinear differential equation arises, one must resort to numerical methods or approximation techniques to solve the equation. Recall that second order ordinary differential equations require two independent constants in the general solution and so the general solution of the Euler-Lagrange equation is called a two-parameter family of solution curves. Under favorable conditions the Euler equation can be solved and the general solution will contain two arbitrary constants. These constants occurring in the general solution must be selected to satisfy the end conditions given by equation (3.12).

It is left as an exercise to show that the Euler-Lagrange equation (3.22) can also be written in the equivalent form

$$\frac{d}{dx}\left(f - \frac{\partial f}{\partial y'}\frac{dy}{dx}\right) - \frac{\partial f}{\partial x} = 0. \tag{3.23}$$

Example 3-1.   Special cases for the Euler-Lagrange equation

We consider special conditions under which the functional $I = \int_{x_1}^{x_2} f(x, y, y')\, dx$ assumes a stationary value. We assume that the solution $y = y(x)$ satisfies the end point conditions $y(x_1) = y_1$ and $y(x_2) = y_2$ for the following special cases.

Case 1: Dependent variable absent

If $f = f(x, y')$, then $\frac{\partial f}{\partial y} = 0$ so that the Euler-Lagrange equation (3.20) reduces to the form $\frac{d}{dx}\left(\frac{\partial f}{\partial y'}\right) = 0$. An integration of this equation gives an equation of the form $\frac{\partial f}{\partial y'} = C_1 = constant$. This equation can now be solved to represent y' as a function of x and C_1 to obtain an equation of the form $y' = \frac{dy}{dx} = G(x, C_1)$. This equation can now be solved by an integration to obtain $y = \int_{x_1}^{x} G(x, C_1)\, dx + C_2$.

Case 2: Independent variable absent

For $f = f(x, y, y')$ one can verify the Beltrami identity

$$\begin{aligned}\frac{d}{dx}\left[y'\frac{\partial f}{\partial y'} - f\right] &= y'\frac{d}{dx}\left(\frac{\partial f}{\partial y'}\right) + y''\frac{\partial f}{\partial y'} - \frac{\partial f}{\partial x} - \frac{\partial f}{\partial y}y' - \frac{\partial f}{\partial y'}y'' \\ &= -y'\left[\frac{\partial f}{\partial y} - \frac{d}{dx}\left(\frac{\partial f}{\partial y'}\right)\right] - \frac{\partial f}{\partial x}.\end{aligned} \tag{3.24}$$

In the special case $f = f(y, y')$ the term inside the brackets of Beltrami's identity is zero because the Euler-Lagrange equation is assumed to be satisfied. The last term $\frac{\partial f}{\partial x} = 0$ because f is assumed to be independent of the variable x. The Beltrami identity implies that in the special case where $f = f(y, y')$, then a first integral of the Euler-Lagrange equation can be written as

$$y'\frac{\partial f}{\partial y'} - f = \alpha = constant. \tag{3.25}$$

This first integral is a first order ordinary differential equation containing only y and y' terms.

Case 3: Exact or total derivative exists

Assume there exists a function $\phi = \phi(x, y)$ such that

$$\frac{d\phi}{dx} = \frac{\partial \phi}{\partial x} + \frac{\partial \phi}{\partial y}y' = f(x, y, y'). \tag{3.26}$$

In this special case the functional can be written in a form which is easily integrated. That is,

$$I = \int_{x_1}^{x_2} f(x, y, y')\, dx = \int_{x_1}^{x_2} d\phi = \phi(x_2, y_2) - \phi(x_1, y_1) \tag{3.27}$$

In this special case the integral is independent of the path connecting the points (x_1, y_1) and (x_2, y_2). This implies that all admissible functions $y(x)$ yield stationary values. Hence, the Euler-Lagrange equation degenerates into an identity. If $f = \phi_x + \phi_y y'$, then

$$\frac{\partial f}{\partial y} - \frac{d}{dx}\left(\frac{\partial f}{\partial y'}\right) = \phi_{xy} + \phi_{yy}y' - \frac{d}{dx}(\phi_y) = \phi_{xy} + \phi_{yy}y' - \phi_{yx} - \phi_{yy}y' = 0$$

is an identity. Conversely, let us assume that

$$\frac{\partial f}{\partial y} - \frac{d}{dx}\left(\frac{\partial f}{\partial y'}\right) = \frac{\partial f}{\partial y} - \frac{\partial^2 f}{\partial y'\partial x} - \frac{\partial^2 f}{\partial y'\partial y}y' - \frac{\partial^2 f}{\partial y'^2}y'' = 0 \tag{3.28}$$

is satisfied for all values of x, y, y', y''. The term y'' does not occur in the first three terms of equation (3.28) and since y'' can be selected arbitrarily, then the coefficient of y'' in equation (3.28) must equal zero in order for this equation to hold for arbitrary functions y. Hence, for arbitrary functions y we require that the following equations are satisfied

$$\frac{\partial^2 f}{\partial y'^2} = 0 \quad \text{and} \quad \frac{\partial f}{\partial y} - \frac{\partial^2 f}{\partial y'\partial x} - \frac{\partial^2 f}{\partial y'\partial y}y' = 0. \tag{3.29}$$

The first of these equations is integrated to obtain $\frac{\partial f}{\partial y'} = N(x, y)$ where $N(x, y)$ is an arbitrary function of x and y. Another integration gives

$$f = N(x, y)y' + M(x, y) \tag{3.30}$$

where $M(x, y)$ is another arbitrary function of x and y. Now substitute the results from equation (3.30) into the second condition from equation (3.29) to obtain

$$\frac{\partial N}{\partial y}y' + \frac{\partial M}{\partial y} - \frac{\partial N}{\partial x} - \frac{\partial N}{\partial y}y' = 0$$

which implies

$$\frac{\partial M}{\partial y} = \frac{\partial N}{\partial x}. \tag{3.31}$$

The equation (3.31) is a necessary condition that the integral

$$I = \int_{x_1}^{x_2} [N(x, y)y' + M(x, y)]\, dx = \int_{x_1}^{x_2} N(x, y)\, dy + M(x, y)\, dx = \int_{x_1}^{x_2} d\phi \tag{3.32}$$

be independent of the path connecting the points (x_1, y_1) and (x_2, y_2). In this case the functional has a constant value for every admissible curve and so the variational problem is of no interest.

Consider the addition of a term to the integrand of a given integral. In order that the added term not affect the resulting Euler-Lagrange equation, we must impose the condition that the added term be the exact or total derivative with respect to x of some function $\phi(x, y)$. This is a necessary and sufficient condition.

Case 4: No solution exists

There can arise situations where the resulting Euler-Lagrange equation does not have a solution. Usually, these situations are not of interest or a trivial solution will exist. ∎

The following is a well known example illustrating how a simple problem can be formulated in different ways. These various formulations can also be achieved by using a change of variable to transform the problem into a new coordinate system.

Example 3-2. **Shortest distance between two points in plane**

Join two points in a plane with the shortest arc.

Solution 1: In Cartesian coordinates a curve connecting the given points has the arc length squared given by

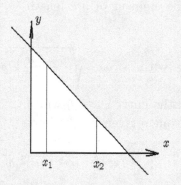
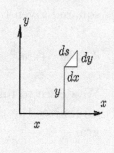

$$ds^2 = dx^2 + dy^2$$

with

$$ds = \sqrt{1 + (y')^2}\, dx.$$

We desire to find the curve $y = y(x)$ which minimizes the distance

$$I = \int_{x_1}^{x_2} ds = \int_{x_1}^{x_2} \sqrt{1 + (y')^2}\, dx. \tag{3.33}$$

Assume that the curve is to pass through the points (x_1, y_1) and (x_2, y_2) with $x_1 < x_2$. Let $f = \sqrt{1 + (y')^2}$ denote the integrand of the above integral and calculate the derivatives

$$\frac{\partial f}{\partial y} = 0, \quad \text{and} \quad \frac{\partial f}{\partial y'} = \frac{y'}{\sqrt{1 + (y')^2}}.$$

The Euler-Lagrange equation for an extremum is given by

$$\frac{\partial f}{\partial y} - \frac{d}{dx}\left(\frac{\partial f}{\partial y'}\right) = 0, \quad \text{or} \quad \frac{d}{dx}\left(\frac{y'}{\sqrt{1 + (y')^2}}\right) = 0.$$

An integration gives the first integral

$$\frac{y'}{\sqrt{1 + (y')^2}} = C = \text{constant}.$$

Solving for y' gives

$$y' = \frac{dy}{dx} = \sqrt{\frac{C^2}{1 - C^2}} = m = \text{a constant}.$$

Another integration gives the straight line $y = y(x) = mx + b$ where b is a constant of integration. The constants m and b are selected so that the line passes through the given end points. This produces the point-slope formula

$$y - y_1 = m(x - x_1), \quad \text{where} \quad m = \frac{y_2 - y_1}{x_2 - x1}. \tag{3.34}$$

Solution 2: In polar coordinates (r, θ), a curve joining two given points (r_1, θ_1) and (r_2, θ_2) has the element of arc length squared given by

$$ds^2 = dr^2 + r^2 d\theta^2 \qquad \text{with} \qquad ds = \sqrt{r^2 + \left(\frac{dr}{d\theta}\right)^2} \, d\theta.$$

We desire to find the curve $r = r(\theta)$ which minimizes the distance given by the integral

$$I = \int_{\theta_1}^{\theta_2} \sqrt{r^2 + (r')^2} \, d\theta, \qquad \text{where} \quad r' = \frac{dr}{d\theta}. \tag{3.35}$$

The desired curve is to satisfy the end point conditions $r_1 = r(\theta_1)$ and $r_2 = r(\theta_2)$. Let

$$F = F(\theta, r, \frac{dr}{d\theta}) = \sqrt{r^2 + (r')^2}$$

denote the integrand of the integral to be minimized and calculate the derivatives

$$\frac{\partial F}{\partial r} = \frac{r}{\sqrt{r^2 + (r')^2}}, \qquad \frac{\partial F}{\partial r'} = \frac{r'}{\sqrt{r^2 + (r')^2}}$$

so that the Euler-Lagrange equation becomes

$$\frac{\partial F}{\partial r} - \frac{d}{d\theta}\left(\frac{\partial F}{\partial r'}\right) = 0 \quad \text{or} \quad \frac{r}{\sqrt{r^2 + (r')^2}} - \frac{d}{d\theta}\left(\frac{r'}{\sqrt{r^2 + (r')^2}}\right) = 0. \tag{3.36}$$

The equation (3.36) simplifies to the ordinary differential equation

$$rr'' - 2(r')^2 - r^2 = 0. \tag{3.37}$$

Make the substitution $\frac{dr}{d\theta} = u = u(r)$ with $\frac{d^2 r}{d\theta^2} = u\frac{du}{dr}$ and then integrate equation (3.37) to obtain, after simplification, the result

$$\frac{dr}{d\theta} = r\sqrt{(r/p)^2 - 1} \tag{3.38}$$

where $1/p^2$ is a constant of integration. Separate the variables in equation (3.38) and then perform another integration to obtain, after simplification, the solution $r = p\sec(\theta - \beta)$ where $-\beta$ is the integration constant. This solution is the polar form for the equation of a straight line. Here p has the physical interpretation of representing the perpendicular distance from the origin to the line and β has the physical interpretation of representing the angle of inclination of the normal line which passes through the origin. The situation is illustrated in the accompanying figure.

Alternatively, introduce the change of variables $x = r\cos\theta$, $y = r\sin\theta$ in equation (3.35) and reduce it to the form of equation (3.33). This substitution is left as an exercise.

Note also that the problem in polar coordinates can be formulated in the form where $I = \int_{r_1}^{r_2} \sqrt{1 + r^2 \left(\frac{d\theta}{dr}\right)^2}\, dr$ where r is the independent variable. It is left as an exercise to show that the resulting Euler-Lagrange equation has the solution $r\sin(\theta + \alpha) = \beta$ where α and β are constants. The solution being another form for the polar equation of a straight line.

■

Invariance under a change of variables

We examine the functional

$$I = \int_{x_1}^{x_2} f(x, y(x), y'(x))\, dx \tag{3.39}$$

which undergoes a transformation from an (x,y) Cartesian coordinate system to a general (u,v) coordinate system. Assume that we are given a set of transformation equations of the form

$$x = x(u,v), \quad y = y(u,v), \quad \text{with Jacobian} \quad \frac{\partial(x,y)}{\partial(u,v)} = \begin{vmatrix} \frac{\partial x}{\partial u} & \frac{\partial x}{\partial v} \\ \frac{\partial y}{\partial u} & \frac{\partial y}{\partial v} \end{vmatrix} \neq 0. \tag{3.40}$$

Here the Jacobian of the transformation $\frac{\partial(x,y)}{\partial(u,v)} \neq 0$ insures that an inverse transformation

$$u = u(x,y), \quad \text{and} \quad v = v(x,y) \tag{3.41}$$

exists. Further assume that the curve $y = y(x)$ which produces a stationary value of the functional is transformed to a curve $v = v(u)$ in the new coordinate system. Substituting the change of variables given by equations (3.40) into the above functional produces the new functional

$$I = I(v) = \int_{u_1}^{u_2} f\left(x(u,v), y(u,v), \frac{\frac{\partial y}{\partial u} + \frac{\partial y}{\partial v}\frac{dv}{du}}{\frac{\partial x}{\partial u} + \frac{\partial x}{\partial v}\frac{dv}{du}}\right)\left(\frac{\partial x}{\partial u} + \frac{\partial x}{\partial v}\frac{dv}{du}\right) du = \int_{u_1}^{u_2} F(u,v,v')\, du \tag{3.42}$$

where

$$F(u,v,v') = f\left(x(u,v), y(u,v), \frac{\frac{\partial y}{\partial u} + \frac{\partial y}{\partial v}\frac{dv}{du}}{\frac{\partial x}{\partial u} + \frac{\partial x}{\partial v}\frac{dv}{du}}\right)\left(\frac{\partial x}{\partial u} + \frac{\partial x}{\partial v}\frac{dv}{du}\right) \tag{3.43}$$

is some new function of the variables u, v, v' where $v' = \dfrac{dv}{du}$.

The invariance of the Euler-Lagrange equation is a property that we state without proof. (For a proof see the reference by R. Courant.) The basic idea behind the proof is that if $y = y(x)$ satisfies the Euler-Lagrange equation

$$\frac{\partial f}{\partial y} - \frac{d}{dx}\left(\frac{\partial f}{\partial y'}\right) = 0,$$

associated with the functional given by equation (3.39), then the transformation equations (3.40) transform $y = y(x)$ to a curve $v = v(u)$ which satisfies the Euler-Lagrange equation

$$\frac{\partial F}{\partial v} - \frac{d}{du}\left(\frac{\partial F}{\partial v'}\right) = 0$$

108

associated with the new functional given by equation (3.42). Courant shows that for a general transformation of coordinates the following equation must hold

$$\left[\frac{\partial f}{\partial y} - \frac{d}{dx}\left(\frac{\partial f}{\partial y'}\right)\right] = \left[\frac{\partial F}{\partial v} - \frac{d}{du}\left(\frac{\partial F}{\partial v'}\right)\right]\frac{\partial(x,y)}{\partial(u,v)} \tag{3.44}$$

By hypothesis we have assumed that the Jacobian of the transformation is different from zero. The equation (3.44) therefore implies that if the Euler-Lagrange equation is satisfied in one coordinate system, then it must be satisfied in the new coordinate system also. The equation (3.44) demonstrates that the Euler-Lagrange equation is an invariant under a general coordinate transformation with nonzero Jacobian. The invariance of the Euler-Lagrange equation is a basic principle that is often used to express mathematical equations in new coordinates. For example, if you know the Euler-Lagrange equation produces the Laplacian equation $\nabla^2 u = 0$ in one coordinate system, then you can use equation (3.44) to find the Laplacian equation in the transformed coordinate system.

Parametric representation

The functional

$$I = \int_{x_1}^{x_2} f(x,y,y')\,dx \quad \text{with Euler-Lagrange equation} \quad \frac{\partial f}{\partial y} - \frac{d}{dx}\left(\frac{\partial f}{\partial y'}\right) = 0 \tag{3.45}$$

assumes that y is a single-valued function of x subject to the end point conditions $y(x_1) = y_1$ and $y(x_2) = y_2$. One can always introduce a parametric representation $x = x(t)$ and $y = y(t)$ for $t_1 \leq t \leq t_2$ where $x(t_i) = x_i$ and $y(t_i) = y_i$ for $i = 1, 2$. The differential and derivative relations $dx = \dot{x}\,dt$ and $y' = \dot{y}/\dot{x}$ allows one to write the functional as

$$I = \int_{x_1}^{x_2} f(x,y,y')\,dx = \int_{t_1}^{t_2} f\left(x,y,\frac{\dot{y}}{\dot{x}}\right)\dot{x}\,dt = \int_{t_1}^{t_2} g(x,y,\dot{x},\dot{y})\,dt \tag{3.46}$$

where $g = f\dot{x}$. Later we will demonstrate that the equations defining the functions $x = x(t)$ and $y = y(t)$ producing a stationary value for the functional (3.46) are determined from the system of Euler-Lagrange differential equations

$$\frac{\partial g}{\partial x} - \frac{d}{dt}\left(\frac{\partial g}{\partial \dot{x}}\right) = 0, \qquad \frac{\partial g}{\partial y} - \frac{d}{dt}\left(\frac{\partial g}{\partial \dot{y}}\right) = 0 \tag{3.47}$$

We can also work backwards and show that the above system of differential equations result because of the following considerations. It is left as an exercise to verify the derivatives

$$\frac{\partial g}{\partial x} = \frac{\partial f}{\partial x}\dot{x}, \qquad \frac{\partial g}{\partial \dot{x}} = f - \dot{x}\frac{\partial f}{\partial y'}\left(\frac{\dot{y}}{\dot{x}^2}\right) = f - y'\frac{\partial f}{\partial y'}, \qquad \frac{d}{dt}\left(\frac{\partial g}{\partial \dot{x}}\right) = \left\{y'\left[\frac{\partial f}{\partial y} - \frac{d}{dx}\left(\frac{\partial f}{\partial y'}\right)\right] + \frac{\partial f}{\partial x}\right\}\dot{x}$$

$$\frac{\partial g}{\partial y} = \frac{\partial f}{\partial y}\dot{x}, \qquad \frac{\partial g}{\partial \dot{y}} = \dot{x}\frac{\partial f}{\partial y'}\left(\frac{1}{\dot{x}}\right) = \frac{\partial f}{\partial y'}, \qquad \frac{d}{dt}\left(\frac{\partial g}{\partial \dot{y}}\right) = \frac{d}{dx}\left(\frac{\partial f}{\partial y'}\right)\dot{x}$$

Then one can substitute these derivatives into the equations (3.47) to obtain the following results

$$\frac{\partial g}{\partial x} - \frac{d}{dt}\left(\frac{\partial g}{\partial \dot{x}}\right) = -\left[\frac{\partial f}{\partial y} - \frac{d}{dx}\left(\frac{\partial f}{\partial y'}\right)\right]\dot{y} = 0, \quad \text{and} \quad \frac{\partial g}{\partial y} - \frac{d}{dt}\left(\frac{\partial g}{\partial \dot{y}}\right) = \left[\frac{\partial f}{\partial y} - \frac{d}{dx}\left(\frac{\partial f}{\partial y'}\right)\right]\dot{x} = 0$$

The above set of equations show that if the Euler-Lagrange equation (3.45) is satisfied, then the equations (3.47) must be satisfied. Thus, one can write variational problems in the form where y as a single-valued function of x, equation (3.45), or alternatively it can be written in the parametric form given by the equation (3.46) having the system of Euler-Lagrange equations (3.47).

The variational notation δ

Given a curve $y = f(x)$ one then has a specific relationship between the x and y variables. When one moves from a point $P(x,y)$ to a near point $Q(x+dx, y+dy)$ there is a definite relationship between the differentials dx and dy, namely that $dy = f'(x)\,dx$. In the calculus of variations one can assign arbitrary and independent changes to the variables x and y so that one can move from a point $P(x,y)$ on the given curve to a point $P^*(x+\delta x, y+\delta y)$ which may or may not lie on the given curve. When we apply the arbitrary independent changes to each point P of the given curve C, we shall restrict ourselves to studying those changes where the variational points P^* are such that they lie on a continuous curve C^*. The independent arbitrary variations are denoted using the notations δx and δy for these changes. We shall at times require that the variations δx and δy be zero at the end points of a curve.

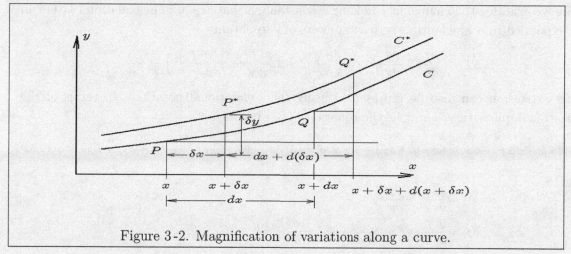

Figure 3-2. Magnification of variations along a curve.

In terms of differentials one can write $d(\delta x) = \delta(dx)$. These changes are illustrated geometrically in the figure 3-2. Assume that two near points P and Q on a continuous curve C get deformed to points P^* and Q^* by using arbitrary and independent variations δx and δy. The variations δx and δy of each point $P \to P^*$ and $Q \to Q^*$ are such that there results a deformed curve C^* which is continuous. We wish to examine the arbitrary change in dx as the point x changes to $x+dx$ and simultaneously the point $x+\delta x$ changes to $x+\delta x+d(x+\delta x)$. The figure 3-2 illustrates an arbitrary change in dx which can be written

$$\delta(dx) = [x + \delta x + d(x+\delta x) - (x+\delta x)] - [dx] = d(\delta x) \tag{3.48}$$

which shows the operators δ and d are commutative.

Let us examine variations in the functional $I = \int_{x_1}^{x_2} f(x, y, y')\, dx$ if $y(x)$ is changed to the comparison function $Y(x) = y(x) + \epsilon \eta(x)$ where ϵ is independent of x. Define the variation in the function y by

$$\delta y = \text{variation in } y = Y - y = \epsilon \eta(x).$$

The variation in y is a function of x and so it can be differentiated to obtain

$$\frac{d}{dx}(\delta y) = (\delta y)' = Y' - y' = \epsilon \eta'(x) = \text{variation in } y' = \delta y' = \delta \frac{dy}{dx}. \tag{3.49}$$

By taking higher derivatives there results

$$(\delta y)'' = \delta y'', \quad (\delta y)''' = \delta y''', \quad \dots, \quad (\delta y)^{(n)} = \delta y^{(n)}.$$

The variational symbol δ and the derivative symbol $\frac{d}{dx}$ are commutative since

$$\frac{d}{dx}(\delta y) = \frac{d}{dx}(\epsilon \eta(x)) = \epsilon \eta'(x) = \delta y' = \delta \frac{dy}{dx}. \tag{3.50}$$

The change in $f = f(x, y, y')$ corresponding to a change in y and y' is represented

$$\Delta f = f(x, y + \epsilon \eta, y' + \epsilon \eta') - f(x, y, y'). \tag{3.51}$$

Here we examine the change in f holding x constant so that $\delta x = 0$. The equation (3.51) can be expanded in a Maclaurin's series in powers of ϵ to obtain

$$\Delta f = \left(\frac{\partial f}{\partial y}\eta + \frac{\partial f}{\partial y'}\eta'\right)\epsilon + \left(\frac{\partial^2 f}{\partial y^2}\eta^2 + 2\frac{\partial^2 f}{\partial y \partial y'}\eta\eta' + \frac{\partial^2 f}{\partial y'^2}\eta'^2\right)\frac{\epsilon^2}{2!} + \cdots$$

This expansion can also be represented using the δ-variational notation. In terms of the variational quantities δy and $\delta y'$ the equation (3.51) becomes

$$\Delta f = f(x, y + \delta y, y' + \delta y') - f(x, y, y') = \delta f_1 + \frac{1}{2!}\delta f_2 + \frac{1}{3!}\delta f_3 + \cdots + \frac{1}{n!}\delta f_n + R_n \tag{3.52}$$

where

$$\delta f_1 = \frac{\partial f}{\partial y}\delta y + \frac{\partial f}{\partial y'}\delta y'$$

$$\delta f_2 = \frac{\partial^2 f}{\partial y^2}(\delta y)^2 + 2\frac{\partial^2 f}{\partial y \partial y'}\delta y \delta y' + \frac{\partial^2 f}{\partial y'^2}(\delta y')^2$$

$$\delta f_3 = \frac{\partial^3 f}{\partial y^3}(\delta y)^3 + 3\frac{\partial^3 f}{\partial y^2 \partial y'}(\delta y)^2 \delta y' + 3\frac{\partial^3 f}{\partial y \partial y'^2}\delta y(\delta y')^2 + \frac{\partial^3 f}{\partial y'^3}(\delta y')^3$$

$$\vdots$$

$$\delta f_n = \left(\delta y \frac{\partial}{\partial y} + \delta y' \frac{\partial}{\partial y'}\right)^n f$$

and R_n is a remainder term. Often times the variations δy and $\delta y'$ are required to be zero at the end points where $x = x_1$ and $x = x_2$. However, these end point conditions can be subject to change later on.

Define variations in the integral I, as

$$\delta I = \int_{x_1}^{x_2} f(x, y + \delta y, y' + \delta y')\, dx - \int_{x_1}^{x_2} f(x, y, y')\, dx = \int_{x_1}^{x_2} \Delta f\, dx$$

and use the equation (3.52) to obtain $\quad \delta I = \delta I_1 + \frac{1}{2!}\delta I_2 + \frac{1}{3!}\delta I_3 + \cdots$ where

$$\delta I_1 = \int_{x_1}^{x_2} \delta f_1 \, dx \quad \text{is called the first variation}$$

$$\delta I_2 = \int_{x_1}^{x_2} \delta f_2 \, dx \quad \text{is called the second variation}$$

$$\vdots$$

$$\delta I_n = \int_{x_1}^{x_2} \delta f_n \, dx \quad \text{is called the } n\text{th variation}$$

Examine the first variation of the integral I to obtain $\delta I_1 = \int_{x_1}^{x_2} \left[\frac{\partial f}{\partial y} \delta y + \frac{\partial f}{\partial y'} \delta y' \right] dx$ and use integration by parts on the second term of this integral with

$$U = \frac{\partial f}{\partial y'}, \qquad dU = \frac{d}{dx}\left(\frac{\partial f}{\partial y'} \right) dx, \qquad dV = \delta y' \, dx, \qquad V = \delta y$$

to obtain

$$\delta I_1 = \left[\frac{\partial f}{\partial y'} \delta y \right]_{x_1}^{x_2} + \int_{x_1}^{x_2} \left[\frac{\partial f}{\partial y} - \frac{d}{dx}\left(\frac{\partial f}{\partial y'} \right) \right] \delta y \, dx.$$

If δy is assumed to be zero at the end points, then the first term is zero.

If I has a stationary value along the curve C, then this first variation must equal zero and the function $y = y(x)$ must satisfy the Euler-Lagrange equation

$$\frac{\partial f}{\partial y} - \frac{d}{dx}\left(\frac{\partial f}{\partial y'} \right) = 0, \quad \text{or} \quad \frac{\partial f}{\partial y} - \frac{\partial^2 f}{\partial y' \partial x} - \frac{\partial^2 f}{\partial y' \partial y} y' - \frac{\partial^2 f}{\partial y'^2} y'' = 0. \qquad (3.53)$$

Solutions of the Euler-Lagrange equation subject to the fixed boundary conditions are referred to as stationary solutions. Further testing is required to determine if the functions produce an extreme value.

Other functionals

We investigate other functionals having fixed end point conditions or fixed boundary conditions and derive the necessary conditions under which these functionals take on a stationary value. Note in the following derivations that integration by parts in some form plays an essential role in simplifying the resulting equations to a representation where the basic lemma can then be applied to obtain the necessary condition for a stationary value to occur. Also associated with the derived equations are boundary conditions. For now we consider these boundary conditions as fixed or specified. Other types of boundary conditions will be considered later in the text. The investigation of the stationary values to determine if they correspond to an extreme value of a maximum or minimum value for the integral I will be considered in the next chapter.

[f2]: Integrand f(x, y, y′, y″)

Consider the functional

$$I = \int_a^b f(x, y, y', y'') \, dx \tag{3.54}$$

where the integrand is a function of the variables x, y, y', y'' involving both the first and second derivatives of a function $y = y(x)$. We assume that $y = y(x)$ and all its derivatives are continuous over the interval (a, b) and that the above integral exists. The function $y = y(x)$ is subject to the boundary conditions

$$y(a) = \alpha_0, \quad y'(a) = \alpha_1, \qquad y(b) = \beta_0, \quad y'(b) = \beta_1 \tag{3.55}$$

where the quantities $\alpha_0, \alpha_1, \beta_0, \beta_1$ are given constants. We use the variational notation to find the differential equation satisfied by the function $y = y(x)$ which produces a stationary value for the integral I. If this stationary value corresponds to an extremal, the first variation in I is required to satisfy

$$\delta I = \int_a^b \left[\frac{\partial f}{\partial y} \delta y + \frac{\partial f}{\partial y'} \delta y' + \frac{\partial f}{\partial y''} \delta y'' \right] dx = 0 \tag{3.56}$$

where $\delta y = \delta y' = 0$ at the end points $x = a$ and $x = b$. In equation (3.56) we integrate the second term by parts to obtain

$$\int_a^b \frac{\partial f}{\partial y'} \delta y' \, dx = \left[\frac{\partial f}{\partial y'} \delta y \right]_a^b - \int_a^b \frac{d}{dx} \left(\frac{\partial f}{\partial y'} \right) \delta y \, dx. \tag{3.57}$$

where the first term of this integral is zero because of the boundary condition assumptions. Now integrate the third term in equation (3.56) by parts to obtain

$$\int_a^b \frac{\partial f}{\partial y''} \delta y'' \, dx = \left[\frac{\partial f}{\partial y''} \delta y' \right]_a^b - \int_a^b \frac{d}{dx} \left(\frac{\partial f}{\partial y''} \right) \delta y' \, dx \tag{3.58}$$

where again the first term in this integral is zero because of the boundary condition assumptions. Performing another integration by parts on the integral in equation (3.58), there results

$$-\int_a^b \frac{d}{dx} \left(\frac{\partial f}{\partial y''} \right) \delta y' \, dx = -\left[\frac{d}{dx} \left(\frac{\partial f}{\partial y''} \right) \delta y \right]_a^b + \int_a^b \frac{d^2}{dx^2} \left(\frac{\partial f}{\partial y''} \right) \delta y \, dx. \tag{3.59}$$

where the first term in equation (3.59) is zero because of the boundary condition assumptions. Therefore, if integration by parts is employed on the second term of equation (3.56) and integration by parts is used twice on the third term of equation (3.56), together with the boundary conditions that $\delta y = \delta y' = 0$ at the end points, then the equation (3.56) is simplified to the form

$$\delta I = \int_a^b \left[\frac{\partial f}{\partial y} - \frac{d}{dx} \left(\frac{\partial f}{\partial y'} \right) + \frac{d^2}{dx^2} \left(\frac{\partial f}{\partial y''} \right) \right] \delta y \, dx = 0. \tag{3.60}$$

Now we can use the basic lemma to obtain the Euler-Lagrange necessary condition for a stationary value

$$\frac{\partial f}{\partial y} - \frac{d}{dx} \left(\frac{\partial f}{\partial y'} \right) + \frac{d^2}{dx^2} \left(\frac{\partial f}{\partial y''} \right) = 0. \tag{3.61}$$

This is the Euler-Lagrange equation which is a necessary condition to be satisfied in order for a stationary value of the functional (3.54) to exist.

Example 3-3.

Find the Euler-Lagrange equation associated with the functional
$$I = \int_{x_1}^{x_2} \left[\omega^4 y^2 - (y'')^2 \right] dx \text{ where } \omega \text{ is a constant.}$$
Solution: Here $f = f(x, y, y', y'') = \omega^4 y^2 - (y'')^2$ with derivatives

$$\frac{\partial f}{\partial y} = 2\omega^4 y, \quad \frac{\partial f}{\partial y'} = 0, \quad \frac{\partial f}{\partial y''} = -2y''$$

The Euler-Lagrange equation $\dfrac{\partial f}{\partial y} - \dfrac{d}{dx}\left(\dfrac{\partial f}{\partial y'}\right) + \dfrac{d^2}{dx^2}\left(\dfrac{\partial f}{\partial y''}\right) = 0$ can be written

$$2\omega^4 y + (-2y^{(iv)}) = 0, \quad \text{or} \quad y^{(iv)} - \omega^4 y = 0.$$

This is a fourth order ordinary differential equation with constant coefficients. Assume an exponential solution $y = e^{mx}$, where m is a constant, and obtain the characteristic equation $m^4 - \omega^4 = (m^2 - \omega^2)(m^2 + \omega^2) = 0$, with characteristic roots $m = \pm\omega, \pm i\omega$. This gives the fundamental set of solutions $\{e^{\omega x}, e^{-\omega x}, \sin \omega x, \cos \omega x\}$. The general solution is therefore any linear combination of these functions. The general solution can be expressed in either of the forms $y = c_1 e^{\omega x} + c_2 e^{-\omega x} + c_3 \sin \omega x + c_4 \cos \omega x$ or $y = c_1 \sinh \omega x + c_2 \cosh \omega x + c_3 \sin \omega x + c_4 \cos \omega x$, where c_1, c_2, c_3, c_4 are constants. Here a four parameter family of curves exists where each curve produces a stationary value associated with the integral I. We must select the particular function from the family of curves which satisfies the given boundary conditions and then determine if this particular curve corresponds to an extreme value associated with I. ∎

[f3]: Integrand $\mathbf{f(x, y, y', y'', y''', \ldots, y^{(n)})}$

Consider functionals of the form

$$I = \int_a^b f(x, y, y', y'', \ldots, y^{(n)}) \, dx \tag{3.62}$$

which involve derivatives through the n-th order. This functional is a generalization of the previous two functionals. We desire to find a function $y = y(x)$ which produces an extremum, if one exists.

Assume that the functional (3.62) undergoes a variation in y and its derivatives where $\delta y = \epsilon \eta, \ \delta y' = \epsilon \eta', \ldots, \delta y^{(n)} = \epsilon \eta^{(n)}$, with primes denoting differentiation with respect to x, then

$$\delta I = \int_a^b f(x, y + \epsilon\eta, y' + \epsilon\eta', y'' + \epsilon\eta'', \ldots, y^{(n)} + \epsilon\eta^{(n)}) \, dx - \int_a^b f(x, y, y', y'', \ldots, y^{(n)}) \, dx$$

Further assume $f \in \mathcal{C}^n$, then one can employ a Taylor series expansion about $\epsilon = 0$ and obtain

$$\delta I = \epsilon \int_a^b (\Delta f) \, dx + \frac{\epsilon^2}{2!} \int_a^b (\Delta^2 f) \, dx + \frac{\epsilon^3}{3!} \int_a^b R \, dx \tag{3.63}$$

where Δ is the differential operator

$$\Delta = \eta \frac{\partial}{\partial y} + \eta' \frac{\partial}{\partial y'} + \eta'' \frac{\partial}{\partial y''} + \cdots + \eta^{(n)} \frac{\partial}{\partial y^{(n)}}$$

and R is the remainder term. One can verify, using integration by parts, the following integrals

$$\int \eta \frac{\partial f}{\partial y}\, dx = \qquad\qquad\qquad\qquad\qquad\qquad \int \eta \frac{\partial f}{\partial y}\, dx$$

$$\int \eta' \frac{\partial f}{\partial y'}\, dx = \qquad\qquad\qquad\qquad \eta \frac{\partial f}{\partial y'} - \int \eta \frac{d}{dx}\left(\frac{\partial f}{\partial y'}\right)\, dx$$

$$\int \eta'' \frac{\partial f}{\partial y''}\, dx = \qquad\qquad \eta' \frac{\partial f}{\partial y'} - \eta \frac{d}{dx}\left(\frac{\partial f}{\partial y''}\right) + \int \eta \frac{d^2}{dx^2}\left(\frac{\partial f}{\partial y''}\right)\, dx$$

$$\vdots \qquad\qquad\qquad\qquad\qquad\qquad\qquad\qquad\qquad \vdots$$

$$\int \eta^{(n)} \frac{\partial f}{\partial y^{(n)}}\, dx = \eta^{(n-1)} \frac{\partial f}{\partial y^{(n)}} - \eta^{(n-2)} \frac{d}{dx}\left(\frac{\partial f}{\partial y^{(n)}}\right) + \cdots + (-1)^{n-1}\eta \frac{d^{(n-1)}}{dx^{(n-1)}}\left(\frac{\partial f}{\partial y^{(n)}}\right) + (-1)^n \int \eta \frac{d^n}{dx^n}\left(\frac{\partial f}{\partial y^{(n)}}\right)\, dx$$

By integrating the first term in equation (3.63) between the limits a and b and utilizing the above integrals, we obtain by adding the left column of integrals and right column of integrals the first variation

$$\delta I_1 = \int_a^b \epsilon\,(\Delta f)\, dx = \left[\text{Boundary terms}\right]_a^b$$

$$+ \int_a^b \eta \left[\frac{\partial f}{\partial y} - \frac{d}{dx}\left(\frac{\partial f}{\partial y'}\right) + \frac{d^2}{dx^2}\left(\frac{\partial f}{\partial y''}\right) - \frac{d^3}{dx^3}\left(\frac{\partial f}{\partial y'''}\right) + \cdots + (-1)^n \frac{d^n}{dx^n}\left(\frac{\partial f}{\partial y^{(n)}}\right)\right]\, dx$$

If we assume that $\delta y = \delta y' = \delta y'' = \cdots = \delta y^{(n-1)} = 0$ at the end points $x = a$ and $x = b$, then the boundary terms are zero. Equate the first variation to zero and employ the basic lemma. One finds the functions $\overset{\ast}{y} = y(x)$ which produce a stationary value must satisfy the Euler-Lagrange equation

$$\frac{\partial f}{\partial y} - \frac{d}{dx}\left(\frac{\partial f}{\partial y'}\right) + \frac{d^2}{dx^2}\left(\frac{\partial f}{\partial y''}\right) - \frac{d^3}{dx^3}\left(\frac{\partial f}{\partial y'''}\right) + \cdots + (-1)^n \frac{d^n}{dx^n}\left(\frac{\partial f}{\partial y^{(n)}}\right) = 0. \qquad (3.64)$$

This is a generalization of the results under the headings [f1] and [f2] previously considered. The Euler-Lagrange equation (3.64) is an ordinary differential equation of order 2n. It is sometimes referred to as the Euler-Poisson equation associated with the functional given by equation (3.62). The general solution of this equation will involve 2n arbitrary constants. To assign values to these constants when the variations $\delta y, \delta y', \ldots, \delta y^{(n-1)}$ are zero, it is customary to employ the 2n boundary conditions

$$y(a) = \alpha_0, \quad y'(a) = \alpha_1, \ldots, y^{(n-1)}(a) = \alpha_{n-1}$$
$$y(b) = \beta_0, \quad y'(b) = \beta_1, \ldots, y^{(n-1)}(b) = \beta_{n-1} \qquad (3.65)$$

where the α_i and β_i values, for $i = 0, 1, 2, \ldots, (n-1)$, are specified constants.

Special case 1

Consider the special case were f does not contain the independent variable x. In this case one can immediately obtain a first integral of the resulting Euler-Lagrange equation as follows. Note that in this special case $f = f(y, y', y'', \ldots, y^{(n)})$ so that one can write

$$\frac{df}{dx} = \frac{\partial f}{\partial y}y' + \frac{\partial f}{\partial y'}y'' + \frac{\partial f}{\partial y''}y''' + \cdots + \frac{\partial f}{\partial y^{(n)}}y^{(n+1)}. \qquad (3.66)$$

Using integration by parts we integrate each term on the right-hand side of equation (3.66) to obtain

$$\int \frac{\partial f}{\partial y} y' \, dx = \boxed{} \qquad \int \frac{\partial f}{\partial y} y' \, dx$$

$$\int \frac{\partial f}{\partial y'} y'' \, dx = \frac{\partial f}{\partial y'} y' \qquad - \int \frac{d}{dx}\left(\frac{\partial f}{\partial y'}\right) y' \, dx$$

$$\int \frac{\partial f}{\partial y''} y''' \, dx = \frac{\partial f}{\partial y''} y'' - \frac{d}{dx}\left(\frac{\partial f}{\partial y''}\right) y' \qquad + \int \frac{d^2}{dx^2}\left(\frac{\partial f}{\partial y''}\right) y' \, dx$$

$$\int \frac{\partial f}{\partial y'''} y^{(iv)} \, dx = \frac{\partial f}{\partial y'''} y''' - \frac{d}{dx}\left(\frac{\partial f}{\partial y'''}\right) y'' + \frac{d^2}{dx^2}\left(\frac{\partial f}{\partial y'''}\right) y' \qquad - \int \frac{d^3}{dx^3}\left(\frac{\partial f}{\partial y'''}\right) y' \, dx$$

$$\vdots$$

$$\int \frac{\partial f}{\partial y^{(n)}} y^{(n+1)} \, dx = \frac{\partial f}{\partial y^{(n)}} y^{(n)} - \frac{d}{dx}\left(\frac{\partial f}{\partial y^{(n)}}\right) y^{(n-1)} + \cdots + (-1)^{n-1} \frac{d^{(n-1)}}{dx^{(n-1)}}\left(\frac{\partial f}{\partial y^{(n)}}\right) y' + (-1)^n \int \frac{d^n}{dx^n}\left(\frac{\partial f}{\partial y^{(n)}}\right) y' \, dx$$

Define the sum of the terms in the white rectangular area above as $T_n = T_n(y, y', \ldots, y^{(n)})$ and then add the above equations. Use equation (3.66) and note that the sum of the integrals on the left-hand side can be integrated to obtain $f + \alpha$ where α is a constant of integration. Also note that the sum of the integrals on the right-hand side gives zero because the resulting integrand is the Euler-Lagrange equation. By adding the above equations one obtains a first integral of the Euler-Lagrange equation as

$$f + \alpha = T_n \tag{3.67}$$

Note the following special cases of equation (3.67) where α denotes a constant of integration.

Integrand f	First integral of Euler-Lagrange equation
$f = f(y, y')$	$f + \alpha = \dfrac{\partial f}{\partial y'} y' \quad$ See also equation (3.25)
$f = f(y, y', y'')$	$f + \alpha = \dfrac{\partial f}{\partial y'} y' + \dfrac{\partial f}{\partial y''} y'' - \dfrac{d}{dx}\left(\dfrac{\partial f}{\partial y''}\right) y'$
$f = f(y, y', y'', y''')$	$f + \alpha = \dfrac{\partial f}{\partial y'} y' + \dfrac{\partial f}{\partial y''} y'' + \dfrac{\partial f}{\partial y'''} y''' - \dfrac{d}{dx}\left(\dfrac{\partial f}{\partial y''}\right) y' - \dfrac{d}{dx}\left(\dfrac{\partial f}{\partial y'''}\right) y'' + \dfrac{d^2}{dx^2}\left(\dfrac{\partial f}{\partial y'''}\right) y'$

The verification of these results is left as an exercise. The general case is given by equation (3.67).

Special case 2

In the special case f does not contain the dependent variable y, then the Euler-Lagrange equation (3.64) reduces to

$$-\frac{d}{dx}\left(\frac{\partial f}{\partial y'}\right) + \frac{d^2}{dx^2}\left(\frac{\partial f}{\partial y''}\right) - \frac{d^3}{dx^3}\left(\frac{\partial f}{\partial y'''}\right) + \cdots + (-1)^n \frac{d^n}{dx^n}\left(\frac{\partial f}{\partial y^{(n)}}\right) = 0. \tag{3.68}$$

and this equation can be immediately integrated to obtain the first integral

$$\frac{\partial f}{\partial y'} - \frac{d}{dx}\left(\frac{\partial f}{\partial y''}\right) + \cdots + (-1)^n \frac{d^{n-1}}{dx^{n-1}}\left(\frac{\partial f}{\partial y^{(n)}}\right) = \alpha^* \tag{3.69}$$

where α^* is a new constant.

[f4]: Integrand $f\left(x, y_1, y_1', y_1'', \ldots, y_1^{(n_1)}, y_2, y_2', y_2'', \ldots, y_2^{(n_2)}, \ldots, y_m, y_m', y_m'', \ldots, y_m^{(n_m)}\right)$

The previous results can be generalized to functionals having more than one dependent variable which contain derivatives of arbitrary order. Consider functionals having the form

$$I(y_1, y_2, \ldots, y_m) = \int_a^b f\left(x, y_1, y_1', \ldots, y_1^{(n_1)}, y_2, y_2', \ldots, y_2^{(n_2)}, \ldots, y_m, y_m', \ldots, y_m^{(n_m)}\right) dx \qquad (3.70)$$

where $n_1, n_2, \ldots, n_m$ are integers. The problem is to find a set of functions $y_1(x), y_2(x), \ldots, y_m(x)$ which make the functional given by equation (3.70) have a stationary value which may or may not correspond to an extreme value. Here one must assume that each function $y_i(x) \in \mathcal{C}^{(n_i)}$ for $x \in [a, b]$ where the index i varies from 1 to m. It is assumed that each function $y_i(x)$, $i = 1, \ldots, m$ satisfies the end point boundary conditions where $x = a$ and $x = b$ of the form

$$
\begin{aligned}
y_1(a) &= \alpha_{1,0}, & y_1'(a) &= \alpha_{1,1}, & \ldots & & y_1^{(n_1-1)}(a) &= \alpha_{1,n_1-1} \\
y_2(a) &= \alpha_{2,0}, & y_2'(a) &= \alpha_{2,1}, & \ldots & & y_2^{(n_2-1)}(a) &= \alpha_{2,n_2-1} \\
&\ \vdots & &\ \vdots & \ldots & & &\ \vdots \\
y_m(a) &= \alpha_{m,0}, & y_m'(a) &= \alpha_{m,1}, & \cdots & & y_m^{(n_m-1)}(a) &= \alpha_{m,n_m-1}
\end{aligned}
\qquad (3.71)
$$

and

$$
\begin{aligned}
y_1(b) &= \beta_{1,0}, & y_1'(b) &= \beta_{1,1}, & \ldots & & y_1^{(n_1-1)}(b) &= \beta_{1,n_1-1} \\
y_2(b) &= \beta_{2,0}, & y_2'(b) &= \beta_{2,1}, & \ldots & & y_2^{(n_2-1)}(b) &= \beta_{2,n_2-1} \\
&\ \vdots & &\ \vdots & \ldots & & &\ \vdots \\
y_m(b) &= \beta_{m,0}, & y_m'(b) &= \beta_{m,1}, & \cdots & & y_m^{(n_m-1)}(b) &= \beta_{m,n_m-1}
\end{aligned}
\qquad (3.72)
$$

where the α and β terms are given constants and $n_1, n_2, \ldots, n_m$ are integers.

Note that by holding $y_2, y_3, \ldots, y_m$ and their derivatives constant, this functional reduces to the functional [f3] previously considered. By letting only one function vary, one obtains an Euler-Lagrange equation to be satisfied. Do this for each function y_i, $i = 1, \ldots, m$. This gives the Euler-Lagrange system of equations which give a necessary condition for an extremum

$$
\begin{aligned}
\frac{\partial f}{\partial y_1} - \frac{d}{dx}\left(\frac{\partial f}{\partial y_1'}\right) + \frac{d^2}{dx^2}\left(\frac{\partial f}{\partial y_1''}\right) + \cdots + (-1)^{n_1}\frac{d^{n_1}}{dx^{n_1}}\left(\frac{\partial f}{\partial y_1^{(n_1)}}\right) &= 0 \\
\frac{\partial f}{\partial y_2} - \frac{d}{dx}\left(\frac{\partial f}{\partial y_2'}\right) + \frac{d^2}{dx^2}\left(\frac{\partial f}{\partial y_2''}\right) + \cdots + (-1)^{n_2}\frac{d^{n_2}}{dx^{n_2}}\left(\frac{\partial f}{\partial y_2^{(n_2)}}\right) &= 0 \\
\vdots \qquad\qquad\qquad & \\
\frac{\partial f}{\partial y_m} - \frac{d}{dx}\left(\frac{\partial f}{\partial y_m'}\right) + \frac{d^2}{dx^2}\left(\frac{\partial f}{\partial y_m''}\right) + \cdots + (-1)^{n_m}\frac{d^{n_m}}{dx^{n_m}}\left(\frac{\partial f}{\partial y_m^{(n_m)}}\right) &= 0
\end{aligned}
\qquad (3.73)
$$

Each of these equations has the same form as equation (3.64). Note that only the last index n is changed in each equation. The final result is the system of Euler-Lagrange equations (3.73) which must be solved.

[f5]: Integrand $f(t, y_1(t), y_2(t), \dot{y}_1(t), \dot{y}_2(t))$

Let t denote the independent variable and let $y_1(t)$ and $y_2(t)$ denote dependent variables and consider a functional of the form

$$I = \int_a^b f(t, y_1(t), y_2(t), \dot{y}_1(t), \dot{y}_2(t))\, dt \tag{3.74}$$

where we use the dot notation $\dot{y}_1(t) = \dfrac{dy_1}{dt}$ and $\dot{y}_2(t) = \dfrac{dy_2}{dt}$ to denote derivatives with respect to the independent variable. One can now use the variational notation to derive the Euler-Lagrange equations for determining the functions which produce an extremal value.

Here $I = I(y_1, y_2)$ and so we let $\delta_1 I$ denote a variation with respect to y_1 holding y_2 constant and let $\delta_2 I$ denote a variation with respect to y_2 while holding y_1 constant. At an extremum we require $\delta_1 I = 0$ and $\delta_2 I = 0$. This requires that the system of equations

$$\delta_1 I = \int_a^b \left[\frac{\partial f}{\partial y_1} \delta y_1 + \frac{\partial f}{\partial \dot{y}_1} \delta \dot{y}_1 \right] dt = 0 \quad \text{and} \quad \delta_2 I = \int_a^b \left[\frac{\partial f}{\partial y_2} \delta y_2 + \frac{\partial f}{\partial \dot{y}_2} \delta \dot{y}_2 \right] dt = 0 \tag{3.75}$$

be satisfied simultaneously. Each equation can be integrated by parts to obtain

$$\delta_1 I = \int_a^b \left[\frac{\partial f}{\partial y_1} - \frac{d}{dt}\left(\frac{\partial f}{\partial \dot{y}_1} \right) \right] \delta y_1\, dt = 0 \quad \text{and} \quad \delta_2 I = \int_a^b \left[\frac{\partial f}{\partial y_2} - \frac{d}{dt}\left(\frac{\partial f}{\partial \dot{y}_2} \right) \right] \delta y_2\, dt = 0 \tag{3.76}$$

One can now employ the basic lemma to obtain the Euler-Lagrange equations

$$\frac{\partial f}{\partial y_1} - \frac{d}{dt}\left(\frac{\partial f}{\partial \dot{y}_1} \right) = 0 \quad \text{and} \quad \frac{\partial f}{\partial y_2} - \frac{d}{dt}\left(\frac{\partial f}{\partial \dot{y}_2} \right) = 0 \tag{3.77}$$

which represents a system of differential equations for determining the functions y_1 and y_2. These solutions are subject to specified end point conditions. Note that this is a special case of the previous functional [f4].

Example 3-4.

Find a stationary value for the functional $I = \int_0^{\pi/2} \left[(\dot{y}_1)^2 + (\dot{y}_2)^2 + 2y_1 y_2 \right] dx$ subject to the end point conditions

$$y_1(0) = 0, \quad y_1(\pi/2) = 1, \quad y_2(0) = 0, \quad y_2(\pi/2) = -1.$$

Solution: Here $f = (\dot{y}_1)^2 + (\dot{y}_2)^2 + 2y_1 y_2$ and the Euler-Lagrange equations given by equations (3.77) are calculated

$$\frac{\partial f}{\partial y_1} - \frac{d}{dt}\left(\frac{\partial f}{\partial \dot{y}_1} \right) = 0 \Rightarrow \quad 2y_2 - \frac{d}{dt}(2\dot{y}_1) = 0, \qquad \frac{\partial f}{\partial y_2} - \frac{d}{dt}\left(\frac{\partial f}{\partial \dot{y}_2} \right) = 0 \Rightarrow \quad 2y_1 - \frac{d}{dt}(2\dot{y}_2) = 0$$

which simplifies to the system of differential equations

$$\ddot{y}_1 - y_2 = 0, \qquad \ddot{y}_2 - y_1 = 0. \tag{3.78}$$

Differentiate the first equation with respect to t twice and use the result from the second equation to obtain the fourth order ordinary differential equation

$$\frac{d^4 y_1}{dt^4} - y_1 = 0.$$

This equation has the general solution $y_1 = y_1(t) = c_1 e^t + c_2 e^{-t} + c_3 \sin t + c_4 \cos t$, where c_1, c_2, c_3, c_4 are constants. The function y_2 is obtained from equation (3.78) and one finds

$$y_2 = y_2(t) = \ddot{y}_1 = c_1 e^t + c_2 e^{-t} - c_3 \sin t - c_4 \cos t.$$

Applying the boundary conditions to these equations produces the solutions

$$y_1 = y_1(t) = \sin t, \quad \text{and} \quad y_2 = y_2(t) = -\sin t$$

for t over the interval $0 \le t \le \pi/2$. One must now investigate this solution to determine if it produces a maximum or minimum value for the integral I.

■

[f6]: Integrand $\mathbf{f(t, y_1, y_2, \ldots, y_n, \dot{y}_1, \dot{y}_2, \ldots, \dot{y}_n)}$

The previous result can be generalized to n-dependent variables $y_1, \ldots, y_n$ which are functions of time t. Consider the stationary values associated with the functional

$$I = \int_a^b f(t, y_1, y_2, \ldots, y_n, \dot{y}_1, \dot{y}_2, \ldots, \dot{y}_n) \, dt \tag{3.79}$$

where $\dot{y}_i = \frac{dy_i(t)}{dt}$ for $i = 1, \ldots, n$. This is a generalization of the functional [f5]. We use the same type of arguments that we used to determine the Euler-Lagrange system for the functional [f5] previously considered. Hold all the $y_i's$ constant except one and derive the Euler-Lagrange equation. Then repeat these arguments for each value of the index i to verify that the Euler-Lagrange equations are a system of ordinary differential equations of the form

$$\frac{\partial f}{\partial y_i} - \frac{d}{dt}\left(\frac{\partial f}{\partial \dot{y}_i}\right) = 0, \qquad \text{for} \quad i = 1, 2, \ldots, n \tag{3.80}$$

The identity

$$\frac{d}{dt}\left(\frac{\partial f}{\partial \dot{y}_1}\dot{y}_1 + \frac{\partial f}{\partial \dot{y}_2}\dot{y}_2 + \cdots + \frac{\partial f}{\partial \dot{y}_n}\dot{y}_n - f\right)$$
$$= -\dot{y}_1\left[\frac{\partial f}{\partial y_1} - \frac{d}{dt}\left(\frac{\partial f}{\partial \dot{y}_1}\right)\right] - \dot{y}_2\left[\frac{\partial f}{\partial y_2} - \frac{d}{dt}\left(\frac{\partial f}{\partial \dot{y}_2}\right)\right] - \cdots - \dot{y}_n\left[\frac{\partial f}{\partial y_n} - \frac{d}{dt}\left(\frac{\partial f}{\partial \dot{y}_n}\right)\right] - \frac{\partial f}{\partial t} \tag{3.81}$$

can be used to obtain a first integral to the system of equations (3.80) whenever the independent variable t is absent from the integrand f. In this special case the last term of the identity (3.81) becomes $\frac{\partial f}{\partial t} = 0$ and so one can integrate equation (3.81) to obtain

$$\frac{\partial f}{\partial \dot{y}_1}\dot{y}_1 + \frac{\partial f}{\partial \dot{y}_2}\dot{y}_2 + \cdots + \frac{\partial f}{\partial \dot{y}_n}\dot{y}_n - f = C = \text{a constant.} \tag{3.82}$$

Sometimes it is convenient to save space and represent the above integral using the notation $I = \int_a^b f(t, y_i, \dot{y}_i)\,dt$ for $i = 1, 2, \ldots, n$. If the system of differential equations (3.80) have nonzero solutions, then the general solutions of the system will involve 2n arbitrary constants $c_1, c_2, \ldots, c_{2n}$ and the solutions will have the form $y_i = y_i(t, c_1, c_2, \ldots, c_{2n})$ for $i = 1, \ldots, n$. The 2n constants are determined from the given boundary conditions. Note that this is also a special case of [f4].

[f7]: Integrand $\mathbf{F}\left(\mathbf{x}, \mathbf{y}, \mathbf{w}, \dfrac{\partial \mathbf{w}}{\partial \mathbf{x}}, \dfrac{\partial \mathbf{w}}{\partial \mathbf{y}}\right)$

We now consider functionals with more than one independent variable and a single dependent variable. Let $w = w(x, y)$ denote a continuous function with partial derivatives which are also continuous. We consider functionals of the form

$$I = \iint_R F(x, y, w, \frac{\partial w}{\partial x}, \frac{\partial w}{\partial y})\,dxdy \tag{3.83}$$

where we desired to find $w = w(x, y)$ which produces an extremal value.

Assume $w = w(x, y)$ is the surface which produces a stationary value for the functional I. Let R denote the region of the x, y-plane where the integration is performed. The boundary of this region is denoted using the notation ∂R. It is further assumed that $w(x, y)$ is prescribed along the boundary curve ∂R. Construct the comparison functions $W(x, y) = w(x, y) + \epsilon\eta(x, y)$, where the function $\eta(x, y)$ is selected to satisfy the boundary condition that $\eta(x, y) = 0$ for $(x, y) \in \partial R$. The situation is illustrated in the figure 3-3.

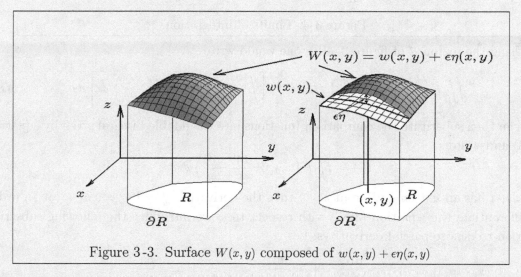

Figure 3-3. Surface $W(x, y)$ composed of $w(x, y) + \epsilon\eta(x, y)$

Note that there are many situations where the integration of an area integral is greatly simplified by a proper choice of coordinates. There may be times when it is necessary to change from an x, y-coordinate system to a u, v-coordinate system, using a change of variables $x = x(u, v)$ and $y = y(u, v)$, in order to simplify the representation of the integral I. In such

120

circumstances the integral I can be changed to the new coordinates using the relation

$$I = \iint_R F \, dxdy = \iint_{R^*} F \, \frac{\partial(x,y)}{\partial(u,v)} \, dudv \qquad (3.84)$$

where $\frac{\partial(x,y)}{\partial(u,v)}$ denotes the Jacobian of the transformation and R^* is such that the bounding curves of the region are expressed in the u,v coordinate system. We assume that such a transformation has been performed and the resulting equation is of the form given by equation (3.83). The notation $\iint_R$ is used to denote a double integral with the appropriate limits of integration over a region R. If the region is rectangular, then the limits of integration are constants. We consider the more general situation where the projected region R has one of the configurations illustrated in the figure 3-4. For the region in figure 3-4(a), one can write

$$\iint_R F\left(x,y,w,\frac{\partial w}{\partial x},\frac{\partial w}{\partial y}\right) dxdy = \int_a^b \left[\int_{y_1(x)}^{y_2(x)} F\left(x,y,w,\frac{\partial w}{\partial x},\frac{\partial w}{\partial y}\right) dy\right] dx \qquad (3.85)$$

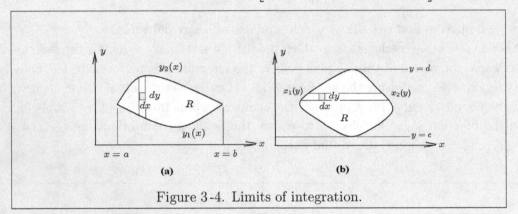

Figure 3-4. Limits of integration.

Using the region R of figure 3-4(b), one would write

$$\iint_R F\left(x,y,w,\frac{\partial w}{\partial x},\frac{\partial w}{\partial y}\right) dxdy = \int_c^d \left[\int_{x_1(y)}^{x_2(y)} F\left(x,y,w,\frac{\partial w}{\partial x},\frac{\partial w}{\partial y}\right) dx\right] dy. \qquad (3.86)$$

One can then substitute the comparison functions into the double integral given by equation (3.83) and write

$$I(\epsilon) = \iint_R F\left(x,y,W,\frac{\partial W}{\partial x},\frac{\partial W}{\partial y}\right) dxdy. \qquad (3.87)$$

If $I = I(\epsilon)$ has an extreme value at $\epsilon = 0$, then the derivative $\frac{dI}{d\epsilon} = I'(\epsilon) = 0$ at $\epsilon = 0$. In order to differentiate the equation (3.87) with respect to ϵ, we introduce the following subscript notation to denote partial derivatives. Let

$$W_x = \frac{\partial W}{\partial x}, \quad W_y = \frac{\partial W}{\partial y}, \quad \eta_x = \frac{\partial \eta}{\partial x}, \quad \eta_y = \frac{\partial \eta}{\partial y},$$

with higher derivatives denoted

$$W_{xx} = \frac{\partial^2 W}{\partial x^2}, \quad W_{xy} = \frac{\partial^2 W}{\partial x \partial y}, \quad W_{xxy} = \frac{\partial^3 W}{\partial x^2 \partial y}, \quad W_{xxxx} = \frac{\partial^4 W}{\partial x^4}, \quad \text{etc.}$$

Employing the subscript notation for denoting partial derivatives, the derivative $I'(\epsilon)$ can be represented in the form

$$\frac{dI}{d\epsilon} = I'(\epsilon) = \iint_R \left(\frac{\partial F}{\partial W}\eta + \frac{\partial F}{\partial W_x}\eta_x + \frac{\partial F}{\partial W_y}\eta_y \right) dxdy.$$

At a stationary value we require that the derivative of I with respect to ϵ satisfy

$$I'(0) = \iint_R \left(\frac{\partial F}{\partial w}\eta + \frac{\partial F}{\partial w_x}\eta_x + \frac{\partial F}{\partial w_y}\eta_y \right) dxdy = 0. \qquad (3.88)$$

One can now employ the Green's theorem

$$\iint_R \left(\frac{\partial Q}{\partial x} - \frac{\partial P}{\partial y} \right) dxdy = \oint_{\partial R} (P\,dx + Q\,dy), \qquad (3.89)$$

where ∂R denotes the boundary curve around the region R. In order to simplify the necessary condition given by equation (3.88) we make the substitutions

$$Q = \eta \frac{\partial F}{\partial w_x} \quad \text{and} \quad P = -\eta \frac{\partial F}{\partial w_y}$$

into the Green's theorem (1.46) to obtain

$$\iint_R \left[\frac{\partial \left(\eta \frac{\partial F}{\partial w_x} \right)}{\partial x} + \frac{\partial \left(\eta \frac{\partial F}{\partial w_y} \right)}{\partial y} \right] dxdy = \oint_{\partial R} \eta(x,y) \left(F_{w_x}\,dy - F_{w_y}\,dx \right). \qquad (3.90)$$

Expand the left-hand side of equation (3.90) to obtain the result

$$\begin{aligned} \iint_R \left(\frac{\partial F}{\partial w_x}\eta_x + \frac{\partial F}{\partial w_y}\eta_y \right) dxdy = &-\iint_R \eta \left[\frac{\partial}{\partial x}\left(\frac{\partial F}{\partial w_x} \right) + \frac{\partial}{\partial y}\left(\frac{\partial F}{\partial w_y} \right) \right] dxdy \\ &+ \oint_{\partial R} \eta \left(F_{w_x}\,dy - F_{w_y}\,dx \right). \end{aligned} \qquad (3.91)$$

Now substitute this result in place of the last two terms on the right-hand side of equation (3.88) to obtain the condition

$$\begin{aligned} I'(0) = &\iint_R \left[\frac{\partial F}{\partial w} - \frac{\partial}{\partial x}\left(\frac{\partial F}{\partial w_x} \right) - \frac{\partial}{\partial y}\left(\frac{\partial F}{\partial w_y} \right) \right] \eta(x,y)\,dxdy \\ &+ \oint_{\partial R} \eta \left(F_{w_x}\,dy - F_{w_y}\,dx \right) = 0. \end{aligned} \qquad (3.92)$$

The line integral term in equation (3.92) is zero because we required that $\eta \big|_{x,y \in \partial R} = 0$. One can employ a form of the basic lemma on the remaining integral to obtain the necessary condition

$$\frac{\partial F}{\partial w} - \frac{\partial}{\partial x}\left(\frac{\partial F}{\partial w_x} \right) - \frac{\partial}{\partial y}\left(\frac{\partial F}{\partial w_y} \right) = 0 \qquad (3.93)$$

which is known as the Euler-Lagrange necessary condition for a stationary value associated with the functional given by equation (3.83). Note that sometimes y is replaced by a time variable t in the representation of the functional [f7].

In the case the boundary condition is not prescribed on ∂R, then in order for the right-hand side of equation (3.92) to be zero we require that

$$\frac{\partial F}{\partial w_x}\frac{dy}{ds} - \frac{\partial F}{\partial w_y}\frac{dx}{ds} \bigg|_{x,y \in \partial R} = 0$$

where s denotes arc length along the boundary curve.

Example 3-5. **Laplace's equation** Find the Euler-Lagrange equation associated with the functional

$$I = \iint_R \left[\left(\frac{\partial w}{\partial x} \right)^2 + \left(\frac{\partial w}{\partial y} \right)^2 \right] dxdy \tag{3.94}$$

Solution: Here $w = w(x,y)$ and $F = F(x,y,w,w_x,w_y) = \left(\frac{\partial w}{\partial x} \right)^2 + \left(\frac{\partial w}{\partial y} \right)^2$ with partial derivatives

$$F_w = \frac{\partial F}{\partial w} = 0, \quad F_{w_x} = \frac{\partial F}{\partial w_x} = 2w_x, \quad F_{w_y} = \frac{\partial F}{\partial w_y} = 2w_y.$$

The Euler-Lagrange equation

$$F_w - \frac{\partial}{\partial x}(F_{w_x}) - \frac{\partial}{\partial y}(F_{w_y}) = 0 \quad \text{becomes} \quad \frac{\partial^2 w}{\partial x^2} + \frac{\partial^2 w}{\partial y^2} = \nabla^2 w = 0$$

which is the well known Laplace equation.

Note that the Euler-Lagrange equation associated with the functional

$$I = \iint_R \left[(w_x)^2 + (w_y)^2 + 2wf(x,y) \right] dxdy$$

is the Poisson equation $\nabla^2 w = f(x,y)$. This result is left as an exercise. ∎

Example 3-6. **Laplace's equation**

Convert the functional in the previous example to polar coordinates and hence determine the Laplace equation in polar coordinates.

Solution: From the transformation equations $x = r\cos\theta$ and $y = r\sin\theta$, solve for r^2 and $\tan\theta$ to obtain $x^2 + y^2 = r^2$ and $\tan\theta = y/x$. Differentiate these equations to obtain the derivatives

$$\frac{\partial r}{\partial x} = \frac{x}{r} = \cos\theta, \qquad \frac{\partial r}{\partial y} = \frac{y}{r} = \sin\theta, \qquad \frac{\partial \theta}{\partial x} = \frac{-y}{r^2} = \frac{-\sin\theta}{r}, \qquad \frac{\partial \theta}{\partial y} = \frac{x}{r^2} = \frac{\cos\theta}{r}$$

Treat w as a function of r and θ and write $w = w(r,\theta)$. Using chain rule differentiation verify the partial derivatives

$$\frac{\partial w}{\partial x} = \frac{\partial w}{\partial r}\frac{\partial r}{\partial x} + \frac{\partial w}{\partial \theta}\frac{\partial \theta}{\partial x} = \frac{\partial w}{\partial r}\cos\theta - \frac{\partial w}{\partial \theta}\frac{\sin\theta}{r},$$

$$\frac{\partial w}{\partial y} = \frac{\partial w}{\partial r}\frac{\partial r}{\partial y} + \frac{\partial w}{\partial \theta}\frac{\partial \theta}{\partial y} = \frac{\partial w}{\partial r}\sin\theta + \frac{\partial w}{\partial \theta}\frac{\cos\theta}{r}$$

The element of area transforms according to the relation $dxdy = \frac{\partial(x,y)}{\partial(r,\theta)} drd\theta$ where the Jacobian of the transformation is

$$\frac{\partial(x,y)}{\partial(r,\theta)} = \begin{vmatrix} \frac{\partial x}{\partial r} & \frac{\partial x}{\partial \theta} \\ \frac{\partial y}{\partial r} & \frac{\partial y}{\partial \theta} \end{vmatrix} = \begin{vmatrix} \cos\theta & -r\sin\theta \\ \sin\theta & r\cos\theta \end{vmatrix} = r.$$

Substitute the above quantities into the functional given by equation (3.94) and verify that it transforms to

$$I = \iint_R \left\{ \left[\frac{\partial w}{\partial r}\cos\theta - \frac{\partial w}{\partial \theta}\frac{\sin\theta}{r} \right]^2 + \left[\frac{\partial w}{\partial r}\sin\theta + \frac{\partial w}{\partial \theta}\frac{\cos\theta}{r} \right]^2 \right\} rdrd\theta.$$

This functional now has the form $I = \iint_R F(r, \theta, w, w_r, w_\theta) \, dr d\theta$ with the Euler-Lagrange equation given by

$$\frac{\partial F}{\partial w} - \frac{\partial}{\partial r}\left(\frac{\partial F}{\partial w_r}\right) - \frac{\partial}{\partial \theta}\left(\frac{\partial F}{\partial w_\theta}\right) = 0.$$

It is left as an exercise to verify that the Euler-Lagrange equation reduces to the polar form of Laplace's equation, given by

$$r^2 \frac{\partial^2 w}{\partial r^2} + r \frac{\partial w}{\partial r} + \frac{\partial^2 w}{\partial \theta^2} = 0.$$

∎

[f8]: Integrand $\mathbf{F}\left(\mathbf{u}, \mathbf{v}, \mathbf{x}, \mathbf{y}, \mathbf{z}, \dfrac{\partial \mathbf{x}}{\partial \mathbf{u}}, \dfrac{\partial \mathbf{x}}{\partial \mathbf{v}}, \dfrac{\partial \mathbf{y}}{\partial \mathbf{u}}, \dfrac{\partial \mathbf{y}}{\partial \mathbf{v}}, \dfrac{\partial \mathbf{z}}{\partial \mathbf{u}}, \dfrac{\partial \mathbf{z}}{\partial \mathbf{v}}\right)$

Let u and v denote independent variables and let $x = x(u,v)$, $y = y(u,v)$ and $z = z(u,v)$ denote continuous functions of these variables with derivatives through the second order which are also continuous. We consider functionals of the form

$$I = I(x,y,z) = \iint_{R_{uv}} F\left(u, v, x, y, z, \frac{\partial x}{\partial u}, \frac{\partial x}{\partial v}, \frac{\partial y}{\partial u}, \frac{\partial y}{\partial v}, \frac{\partial z}{\partial u}, \frac{\partial z}{\partial v}\right) du dv \tag{3.95}$$

where R_{uv} denotes the region of integration with respect to the parameters u, v. If we hold y and z constant, then the integrand given by equation (3.95) reduces to a form of the previous integrand given by equation (3.83) with the symbols changed. Analogous results are obtained when holding x and z constant, or when holding x and y constant. One finds that for an extremal to occur a necessary condition to be satisfied is for x, y and z to satisfy the system of simultaneous equations

$$\begin{aligned}
\frac{\partial F}{\partial x} - \frac{\partial}{\partial u}\left(\frac{\partial F}{\partial x_u}\right) - \frac{\partial}{\partial v}\left(\frac{\partial F}{\partial x_v}\right) &= 0 \\
\frac{\partial F}{\partial y} - \frac{\partial}{\partial u}\left(\frac{\partial F}{\partial y_u}\right) - \frac{\partial}{\partial v}\left(\frac{\partial F}{\partial y_v}\right) &= 0 \\
\frac{\partial F}{\partial z} - \frac{\partial}{\partial u}\left(\frac{\partial F}{\partial z_u}\right) - \frac{\partial}{\partial v}\left(\frac{\partial F}{\partial z_v}\right) &= 0
\end{aligned} \tag{3.96}$$

[f9]: Integrand $\mathbf{F}\left(\mathbf{x}, \mathbf{y}, \mathbf{z}, \mathbf{w}, \dfrac{\partial \mathbf{w}}{\partial \mathbf{x}}, \dfrac{\partial \mathbf{w}}{\partial \mathbf{y}}, \dfrac{\partial \mathbf{w}}{\partial \mathbf{z}}\right)$

For x, y, z independent variables and $w = w(x,y,z)$ a continuous function with derivatives through the second order which are also continuous, we consider functionals having the form

$$I = \iiint_V F(x, y, z, w, w_x, w_y, w_z) \, dx dy dz \tag{3.97}$$

where V is a volume enclosed by a surface S. The surface S is the boundary of the region V and one can express this using the notation $S = \partial V$. Here we treat x, y and z as independent variables with w a function of x, y, z to be determined such that I is an extremum. It is further assumed that $w(x,y,z)$ is specified for $(x,y,z) \in S = \partial V$. Assume $w = w(x,y,z)$ is the function which produces an extremum for the given integral I and then construct the comparison functions $W = W(x, y, z) = w(x, y, z) + \epsilon\eta(x, y, z)$, where $\eta = 0$ for $(x, y, z) \in S$, and substitute this function into the integral (3.97) to obtain

$$I = I(\epsilon) = \iiint_V F(x, y, z, W, W_x, W_y, W_z) \, dx dy dz. \tag{3.98}$$

One can verify the derivative

$$I'(\epsilon) = \iiint_V \left(\frac{\partial F}{\partial W}\eta + \frac{\partial F}{\partial W_x}\eta_x + \frac{\partial F}{\partial W_y}\eta_y + \frac{\partial F}{\partial W_z}\eta_z \right) dx dy dz.$$

At a stationary value $I'(0) = 0$ which requires that

$$I'(0) = \iiint_V \left(\frac{\partial F}{\partial w}\eta + \frac{\partial F}{\partial w_x}\eta_x + \frac{\partial F}{\partial w_y}\eta_y + \frac{\partial F}{\partial w_z}\eta_z \right) dx dy dz = 0. \tag{3.99}$$

The Gauss divergence theorem can be used to simplify this integral. Recall the Gauss divergence theorem can be written

$$\iint_V \operatorname{div} \vec{F} \, d\tau = \iint_S \vec{F} \cdot \hat{n} \, d\sigma \tag{3.100}$$

where $\vec{F} = P(x, y, z)\,\hat{e}_1 + Q(x, y, z)\,\hat{e}_2 + R(x, y, z)\,\hat{e}_3$ represents a continuous vector field, $d\tau = dx dy dz$ represents an element of volume, $\hat{n} = n_1\,\hat{e}_1 + n_2\,\hat{e}_2 + n_3\,\hat{e}_3$ represents an outward unit normal vector to the bounding surface S, and $d\sigma$ represents an element of surface area. An equivalent form for the Gauss divergence theorem is given by

$$\iiint_V \left(\frac{\partial P}{\partial x} + \frac{\partial Q}{\partial y} + \frac{\partial R}{\partial z} \right) d\tau = \iint_S (Pn_1 + Qn_2 + Rn_3) \, d\sigma \tag{3.101}$$

In the Gauss divergence theorem, given by the equation (3.101), let

$$P = \eta\frac{\partial F}{\partial w_x}, \qquad Q = \eta\frac{\partial F}{\partial w_y}, \qquad R = \eta\frac{\partial F}{\partial w_z}$$

to obtain the result

$$\iiint_V \left[\frac{\partial\left(\eta\frac{\partial F}{\partial w_x}\right)}{\partial x} + \frac{\partial\left(\eta\frac{\partial F}{\partial w_y}\right)}{\partial y} + \frac{\partial\left(\eta\frac{\partial F}{\partial w_x}\right)}{\partial z} \right] d\tau = \iint_S \eta \left[\frac{\partial F}{\partial w_x}n_1 + \frac{\partial F}{\partial w_y}n_2 + \frac{\partial F}{\partial w_z}n_3 \right] d\sigma. \tag{3.102}$$

Expanding the left-hand side of the equation (3.102) produces the result

$$\iiint_V \left[\frac{\partial F}{\partial w_x}\eta_x + \frac{\partial F}{\partial w_y}\eta_y + \frac{\partial F}{\partial w_z}\eta_z \right] dx dy dz =$$
$$- \iint_V \left[\frac{\partial}{\partial x}\left(\frac{\partial F}{\partial w_x}\right) + \frac{\partial}{\partial y}\left(\frac{\partial F}{\partial w_y}\right) + \frac{\partial}{\partial z}\left(\frac{\partial F}{\partial w_z}\right) \right] \eta \, dx dy dz \tag{3.103}$$
$$+ \iint_S \eta \left[\frac{\partial F}{\partial w_x}n_1 + \frac{\partial F}{\partial w_y}n_2 + \frac{\partial F}{\partial w_z}n_3 \right] d\sigma$$

Now substitute the result from equation (3.103) into the equation (3.99) to obtain the necessary condition for a stationary value that

$$
\begin{aligned}
I'(0) = \iiint_V \eta &\left[\frac{\partial F}{\partial w} - \frac{\partial}{\partial x}\left(\frac{\partial F}{\partial w_x} \right) - \frac{\partial}{\partial y}\left(\frac{\partial F}{\partial w_y} \right) - \frac{\partial}{\partial z}\left(\frac{\partial F}{\partial w_z} \right) \right] dx\,dy\,dz \\
&+ \iint_S \eta \left[\frac{\partial F}{\partial w_x}n_1 + \frac{\partial F}{\partial w_y}n_2 + \frac{\partial F}{\partial w_z}n_3 \right] d\sigma = 0.
\end{aligned}
\tag{3.104}
$$

The surface integral term is zero because we require that $\eta = 0$ for $(x, y, z) \in S$. For arbitrary values of η one can employ a form of the basic lemma and obtain the necessary condition for an extremal

$$
\frac{\partial F}{\partial w} - \frac{\partial}{\partial x}\left(\frac{\partial F}{\partial w_x} \right) - \frac{\partial}{\partial y}\left(\frac{\partial F}{\partial w_y} \right) - \frac{\partial}{\partial z}\left(\frac{\partial F}{\partial w_z} \right) = 0.
\tag{3.105}
$$

[f10]: Integrand $F\left(x_1, x_2, \ldots, x_n, w, w_{x_1}, w_{x_2}, \ldots, w_{x_n}\right)$

The previous two functionals can be generalized to multiple integrals over n-dimensional regions V_n with boundary surface ∂V_n. Consider the functional

$$
I = \int \cdots \int_{V_n} F\left(x_1, x_2, \ldots, x_n, w, w_{x_1}, w_{x_2}, \ldots, w_{x_n}\right) dx_1 dx_2 \cdots dx_n
\tag{3.106}
$$

where w is specified for $(x_1, x_2, \ldots, x_n) \in \partial V_n$. It can be demonstrated that in order for this functional to have an extremal value it is necessary for the Euler-Lagrange condition

$$
\frac{\partial F}{\partial w} - \frac{\partial}{\partial x_1}\left(\frac{\partial F}{\partial w_{x_1}} \right) - \frac{\partial}{\partial x_2}\left(\frac{\partial F}{\partial w_{x_2}} \right) - \cdots - \frac{\partial}{\partial x_n}\left(\frac{\partial F}{\partial w_{x_n}} \right) = 0
\tag{3.107}
$$

to be satisfied. The proof can be accomplished by using a generalized form of Green's theorem applied to n-dimensions, or alternatively, one can use integration by parts to derive the above result.

Example 3-7. Find the equation that $w = w(x, y, z)$ must satisfy in order for the functional

$$
I = \iiint_V \left[\left(\frac{\partial w}{\partial x} \right)^2 + \left(\frac{\partial w}{\partial y} \right)^2 + \left(\frac{\partial w}{\partial z} \right)^2 \right] dx\,dy\,dz
$$

to have an extremum.

Solution:

The Euler-Lagrange necessary condition for w to produce a stationary value is given by the equation (3.107). Boundary conditions are assumed to be specified on the boundary ∂V. Here the integrand is given by

$$
F = \left(\frac{\partial w}{\partial x} \right)^2 + \left(\frac{\partial w}{\partial y} \right)^2 + \left(\frac{\partial w}{\partial z} \right)^2
$$

from which the following derivatives are obtained

$$
\frac{\partial F}{\partial w} = 0, \quad \frac{\partial}{\partial x}\left(\frac{\partial F}{\partial w_x} \right) = 2w_{xx}, \quad \frac{\partial}{\partial y}\left(\frac{\partial F}{\partial w_y} \right) = 2w_{yy}, \quad \frac{\partial}{\partial z}\left(\frac{\partial F}{\partial w_z} \right) = 2w_{zz}.
$$

126

The Euler-Lagrange equation then takes the form of Laplace's equation

$$\nabla^2 w = \frac{\partial^2 w}{\partial x^2} + \frac{\partial^2 w}{\partial y^2} + \frac{\partial^2 w}{\partial z^2} = 0.$$

Hence a necessary condition that the function w produce an extremum is that w must satisfy the Laplace equation over the region of integration and satisfy the given boundary conditions.

■

[f11]: Integrand $F\left(t, x, y, z, u, v, w, u_t, u_x, u_y, u_z, v_t, v_x, v_y, v_z, w_t, w_x, w_y, w_z\right)$

The functional

$$I = \int_{t_1}^{t_2} \iiint_V F\left(t, x, y, z, u, v, w, u_t, u_x, u_y, u_z, v_t, v_x, v_y, v_z, w_t, w_x, w_y, w_z\right) dx dy dz\, dt \tag{3.108}$$

can be reduced to the previous problem by say holding v and w constant. This functional then becomes a special case of the previous functional [f10]. One can then consider the separate cases where only u varies, only v varies and only w varies to obtain the Euler-Lagrange system of partial differential equations.

$$\frac{\partial F}{\partial u} - \frac{\partial}{\partial t}\left(\frac{\partial F}{\partial u_t}\right) - \frac{\partial}{\partial x}\left(\frac{\partial F}{\partial u_x}\right) - \frac{\partial}{\partial y}\left(\frac{\partial F}{\partial u_y}\right) - \frac{\partial}{\partial z}\left(\frac{\partial F}{\partial u_z}\right) = 0$$

$$\frac{\partial F}{\partial v} - \frac{\partial}{\partial t}\left(\frac{\partial F}{\partial v_t}\right) - \frac{\partial}{\partial x}\left(\frac{\partial F}{\partial v_x}\right) - \frac{\partial}{\partial y}\left(\frac{\partial F}{\partial v_y}\right) - \frac{\partial}{\partial z}\left(\frac{\partial F}{\partial v_z}\right) = 0$$

$$\frac{\partial F}{\partial w} - \frac{\partial}{\partial t}\left(\frac{\partial F}{\partial w_t}\right) - \frac{\partial}{\partial x}\left(\frac{\partial F}{\partial w_x}\right) - \frac{\partial}{\partial y}\left(\frac{\partial F}{\partial w_y}\right) - \frac{\partial}{\partial z}\left(\frac{\partial F}{\partial w_z}\right) = 0$$

[f12]: Integrand $F\left(x, y, z, \dfrac{\partial z}{\partial x}, \dfrac{\partial z}{\partial y}, \dfrac{\partial^2 z}{\partial x^2}, \dfrac{\partial^2 z}{\partial x \partial y}, \dfrac{\partial^2 z}{\partial y^2}\right)$

We consider cases where z is a function of x and y which has continuous derivatives through the second order. We examine a functional which depends upon these higher ordered derivatives. We proceed exactly as we have done for functionals with lower ordered derivatives. Assume that $z = z(x, y)$ is the function which produces an extremum for the integral

$$\iint_R F\left(x, y, z, \frac{\partial z}{\partial x}, \frac{\partial z}{\partial y}, \frac{\partial^2 z}{\partial x^2}, \frac{\partial^2 z}{\partial x \partial y}, \frac{\partial^2 z}{\partial y^2}\right) dx dy \tag{3.109}$$

where the integrations over R are similar to the equations (3.85) and (3.86) previously considered. Assume that $z(x, y)$ is prescribed along the boundary curve ∂R and then construct a comparison function $Z = z(x, y) + \epsilon \eta(x, y)$ and substitute this comparison function into the functional (3.109) to obtain

$$I = I(\epsilon) = \iint_R F(x, y, Z, Z_x, Z_y, Z_{xx}, Z_{xy}, Z_{yy})\, dx dy \tag{3.110}$$

By assumption $I(\epsilon)$ has an extremum at $\epsilon = 0$ and so we require that $\left.\frac{dI}{d\epsilon}\right|_{\epsilon=0} = 0$. We find

$$\frac{dI}{d\epsilon} = I'(\epsilon) = \iint_R \left(\frac{\partial F}{\partial Z}\eta + \frac{\partial F}{\partial Z_x}\eta_x + \frac{\partial F}{\partial Z_y}\eta_y + \frac{\partial F}{\partial Z_{xx}}\eta_{xx} + \frac{\partial F}{\partial Z_{xy}}\eta_{xy} + \frac{\partial F}{\partial Z_{yy}}\eta_{yy}\right) dx dy \tag{3.111}$$

so that a necessary condition for $z = z(x, y)$ to produce an extremum is for

$$I'(0) = \iint_R \left(\frac{\partial F}{\partial z}\eta + \frac{\partial F}{\partial z_x}\eta_x + \frac{\partial F}{\partial z_y}\eta_y + \frac{\partial F}{\partial z_{xx}}\eta_{xx} + \frac{\partial F}{\partial z_{xy}}\eta_{xy} + \frac{\partial F}{\partial z_{yy}}\eta_{yy} \right) dxdy = 0. \tag{3.112}$$

One can employ the Green's theorem to represent this necessary condition in an alternative form. The second and third terms in the integrand of equation (3.112) can be integrated by substituting

$$Q = \eta\frac{\partial F}{\partial z_x} \quad \text{and} \quad P = -\eta\frac{\partial F}{\partial z_y}$$

into the Green's theorem equation (1.45) to obtain

$$\iint_R \left[\frac{\partial}{\partial x}\left(\eta\frac{\partial F}{\partial z_x} \right) + \frac{\partial}{\partial y}\left(\eta\frac{\partial F}{\partial z_y} \right) \right] dxdy = \oint_{\partial R} \eta\left[\frac{\partial F}{\partial z_x}dy - \frac{\partial F}{\partial z_y}dx \right] \tag{3.113}$$

The left-hand side of equation (3.113) can then be expanded to obtain the result

$$\iint_R \left(\frac{\partial F}{\partial z_x}\eta_x + \frac{\partial F}{\partial z_y}\eta_y \right) dxdy =$$
$$- \iint_R \eta\left[\frac{\partial}{\partial x}\left(\frac{\partial F}{\partial z_x} \right) + \frac{\partial}{\partial y}\left(\frac{\partial F}{\partial z_y} \right) \right] dxdy + \oint_{\partial R} \eta\left[\frac{\partial F}{\partial z_x}dy - \frac{\partial F}{\partial z_y}dx \right] \tag{3.114}$$

The Green's theorem can also be used to evaluate the integral of the last three terms in the integrand of equation (3.112). Consider the following special case of the Green's theorem

$$\iint_R \left(\frac{\partial Q}{\partial x} - \frac{\partial P}{\partial y} \right) dxdy = \oint_{\partial R} P\,dx + Q\,dy \tag{3.115}$$

where Q and P have the following special values

$$Q = \frac{\partial F}{\partial z_{xx}}\eta_x - \frac{\partial}{\partial x}\left(\frac{\partial F}{\partial z_{xx}} \right)\eta + \frac{\partial F}{\partial z_{xy}}\eta_y$$
$$P = \frac{\partial}{\partial y}\left(\frac{\partial F}{\partial z_{yy}} \right)\eta - \frac{\partial F}{\partial z_{yy}}\eta_y + \frac{\partial}{\partial x}\left(\frac{\partial F}{\partial z_{xy}} \right)\eta \tag{3.116}$$

Performing the necessary derivatives and simplifying one obtains the result

$$\iint_R \left[\frac{\partial F}{\partial z_{xx}}\eta_{xx} + \frac{\partial F}{\partial z_{xy}}\eta_{xy} + \frac{\partial F}{\partial z_{yy}}\eta_{yy} \right] dxdy =$$
$$\iint_R \eta\left[\frac{\partial^2}{\partial x^2}\left(\frac{\partial F}{\partial z_{xx}} \right) + \frac{\partial^2}{\partial x\partial y}\left(\frac{\partial F}{\partial z_{xy}} \right) + \frac{\partial^2}{\partial y^2}\left(\frac{\partial F}{\partial z_{yy}} \right) \right] dxdy$$
$$+ \oint_{\partial R} \left[\frac{\partial}{\partial y}\left(\frac{\partial F}{\partial z_{yy}} \right)\eta - \frac{\partial F}{\partial z_{yy}}\eta_y + \frac{\partial}{\partial x}\left(\frac{\partial F}{\partial z_{xy}} \right)\eta \right] dx$$
$$+ \oint_{\partial R} \left[\frac{\partial F}{\partial z_{xx}}\eta_x - \frac{\partial}{\partial x}\left(\frac{\partial F}{\partial z_{xx}} \right)\eta + \frac{\partial F}{\partial z_{xy}}\eta_y \right] dy \tag{3.117}$$

Substitute the results from equations (3.114) and (3.117) into the necessary condition given by equation (3.112) and simplify the results to obtain

$$I'(0) = \iint_R \eta\left[\frac{\partial F}{\partial z} - \frac{\partial}{\partial x}\left(\frac{\partial F}{\partial z_x} \right) - \frac{\partial}{\partial y}\left(\frac{\partial F}{\partial z_y} \right) + \frac{\partial^2}{\partial x^2}\left(\frac{\partial F}{\partial z_{xx}} \right) + \frac{\partial^2}{\partial x\partial y}\left(\frac{\partial F}{\partial z_{xy}} \right) + \frac{\partial^2}{\partial y^2}\left(\frac{\partial F}{\partial z_{yy}} \right) \right] dxdy$$
$$+ \oint_{\partial R} \left[\eta\frac{\partial F}{\partial z_x} + \frac{\partial F}{\partial z_{zz}}\eta_x - \frac{\partial}{\partial x}\left(\frac{\partial F}{\partial z_{xx}} \right)\eta + \frac{\partial F}{\partial z_{xy}}\eta_y \right] dy$$
$$+ \oint_{\partial R} \left[-\frac{\partial F}{\partial z_y}\eta + \frac{\partial}{\partial y}\left(\frac{\partial F}{\partial z_{yy}} \right)\eta - \frac{\partial F}{\partial z_{yy}}\eta_y + \frac{\partial}{\partial x}\left(\frac{\partial F}{\partial z_{xy}} \right)\eta \right] dx = 0$$
$$\tag{3.118}$$

To make the line integral terms vanish for arbitrary η, we require $\eta = 0$ on the boundary together with $\eta_x = 0$ and $\eta_y = 0$ on the boundary. The final result that we obtain, after employing a form of the basic lemma, is that to have an extreme value for the functional given by equation (3.109), the function $z = z(x, y)$ must satisfy the Euler-Lagrange equation necessary condition

$$\frac{\partial F}{\partial z} - \frac{\partial}{\partial x}\left(\frac{\partial F}{\partial z_x}\right) - \frac{\partial}{\partial y}\left(\frac{\partial F}{\partial z_y}\right) + \frac{\partial^2}{\partial x^2}\left(\frac{\partial F}{\partial z_{xx}}\right) + \frac{\partial^2}{\partial x \partial y}\left(\frac{\partial F}{\partial z_{xy}}\right) + \frac{\partial^2}{\partial y^2}\left(\frac{\partial F}{\partial z_{yy}}\right) = 0 \qquad (3.119)$$

The resulting equation is a fourth-order partial differential equation which defines a family of functions $z = z(x, y)$ which produce an extremum.

Example 3-8. Find the Euler-Lagrange equation associated with the functional

$$I = \iint_R \left(\left(\frac{\partial^2 z}{\partial x^2}\right)^2 + \left(\frac{\partial^2 z}{\partial y^2}\right)^2 + 2\left(\frac{\partial^2 z}{\partial x \partial y}\right)^2\right) dx\, dy,$$

Solution: Here the integrand is given by

$$F = \left(\frac{\partial^2 z}{\partial x^2}\right)^2 + \left(\frac{\partial^2 z}{\partial y^2}\right)^2 + 2\left(\frac{\partial^2 z}{\partial x \partial y}\right)^2.$$

One can substitute into the Euler-Lagrange equation (3.119) the following derivatives

$$\frac{\partial F}{\partial z} = 0, \qquad \frac{\partial F}{\partial z_x} = 0, \qquad \frac{\partial F}{\partial z_y} = 0$$

$$\frac{\partial F}{\partial z_{xx}} = 2z_{xx}, \qquad \frac{\partial F}{\partial z_{xy}} = 4z_{xy}, \qquad \frac{\partial F}{\partial z_{yy}} = 2z_{yy}$$

The resulting Euler-Lagrange equation (3.119) becomes

$$\frac{\partial^4 z}{\partial x^4} + 2\frac{\partial^4 z}{\partial x^2 \partial y^2} + \frac{\partial^4 z}{\partial y^4} = 0$$

which is known as the biharmonic equation. It can be denoted using the alternative notations

$$\nabla^4 z = \nabla^2 \nabla^2 z = 0.$$

■

[f13]: Integrand $\mathbf{F}\left(\mathbf{x, y, u, v}, \dfrac{\partial \mathbf{u}}{\partial \mathbf{x}}, \dfrac{\partial \mathbf{u}}{\partial \mathbf{y}}, \dfrac{\partial \mathbf{v}}{\partial \mathbf{x}}, \dfrac{\partial \mathbf{v}}{\partial \mathbf{y}}\right)$

Let $u = u(x, y)$ and $v = v(x, y)$ denote functions of x and y which possess derivatives through the second order. We consider functionals of the form

$$I = \iint_R F\left(x, y, u, v, \frac{\partial u}{\partial x}, \frac{\partial u}{\partial y}, \frac{\partial v}{\partial x}, \frac{\partial v}{\partial y}\right) dx\, dy \qquad (3.120)$$

where R is a region of the x, y-plane. We desire to find functions $u = u(x, y)$ and $v = v(x, y)$ such that I is maximized or minimized. Further assume, that the values of u and v are prescribed upon the boundary curve $C = \partial R$ associated with the region R. One can construct the comparison functions

$$U(x,y) = u(x,y) + \epsilon\eta(x,y), \qquad \text{and} \qquad V(x,y) = v(x,y) + \epsilon\zeta(x,y)$$

and form the functional

$$I(\epsilon) = \iint_R F\left(x,y,U,V,\frac{\partial U}{\partial x},\frac{\partial U}{\partial y},\frac{\partial V}{\partial x},\frac{\partial V}{\partial y}\right)\,dxdy.$$

If one assumes this functional has an extremum when $\epsilon = 0$, then it is necessary that $I'(\epsilon)$ equal zero at $\epsilon = 0$. One can verify the derivative

$$\frac{dI}{d\epsilon} = I'(\epsilon) = \iint_R \left[\left(\frac{\partial F}{\partial U}\eta + \frac{\partial F}{\partial U_x}\eta_x + \frac{\partial F}{\partial U_y}\eta_y\right) + \left(\frac{\partial F}{\partial V}\zeta + \frac{\partial F}{\partial V_x}\zeta_x + \frac{\partial F}{\partial V_y}\zeta_y\right)\right]\,dxdy \qquad (3.121)$$

so that when $\epsilon = 0$ one obtains

$$\frac{dI}{d\epsilon}\bigg|_{\epsilon=0} = I'(0) = \iint_R \left[\left(\frac{\partial F}{\partial u}\eta + \frac{\partial F}{\partial u_x}\eta_x + \frac{\partial F}{\partial u_y}\eta_y\right) + \left(\frac{\partial F}{\partial v}\zeta + \frac{\partial F}{\partial v_x}\zeta_x + \frac{\partial F}{\partial v_y}\zeta_y\right)\right]\,dxdy \qquad (3.122)$$

We can apply the Green's theorem to simplify this expression to the form

$$I'(0) = \iint_R \left\{\left[\frac{\partial F}{\partial u} - \frac{\partial}{\partial x}\left(\frac{\partial F}{\partial u_x}\right) - \frac{\partial}{\partial y}\left(\frac{\partial F}{\partial u_y}\right)\right]\eta + \left[\frac{\partial F}{\partial v} - \frac{\partial}{\partial x}\left(\frac{\partial F}{\partial v_x}\right) - \frac{\partial}{\partial y}\left(\frac{\partial F}{\partial v_y}\right)\right]\zeta\right\}\,dxdy = 0$$

The variations η and ζ are independent of each other and so each term within the brackets must equal zero. This gives the Euler-Lagrange system of equations

$$\frac{\partial F}{\partial u} - \frac{\partial}{\partial x}\left(\frac{\partial F}{\partial u_x}\right) - \frac{\partial}{\partial y}\left(\frac{\partial F}{\partial u_y}\right) = 0, \qquad \frac{\partial F}{\partial v} - \frac{\partial}{\partial x}\left(\frac{\partial F}{\partial v_x}\right) - \frac{\partial}{\partial y}\left(\frac{\partial F}{\partial v_y}\right) = 0 \qquad (3.123)$$

which represent a system of two partial differential equations to be solved over the region R subject to the prescribed conditions for u and v on the boundary of the region R.

Note that x and y are independent variables and the terms u,v,u_x,v_x,u_y,v_y are treated as functions of x and y. The derivative terms in the equation (3.123) are found using chain rule differentiation. For example,

$$\frac{\partial}{\partial x}\left(\frac{\partial F}{\partial u_x}\right) = \frac{\partial^2 F}{\partial u_x \partial x} + \frac{\partial^2 F}{\partial u_x \partial u}\frac{\partial u}{\partial x} + \frac{\partial^2 F}{\partial u_x \partial v}\frac{\partial v}{\partial x} + \frac{\partial^2 F}{\partial u_x{}^2}\frac{\partial u_x}{\partial x}$$
$$+ \frac{\partial^2 F}{\partial u_x \partial u_y}\frac{\partial u_y}{\partial x} + \frac{\partial^2 F}{\partial u_x \partial v_x}\frac{\partial v_x}{\partial x} + \frac{\partial^2 F}{\partial u_x \partial v_y}\frac{\partial v_y}{\partial x}$$

which simplifies to

$$\frac{\partial}{\partial x}\left(\frac{\partial F}{\partial u_x}\right) = \frac{\partial^2 F}{\partial u_x{}^2}\frac{\partial^2 u}{\partial x^2} + \frac{\partial^2 F}{\partial u_x \partial u_y}\frac{\partial^2 u}{\partial x \partial y} + \frac{\partial^2 F}{\partial u_x \partial v_x}\frac{\partial^2 v}{\partial x^2} + \frac{\partial^2 F}{\partial u_x \partial v_y}\frac{\partial^2 v}{\partial x \partial y}$$
$$+ \frac{\partial^2 F}{\partial u \partial u_x}\frac{\partial u}{\partial x} + \frac{\partial^2 F}{\partial v \partial u_x}\frac{\partial v}{\partial x} + \frac{\partial^2 F}{\partial x \partial u_x} \qquad (3.124)$$

Expressions for the other derivative terms are calculated in a similar manner.

Differential constraint conditions

The problem of finding an extreme value for the functional

$$I = \int_{t_1}^{t_2} f(t, x_1, x_2, \ldots, x_n, \dot{x}_1, \dot{x}_2, \ldots, \dot{x}_n)\, dt$$

where $x_i = x_i(t)$ and $\dot{x}_i = \frac{dx_i}{dt}$ for $i = 1, \ldots, n$, occurs in many engineering and scientific applications. Related to this problem is the more complicated problem of finding an extreme value to the above functional I subject to constraint conditions that the functions $x_i(t)$ must satisfy given differential equations having the general form

$$\phi_k(x_1(t), x_2(t), \ldots, x_n(t), \dot{x}_1(t), \dot{x}_2(t), \ldots, \dot{x}_n(t)) = 0 \tag{3.125}$$

for the index k ranging over the values $k = 1, 2, \ldots, m$, with $m < n$. This type of variational problem with constraints can be solved by introducing Lagrange multipliers $\lambda_1(t), \ldots, \lambda_m(t)$ which are functions of time t. The problem is then reformulated to that of finding an extreme value for the functional

$$I^* = \int_{t_1}^{t_2} \left(f + \sum_{k=1}^{m} \lambda_k \phi_k \right) dt \tag{3.126}$$

without regard to the constraint conditions. This is sometimes referred to as the Lagrange multiplier rule. Note that if the constraint conditions are independent of derivatives, then they become algebraic equations

$$\phi_k(x_1(t), x_2(t), \ldots, x_n(t)) = 0 \tag{3.127}$$

In this case the equation (3.126) is still used to find an extreme value. In the next chapter we will investigate this use of the Lagrange multiplier and the derivation of equation (3.126) as a functional to solve the variational problem with either algebraic or differential equation constraints. In addition we consider other complications associated with the basic variational problem.

Weierstrass criticism

Weierstrass was the first to criticize the Euler-Lagrange method on logical grounds. The Euler-Lagrange equations with assumed end point conditions are derived by first assuming the existence of an extremal solution and then considering variations from this solution. For the class of problems where this original assumption is valid, there is nothing wrong with this approach. Problems arise when a given problem does not satisfy the original assumption. In engineering applications one usually assumes all necessary and sufficient conditions are satisfied in order for a maximum or minimum value to be produced by a solution. This assumption can sometimes be justified from a physical understanding of the problem to be solved. One should be aware that there may arise an occasion when this original assumption is false, with the consequences leading to unforeseen results.

"When the mind grapples with a great and intricate problem, it makes its advances, it secures its positions step by step, with but little realization of the gains it has made, until suddenly, with an effect of abrupt illumination, it realizes its victory."

H.G. Wells (1866-1946)

Exercises Chapter 3

▶ **1.**

(a) Apply Green's theorem to $\oint_C P\,dx + Q\,dy$ with $P = 0$ and $Q = f(x,y)g(x,y)$ and show that

$$\iint_R f\frac{\partial g}{\partial x}\,dxdy = \oint_C fg\,(\widehat{n}\cdot\widehat{\mathbf{e}}_1)\,ds - \iint_R g\frac{\partial f}{\partial x}\,dxdy$$

where $\widehat{n} = \frac{dy}{ds}\widehat{\mathbf{e}}_1 - \frac{dx}{ds}\widehat{\mathbf{e}}_2$ is the outer normal to the simple closed curve C which bounds the region R. This is a two-dimensional form of integration by parts.

(b) Apply Green's theorem to $\oint_C P\,dx + Q\,dy$ with $P = -f(x,y)h(x,y)$ and $Q = 0$ and show that

$$\iint_R f\frac{\partial h}{\partial y}\,dxdy = \oint_C fh\,(\widehat{n}\cdot\widehat{\mathbf{e}}_2)\,ds - \iint_R h\frac{\partial f}{\partial y}\,dxdy$$

This is another form of the two-dimensional integration by parts.

(c) Add the results from parts (a) and (b) and show that

$$\iint_R f\left(\frac{\partial g}{\partial x} + \frac{\partial h}{\partial y}\right)\,dxdy = \oint_C f(g\,\widehat{\mathbf{e}}_1 + h\,\widehat{\mathbf{e}}_2)\cdot\widehat{n}\,ds - \iint_R \left(g\frac{\partial f}{\partial x} + h\frac{\partial f}{\partial y}\right)\,dxdy$$

which is still another form of two-dimensional integration by parts.

▶ **2.** Show that the Euler-Lagrange equation $\dfrac{\partial f}{\partial y} - \dfrac{d}{dx}\left(\dfrac{\partial f}{\partial y'}\right) = 0$ can be written in the alternate form $\dfrac{\partial f}{\partial x} - \dfrac{d}{dx}\left(f - y'\dfrac{\partial f}{\partial y'}\right) = 0$.

▶ **3.**

(a) Find the shortest distance between two points (x_0, y_0) and (x_1, y_1) lying in the x,y-plane.

(b) Show that the resulting straight line path satisfies the condition $\begin{vmatrix} x & y & 1 \\ x_0 & y_0 & 1 \\ x_1 & y_1 & 1 \end{vmatrix} = 0.$

▶ **4.** If $f = f(x, y, y')$, then calculate $\dfrac{d}{dx}\left(\dfrac{\partial f}{\partial y'}\right)$.

▶ **5.** If $f = f(x, y, y', y'')$, then calculate the following derivatives

$$(a)\quad \frac{d}{dx}\left(\frac{\partial f}{\partial y'}\right) \qquad (b)\quad \frac{d}{dx}\left(\frac{\partial f}{\partial y''}\right)$$

▶ **6.** Find the Euler-Lagrange equations for y_1 and y_2 such that the functional

$$I = I(y_1, y_2) = \int_{t_0}^{t_1} \left[2y_1 y_2 - 2(y_1)^2 + (\dot{y}_1)^2 - (\dot{y}_2)^2 \right] dt$$

has a stationary value.

▶ **7.**

(a) Find the Euler-Lagrange equation associated with the functional $I = I(y) = \int_a^b f(x, y, y') \, dx$
where $f = f(x, y, y') = \alpha(x, y) + \beta(x, y)y' + \gamma(x, y)(y')^2$

(b) What happens in the special case $\gamma(x, y) = 0$?
Assume $y(x)$ is prescribed at the end points.

▶ **8.**

(a) Find the Euler-Lagrange equation associated with the functional $I = \int_{x_1}^{x_2} f(x, y, y', y'') \, dx$

(b) Find a first integral of the Euler-Lagrange equation in the special case $f = f(y, y', y'')$

(c) Find a first integral of the Euler-Lagrange equation in the special case $f = f(x, y', y'')$

▶ **9.**

(a) Find the Euler-Lagrange equation associated with the functional $I = \int_{x_1}^{x_2} f(x, y, y', y'', y''') \, dx$

(b) Find a first integral of the Euler-Lagrange equation in the special case $f = f(y, y', y'', y''')$

(c) Find a first integral of the Euler-Lagrange equation in the special case $f = f(x, y', y'', y''')$

▶ **10.** Find the curve which produces an extreme value for the functional

$$I = \int_{(0,0)}^{(\ell,1)} \left(y'^2 + \omega^2 y^2 \right) dx$$

where ℓ and ω are given positive constants.

▶ **11.** Consider all parabolas of the form $y = x + \alpha x(1 - x)$ which pass through the points $(0, 0)$ and $(1, 1)$. Each parabola in the family is rotated about the x-axis to form a solid of revolution for the region between $x = 0$ and $x = 1$. Which parabola produces a solid of revolution with the minimum volume?

▶ **12.**

(a) Show the element of arc length squared on the surface of a sphere of radius r_0 is given by $ds^2 = r_0^2 \, d\theta^2 + r_0^2 \sin^2 \theta \, d\phi^2$

(b) On the curve of shortest distance on the surface of a sphere show that $\phi = \alpha - \sin^{-1}(\beta \cot \theta)$ where α and β are constants.

▶ **13.** Find the extremal for the functional $I = \int_1^2 (y' + x^2 y'^2) \, dx$ subject to the end conditions $y(1) = 1$ and $y(2) = 2$.

▶ **14.** Find the Euler-Lagrange equation associated with the functional

$$I = \int_{x_0}^{x_1} \left[p(x) \left(\frac{dy}{dx} \right)^2 + q(x) y^2 \right] dx$$

subject to the boundary conditions $y(x_0) = y_0$ and $y(x_1) = y_1$. Assume $p = p(x) > 0$ and $q = q(x) > 0$ for $x_0 \leq x \leq x_1$ are well behaved differentiable functions.

▶ **15.** Consider the problem to minimize the functional

$$I = \int_{x_1}^{x_2} f(y_1, y_2, \ldots, y_n, y_1', y_2', \ldots, y_n') \, dx$$

where f is independent of the variable x.
(a) Show the Euler-Lagrange equations is the system of differential equations

$$\frac{\partial f}{\partial y_i} - \frac{d}{dx} \left(\frac{\partial f}{\partial y_i'} \right) = 0, \qquad i = 1, 2, \ldots, n$$

(b) Because f is independent of the variable x show that $f - \sum_{i=1}^{n} y_i' \frac{\partial f}{\partial y_i'} = constant$

(c) Show that an integration of the Euler-Lagrange equations in part (a) gives the Euler-Lagrange equations in the integral form

$$\frac{\partial f}{\partial y_i'} = \int_{x_1}^{x} \frac{\partial f}{\partial y_i} \, dx + C_i, \qquad i = 1, 2, \ldots, n$$

where C_i for $i = 1, 2, \ldots, n$ denote arbitrary constants.

▶ **16.** Find the curve $\vec{r} = x \, \hat{e}_1 + y(x) \, \hat{e}_2 + z(x) \, \hat{e}_3$ which passes through the points $(0, 0, 0)$ and $(1, 1, 1)$ which minimizes the functional $I = \int_0^1 \left[y^2 + \left(\frac{dy}{dx} \right)^2 + \left(\frac{dz}{dx} \right)^2 \right] dx$

▶ **17.**
(a) Find the general family of curves such that $I = \int_0^{x_0} \left(\frac{d^2y}{dx^2} \right)^2 dx$ is an extremum.
(b) Find the particular curve satisfying the end point conditions

$$y(0) = 0, \quad y(x_0) = 0, \quad y'(0) = \alpha, \quad y'(x_0) = \beta$$

where x_0, α and β are constants.
(c) What kind of curve results in the special case $\beta = -\alpha$?
(d) Test your results against some other curves and determine if the extremum is a maximum or minimum.

▶ **18.** Find an extremum for the functional $I = I(y) = \int_0^{\pi} \left[(y')^2 + y^2 + 2xy \right] dx$ subject to the boundary conditions that $y(0) = 1$ and $y(\pi) = 0$.

► **19.**

(a) Expand the Euler-Lagrange equations (3.77)

(b) Obtain a solution to the Euler-Lagrange equations associated with the functional

$$I = \int_{t_0}^{t_1} \frac{x^2 \dot{x}^2}{\dot{y}} \, dt$$

► **20.** Find the extremal for the functional $I = \int_0^1 \left[1 + (y'')^2\right] dx$ satisfying the end conditions $y(0) = 0$, $y'(0) = 1$, $y(1) = 1$, $y'(1) = 0$.

► **21.** Determine the Euler-Lagrange equation associated with the functional

$$I = I(u) = \iint_R \left[\left(\frac{\partial u}{\partial x}\right)^2 + \left(\frac{\partial u}{\partial y}\right)^2 + 2q(x,y)u\right] dxdy$$

where u is subject to the boundary condition that $u(x,y) = g(x,y)$ for $x, y \in C = \partial R$.

► **22.** Determine the Euler-Lagrange equation associated with the functional

$$I = I(u) = \iiint_V \left[\left(\frac{\partial u}{\partial x}\right)^2 + \left(\frac{\partial u}{\partial y}\right)^2 + \left(\frac{\partial u}{\partial z}\right)^2 + 2q(x,y,z)u\right] dxdydz$$

where u is subject to the boundary condition that $u(x,y,z) = g(x,y,z)$ for $x, y, z \in S = \partial V$. Give a physical interpretation associated with the resulting Euler-Lagrange equation.

► **23.**

(a) Show the problem 22 can be written in the form $I = I(u) = \iiint_V (\nabla u \cdot \nabla u + 2q(x,y,z)u) \, d\tau$ where $d\tau$ is an element of volume and ∇ is the gradient operator.

(b) Express the integral in part (a) in cylindrical coordinates (r, θ, z) and find the Euler-Lagrange equation in cylindrical coordinates.

(c) Express the integral in part (a) in spherical coordinates (ρ, θ, ϕ) and find the Euler-Lagrange equation in spherical coordinates.

(d) Give a physical interpretation associated with the resulting Euler-Lagrange equations.

► **24.** Determine the Euler-Lagrange equation associated with the functional

$$I = I(u) = \iint_R \left[\frac{\alpha(x,y)}{2}\left(\frac{\partial u}{\partial x}\right)^2 + \frac{\alpha(x,y)}{2}\left(\frac{\partial u}{\partial y}\right)^2 + q(x,y)u(x,y)\right] dxdy$$

subject to the boundary condition that $u(x,y) = g(x,y)$ for $x, y \in C = \partial R$.

► **25.** Find the Euler-Lagrange equation associated with the functional

$$I = \int_{x_1}^{x_2} f\left(x, y, y', \int_{x_1}^x y(\xi) \, d\xi\right) dx$$

▶ **26.** The Du Bois-Reymond form of the Euler-Lagrange equation associated with the functional $I = \int_{x_1}^{x_2} f(x,y,y')\,dx$ is given by

$$\frac{\partial f}{\partial y'} = \int_{x_1}^{x} \frac{\partial f}{\partial y}\,dx + C$$

where C is a constant. Derive this necessary condition. Hint: Integrate the first term in equation (3.17) using integration by parts.

▶ **27.** Find the function which creates a stationary value for the functional

$$I = \int_0^1 (y'^2 + 2xyy')\,dx$$

satisfying the end conditions $y(0) = y_0$ and $y(1) = y_1$.

▶ **28.** For f, g, h, q continuous functions of x and y, find the Euler-Lagrange equation associated with the functional

$$I = I(u) = \frac{1}{2} \iint_R \left[f(x,y) \left(\frac{\partial u}{\partial x}\right)^2 + 2g(x,y) \left(\frac{\partial u}{\partial x}\right) \left(\frac{\partial u}{\partial y}\right) + h(x,y) \left(\frac{\partial u}{\partial y}\right)^2 - 2q(x,y)u\frac{\partial u}{\partial y} \right] dxdy$$

subject to the boundary condition that $u(x,y) = F(x,y)$ for $x,y \in C = \partial R$, where the region R and boundary curve C are known. The resulting Euler-Lagrange equation is called an elliptic boundary value problem if $fh - g^2 > 0$. If y is replaced by t and $fh - g^2 = 0$, then the resulting Euler-Lagrange equation is called a parabolic initial-value problem. If y is replaced by t and $fh - g^2 < 0$, then the resulting Euler-Lagrange equation is called a hyperbolic initial-value problem. Determine values for f, g, h and q such that the Euler-Lagrange equation is
(a) the Laplace equation $\frac{\partial^2 u}{\partial x^2} + \frac{\partial^2 u}{\partial y^2} = 0$. Show this equation is an elliptic equation.
(b) the heat or diffusion equation $\frac{\partial^2 u}{\partial x^2} - \frac{\partial u}{\partial t} = 0$. Show this equation is a parabolic equation.
(c) the wave equation $\frac{\partial^2 u}{\partial x^2} - \frac{\partial^2 u}{\partial t^2} = 0$. Show this equation is a hyperbolic equation.

▶ **29.**

(a) Consider the functional $I = I(y) = \frac{\int_{x_0}^{x_1} F(x,y,y')\,dx}{\int_{x_0}^{x_1} G(x,y,y')\,dx}$ which is a ratio of integrals where the integrands $F \in \mathcal{C}^{(2)}$ and $G \in \mathcal{C}^{(2)}$ over the interval $x_0 \leq x \leq x_1$. Show that the Euler-Lagrange necessary condition for the function $y = y(x)$ to be and extreme value for this ratio, and satisfy the boundary conditions $y(x_0) = y_0$ and $y(x_1) = y_1$, is given by the equation

$$\frac{\partial F}{\partial y} - \frac{d}{dx}\left(\frac{\partial F}{\partial y'}\right) - I(y)\left[\frac{\partial G}{\partial y} - \frac{d}{dx}\left(\frac{\partial G}{\partial y'}\right)\right] = 0.$$

▶ **30.** Assume that $r(x)y'(x)$ is zero at $x = a$ and $x = b$ and show that the stationary values of the ratio

$$\lambda = \frac{\int_a^b \left[r(x)y'^2 - q(x)y^2\right]\,dx}{\int_a^b p(x)y^2\,dx}$$

are found by solving the Sturm-Liouville equation $\frac{d}{dx}\left(r(x)\frac{dy}{dx}\right) + [q(x) + \lambda p(x)]\,y = 0.$

▶ **31.** Consider the special case $I = \int_{x_1}^{x_2} f(y, y') \, dx$ where f is independent of x.

(a) Show that $\dfrac{d}{dx}\left(f - \dfrac{\partial f}{\partial y'} y'\right) = y'\left[\dfrac{\partial f}{\partial y} - \dfrac{d}{dx}\left(\dfrac{\partial f}{\partial y'}\right)\right]$

(b) Show that a first integral of the Euler-Lagrange equation is given by $f - \dfrac{\partial f}{\partial y'} y' = constant$

▶ **32.** Consider the functional $I = \lambda \int_{x_1}^{x_2} \int_{x_1}^{x_2} G(\xi, x) y(\xi) y(x) \, d\xi dx + \int_{x_1}^{x_2} \left[2H(x)y(x) - y^2(x)\right] dx$ where λ is constant, $H(x)$ is a known function and $G(x, \xi) = G(\xi, x)$ is a given symmetric function called a kernel function. Show that the Euler-Lagrange equation associated with this functional is the integral equation $y(x) = H(x) + \lambda \int_{x_1}^{x_2} G(\xi, x) y(\xi) \, d\xi$

▶ **33.** Find the family of curves which satisfy the Euler-Lagrange equation associated with the functional $I = \int_{x_0}^{x_1} f(y) \left[1 + y'^2\right] dx$ where $f \in C^{(2)}$.

▶ **34.** Join two points (x_0, y_0) and (x_1, y_1) in the plane with the curve of shortest length. Solve this problem by representing the curve in parametric form $x = x(t)$ and $y = y(t)$. Here the functional to be minimized is

$$L = \int_{t_0}^{t_1} \sqrt{x'^2 + y'^2} \, dt \qquad \text{where} \quad x' = \frac{dx}{dt} \quad \text{and} \quad y' = \frac{dy}{dt}.$$

Assume that $x(t_0) = x_0$, $y(t_0) = y_0$, $x(t_1) = x_1$, and $y(t_1) = y_1$.

▶ **35.** Find the minimum value of the functional $I = \int_{x_0}^{x_1} (\alpha - x)^2 \left(\dfrac{dy}{dx}\right)^2 dx, \qquad x_1 > x_0$ where α is a constant and $y = y(x)$ is bounded over the interval (x_0, x_1) and required to satisfy the end point conditions $y(x_0) = y_0$ and $y(x_1) = y_1$.

▶ **36.** Calculate the biharmonic equation $\nabla^4 z = \nabla^2 \nabla^2 z = 0$.

(a) In rectangular coordinates using $\left(\dfrac{\partial^2}{\partial x^2} + \dfrac{\partial^2}{\partial y^2}\right)\left(\dfrac{\partial^2 z}{\partial x^2} + \dfrac{\partial^2 z}{\partial y^2}\right) = 0$

(b) In polar coordinates using $\left(\dfrac{\partial^2}{\partial r^2} + \dfrac{1}{r}\dfrac{\partial}{\partial r} + \dfrac{1}{r^2}\dfrac{\partial^2}{\partial \theta^2}\right)\left(\dfrac{\partial^2 z}{\partial r^2} + \dfrac{1}{r}\dfrac{\partial z}{\partial r} + \dfrac{1}{r^2}\dfrac{\partial^2 z}{\partial \theta^2}\right) = 0$

▶ **37.** Find the Euler-Lagrange equation associated with the functional

$$I = \iint_R \left(\frac{\partial^2 z}{\partial x^2} + \frac{\partial^2 z}{\partial y^2}\right)^2 dxdy$$

▶ **38.** Find the Euler-Lagrange equation associated with the functional

$$I = \iint_R \left(\left(\frac{\partial^2 z}{\partial x^2} + \frac{\partial^2 z}{\partial y^2}\right)^2 - 2(1 - \nu)\left[\frac{\partial^2 z}{\partial x^2}\frac{\partial^2 z}{\partial y^2} - \left(\frac{\partial^2 z}{\partial x \partial y}\right)^2\right]\right) dxdy$$

where ν is a constant.

▶ **39.** Find the Euler-Lagrange equation associated with the functional

$$I = \iint_R \left[\left(\frac{\partial^2 z}{\partial x^2}\right)^2 + \left(\frac{\partial^2 z}{\partial y^2}\right)^2 + 2\left(\frac{\partial^2 z}{\partial x \partial y}\right)^2 - 2z \, q(x, y)\right] dxdy,$$

Chapter 4
Additional Variational Concepts

In the previous chapter we considered calculus of variation problems which had fixed boundary conditions. That is, in one dimension the end point conditions were specified. In two and three dimensions, a boundary condition or surface condition was specified. In this chapter we shall consider other types of boundary conditions together with subsidiary conditions or constraints imposed upon the class of admissible solutions which produce an extremal value for a given functional. We shall also examine necessary conditions for a maximum or minimum value of a functional to exist. Let us begin by examining the one-dimensional case where either one or both boundary conditions are not prescribed.

Natural boundary conditions

Consider the problem of finding a function $y = y(x)$ such that the functional

$$I = I(y) = \int_{x_1}^{x_2} f(x, y, y') \, dx \tag{4.1}$$

has a stationary value where either one or both of the end point conditions $y(x_1) = y_1$ and $y(x_2) = y_2$ are not prescribed. One proceeds exactly as we have done previously. Assume that $y = y(x)$ produces an extremal value and construct the class of comparison functions $Y(x) = y(x) + \epsilon \eta(x)$. Substitute $Y(x)$ into the functional given by equation (4.1) to obtain

$$I = I(\epsilon) = \int_{x_1}^{x_2} f(x, y + \epsilon \eta, y' + \epsilon \eta') \, dx.$$

By hypothesis, the functional I has a stationary value at $\epsilon = 0$ so that

$$\frac{dI}{d\epsilon}\bigg|_{\epsilon=0} = I'(0) = \int_{x_1}^{x_2} \left(\frac{\partial f}{\partial y} \eta + \frac{\partial f}{\partial y'} \eta' \right) dx = 0. \tag{4.2}$$

Now integrate the second term in equation (4.2) using integration by parts to obtain

$$\frac{dI}{d\epsilon}\bigg|_{\epsilon=0} = I'(0) = \frac{\partial f}{\partial y'} \eta \bigg|_{x_1}^{x_2} + \int_{x_1}^{x_2} \left[\frac{\partial f}{\partial y} - \frac{d}{dx}\left(\frac{\partial f}{\partial y'} \right) \right] \eta \, dx = 0. \tag{4.3}$$

Assume that the Euler-Lagrange equation

$$\frac{\partial f}{\partial y} - \frac{d}{dx}\left(\frac{\partial f}{\partial y'} \right) = 0 \tag{4.4}$$

is satisfied over the interval $x_1 \leq x \leq x_2$, then in order for equation (4.3) to be satisfied it is necessary that the first term of equation (4.3) also equal zero. This requires that

$$\frac{\partial f}{\partial y'} \eta \bigg|_{x_1}^{x_2} = \frac{\partial f}{\partial y'}\bigg|_{x=x_2} \eta(x_2) - \frac{\partial f}{\partial y'}\bigg|_{x=x_1} \eta(x_1) = 0. \tag{4.5}$$

In the case of fixed end points, the Euler-Lagrange equation together with the end conditions $\eta(x_1) = 0$ and $\eta(x_2) = 0$ guarantees that the equation (4.5) is satisfied. In the case both end points are variable, then the terms $\eta(x_1)$ and $\eta(x_2)$ are arbitrary, and so in order for equation (4.5) to be satisfied one must require in addition to the Euler-Lagrange equation, the end point conditions

$$\frac{\partial f}{\partial y'}\bigg|_{x=x_1} = 0, \qquad \text{and} \qquad \frac{\partial f}{\partial y'}\bigg|_{x=x_2} = 0. \qquad (4.6)$$

These conditions are called the natural boundary conditions or transversality conditions associated with the extremum problem. The case of mixed boundary conditions occurs when one end of the curve is fixed and the other end can vary. The above considerations give rise to the following boundary cases.

Case 1: (Fixed end points) If $y(x_1) = y_1$ and $y(x_2) = y_2$ are prescribed, then there is no variation of the end points so that $\eta(x_1) = 0$ and $\eta(x_1) = 0$.

Case 2: (Mixed end condition) If $y(x_1) = y_1$ is given and $y(x_2)$ is unknown, then one must impose the natural boundary condition $\frac{\partial f}{\partial y'}\bigg|_{x=x_2} = 0$ and $\eta(x_1) = 0$.

Case 3: (Mixed end condition) If $y(x_2) = y_2$ is given and $y(x_1)$ is unknown, then one must impose the natural boundary condition $\frac{\partial f}{\partial y'}\bigg|_{x=x_1} = 0$ and $\eta(x_2) = 0$.

Case 4: (Variable end points) If $y = y(x)$ is not prescribed at the end points of the given interval $x_1 \leq x \leq x_2$, then the partial derivative $\frac{\partial f}{\partial y'}$ must satisfy the natural boundary conditions given by equations (4.6).

Note that the requirement that $\eta = 0$ at an end point is equivalent to specifying the value of the dependent variable y at an end point. Boundary conditions where $\eta = 0$ or $\delta y = 0$, where the values of y are specified at a boundary, are called essential boundary conditions sometimes referred to as Dirichlet boundary conditions or geometric boundary conditions. Natural boundary conditions are sometimes referred to as Neumann boundary conditions or dynamic boundary conditions. Mixed boundary value problems, sometimes referred to as Robin boundary value problems, occur when both essential and natural boundary are specified on portions of the boundary. A general rule is that the vanishing of the variation η on a boundary is an essential boundary condition and the vanishing of the term that multiplies η is called a natural boundary condition. This general rule, and the above terminology, with slight modifications, can be applied to one, two and three dimensional problems.

Example 4-1.   Natural boundary condition

Find the curve $y = y(x)$ producing the shortest distance between the points x_0 and x_1 subject to natural boundary conditions.

Solution: Let $ds^2 = dx^2 + dy^2$ denote an element of distance squared in Cartesian coordinates. One can then form the functional

$$I = \int_{x_0}^{x_1} ds = \int_{x_0}^{x_1} \sqrt{1 + \left(\frac{dy}{dx}\right)^2}\, dx.$$

The integrand for this functional is $f = \sqrt{1 + (y')^2}$ with partial derivatives

$$\frac{\partial f}{\partial y} = 0, \qquad \frac{\partial f}{\partial y'} = \frac{y'}{\sqrt{1 + (y')^2}}.$$

The Euler-Lagrange equation associated with this functional is $y'' = 0$ which is subject to the natural boundary conditions

$$\frac{\partial f}{\partial y'}\bigg|_{x=x_0} = \frac{y'(x_0)}{\sqrt{1 + (y'(x_0))^2}} = 0$$
$$\frac{\partial f}{\partial y'}\bigg|_{x=x_1} = \frac{y'(x_1)}{\sqrt{1 + (y'(x_1))^2}} = 0$$

The Euler-Lagrange equation has the solution $y = y(x) = Ax + B$ where A, B are constants. The natural boundary conditions require that $A = 0$ and so the solution is $y = B = constant$. ∎

Natural boundary conditions for other functionals

Consider the functional

$$I = \int_a^b f(x, y, y', y'') \, dx \tag{4.7}$$

where we hold the variation in x constant and consider only the variation in y. This functional has first variation

$$\delta I = \int_a^b \left(\frac{\partial f}{\partial y} \delta y + \frac{\partial f}{\partial y'} \delta y' + \frac{\partial f}{\partial y''} \delta y'' \right) dx = 0. \tag{4.8}$$

Use integration by parts on the second term of equation (4.8) using

$$U = \frac{\partial f}{\partial y'}, \quad dU = \frac{d}{dx}\left(\frac{\partial f}{\partial y'} \right) dx, \qquad dV = \delta y' \, dx, \quad V = \delta y$$

and then use integration by parts on the third term in equation (4.8) using

$$U = \frac{\partial f}{\partial y''}, \quad dU = \frac{d}{dx}\left(\frac{\partial f}{\partial y''} \right) dx, \qquad dV = \delta y'' \, dx, \quad V = \delta y'$$

and verify that equation (4.8) can be expressed in the form

$$\delta I = \int_a^b \left[\frac{\partial f}{\partial y} \delta y - \frac{d}{dx}\left(\frac{\partial f}{\partial y'} \right) \delta y - \frac{d}{dx}\left(\frac{\partial f}{\partial y''} \right) \delta y' \right] dx + \frac{\partial f}{\partial y'} \delta y \bigg|_a^b + \frac{\partial f}{\partial y''} \delta y' \bigg|_a^b = 0 \tag{4.9}$$

Now use integration by parts again on the third term under the integral in equation (4.9) to show

$$\delta I = \int_a^b \left[\frac{\partial f}{\partial y} - \frac{d}{dx}\left(\frac{\partial f}{\partial y'} \right) + \frac{d^2}{dx^2}\left(\frac{\partial f}{\partial y''} \right) \right] \delta y \, dx + \left[\frac{\partial f}{\partial y'} - \frac{d}{dx}\left(\frac{\partial f}{\partial y''} \right) \right] \delta y \bigg|_a^b + \frac{\partial f}{\partial y''} \delta y' \bigg|_a^b = 0. \tag{4.10}$$

The Euler-Lagrange equation associated with the functional given by equation (4.7) is

$$\frac{\partial f}{\partial y} - \frac{d}{dx}\left(\frac{\partial f}{\partial y'} \right) + \frac{d^2}{dx^2}\left(\frac{\partial f}{\partial y''} \right) = 0.$$

140

This is a fourth order ordinary differential equation so that the general solution will contain four arbitrary constants. The boundary conditions come from the requirement that the remaining terms on the right-hand side of equation (4.10) must equal zero. This requires

$$\left[\frac{\partial f}{\partial y'} - \frac{d}{dx}\left(\frac{\partial f}{\partial y''}\right)\right]\delta y\,\Big|_a^b = 0 \quad \text{and} \quad \frac{\partial f}{\partial y''}\delta y'\,\Big|_a^b = 0. \tag{4.11}$$

This produces the essential boundary conditions δy and $\delta y'$ equal to zero at $x = a$ and $x = b$ producing the conditions that $y(a), y(b)$ and $y'(a), y'(b)$ are specified at the end points. Alternatively, one can use the natural boundary conditions that $\frac{\partial f}{\partial y'} - \frac{d}{dx}\left(\frac{\partial f}{\partial y''}\right)$ and $\frac{\partial f}{\partial y''}$ are zero at the end points $x = a$ and $x = b$, or one can employ some combination of mixed boundary conditions determined from the equation (4.11). For example, one can select $\delta y = 0$ at $x = a$ and $x = b$ together with $\frac{\partial f}{\partial y''} = 0$ and $x = a$ and $x = b$ as a mixed boundary condition involving both essential and natural boundary conditions. The other combination is to assume $\delta y' = 0$ at $x = a$ and $x = b$ together with $\frac{\partial f}{\partial y'} - \frac{d}{dx}\left(\frac{\partial f}{\partial y''}\right) = $ at the end points. Note that the conditions $\delta y = 0$ and $\delta y' = 0$ implies that $y(a), y(b), y'(a), y'(b)$ are all specified so that there is no variation of these quantities at the end points.

Example 4-2. Natural boundary condition

Find the function $y = y(x)$ such that the functional

$$I = \int_a^b [a(x)(y'')^2 - b(x)(y')^2 + c(x)y^2]\,dx$$

is an extremum.

Solution: Here $f = a(x)(y'')^2 - b(x)(y')^2 + c(x)y^2$ with partial derivatives

$$\frac{\partial f}{\partial y} = 2c(x)y, \qquad \frac{\partial f}{\partial y'} = -2b(x)y', \qquad \frac{\partial f}{\partial y''} = 2a(x)y''$$

The Euler-Lagrange equation associated with the above functional is the fourth order ordinary differential equation

$$c(x)y + \frac{d}{dx}\left(b(x)y'\right) + \frac{d^2}{dx^2}\left(a(x)y''\right) = 0, \qquad a \le x \le b$$

Various types of boundary conditions can be applied to obtain a solution to this differential equation.

Case 1 If $\delta y = 0$ and $\delta y' = 0$ at the end points, then $y(a), y'(a), y(b), y'(b)$ must be prescribed.

Case 2 If $\delta y = 0$ and $\delta y' \ne 0$ at the end points, then $y(a)$ and $y(b)$ must be prescribed and the natural boundary conditions

$$\frac{\partial f}{\partial y''}\,\Big|_{x=a} = 0, \qquad \text{and} \qquad \frac{\partial f}{\partial y''}\,\Big|_{x=b} = 0$$

must be satisfied. This implies $y''(a)$ and $y''(b)$ have prescribed values.

Case 3 If $\delta y \neq 0$ and $\delta y' = 0$ at the end points, then $y'(a)$ and $y'(b)$ are prescribed and the natural boundary conditions

$$\left[\frac{\partial f}{\partial y'} - \frac{d}{dx}\left(\frac{\partial f}{\partial y''}\right)\right]_{x=a} = 0, \quad \text{and} \quad \left[\frac{\partial f}{\partial y'} - \frac{d}{dx}\left(\frac{\partial f}{\partial y''}\right)\right]_{x=b} = 0,$$

must be satisfied.

Case 4 If $\delta y \neq 0$ and $\delta y' \neq 0$ at the end points, then the natural boundary conditions

$$\frac{\partial f}{\partial y''} = 0 \quad \text{and} \quad \frac{\partial f}{\partial y'} - \frac{d}{dx}\left(\frac{\partial f}{\partial y''}\right) = 0$$

at the end points $x = a$ and $x = b$ must be satisfied.

$\blacksquare$

More natural boundary conditions

Consider the functional

$$I = \iint_R f(x, y, w, w_x, w_y) \, dxdy \tag{4.12}$$

where $w = w(x, y)$ is a function of x and y and R is a region enclosed by a simple closed curve $C = \partial R$. We use the variational notation and require that at an extremum the first variation is zero so that

$$\delta I = \iint_R \left[\frac{\partial f}{\partial w}\delta w + \frac{\partial f}{\partial w_x}\delta w_x + \frac{\partial f}{\partial w_y}\delta w_y\right] dxdy = 0. \tag{4.13}$$

Now use the Green's theorem in the plane

$$\iint_R \left(\frac{\partial N}{\partial x} - \frac{\partial M}{\partial y}\right) dxdy = \oint_C M\, dx + N\, dy \tag{4.14}$$

with $N = \frac{\partial f}{\partial w_x}\delta w$ and $M = -\frac{\partial f}{\partial w_y}\delta w$ to integrate the last two terms under the integral on the right-hand side of equation (4.13). Note that for the above choices for M and N we have

$$\frac{\partial N}{\partial x} = \frac{\partial f}{\partial w_x}\delta w_x + \frac{\partial}{\partial x}\left(\frac{\partial f}{\partial w_x}\right)\delta w \quad \text{and} \quad \frac{\partial M}{\partial y} = -\frac{\partial f}{\partial w_y}\delta w_y - \frac{\partial}{\partial y}\left(\frac{\partial f}{\partial w_y}\right)\delta w$$

and so the use of the Green's theorem produces the result

$$\iint_R \left(\frac{\partial f}{\partial w_x}\delta w_x + \frac{\partial f}{\partial w_y}\delta w_y\right) dxdy = \oint_C \left(-\frac{\partial f}{\partial w_y}\delta w\, dx + \frac{\partial f}{\partial w_x}\delta w\, dy\right)$$
$$- \iint_R \left[\frac{\partial}{\partial x}\left(\frac{\partial f}{\partial w_x}\right)\delta w + \frac{\partial}{\partial y}\left(\frac{\partial f}{\partial w_y}\right)\delta w\right] dxdy \tag{4.15}$$

Substitute the result from equation (4.15) into the equation (4.13) and simplify to obtain

$$\delta I = \iint_R \left[\frac{\partial f}{\partial w} - \frac{\partial}{\partial x}\left(\frac{\partial f}{\partial w_x}\right) - \frac{\partial}{\partial y}\left(\frac{\partial f}{\partial w_y}\right)\right]\delta w\, dxdy$$
$$+ \oint_C \delta w \left(-\frac{\partial f}{\partial w_y}\, dx + \frac{\partial f}{\partial w_x}\, dy\right) = 0. \tag{4.16}$$

If the equation (4.16) is to be zero, then $w = w(x, y)$ must satisfy the Euler-Lagrange equation

$$\frac{\partial f}{\partial w} - \frac{\partial}{\partial x}\left(\frac{\partial f}{\partial w_x}\right) - \frac{\partial}{\partial y}\left(\frac{\partial f}{\partial w_y}\right) = 0$$

and the boundary conditions come from an analysis of the remaining line integral term. If $\delta w = 0$, then the boundary conditions required are for the function $w = w(x, y)$ to be specified everywhere along the boundary curve $C = \partial R$. If $\delta w \neq 0$, then we require that

$$-\frac{\partial f}{\partial w_y}\, dx + \frac{\partial f}{\partial w_x}\, dy \, \bigg|_{(x,y)\in\partial R} = 0. \tag{4.17}$$

This condition can be written in terms of the unit normal vector to the boundary curve $C = \partial R$. Note that if $\vec{r} = x(s)\,\widehat{\mathbf{e}}_1 + y(s)\,\widehat{\mathbf{e}}_2$ is the position vector defining the boundary curve C, then $\frac{d\vec{r}}{ds} = \hat{t} = \frac{dx}{ds}\,\widehat{\mathbf{e}}_1 + \frac{dy}{ds}\,\widehat{\mathbf{e}}_2$ is the unit tangent vector to a point on the boundary curve. The cross product

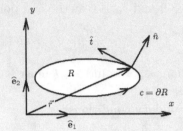

$$\hat{n} = \hat{t} \times \widehat{\mathbf{e}}_3 = \begin{vmatrix} \widehat{\mathbf{e}}_1 & \widehat{\mathbf{e}}_2 & \widehat{\mathbf{e}}_3 \\ \frac{dx}{ds} & \frac{dy}{ds} & 0 \\ 0 & 0 & 1 \end{vmatrix} = \frac{dy}{ds}\,\widehat{\mathbf{e}}_1 - \frac{dx}{ds}\,\widehat{\mathbf{e}}_2$$

gives the unit normal vector to the boundary curve. The boundary condition given by equation (4.17) can then be written in the form

$$\hat{n} \cdot \left(\frac{\partial f}{\partial w_x}\,\widehat{\mathbf{e}}_1 + \frac{\partial f}{\partial w_y}\,\widehat{\mathbf{e}}_2 \right) \bigg|_{(x,y)\in\partial R} = 0 \tag{4.18}$$

where $\hat{n} = \frac{dy}{ds}\,\widehat{\mathbf{e}}_1 - \frac{dx}{ds}\,\widehat{\mathbf{e}}_2$ is a unit normal vector to the boundary curve $C = \partial R$. The condition (4.18), or (4.17), is called the natural boundary condition associated with the functional given by equation (4.12).

Tests for maxima and minima

So far we have been solving the Euler-Lagrange equation and stating that either a maximum or minimum value exists. We have not actually proved these results. The following are two tests for extremum problems associated with the functional given by equation (4.1). These tests are known as Legendre's test and Jacobi's test.

The Legendre and Jacobi analysis

Consider the functional

$$I = I(y) = \int_{x_1}^{x_2} f(x, y, y')\, dx \tag{4.19}$$

and assume that $y = y(x)$ produces an extremum. If we replace y by the set of comparison functions $Y = y + \epsilon\eta$, where we assume weak variations with $\eta(x_1) = 0$ and $\eta(x_2) = 0$. One can then treat I as a function of ϵ and write $I = I(\epsilon)$ which can be expanded in a Taylor series about $\epsilon = 0$. This produces the result

$$I = I(\epsilon) = \int_{x_1}^{x_2} f(x, y + \epsilon\eta, y' + \epsilon\eta')\, dx = I(0) + I'(0)\epsilon + I''(0)\frac{\epsilon^2}{2!} + I'''(0)\frac{\epsilon^3}{3!} + \cdots \tag{4.20}$$

The change in the value of the functional I is given by

$$\Delta I = \int_{x_1}^{x_2} \left[f(x, y + \epsilon\eta, y' + \epsilon\eta') - f(x, y, y') \right] dx \tag{4.21}$$

Expanded the integrand of equation (4.21) in a Taylor series about $\epsilon = 0$ to produced

$$f(x, y + \epsilon\eta, y' + \epsilon\eta') - f(x, y, y') =$$
$$\frac{\epsilon}{1!} \frac{\partial f(x, y + \epsilon\eta, y' + \epsilon\eta')}{\partial \epsilon} \bigg|_{\epsilon=0} + \frac{\epsilon^2}{2!} \frac{\partial^2 f(x, y + \epsilon\eta, y' + \epsilon\eta')}{\partial \epsilon^2} \bigg|_{\epsilon=0} + \cdots$$
$$+ \frac{\epsilon^{n-1}}{(n-1)!} \frac{\partial^{(n-1)} f(x, y + \epsilon\eta, y' + \epsilon\eta')}{\partial \epsilon^{n-1}} \bigg|_{\epsilon=0} + R_n \tag{4.22}$$

An integration of both sides of equation (4.22) from x_1 to x_2 produces

$$\Delta I = \frac{\epsilon}{1!} \int_{x_1}^{x_2} \frac{\partial f(x, y + \epsilon\eta, y' + \epsilon\eta')}{\partial \epsilon} \bigg|_{\epsilon=0} dx + \frac{\epsilon^2}{2!} \int_{x_1}^{x_2} \frac{\partial^2 f(x, y + \epsilon\eta, y' + \epsilon\eta')}{\partial \epsilon^2} \bigg|_{\epsilon=0} dx + \cdots$$
$$+ \frac{\epsilon^{n-1}}{(n-1)!} \int_{x_1}^{x_2} \frac{\partial^{(n-1)} f(x, y + \epsilon\eta, y' + \epsilon\eta')}{\partial \epsilon^{(n-1)}} \bigg|_{\epsilon=0} dx + \int_{x_1}^{x_2} R_n \, dx \tag{4.23}$$

The coefficients of quantities $\epsilon/1!, \epsilon^2/2!, \ldots, \epsilon^{(n-1)}/(n-1)!$ are referred to as the first, second, ..., (n-1)st variations of the functional I and written as $\delta I_1, \delta I_2, \ldots \delta I_{(n-1)}$. One can calculate the derivatives in equation (4.23) and verify that these variations are given by

$$\delta I_1 = \int_{x_1}^{x_2} \left[\frac{\partial f}{\partial y}\eta + \frac{\partial f}{\partial y'}\eta' \right] dx \tag{4.24}$$

$$\delta I_2 = \int_{x_1}^{x_2} \left[\frac{\partial^2 f}{\partial y^2}\eta^2 + 2\frac{\partial^2 f}{\partial y \partial y'}\eta\eta' + \frac{\partial^2 f}{\partial y'^2}(\eta')^2 \right] dx \tag{4.25}$$

$$\delta I_3 = \int_{x_1}^{x_2} \left[\frac{\partial^3 f}{\partial y^3}\eta^3 + 3\frac{\partial^3 f}{\partial y^2 \partial y'}\eta^2\eta' + 3\frac{\partial^3 f}{\partial y \partial y'^2}\eta\eta'^2 + \frac{\partial^3 f}{\partial y'^3}\eta'^3 \right] dx \tag{4.26}$$

$$\vdots$$

Then the change in the value of the functional I can be written

$$\Delta I = \frac{\epsilon}{1!}\delta I_1 + \frac{\epsilon^2}{2!}\delta I_2 + \frac{\epsilon^3}{3!}\delta I_3 + \cdots + \frac{\epsilon^{(n-1)}}{(n-1)!}\delta I_{n-1} + \int_{x_1}^{x_2} R_n \, dx \tag{4.27}$$

Some textbooks write these variations using the notations

$$\Delta I = \delta I + \frac{1}{2!}\delta^2 I + \frac{1}{3!}\delta^3 I + \cdots \tag{4.28}$$

The relation between these different notations is

$$\delta^m I = \epsilon^m \delta I_m = I^{(m)}(0)\epsilon^m \quad \text{for} \quad m = 1, 2, 3, \ldots \tag{4.29}$$

Examine the first variation and assume that $\eta = 0$ at the end points of the integration interval. Use integration by parts on the second term in equation (4.24) to obtain

$$\delta I_1 = \left[\frac{\partial f}{\partial y'}\eta \right]_{x_1}^{x_2} + \int_{x_1}^{x_2} \left[\frac{\partial f}{\partial y} - \frac{d}{dx}\left(\frac{\partial f}{\partial y'} \right) \right] \eta \, dx. \tag{4.30}$$

The end point conditions insures that the first term in equation (4.30) is zero. In order for the second term in equation (4.30) to be zero the function $y = y(x)$ must satisfy the Euler-Lagrange equation subject to boundary conditions $y(x_1) = y_1$ and $y(x_2) = y_2$. If these conditions are satisfied one can write that $\frac{dI}{d\epsilon}\Big|_{\epsilon=0} = 0$ is a necessary condition that the functional assume a stationary value.

We assume that the extremal curves y which are solutions of the Euler-Lagrange equations and the derivatives of y are continuous functions. However, this is not always the case. Discontinuous solutions to the Euler-Lagrange equation are also possible. In these cases the function y must be continuous but the first or higher derivatives may be discontinuous functions of the independent variable x. Discontinuous solutions are represented by calculating continuous curves over different sub-intervals of the solution domain and then requiring continuity of these solutions at the end points of the sub-intervals. In this way a continuous curve is constructed by joining the solution pieces at the end points of the sub-intervals so that the resulting curve is continuous. In this type of construction one is confronted with corners at the junction points where the left and right-hand derivatives are not always the same. At all corner points the quantities $\frac{\partial f}{\partial y'}$ and $f - \frac{\partial f}{\partial y'}y'$ must be continuous. These conditions are associated with the Weirstrass-Erdmann corner conditions to be discussed later in this chapter. For present we assume continuity for the functions y and its derivative $\frac{dy}{dx}$, where y is a solution of the Euler-Lagrange equation.

If the first variation is zero, $\delta I_1 = 0$, then the sign of the functional change ΔI is determined by the sign of δI_2. If $\delta I_2 > 0$ is positive and of constant sign at all points of the extremal arc $y(x)$, then a relative minimum exists. If $\delta I_2 < 0$, is negative and of constant sign at all points of the extremal arc $y(x)$, then a relative maximum exists. Note that the maximum or minimum value assigned to the functional must be independent of the value of η and ϵ. If δI_2 changes sign over the extremal arc, then the stationary value is neither a maximum or minimum. Examine equation (4.27) and note that if both δI_1 and δI_2 are zero, then the sign of ΔI depends upon the sign of $\epsilon^3 \delta I_3$ which changes sign for $\epsilon > 0$ and $\epsilon < 0$, so that there can be no maximum or minimum value unless δI_3 is also zero. Consequently, for these conditions the sign of ΔI is determined by an examination of the sign of the δI_4 variation, if it is different from zero.

We wish to examine the second variation

$$\delta I_2 = \int_{x_1}^{x_2} \left[\frac{\partial^2 f}{\partial y^2} \eta^2 + 2 \frac{\partial^2 f}{\partial y \partial y'} \eta \eta' + \frac{\partial^2 f}{\partial y'^2} (\eta')^2 \right] dx \tag{4.31}$$

under the conditions of a weak variation where $\eta(x_1) = 0$, $\eta(x_2) = 0$ and the Euler-Lagrange equation $f_y - \frac{d}{dx}(f_{y'}) = 0$ is satisfied by a function $y(x)$ which satisfies end point conditions $y(x_1) = y_1$ and $y(x_2) = y_2$. If $y(x)$ is a solution of the Euler-Lagrange equation and the functional I has a minimum value, then δI_2 must be positive for all values of η over the interval $x_1 \leq x \leq x_2$. If $y(x)$ is a solution of the Euler-Lagrange equation and the functional I has a maximum value, then δI_2 must be negative for all values of η over the interval $x_1 \leq x \leq x_2$.

The second variation can be written in several different forms for analysis. One form is to integrate the middle term in equation (4.31) using integration by parts to obtain

$$\delta I_2 = \left[\eta^2 \frac{\partial^2 f}{\partial y \partial y'}\right]_{x_1}^{x_2} + \int_{x_1}^{x_2} \left\{\eta^2 \left[\frac{\partial^2 f}{\partial y^2} - \frac{d}{dx}\left(\frac{\partial^2 f}{\partial y \partial y'}\right)\right] + (\eta')^2 \left[\frac{\partial^2 f}{\partial y'^2}\right]\right\} dx \tag{4.32}$$

Assume η is zero at the end points, then one can say that a necessary and sufficient condition for I to be a minimum is for $\delta I_2 > 0$ which requires that

$$\frac{\partial^2 f}{\partial y^2} - \frac{d}{dx}\left(\frac{\partial^2 f}{\partial y \partial y'}\right) > 0 \quad \text{and} \quad \frac{\partial^2 f}{\partial y'^2} > 0 \tag{4.33}$$

for all smooth weak variations η.

One can also write the second variation in the form

$$\delta I_2 = \int_{x_1}^{x_2} \left\{[\eta^2 f_{yy} + \eta \eta' f_{yy'}] + \eta'[\eta f_{yy'} + \eta' f_{y'y'}]\right\} dx \tag{4.34}$$

where subscripts denote partial differentiation. Integrate the second half of the integral (4.34) using integration by parts to obtain after simplification

$$\delta I_2 = \left[\eta^2 f_{yy'} + \eta \eta' f_{y'y'}\right]_{x_1}^{x_2} + \int_{x_1}^{x_2} \left[\eta^2 f_{yy} - \eta^2 \frac{d}{dx}(f_{yy'}) - \eta \frac{d}{dx}(\eta' f_{y'y'})\right] dx \tag{4.35}$$

By assumption the first term is zero at the end points. We now express the equation (4.35) using a differential operator $L(\ )$ and write

$$\delta I_2 = -\int_{x_1}^{x_2} \eta L(\eta)\, dx \tag{4.36}$$

where $L(\eta)$ is the differential operator

$$L(\eta) = \frac{d}{dx}\left[f_{y'y'} \frac{d\eta}{dx}\right] - \left[f_{yy} - \frac{d}{dx}(f_{yy'})\right]\eta \tag{4.37}$$

The differential operator (4.37) has the basic form

$$L(u) = \frac{d}{dx}\left[p(x)\frac{du}{dx}\right] - q(x)u \tag{4.38}$$

where

$$p(x) = f_{y'y'} \quad \text{and} \quad q(x) = f_{yy} - \frac{d}{dx}(f_{yy'}) \tag{4.39}$$

and the differential equation

$$L(u) = \frac{d}{dx}\left[p(x)\frac{du}{dx}\right] - q(x)u = 0, \qquad x_1 \le x \le x_2 \tag{4.40}$$

is known as Jacobi's differential equation. The operator $L(\ )$ is a self-adjoint Sturm-Liouville operator which satisfies the Lagrange identity

$$u L(\eta) - \eta L(u) = \frac{d}{dx}[p(x)(u\eta' - \eta u')] \tag{4.41}$$

146

We make use of the Lagrange identity and write the second variation (4.36) in the form

$$\delta I_2 = -\int_{x_1}^{x_2} \frac{\eta}{u} u L(\eta)\, dx = -\int_{x_1}^{x_2} \frac{\eta}{u}\left(\eta L(u) + \frac{d}{dx}\left[p(x)(u\eta' - \eta u')\right]\right) dx \qquad (4.42)$$

This form can be simplified by using integration by parts on the last integral with

$$
\begin{aligned}
U &= \eta/u & dV &= \frac{d}{dx}\left[p(x)(u\eta' - \eta u')\right] dx \\
dU &= \left(\frac{u\eta' - \eta u'}{u^2}\right) dx & V &= p(x)(u\eta' - \eta u')
\end{aligned}
\qquad (4.43)
$$

The second variation can now be expressed in the form

$$\delta I_2 = -\int_{x_1}^{x_2} \frac{\eta^2}{u} L(u)\, dx - \frac{\eta}{u} p(x)(u\eta' - \eta u')\Big|_{x_1}^{x_2} + \int_{x_1}^{x_2} p(x)\left(\frac{u\eta' - \eta u'}{u}\right)^2 dx$$

We substitute the value $p(x) = f_{y'y'}$ from equation (4.39) and rearrange terms to write the second variation as

$$\delta I_2 = -\int_{x_1}^{x_2} \frac{\eta^2}{u} L(u)\, dx + \frac{\eta}{u} f_{y'y'}(\eta u' - u\eta')\Big|_{x_1}^{x_2} + \int_{x_1}^{x_2} f_{y'y'}\left(\eta' - \eta\frac{u'}{u}\right)^2 dx \qquad (4.44)$$

Let us analyze the second variation given by equation (4.44). For the time being we shall assume that the following conditions are satisfied.

Condition (i)

Assume u is a nonzero solution to the Jacobi differential equation

$L(u) = 0$ for $x_1 \le x \le x_2$, with initial conditions $u(x_1) = 0$ and $u'(x_1) \ne 0$.

Condition (ii)

Assume the end point conditions $\eta(x_1) = 0$ and $\eta(x_2) = 0$ so that we are assured that the middle term in equation (4.44) is zero.

Condition (iii)

Assume $\left(\eta' - \eta\frac{u'}{u}\right)^2 \ne 0$ so that the sign of the third term in equation (4.44) is determined by the sign of $f_{y'y'}$.

then one can say that

The condition $f_{y'y'} > 0$ for $x_1 \le x \le x_2$ is a necessary
condition for the functional $I(y)$ to be a minimum.

The condition $f_{y'y'} < 0$ for $x_1 \le x \le x_2$ is a necessary
condition for the functional $I(y)$ to be a maximum.

These are known as Legendre's necessary conditions for an extremal to exist. We shall return to analyze the above conditions after first having developed some necessary background material associated with the Jacobi differential equation.

Background material for the Jacobi differential equation

We assume that $y = y(x, c_1, c_2)$ is a two-parameter solution of the Euler-Lagrange equation

$$\frac{\partial f(x, y, y')}{\partial y} - \frac{d}{dx}\left(\frac{\partial f(x, y, y')}{\partial y'}\right) = 0, \qquad x_1 \le x \le x_2 \qquad (4.45)$$

where c_1 and c_2 are constants. That is, $y = y(x, c_1, c_2)$ satisfies the Euler-Lagrange equation (4.45) so that one can write this equation in the form

$$\frac{\partial f(x, y(x, c_1, c_2), y'(x, c_1, c_2))}{\partial y} - \frac{d}{dx}\left(\frac{\partial f(x, y(x, c_1, c_2), y'(x, c_1, c_2))}{\partial y'}\right) = 0 \tag{4.46}$$

which is treated as an identity for all values of x in the interval (x_1, x_2). If we take the derivative of equation (4.46) with respect to the parameter c_1, we find

$$f_{yy}\frac{\partial y}{\partial c_1} + f_{yy'}\frac{\partial y'}{\partial c_1} - \frac{d}{dx}\left(f_{y'y}\frac{\partial y}{\partial c_1}\right) - \frac{d}{dx}\left(f_{y'y'}\frac{\partial y'}{\partial c_1}\right) = 0 \tag{4.47}$$

which simplifies to

$$\left(f_{yy} - \frac{d}{dx}f_{yy'}\right)\frac{\partial y}{\partial c_1} - \frac{d}{dx}\left(f_{y'y'}\frac{\partial y'}{\partial c_1}\right) = 0 \tag{4.48}$$

Let $u_1 = \frac{\partial y}{\partial c_1}$ and write the equation (4.48) in the form

$$L(u_1) = q(x)u_1 - \frac{d}{dx}\left(p(x)\frac{du_1}{dx}\right) = 0 \tag{4.49}$$

where $p(x)$ and $q(x)$ are defined in equation (4.39). Note that $u_1 = \frac{\partial y}{\partial c_1}$ is a solution of the Jacobi differential equation $L(u) = 0$ over the interval (x_1, x_2). Similarly, if you differentiate the Euler-Lagrange equation (4.46) with respect to the parameter c_2 you can show that $u_2 = \frac{\partial y}{\partial c_2}$ is a solution of the Jacobi differential equation

$$L(u_2) = q(x)u_2 - \frac{d}{dx}\left(p(x)\frac{du_2}{dx}\right) = 0.$$

The functions u_1 and u_2 are solutions of the Jacobi differential equation and so we can write

$$\frac{d}{dx}\left(p(x)\frac{du_1}{dx}\right) - q(x)u_1 = 0, \qquad x_1 \leq x \leq x_2 \tag{4.50}$$

$$\frac{d}{dx}\left(p(x)\frac{du_2}{dx}\right) - q(x)u_2 = 0, \qquad x_1 \leq x \leq x_2 \tag{4.51}$$

where $p(x) = f_{y'y'}$ and $q(x) = f_{yy} - \frac{d}{dx}f_{yy'}$ are continuous functions over the interval (x_1, x_2). Multiply equation (4.50) by u_2 and equation (4.51) by u_1 and then subtract the resulting equations to find

$$u_2\frac{d}{dx}\left(p(x)u_1'\right) - u_1\frac{d}{dx}\left(p(x)u_2'\right) = \frac{d}{dx}\left[p(x)\left(u_2u_1' - u_1u_2'\right)\right] = 0 \tag{4.52}$$

An integration of this equation shows that the solutions u_1 and u_2 of the Jacobi differential equation must satisfy

$$p(x)\left(u_2u_1' - u_1u_2'\right) = K = \text{a constant.} \tag{4.53}$$

This implies that the functions u_1 and u_2 are linearly independent solutions of the Jacobi differential equation and so the Wronskian determinant must be different from zero. Consequently, one can write

$$u_1(x)u_2'(x) - u_1'(x)u_2(x) = \begin{vmatrix} u_1(x) & u_2(x) \\ u_1'(x) & u_2'(x) \end{vmatrix} \neq 0 \tag{4.54}$$

The two solutions $u_1 = \frac{\partial y}{\partial c_1}$ and $u_2 = \frac{\partial y}{\partial c_2}$, where $y = y(x, c_1, c_2)$ is a solution of the Euler-Lagrange equation, are linearly independent solutions of the Jacobi differential equation so that the general solution can be written

$$u = u(x) = \beta_1 u_1(x) + \beta_2 u_2(x) \tag{4.55}$$

where β_1 and β_2 are constants.

Observe that if $f_{y'y'} > 0$ is positive for all values of x in the interval (x_1, x_2), then the zero's of any linear independent solutions $u_1(x)$ and $u_2(x)$ of the Jacobi differential equation must separate one another. This result is known as the Sturm separation theorem. One can prove this result as follows. Let $u_1(x)$ have two consecutive zeros at the points x_1 and x_2 so that $u_1(x_1) = 0$ and $u_1(x_2) = 0$, then the second independent solution $u_2(x)$ cannot equal zero at these points because the Wronskian determinant must satisfy the equation (4.54). To show that the zeros of $u_2(x)$ must lie between the consecutive zeros of $u_1(x)$ we use a proof by contradiction. Assume that $u_2(x)$ does not equal zero in the interval (x_1, x_2) where $u_1(x_1) = 0$ and $u_1(x_2) = 0$, then the function defined by $z = z(x) = \frac{u_1(x)}{u_2(x)}$ is continuous over the interval (x_1, x_2) with z being equal to zero at the end points of the interval.

By Rolle's theorem, if $z = z(x)$ for $x_1 \le x \le x_2$ is a function which is continuous with continuous derivative satisfying $z(x_1) = 0$ and $z(x_2) = 0$, then there must exist at least one point ξ in the interval $(x_1 < \xi < x_2)$ such that $z'(\xi) = 0$.

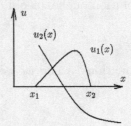

However, the derivative

$$\frac{dz}{dx} = z'(x) = \frac{d}{dx}\left(\frac{u_1(x)}{u_2(x)}\right) = \frac{u_2(x)u_1'(x) - u_1(x)u_2'(x)}{u_2^2(x)} \tag{4.56}$$

is a function with numerator equal to the Wronskian of the functions u_1 and u_2 and cannot equal zero over the interval (x_1, x_2). This is a contradiction to our assumption that $u_2(x)$ does not equal zero in the interval. Hence, $u_2(x)$ must have at least one zero in the interval (x_1, x_2). The function $u_2(x)$ cannot have more than one zero in the interval because if it had two zeros then $u_1(x)$ would have a zero between the consecutive zeros of $u_2(x)$ by the same type of argument just presented. Consequently, there results the Sturm-separation theorem. *The zeros of two real linearly distinct solutions of the linear second order Jacobi differential equation must separate one another.* This theorem does not hold if the solutions u_1 and u_2 of the Jacobi differential equation are not real solutions. The roots in question are sometimes referred to as kinetic foci or conjugate points.

Definition (Conjugate point)

Let $u = u(x)$ denote a nonzero solution to the Jacobi differential equation

$$L(u) = \frac{d}{dx}\left(f_{y'y'}\frac{du}{dx}\right) - \left(f_{yy} - \frac{d}{dx}f_{yy'}\right)u = 0$$

satisfying $u(x_1) = 0$ and $u'(x_1) \neq 0$ where the coefficients $f_{y'y'}$ and $f_{yy} - \frac{d}{dx}f_{yy'}$ are determined by the solution (stationary curve $y(x)$) of the Euler-Lagrange equation (4.45). All points x_c along the stationary curve $y(x)$, which satisfy $u(x_c) = 0$, are called points conjugate to x_1.

We have shown that the general solution of the Jacobi differential equation is given by equation (4.55) so that if $u(x_1) = 0$, then $\beta_1 u_1(x_1) + \beta_2 u_2(x_1) = 0$ with ratio

$$\frac{u_1(x_1)}{u_2(x_1)} = -\frac{\beta_2}{\beta_1}, \qquad \beta_1, \beta_2 \text{ constants} \tag{4.57}$$

Consequently, if x_c is a point conjugate to x_1, then $u(x_c) = \beta_1 u_1(x_c) + \beta_2 u_2(x_c) = 0$ which implies

$$\frac{u_1(x_c)}{u_2(x_c)} = -\frac{\beta_2}{\beta_1} = \frac{u_1(x_1)}{u_2(x_1)} \tag{4.58}$$

This shows that at all conjugate points the ratio of the independent solutions $\frac{u_1(x)}{u_2(x)}$ has the same value. This gives a test for determining conjugate points. Note that if one solves the determinant equation

$$\Delta(x_c, x_1) = \begin{vmatrix} u_1(x_c) & u_2(x_c) \\ u_1(x_1) & u_2(x_1) \end{vmatrix} = 0 = u_1(x_c)u_2(x_1) - u_2(x_c)u_1(x_1) \tag{4.59}$$

for the unknown(s) x_c, then the condition (4.58) is satisfied.

Let us now return to the previous conditions (i), (ii) and (iii) given on page 146.

Comments on Condition (i) Note that we desire a nonzero solution to the Jacobi differential equation. We cannot have the initial conditions $u(x_1) = 0$ and $u'(x_1) = 0$ as this would lead to the trivial solution which we don't want. Consequently, we assumed that $u'(x_1) \neq 0$. Also note that both $\eta(x)$ and $u(x)$ can be expanded in a Taylor series about the point x_1 to obtain

$$\eta(x) = \eta(x_1) + \eta'(x_1)(x - x_1) + \frac{\eta''(x_1)}{2!}(x - x_1)^2 + \cdots$$

$$u(x) = u(x_1) + u(x_1)(x - x_1) + \frac{u''(x_1)}{2!}(x - x_1)^2 + \cdots$$

If $\eta(x_1) = 0$ and $u(x_1) = 0$, then the ratio η/u can be expressed

$$\frac{\eta}{u} = \frac{\eta'(x_1) + \frac{\eta''(x_1)}{2!}(x - x_1) + \cdots}{u'(x_1) + \frac{u''(x_1)}{2!}(x - x_1)} = \cdots$$

so that the indeterminant form $\eta(x_1)/u(x_1)$ has a value in the limit as $x \to x_1$. Now re-examine the equation (4.44) and note that if $L(u) = 0$ and $u = u(x)$ has no conjugate points in the interval (x_1, x_2), then the first term is zero.

Comments on Condition (ii) Re-examine the second term in equation (4.44) and note that the only concern we would have is if x_2 is a conjugate point to x_1. Hence, one can say that the second term in equation (4.44) is zero provided x_2 is not a conjugate point to x_1.

Comments on Condition (iii) In the special case $\eta' - \eta\frac{u'}{u} = 0$ one can write $\frac{\eta'}{\eta} = \frac{u'}{u}$ which can be integrated to obtain $\eta(x) = c_3 u(x)$ where c_3 is a nonzero constant. In this special case the conditions $\eta(x_1) = 0$ and $\eta(x_2) = 0$ imply that $u(x_1) = 0$ and $u(x_2) = 0$. Hence if the roots or zeros of $u(x)$ (conjugate points) are outside the interval (x_1, x_2), then we are insured that $\eta' - \eta\frac{u'}{u} \neq 0$. That is, if the point x_2 is not conjugate to x_1, then $u(x_2)$ cannot equal zero and so η cannot be proportional to u everywhere along the stationary curve. Consequently, the sign of the second derivative term $f_{y'y'}$ in the third term of equation (4.44) controls the sign of the second variation.

In summary, the following, by themselves, are all necessary conditions for an extremal to exist for the variational problem $I = \int_{x_1}^{x_2} f(x, y, y')\, dx$

(i) $y = y(x)$ must be a smooth solution of the Euler-Lagrange equation.

(ii) The Legendre condition must be satisfied. Here $\frac{\partial^2 f}{\partial y'^2}$ must be of constant sign throughout the interval (x_1, x_2) and

    (a) if $\frac{\partial^2 f}{\partial y'^2} < 0$, then I is a maximum.

    (b) if $\frac{\partial^2 f}{\partial y'^2} > 0$, then I is a minimum.

(iii) The Jacobi test must be satisfied. That is, $u(x)$ must have no conjugate points to x_1 in the interval (x_1, x_2).

If all three of the above conditions are satisfied simultaneously, then they are sufficient conditions for a local weak maximum or minimum value to exist.

Example 4-3. **Jacobi test**

Find the extremals associated with the functional

$$I = \int_{(0,1)}^{(\pi/2,0)} f(x, y, y')\, dx = \int_{(0,1)}^{(\pi/2,0)} \left(y' \arcsin y' + \sqrt{1 - y'^2} + xy' \right) dx$$

Solution: First we calculate the Euler-Lagrange equation associated with this functional. We find

$$\frac{\partial f}{\partial y} - \frac{d}{dx}\left(\frac{\partial f}{\partial y'} \right) = 0 \quad \Rightarrow \quad \frac{d}{dx}(\arcsin y' + x) = 0$$

Two integrations of the Euler-Lagrange equation gives

$$\arcsin y' + x = \alpha$$
$$y' = \sin(\alpha - x) = -\sin(x - \alpha)$$
$$y = \cos(x - \alpha) + \beta$$

where α and β are constants of integration. This represents a two parameter family of curves. We desire those curves that pass through the points $(0, 1)$ and $(\pi/2, 0)$. Therefore we must select the constants α and β such that

$$\text{for } x = 0 \text{ and } y = 1 \text{ we require} \qquad 1 = \cos(-\alpha) + \beta$$

$$\text{for } x = \pi/2 \text{ and } y = 0 \text{ we require} \qquad 0 = \cos(\pi/2 - \alpha) + \beta$$

These equations imply that α and β must be chosen so that

$$\sin\alpha = -\beta$$

$$\text{and} \qquad \cos\alpha - \sin\alpha = 1 \tag{4.60}$$

Multiply the bottom equation (4.60) by $1/\sqrt{2}$ and use the cosine addition formula from trigonometry to solve for α as follows

$$\frac{1}{\sqrt{2}}\cos\alpha - \frac{1}{\sqrt{2}}\sin\alpha = \frac{1}{\sqrt{2}}$$

$$\cos\left(\alpha + \frac{\pi}{4}\right) = \frac{\sqrt{2}}{2}$$

$$\alpha + \frac{\pi}{4} = \frac{\pi}{4}, \frac{7\pi}{4}, \frac{9\pi}{4}, \frac{15\pi}{4}, \ldots$$

$$\alpha = 0, \frac{3\pi}{2}, 2\pi, \frac{7\pi}{2}, \ldots$$

and consequently we have $\qquad \beta = 0, \ 1, \ 0, \ 1, \ldots$

so that there are an infinite number of stationary curves to test. We will label these stationary curves $y = y_i(x)$ for $i = 0, 1, 2, \ldots$.

We next calculate two linearly independent solutions of the Jacobi differential equation. We find

$$y = \cos(x - \alpha) + \beta$$

$$\frac{\partial y}{\partial \alpha} = u(x) = \sin(x - \alpha)$$

$$\frac{\partial y}{\partial \beta} = v(x) = 1$$

We test for conjugate points inside the interval $0 < x < \pi/2$

For $\alpha = 0$ and $\beta = 0$ we find

$$\begin{vmatrix} u(x) & v(x) \\ u(0) & v(0) \end{vmatrix} = \begin{vmatrix} \sin x & 1 \\ 0 & 1 \end{vmatrix} = 0$$

or $\sin x = 0$. This gives the points

$$x = 0, \pi, 2\pi, 3\pi, \ldots$$

For $\alpha = 3\pi/2$ and $\beta = 1$ we find

$$\begin{vmatrix} u(x) & v(x) \\ u(0) & v(0) \end{vmatrix} = \begin{vmatrix} \sin(x - 3\pi/2) & 1 \\ 1 & 1 \end{vmatrix} = 0$$

or $\cos x = 1$. This gives the points

$$x = 0, 2\pi, 4\pi, 6\pi, \ldots$$

The above analysis shows that all the conjugate points are outside the interval $0 < x < \pi/2$.

The solutions of the Euler-Lagrange equation can be written

$$\alpha - 0, \quad \beta = 0, \qquad y = y_0(x) = \cos x$$

$$\alpha = 3\pi/2, \quad \beta = 1, \qquad y = y_1(x) = -\sin x + 1$$

$$\alpha = 2\pi, \quad \beta = 0, \qquad y = y_2(x) = \cos x$$

$$\alpha = 7\pi/2, \quad \beta = 1, \qquad y = y_3(x) = -\sin x + 1$$

and the solutions $y_i(x)$, for $i = 4, 5, 6, \ldots$ start to repeat themselves.

We use the Legendre test and calculate $f_{y'y'} = \frac{1}{\sqrt{1-y'^2}}$ and find that for $y = y_0 = \cos x$ the quantity $f_{y'y'}$ is positive everywhere within the interval $0 < x < \pi/2$ and hence y_0 produces a minimum value for I. For $y = y_1 = -\sin x + 1$ the quantity $f_{y'y'}$ is everywhere positive so that y_1 also produces a minimum value for I.

∎

A more general functional

Consider the variational problem

$$I = \int_{x_1}^{x_2} f(x, y, y', y'', \ldots, y^{(n)}) \, dx$$

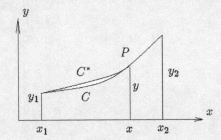

where (x_1, y_1) and point P are two points on a stationary curve C through the points (x_1, y_1) and (x_2, y_2). If P is the first point along the curve where it is possible to construct another stationary curve C^*, then the points (x_1, y_1) and P are said to be conjugate to one another.

If both paths along C and C^* are stationary paths, then each curve must be a solution of the Euler-Lagrange equation. In this general case, the solution of the Euler-Lagrange equation is a 2n-parameter family of curves denoted by $y = y(x, c_1, c_2, \ldots, c_{2n-1}, c_{2n})$. The equation of the stationary curve along C^* must come from this general form. Consequently, we must require that the variations $\delta y, \delta y', \ldots, \delta y^{(n-1)}$ must equal zero at the points (x_1, y_1) and at the conjugate point P. This requires that the system of equations

$$\delta y = \frac{\partial y}{\partial c_1} \delta c_1 + \frac{\partial y}{\partial c_2} \delta c_2 + \cdots + \frac{\partial y}{\partial c_{2n}} \delta c_{2n} = 0$$

$$\delta y' = \frac{\partial y'}{\partial c_1} \delta c_1 + \frac{\partial y'}{\partial c_2} \delta c_2 + \cdots + \frac{\partial y'}{\partial c_{2n}} \delta c_{2n} = 0$$

$$\vdots$$

$$\delta y^{(n-1)} = \frac{\partial y^{(n-1)}}{\partial c_1} \delta c_1 + \frac{\partial y^{(n-1)}}{\partial c_2} \delta c_2 + \cdots + \frac{\partial y^{(n-1)}}{\partial c_{2n}} \delta c_{2n} = 0$$

be satisfied at both the points (x_1, y_1) and (x, y). In order for a nonzero solution to the resulting system of 2n-equations to exist, it is required that the determinant

$$\begin{vmatrix} \frac{\partial y}{\partial c_1} & \frac{\partial y}{\partial c_2} & \cdots & \frac{\partial y}{\partial c_{2n}} \\ \frac{\partial y'}{\partial c_1} & \frac{\partial y'}{\partial c_2} & \cdots & \frac{\partial y'}{\partial c_{2n}} \\ \vdots & \vdots & \ddots & \vdots \\ \frac{\partial y^{(n-1)}}{\partial c_1} & \frac{\partial y^{(n-1)}}{\partial c_2} & \cdots & \frac{\partial y^{(n-1)}}{\partial c_{2n}} \\ \left(\frac{\partial y}{\partial c_1}\right)_1 & \left(\frac{\partial y}{\partial c_2}\right)_1 & \cdots & \left(\frac{\partial y}{\partial c_{2n}}\right)_1 \\ \left(\frac{\partial y'}{\partial c_1}\right)_1 & \left(\frac{\partial y'}{\partial c_2}\right)_1 & \cdots & \left(\frac{\partial y'}{\partial c_{2n}}\right)_1 \\ \vdots & \vdots & \ddots & \vdots \\ \left(\frac{\partial y^{(n-1)}}{\partial c_1}\right)_1 & \left(\frac{\partial y^{(n-1)}}{\partial c_2}\right)_1 & \cdots & \left(\frac{\partial y^{(n-1)}}{\partial c_{2n}}\right)_1 \end{vmatrix} = 0$$

be satisfied, where the top n rows of the above determinant denote the various partial derivatives evaluated at the point (x, y) and the bottom n rows of the above determinant have a suffix 1 attached to each element to denote that the various partial derivatives are to be evaluated at the point (x_1, y_1). Note that the previous example used a special case of this determinant. Solutions for x from this determinant give the points P where a conjugate point exists.

General variation

Previously we have held x constant and have considered only variations in y. Let us consider the functional [$f2$]

$$I = \int_{x_1}^{x_2} f(x, y, y', y'') \, dx \tag{4.61}$$

where we consider changes in both x and y as this is a more general type of variation. Let

$$\delta f = \frac{\partial f}{\partial x} \delta x + \frac{\partial f}{\partial y} \delta y + \frac{\partial f}{\partial y'} \delta y' + \frac{\partial f}{\partial y''} \delta y''$$
$$df = \frac{\partial f}{\partial x} dx + \frac{\partial f}{\partial y} dy + \frac{\partial f}{\partial y'} dy' + \frac{\partial f}{\partial y''} dy'' \tag{4.62}$$

denote two different changes in the integrand $f = f(x, y, y', y'')$. The commutative property of the d and δ operators, $d(\delta x) = \delta(dx)$, is employed to write

$$\delta I = \delta \int_{x_1}^{x_2} f \, dx = \int_{x_1}^{x_2} (\delta f \, dx + f \delta \, dx) = \int_{x_1}^{x_2} (\delta f \, dx + f d \, \delta x)$$

Integrate the last term by parts to obtain

$$\delta I = [f \, \delta x]_{x_1}^{x_2} + \int_{x_1}^{x_2} (\delta f \, dx - df \, \delta x) \tag{4.63}$$

The equations (4.62) are employed to obtain

$$\delta f \, dx - df \, \delta x = \frac{\partial f}{\partial y}(\delta y - y' \delta x) \, dx + \frac{\partial f}{\partial y'}(\delta y' - y'' \delta x) \, dx + \frac{\partial f}{\partial y''}(\delta y'' - y''' \delta x) \, dx$$

so that the equation (4.63) can be expressed in the form

$$\delta I = \delta \int_{x_1}^{x_2} f(x, y, y', y'') \, dx = [f \, \delta x]_{x_1}^{x_2} + \int_{x_1}^{x_2} \left[\frac{\partial f}{\partial y} \omega \, dx + \frac{\partial f}{\partial y'} \omega' \, dx + \frac{\partial f}{\partial y''} \omega'' \, dx \right]. \tag{4.64}$$

where we have made the substitution $\omega = \delta y - y' \delta x$ along with its derivatives $\omega' = \delta y' - y'' \delta x$ and $\omega'' = \delta y'' - y''' \delta x$. An integration by parts of the middle term in equation (4.64) gives us

$$\int_{x_1}^{x_2} \frac{\partial f}{\partial y'} \omega' \, dx = \left[\frac{\partial f}{\partial y'} \omega \right]_{x_1}^{x_2} - \int_{x_1}^{x_2} \frac{d}{dx} \left(\frac{\partial f}{\partial y'} \right) \omega \, dx. \tag{4.65}$$

Use integration by parts twice on the last term in equation (4.64) to obtain

$$\int_{x_1}^{x_2} \frac{\partial f}{\partial y''} \omega'' \, dx = \left[\frac{\partial f}{\partial y''} \omega' - \frac{d}{dx} \left(\frac{\partial f}{\partial y''} \right) \omega \right]_{x_1}^{x_2} + \int_{x_1}^{x_2} \frac{d^2}{dx^2} \left(\frac{\partial f}{\partial y''} \right) \omega \, dx \tag{4.66}$$

The integrals given by equations (4.66) and (4.65) simplifies the equation (4.64) to the form

$$\delta I = \delta \int_{x_1}^{x_2} f(x, y, y', y'') \, dx = \int_{x_1}^{x_2} \left[\frac{\partial f}{\partial y} - \frac{d}{dx}\left(\frac{\partial f}{\partial y'}\right) + \frac{d^2}{dx^2}\left(\frac{\partial f}{\partial y''}\right) \right] \omega \, dx$$
$$+ \left[f\delta x + \frac{\partial f}{\partial y'}\omega + \frac{\partial f}{\partial y''}\omega' - \frac{d}{dx}\left(\frac{\partial f}{\partial y''}\right)\omega \right]_{x_1}^{x_2} \tag{4.67}$$

The equation (4.67) gives a general representation for the variation of the functional I given by equation (4.61). At an extremum we require that $\delta I = 0$ so that the term in brackets under the integral in equation (4.67) must equal zero and the boundary conditions must also equal zero or

$$\frac{\partial f}{\partial y} - \frac{d}{dx}\left(\frac{\partial f}{\partial y'}\right) + \frac{d^2}{dx^2}\left(\frac{\partial f}{\partial y''}\right) = 0$$

$$\left(f\delta x + \left[\frac{\partial f}{\partial y'} - \frac{d}{dx}\left(\frac{\partial f}{\partial y''}\right)\right](\delta y - y'\delta x) \right) + \frac{\partial f}{\partial y''}(\delta y' - y''\delta x)\Bigg|_{x=x_1} = 0 \tag{4.68}$$

$$\left(f\delta x + \left[\frac{\partial f}{\partial y'} - \frac{d}{dx}\left(\frac{\partial f}{\partial y''}\right)\right](\delta y - y'\delta x) \right) + \frac{\partial f}{\partial y''}(\delta y' - y''\delta x)\Bigg|_{x=x_2} = 0$$

A similar type of analysis can be performed on the functional $[f3]$, however the details become more cumbersome.

In the special case where f is not a function of y'' the equation (4.61) reduces to the functional $[f1]$

$$I = \int_{x_1}^{x_2} f(x, y, y') \, dx \tag{4.69}$$

and the equations (4.68) reduce to the form

$$\frac{\partial f}{\partial y} - \frac{d}{dx}\left(\frac{\partial f}{\partial y'}\right) = 0$$

$$f\delta x + \frac{\partial f}{\partial y'}(\delta y - y'\delta x)\Bigg|_{x=x_1} = 0 \tag{4.70}$$

$$f\delta x + \frac{\partial f}{\partial y'}(\delta y - y'\delta x)\Bigg|_{x=x_2} = 0$$

Let us analyze this simpler set of equations.

Case 1: If at the end points we have the prescribed fixed conditions that $y_1 = y(x_1)$ and $y_2 = y(x_2)$, then $\delta x = 0$ and $\delta y = 0$ so that the equations (4.70) reduce to just the Euler-Lagrange equation to be satisfied.

Case 2: If the end points are required to lie on given curves, say $y = g(x)$ at x_1 and $y = h(x)$ at x_2, then the variations δy in the equations (4.70) cease to be arbitrary. In this case we would require that $\delta y\Big|_{x=x_1} = g'(x_1)\delta x$ and $\delta y\Big|_{x=x_2} = h'(x_2)\delta x$.

In the previous chapter 3 we have assumed that the case 1 above was always implied. Let us now investigate the case 2.

Movable boundaries

Another kind of free end condition associated with a calculus of variation problem is the situation where the solution curve is required to satisfy the condition that one or both end points move along given curves.

Consider for example a calculus of variations problem that the functional

$$I = \int_{x_1}^{x} f(x, y, y') \, dx$$

have an extremum, where the left-hand end point (x_1, y_1) is fixed and the right-hand end point x is unknown, but must lie on a given curve $y = h(x)$.

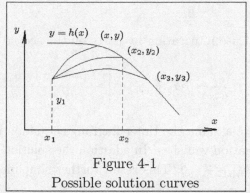

Figure 4-1
Possible solution curves

Here one end point is fixed and the other end point is variable as it can move along the given curve $y = h(x)$. The situation is illustrated in the figure 4-1. We know that the Euler-Lagrange equation is a second order ordinary differential equation with general solution family containing two arbitrary constants. We express this general solution family in the form $y = \phi(x, \alpha, \beta)$ where α, β are constants. This representation is called a two-parameter family of solution curves. For the given problem we want only those curves from the solution family which satisfy the left-end condition $y_1 = \phi(x_1, \alpha, \beta)$. This end point condition allows one to solve for one of the constants in terms of the other constant. Say we solve for β as a function of α. Consequently, the desired solution curve will be of the form $y = \phi(x, \alpha)$ where α is an unknown parameter. If this curve intersects the given curve $y = h(x)$, say at the point (x_3, y_3), then α has some specific value, say $\alpha = \alpha_3$. That is, at the point where the two curves intersect one can write the equation

$$y_3 = \phi(x_3, \alpha) = h(x_3) \tag{4.71}$$

which implies that the constant α depends upon the value x_3. One can then introduce the notation $\alpha = \alpha(x_3) = \alpha_3$ to emphasize this dependence. Substituting this solution curve into the given functional gives

$$I = I(x_3) = \int_{x_1}^{x_3} f\left(x, \phi(x, \alpha_3), \frac{\partial \phi(x, \alpha_3)}{\partial x}\right) dx. \tag{4.72}$$

We wish to examine the variation of this integral as x_3 moves along the given curve $y = h(x)$. Assume that $y = \phi(x, \alpha_2)$, satisfying $y_2 = \phi(x_2, \alpha_2)$, is the curve which produces an extremum and consider the curve $y = \phi(x, \alpha_3)$ a variation away from the true solution curve. At an extremum we require that $\frac{dI}{dx_3}\Big|_{x_3=x_2} = 0$. Differentiating the equation (4.72) using the Leibnitz rule, one finds

$$\frac{dI}{dx_3} = f\left(x_3, \phi(x_3, \alpha_3), \frac{\partial \phi(x_3, \alpha_3)}{\partial x}\right) + \int_{x_1}^{x_3} \left[\frac{\partial f}{\partial y}\frac{\partial y}{\partial \alpha_3} + \frac{\partial f}{\partial y'}\frac{\partial y'}{\partial \alpha_3}\right]\frac{d\alpha_3}{dx_3} \, dx. \tag{4.73}$$

In the equation (4.73) $\frac{\partial y}{\partial \alpha_3} = \frac{\partial \phi}{\partial \alpha_3}$ and $\frac{\partial y'}{\partial \alpha_3} = \frac{\partial^2 \phi}{\partial x \partial \alpha_3}$. Let

$$U = \frac{\partial f}{\partial y'} \qquad\qquad dV = \frac{\partial^2 \phi}{\partial x \partial \alpha_3} \frac{d\alpha_3}{dx_3} dx$$

$$dU = \frac{d}{dx}\left(\frac{\partial f}{\partial y'}\right) dx \qquad V = \frac{\partial \phi(x, \alpha_3)}{\partial \alpha_3} \frac{d\alpha_3}{dx_3}$$

and use integration by parts on the second term in equation (4.73) to obtain

$$\frac{dI}{dx_3} = f(x_3, y_3, y_3') + \left[\frac{\partial f\left(x, \phi, \frac{\partial \phi}{\partial x}\right)}{\partial y'} \frac{\partial \phi(x, \alpha_3)}{\partial \alpha_3} \frac{d\alpha_3}{dx_3}\right]_{x_1}^{x_3} + \int_{x_1}^{x_3}\left(\frac{\partial f}{\partial y} - \frac{d}{dx}\left[\frac{\partial f}{\partial y'}\right]\right)\frac{\partial \phi}{\partial \alpha_3} \frac{d\alpha_3}{dx_3} dx$$

If we assume the Euler-Lagrange equation is satisfied, then the right-hand side of the above equation vanishes. In addition the solution is assumed to satisfy $y_1 = \phi(x_1, \alpha_3) = constant$, and so $\frac{\partial \phi(x_1, \alpha_3)}{\partial \alpha_3} = 0$. This result further simplifies the above equation to the form

$$\frac{dI}{dx_3} = f(x_3, y_3, y_3') + \frac{\partial f(x_3, y_3, y_3')}{\partial y'} \frac{\partial \phi(x_3, \alpha_3)}{\partial \alpha_3} \frac{d\alpha_3}{dx_3}. \tag{4.74}$$

At an extremum we require that

$$\frac{dI}{dx_3}\bigg|_{x_3 = x_2} = f(x_2, y_2, y_2') + \frac{\partial f(x_2, y_2, y_2')}{\partial y'}\left(\frac{\partial \phi}{\partial \alpha_3} \frac{d\alpha_3}{dx_3}\right)\bigg|_{x_3 = x_2} = 0. \tag{4.75}$$

Using our previous assumption that $y_3 = h(x_3) = \phi(x_3, \alpha_3)$ one can calculate the derivative

$$h'(x_3) = \frac{\partial \phi}{\partial x_3} + \frac{\partial \phi}{\partial \alpha_3} \frac{d\alpha_3}{dx_3}$$

so that in the limit as $x_3 \to x_2$ and $\alpha_3 \to \alpha_2$ one obtains

$$h'(x_2) - \frac{\partial \phi(x_2, \alpha_2)}{\partial x_2} = \frac{\partial \phi}{\partial \alpha_3} \frac{d\alpha_3}{dx_3}\bigg|_{x_3 = x_2}.$$

Substitute this result into equation (4.75) to produce the transversality condition that the solution y must satisfy

$$f(x_2, y_2, y_2') + (h'(x_2) - y_2')\frac{\partial f(x_2, y_2, y_2')}{\partial y'} = 0 \tag{4.76}$$

at the end point x_2 where I is an extremum.

This same result can be obtained from the equation (4.70). In the equations (4.70) the first equation is the Euler-Lagrange equation. The second equation reduces to zero because $\delta x = \delta y = 0$ at $x = x_1$. In the third equation, of the set, we have at the point $x = x_2$ the relation $\delta y\big|_{x = x_2} = h'(x_2)\delta x$ which reduces the equation to the transversality condition

$$f(x_2, y_2, y_2') + (h'(x_2) - y_2')\frac{\partial f(x_2, y_2, y_2')}{\partial y'} = 0.$$

Example 4-4.   Transversality condition

Find an extremum for the functional $I = \int_{(0,1)}^{(x,y)} \dfrac{x^2}{y'}\, dx$ where the point (x, y) is to lie on the curve $y = 9 - x^2$ for $x > 0$. The situation is illustrated in the figure 4-2.

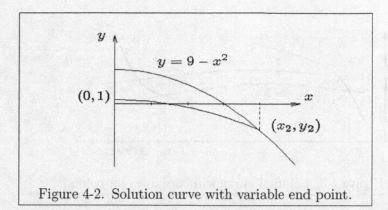

Figure 4-2. Solution curve with variable end point.

Solution: Here $f = \frac{x^2}{y'}$, $\frac{\partial f}{\partial y'} = \frac{-x^2}{(y')^2}$ and $\frac{\partial f}{\partial y} = 0$ so that the Euler-Lagrange equation becomes $\frac{d}{dx}\left(\frac{-x^2}{(y')^2}\right) = 0$. This equation is easily integrated to obtain the first integral $\frac{-x^2}{(y')^2} = C$, where C is a constant of integration. We select $C = \frac{-1}{4\alpha^2}$, where α is some new constant. This selection greatly simplifies the resulting algebra in solving for the derivative y'. One finds that $y' = 2\alpha x$ which is a differential equation that can be integrated directly to obtain the two-parameter solution family

$$y = \alpha x^2 + \beta = \phi(x, \alpha, \beta) \qquad (4.77)$$

The two-parameter family of curves given by equation (4.77) is to pass through the point $(0, 1)$ and consequently, $\beta = 1$. This value for β reduces the equation (4.77) to the one-parameter solution family

$$y = \alpha x^2 + 1 \quad \text{with derivative} \quad y' = 2\alpha x. \qquad (4.78)$$

At a point $x_2 > 0$ we find $y_2 = \alpha x_2^2 + 1$ and $y_2' = 2\alpha x_2$ so that the transversality condition

$$f(x_2, y_2, y_2') + (h'(x_2) - y_2')\frac{\partial f(x_2, y_2, y_2')}{\partial y'} = 0$$

becomes $\qquad \dfrac{x_2^2}{y_2'} + (-2x_2 - y_2')\left[\dfrac{-x_2^2}{(y_2')^2}\right] = 0.$
$$\qquad (4.79)$$

In order for the equation (4.79) to be satisfied either $x_2 = 0$ or $y_2' = -x_2$. We desire $x_2 > 0$ and so we set $y_2' = -x_2$. This gives $\alpha = \frac{y_2'}{2x_2} = -\frac{1}{2}$ and so the desired solution curve is $y = -\frac{1}{2}x^2 + 1$. This curve intersects the given curve $y = h(x) = 9 - x^2$ at the point $x_2 = 4$ and $y_2 = -7$. One finds that the functional I has a maximum value under these conditions.

■

End points on two different curves

Consider the extreme values associated with the functional

$$I = I(y) = \int_{x_1}^{x_2} f(x, y, y')\, dx \qquad (4.80)$$

where we restrict our study to all smooth curves $y(x)$ with end point (x_1, y_1) required to lie on a given curve $y = g(x)$ and the other end point (x_2, y_2) is required to lie on another given curve $y = h(x)$ as illustrated in the figure 4-3.

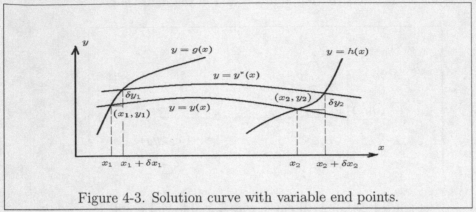

Figure 4-3. Solution curve with variable end points.

The general variation of the functional $I = I(y)$ is given by the equation (4.67) from which we obtain, by removing the y'' terms, the following reduced form for analysis.

$$\delta I = \int_{x_1}^{x_2} \left[\frac{\partial f}{\partial y} - \frac{d}{dx}\left(\frac{\partial f}{\partial y'} \right) \right] dx$$
$$+ \left(\frac{\partial f}{\partial y'} \delta y + \left[f - y' \frac{\partial f}{\partial y'} \right] \delta x \right) \Big|_{x=x_2} - \left(\frac{\partial f}{\partial y'} \delta y + \left[f - y' \frac{\partial f}{\partial y'} \right] \delta x \right) \Big|_{x=x_1} \tag{4.81}$$

If $I = I(y)$ has an extremum for the function $y = y(x)$, then

(i) The function $y = y(x)$ must be a solution of the Euler-Lagrange equation

$$\frac{\partial f}{\partial y} - \frac{d}{dx}\left(\frac{\partial f}{\partial y'} \right) = 0.$$

(ii) $\delta I = 0$ at an extremum, so that

$$\left(\frac{\partial f}{\partial y'} \delta y + \left[f - y' \frac{\partial f}{\partial y'} \right] \delta x \right) \Big|_{x=x_2} - \left(\frac{\partial f}{\partial y'} \delta y + \left[f - y' \frac{\partial f}{\partial y'} \right] \delta x \right) \Big|_{x=x_1} = 0. \tag{4.82}$$

Along the curve $y = g(x)$ we have $\delta y = g'(x)\delta x$ and along the curve $y = h(x)$ we have $\delta y = h'(x)\delta x$ so that the condition given by equation (4.82) can be written as

$$\left(h'(x)\frac{\partial f}{\partial y'} + f - y' \frac{\partial f}{\partial y'} \right) \delta x \Big|_{x=x_2} - \left(g'(x)\frac{\partial f}{\partial y'} + f - y' \frac{\partial f}{\partial y'} \right) \delta x \Big|_{x=x_1} = 0 \tag{4.83}$$

Here the variations in δx_1 and δx_2 are assumed to be independent so that the equation (4.83) implies the boundary conditions

$$\left(h'(x)\frac{\partial f}{\partial y'} + f - y' \frac{\partial f}{\partial y'} \right) \Big|_{(x_2, y_2)} = 0$$
$$\left(g'(x)\frac{\partial f}{\partial y'} + f - y' \frac{\partial f}{\partial y'} \right) \Big|_{(x_1, y_1)} = 0 \tag{4.84}$$

These are the transversality conditions and the curve $y = y(x)$ which satisfies these conditions is called a transversal of the given curves $y = g(x)$ and $y = h(x)$.

Example 4-5. Transversality condition

Find the shortest distance between the curves $y = 1 + \frac{x^2}{4}$ and $y = \frac{x}{2}$.

Solution:

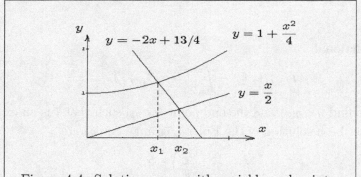

Figure 4-4. Solution curve with variable end points.

Let (x_1, y_1) and (x_2, y_2) denote the points on the given curves which are closest to one another. The arc length between these points is determined from the integral

$$I = \int_{x_1}^{x_2} \sqrt{1 + (y')^2}\, dx.$$

Here $f = \sqrt{1 + (y')^2}$ with $\frac{\partial f}{\partial y'} = \frac{y'}{\sqrt{1+(y')^2}}$ so that the Euler-Lagrange equation can be written

$$\frac{d}{dx}\left(\frac{y'}{\sqrt{1+(y')^2}}\right) = 0 \qquad \text{which implies} \quad y' = \alpha = \text{a constant.}$$

Another integration gives the two parameter solution family

$$y = y(x, \alpha, \beta) = \alpha x + \beta \quad \text{where} \quad \alpha, \ \beta \quad \text{are constants.}$$

We apply the transversality condition given by equations (4.84) at the end points x_1 and x_2. At the right end point x_2 we require that

$$\sqrt{1 + (y_2')^2} + \frac{y_2'}{\sqrt{1+(y_2')^2}}\left(\frac{1}{2} - y_2'\right) = 0$$

and since $y_2' = \alpha$ this reduces to

$$1 + \alpha^2 + \alpha\left(\frac{1}{2} - \alpha\right) = 0$$

which implies that $\alpha = -2$. At the left end point x_1 the transversality condition becomes

$$\sqrt{1 + (y_1')^2} + \frac{y_1'}{\sqrt{1+(y_1')^2}}\left(\frac{x_1}{2} - y_1'\right) = 0.$$

Using the result $y_1' = \alpha$ this equation reduces to

$$1 + \alpha^2 + \alpha\left(\frac{x_1}{2} - \alpha\right) = 0, \quad \text{which implies that} \quad x_1 = \frac{-2}{\alpha} = 1.$$

At $x_1 = 1$ we have $y_1 = 1 + \frac{1}{4} = \frac{5}{4} = -2(1) + \beta$ so that $\beta = \frac{13}{4}$. At x_2 one finds $y_2 = \frac{x_2}{2} = -2x_2 + \frac{13}{4}$ which implies $x_2 = \frac{13}{10}$. Therefore, $(x_1, y_1) = (1, 5/4)$ and $(x_2, y_2) = (13/10, 13/20)$ and the straight line through these points is given by $y = -2x + 13/4$. The minimum distance is $3\sqrt{5}/10$.

∎

Free end points

Consider the functional

$$I = I(x_2, y_2) = \int_{(x_0, y_0)}^{(x_2, y_2)} f(x, y, y') \, dx \tag{4.85}$$

where it is desired to find $y = y(x)$ and the end point (x_2, y_2) such that I is an extremum. The function $y = y(x)$ must be a solution of the Euler-Lagrange equation

$$\frac{\partial f}{\partial y} - \frac{d}{dx} \left(\frac{\partial f}{\partial y'} \right) = 0$$

which is a second-order ordinary differential equation. Second order ordinary differential equations have general solutions containing two parameters, say, α and β and so the general solution can be expressed in the form $y = y(x, \alpha, \beta)$. The initial condition $y_0 = y(x_0, \alpha, \beta)$ allows one to solve for β in terms of α or α in terms of β so that the general solution now contains only one parameter. We therefore consider the one-parameter family of solutions $y = \phi(x, \gamma)$ which satisfy the Euler-Lagrange equation and pass through the end point (x_0, y_0). That is, each member of the family must satisfy the end condition $y_0 = \phi(x_0, \gamma)$. Selected members from this family of curves are illustrated in the figure 4-4.

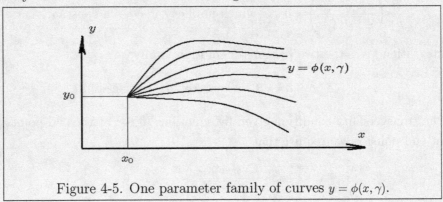

Figure 4-5. One parameter family of curves $y = \phi(x, \gamma)$.

Note that the parameter γ is a function of the coordinates (x_2, y_2) since the solution curve must satisfy $y_2 = \phi(x_2, \gamma)$. This functional dependence is denoted $\gamma = \gamma(x_2, y_2)$. We desire to find (x_2, y_2) such that

$$I = I(x_2, y_2) = \int_{(x_0, y_0)}^{(x_2, y_2)} f(x, \phi(x, \gamma(x_2, y_2)), \phi_x(x, \gamma(x_2, y_2))) \, dx \tag{4.86}$$

is an extremum. Here I is treated as a function of the two-variables x_2 and y_2. We desire to find the critical point (x_2, y_2) such that

$$\frac{\partial I}{\partial x_2} = 0, \qquad \text{and} \qquad \frac{\partial I}{\partial y_2} = 0, \tag{4.87}$$

because this is where an extremum for I occurs. Calculating the partial derivatives of the equation (4.86) one obtains

$$\frac{\partial I}{\partial x_2} = f(x_2, y_2, y_2') + \int_{x_0}^{x_2} \left[f_y(x, \phi, \phi_x) \frac{\partial \phi}{\partial \gamma} \frac{\partial \gamma}{\partial x_2} + f_{y'}(x, \phi, \phi_x) \frac{\partial^2 \phi}{\partial x \partial \gamma} \frac{\partial \gamma}{\partial x_2} \right] dx \tag{4.88}$$

$$\frac{\partial I}{\partial y_2} = \int_{x_0}^{x_2} \left(f_y(x, \phi, \phi_x) \frac{\partial \phi}{\partial \gamma} \frac{\partial \gamma}{\partial y_2} + f_{y'}(x, \phi, \phi_x) \frac{\partial^2 \phi}{\partial x \partial \gamma} \frac{\partial \gamma}{\partial y_2} \right) dx \tag{4.89}$$

Integrate the second term within the integral in equation (4.88) using integration by parts with

$$U = f_{y'} \qquad\qquad dV = \frac{\partial^2 \phi}{\partial x \partial \gamma} \frac{\partial \gamma}{\partial x_2} dx$$
$$dU = \frac{d}{dx}(f_y') \, dx \qquad\qquad V = \frac{\partial \phi}{\partial y} \frac{\partial \gamma}{\partial x_2}$$

Also integrate the second term in equation (4.89) using integration by parts with

$$U = f_{y'} \qquad\qquad dV = \frac{\partial^2 \phi}{\partial x \partial \gamma} \frac{\partial \gamma}{\partial y_2} dx$$
$$dU = \frac{d}{dx}(f_y') \, dx \qquad\qquad V = \frac{\partial \phi}{\partial \gamma} \frac{\partial \gamma}{\partial y_2}$$

One can then verify that the equations (4.88) and (4.89) reduce to

$$\frac{\partial I}{\partial x_2} = f(x_2, y_2, y_2') + \left[\frac{\partial f}{\partial y'} \frac{\partial \phi}{\partial \gamma} \frac{\partial \gamma}{\partial x_2} \right]_{x_0}^{x_2} + \int_{x_0}^{x_2} \left[\frac{\partial f}{\partial y} - \frac{d}{dx} \left(\frac{\partial f}{\partial y'} \right) \right] \frac{\partial \phi}{\partial \gamma} \frac{\partial \gamma}{\partial x_2} dx \tag{4.90}$$

$$\frac{\partial I}{\partial y_2} = \frac{\partial f}{\partial y'} \frac{\partial \phi}{\partial \gamma} \frac{\partial \gamma}{\partial y_2} \Big|_{x_0}^{x_2} + \int_{x_0}^{x_2} \left[\frac{\partial f}{\partial y} - \frac{d}{dx} \left(\frac{\partial f}{\partial y'} \right) \right] \frac{\partial \phi}{\partial \gamma} \frac{\partial \gamma}{\partial y_2} dx \tag{4.91}$$

The terms within the brackets under the integral sign is zero because the solution is assumed to satisfy the Euler-Lagrange equation. Also note that since $y_0 = \phi(x_0, \gamma)$, and $y_2 = \phi(x_2, \gamma)$, then one can write

$$\frac{\partial \phi}{\partial \gamma} \Big|_{x=x_0} = 0, \qquad 0 = \frac{\partial \phi}{\partial x_2} + \frac{\partial \phi}{\partial \gamma} \frac{\partial \gamma}{\partial x_2}, \qquad 1 = \frac{\partial \phi}{\partial \gamma} \frac{\partial \gamma}{\partial y_2} \tag{4.92}$$

Therefore, at an extremum of I, we must require that x_2 and y_2 are selected such that

$$\frac{\partial I}{\partial x_2} = f(x_2, y_2, y_2') + f_{y'}(x_2, y_2, y_2')(-y_2') = 0 \tag{4.93}$$

$$\frac{\partial I}{\partial y_2} = f_{y'}(x_2, y_2, y_2') \frac{\partial \phi}{\partial \gamma} \frac{\partial \gamma}{\partial y_2} = f_{y'}(x_2, y_2, y_2') = 0 \tag{4.94}$$

The equation (4.94) simplifies the condition (4.93) to the form $f(x_2, y_2, y_2') = 0$. The conditions (4.93) and (4.94) together with the initial conditions enable one to solve for the free coordinates (x_2, y_2). In summary, if $y = y(x, \alpha, \beta)$ is a solution of the second order Euler-Lagrange equation $\frac{\partial f}{\partial y} - \frac{d}{dx} \left(\frac{\partial f}{\partial y'} \right) = 0$, then the constants α, β, x_2 and y_2 are determined from solving the simultaneous system of equations

$$y_0 = y(x_0, \alpha, \beta)$$
$$y_2 = y(x_2, \alpha, \beta)$$
$$f(x_2, y_2, y_2') = 0$$
$$f_{y'}(x_2, y_2, y_2') = 0 \tag{4.95}$$

Example 4-6.   Free end conditions

Find the arc $y = y(x)$ such that $I = \int_{(0,1)}^{(x_2,y_2)} y^2[1+(y')^2]\,dx$ is an extremum, where the upper end point is free to be selected.

Solution: Here $f = y^2[1+(y')^2]$ is independent of the variable x and has the partial derivatives

$$\frac{\partial f}{\partial y} = 2y[1+(y')^2], \qquad \frac{\partial f}{\partial y'} = 2y^2 y' \tag{4.96}$$

The Euler-Lagrange equation is calculated and one finds

$$f_y - \frac{d}{dx}(f'_y) = 0 \quad \Rightarrow \quad 2y[1+(y')^2] - \frac{d}{dx}\left(2y^2 y'\right) = 0. \tag{4.97}$$

The first integral can be expressed $2y^2(y')^2 - y^2[1+(y')^2] = \alpha^2$ which implies

$$y^2(y')^2 - y^2 = \alpha^2 \quad \text{or} \quad \frac{2y\,dy}{\sqrt{y^2+\alpha^2}} = 2\,dx.$$

A second integration produces the result $2\sqrt{y^2+\alpha^2} = 2x + 2\beta$ or $y^2 = (x+\beta)^2 - \alpha^2$, where α and β are constants. We require that at $x = 0$, $y = 1$ and at $x = x_2$, $y = y_2$ together with the conditions $f(x_2, y_2, y'_2) = 0$ and $f_{y'}(x_2, y_2, y'_2) = 0$. This gives the system of equations

$$x = 0,\ y = 1 \text{ condition} \quad \Rightarrow \quad 1 = \beta^2 - \alpha^2$$

$$x = x_2,\ y = y_2 \text{ condition} \quad \Rightarrow \quad y_2^2 = (x_2 + \beta)^2 - \alpha^2$$

$$f(x_2, y_2, y'_2) = 0 \text{ condition} \quad \Rightarrow \quad y_2^2[1 + (y'_2)^2] = 0$$

$$f_{y'}(x_2, y_2, y'_2) = 0 \text{ condition} \quad \Rightarrow \quad 2y_2^2 y'_2 = 0$$

Solving this system of equations we find $y_2 = 0$, so that from the first integral $\alpha = 0$, which implies $\beta = 1$ and $x_2 = -1$. This gives the solution curve $y = x + 1$ which makes I have a minimum value.

■

Example 4-7. In the special case one is to find an extremum for the integral $I = \int_{x_1}^{x_2} f(y, y')\,dx$ we note that x does not explicitly occur in the problem and therefore a first integral of the Euler-Lagrange equation is written

$$f - y'\frac{\partial f}{\partial y'} = \alpha_0 = \text{a constant} \tag{4.98}$$

The end conditions from the equations (4.82)

$$\frac{\partial f}{\partial y'}\delta y + \left(f - y'\frac{\partial f}{\partial y'}\right)\delta x \bigg|_{x=x_1} = 0, \qquad \text{and} \qquad \frac{\partial f}{\partial y'}\delta y + \left(f - y'\frac{\partial f}{\partial y'}\right)\delta x \bigg|_{x=x_2} = 0 \tag{4.99}$$

then become

$$\alpha_0\,\delta x + \frac{\partial f}{\partial y'}\delta y \bigg|_{x=x_1} = 0, \qquad \text{and} \qquad \alpha_0\,\delta x + \frac{\partial f}{\partial y'}\delta y \bigg|_{x=x_2} = 0 \tag{4.100}$$

Case 1: If the terminal points are fixed, then $\delta x = 0$ and $\delta y = 0$ so that the conditions given by equation (4.100) are satisfied. A second integration of equation (4.98) will give a solution $y = y(x, \alpha_0, \beta_0)$ where α_0 and β_0 are constants selected so that the solution passes through the given end points (x_1, y_1) and (x_2, y_2).

Case 2: If the end points are to lie on given curves

$$y = g(x) \qquad \text{and} \qquad y = h(x)$$
$$\text{with} \qquad \delta y = g'(x)\,\delta x \qquad \text{and} \qquad \delta y = h'(x)\,\delta x \tag{4.101}$$

then the equations (4.100) become

$$\left[\alpha_0 + \frac{\partial f}{\partial y'}g'(x)\right]_{x=x_1} = 0, \qquad \text{and} \qquad \left[\alpha_0 + \frac{\partial f}{\partial y'}h'(x)\right]_{x=x_2} = 0 \tag{4.102}$$

because δx_1 and δx_2 are assumed to be independent variations. The equations (4.102) together with the requirements

$$g(x_1) = y(x_1, \alpha_0, \beta_0) \qquad \text{and} \qquad h(x_2) = y(x_2, \alpha_0, \beta_0) \tag{4.103}$$

are four equations from which the quantities x_1, x_2, α_0 and β_0 can be determined. ∎

Functional containing end points

Functionals of the form $I = \int_{x_1}^{x_2} f(x, y, y', \ldots; x_1, y_1, x_2, y_2, y_1', y_2', \ldots)\,dx$, which contain end point information, need to be examined carefully for additional end point conditions. For example, a variation of the functional $I = \int_{x_1}^{x_2} f(x, y, y', x_1, x_2, y_1, y_2)\,dx$ is written

$$\delta I = \int_{x_1}^{x_2} \left(\frac{\partial f}{\partial x}\delta x + \frac{\partial f}{\partial y}\delta y + \frac{\partial f}{\partial y'}\delta y' + \frac{\partial f}{\partial x_1}\delta x_1 + \frac{\partial f}{\partial x_2}\delta x_2 + \frac{\partial f}{\partial y_1}\delta y_1 + \frac{\partial f}{\partial y_2}\delta y_2\right) dx \tag{4.104}$$

The additional variations lead to integrals of the form

$$\delta x_1 \int_{x_1}^{x_2} \frac{\partial f}{\partial x_1}\,dx + \delta x_2 \int_{x_1}^{x_2} \frac{\partial f}{\partial x_2}\,dx + \delta y_1 \int_{x_1}^{x_2} \frac{\partial f}{\partial y_1}\,dx + \delta y_2 \int_{x_1}^{x_2} \frac{\partial f}{\partial y_2}\,dx \tag{4.105}$$

where the variation $\delta x_1, \delta x_2, \delta y_2$ are not functions of x. These extra terms provide additional conditions to be satisfied at the end points.

Broken extremal (weak variations)

Recall that a parametric curve C defined by $x = x(t)$, $y = y(t)$ for $t_1 \leq t \leq t_2$ is called a smooth curve if both $x(t)$ and $y(t)$ are well defined over the given interval (t_1, t_2) and have continuous first derivatives $x'(t)$ and $y'(t)$ over the interval where the derivatives satisfy $(x')^2 + (y')^2 > 0$. A curve $y = y(x)$ is called a smooth curve over an interval $a \leq x \leq b$ if the curve is continuous everywhere in the interval and it also has a derivative which exists and is continuous over the interval. A curve $y = y(x)$ is called piecewise smooth over the interval $a \leq x \leq b$ if the curve is continuous everywhere in $[a, b]$ and its derivative is continuous everywhere except for a finite number of points lying in the interval $[a, b]$.

164

In our previous discussions we have constructed solutions to variational problems which are smooth functions. There exists many variational problems which have no smooth solutions but they do have piecewise smooth solutions. For example, consider the problem of finding a piecewise smooth solution for which the functional

$$I(y) = \int_a^b f(x, y, y')\, dx, \qquad y(a) = y_a, \qquad y(b) = y_b \tag{4.106}$$

has an extremum. Further assume the desired solution has only one point $x = c$, lying in the interval (a, b), where the derivative of the solution is not continuous. The figure 4-6 illustrates a representative function from the class of piecewise functions from which a solution is to be extracted.

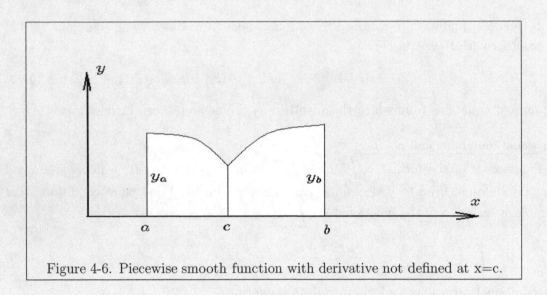

Figure 4-6. Piecewise smooth function with derivative not defined at x=c.

In this case one can write the functional given by equation (4.106) in the form

$$I(y) = \int_a^{c^-} f(x, y, y')\, dx + \int_{c^+}^b f(x, y, y')\, dx = I_1(y) + I_2(y). \tag{4.107}$$

At an extremum we require that $\delta I = \delta I_1 + \delta I_2 = 0$. One can now consider variations in y only over the interval (a, c^-) and require the variations over the interval (c^+, b) to be zero so that the Euler-Lagrange equations must then be satisfied over this interval (a, c^-). Alternatively, require the variations to be zero over the interval (a, c^-) and consider only variations over the interval (c^+, b) where again the Euler-Lagrange equations must apply.

Using the result from equation (4.67), with $\omega = \delta y - y'\, \delta x$, one can write

$$\begin{aligned}
\delta I(y) &= \int_a^{c^-} \left[\frac{\partial f}{\partial y} - \frac{d}{dx}\left(\frac{\partial f}{\partial y'} \right) \right] \omega\, dx + \left[f\, \delta x + \frac{\partial f}{\partial y'}(\delta y - y'\delta x) \right]_a^{c^-} \\
&\quad + \int_{c^+}^b \left[\frac{\partial f}{\partial y} - \frac{d}{dx}\left(\frac{\partial f}{\partial y'} \right) \right] \omega\, dx + \left[f\, \delta x + \frac{\partial f}{\partial y'}(\delta y - y'\, \delta x) \right]_{c^+}^b
\end{aligned} \tag{4.108}$$

Observe that if $y = y(x)$ is a piecewise smooth function which makes I have an extremum, then it must be a solution of the Euler equation and satisfy

$$\frac{\partial f}{\partial y} - \frac{d}{dx}\left(\frac{\partial f}{\partial y'}\right) = 0, \qquad a \leq x \leq c^-$$
$$\frac{\partial f}{\partial y} - \frac{d}{dx}\left(\frac{\partial f}{\partial y'}\right) = 0, \qquad c^+ \leq x \leq b$$

(4.109)

This gives solutions $y = y(x, c_1, c_2)$ over the interval (a, c^-) and solutions $y = y(x, c_3, c_4)$ over the interval (c^+, b) where c_1, c_2, c_3, c_4 are constants. Here the end points at $x = a$ and $x = b$ are fixed so that $\delta x = \delta y = 0$ at the points where $x = a$ and $x = b$. We further require that the piecewise smooth solution be continuous at the point $x = c$ so that the point $x = c$ cannot move freely. The terms in equation (4.108) can be employed to define the quantities

$$\delta I_1 = \left.\frac{\partial f}{\partial y'}\right|_{x=c^-} \delta y + \left.\left(f - y'\frac{\partial f}{\partial y'}\right)\right|_{x=c^-} \delta x$$
$$\delta I_2 = -\left.\frac{\partial f}{\partial y'}\right|_{x=c^+} \delta y - \left.\left(f - y'\frac{\partial f}{\partial y'}\right)\right|_{x=c^+} \delta x$$

At an extremum we require that $\delta I_1 + \delta I_2 = 0$, hence

$$\delta I = \delta I_1 + \delta I_2 = \left(\left.\frac{\partial f}{\partial y'}\right|_{x=c^-} - \left.\frac{\partial f}{\partial y'}\right|_{x=c^+}\right)\delta y + \left[\left.\left(f - y'\frac{\partial f}{\partial y'}\right)\right|_{x=c^-} - \left.\left(f - y'\frac{\partial f}{\partial y'}\right)\right|_{x=c^+}\right]\delta x = 0.$$

Now δx and δy are arbitrary so that in order for the above equation to hold one must require that the following conditions are satisfied

$$\left.\frac{\partial f}{\partial y'}\right|_{x=c^-} = \left.\frac{\partial f}{\partial y'}\right|_{x=c^+}$$
$$\left.\left(f - y'\frac{\partial f}{\partial y'}\right)\right|_{x=c^-} = \left.\left(f - y'\frac{\partial f}{\partial y'}\right)\right|_{x=c^+}.$$

(4.110)

These conditions are known as the Weierstrass-Erdmann corner conditions.

Here one must solve the Euler equations (4.109) over each of the intervals $a \leq x \leq c^-$ and $c^+ \leq x \leq b$. This will produce two equations with four constants to be determined. Two of the four constants are determined from the end conditions $y(a) = y_a$ and $y(b) = y_b$. The remaining two constants are determined from Weierstrass-Erdmann conditions given by equation (4.110).

Many variational problems have discontinuities because of the physical problem being considered. For example, the problem of finding the shortest walking distance between two points A and B located within a city which has streets which run North-South and East-West. Here the solution path will have points where the derivative is discontinuous unless the points A and B are in special locations. A representation of this problem is illustrated in the figure 4-7.

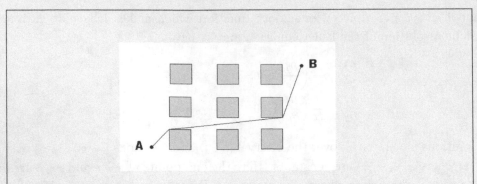

Figure 4-7. Finding shortest distance from point A to point B in a city.

Whenever the geometry of a problem suggests that a discontinuity in the solution might occur, then modifications to the calculus of variation problem can be introduced.

In the case the functional which is to have an extreme value has the form

$$I = \int_{x_1}^{x_2} f(x, y_1, y_2, \ldots, y_n, , y_1', y_2', \ldots, y_n') \, dx$$

the Euler-Lagrange equations are a system of differential equations

$$\frac{\partial f}{\partial y_i} - \frac{d}{dx}\left(\frac{\partial f}{\partial y_i'}\right) = 0, \qquad x_1 \le x \le x_2, \qquad i = 1, 2, \ldots n$$

An integration from x_1 to x gives the Euler-Lagrange equations in the integral form

$$\frac{\partial f}{\partial y_i'} = \int_{x_1}^{x} \frac{\partial f}{\partial y_i} \, dx + C_i, \qquad i = 1, 2, \ldots, n \tag{4.111}$$

where C_i are constants of integration. If there exists points in the interval $x_1 \le x \le x_2$ where any of the derivatives $y_i'(x)$ are discontinuous, then these points are called corners. The Weierstrass-Erdmann corner conditions can be applied whenever there exists a finite number of corners. The first corner condition can be obtained directly from the equation (4.111). If ξ is a point on an solution trajectory where a corner occurs, then one or more of the solution trajectories $y_i(x)$ must have derivatives $y_i'(x)$ which satisfy

$$\frac{\partial f}{\partial y_i'}\bigg|_{x=\xi^-} = \frac{\partial f}{\partial y_i'}\bigg|_{x=\xi^+} \qquad i = 1, 2, \ldots, n \tag{4.112}$$

The second corner condition for the system of equations can be written

$$\left(f - \sum_{i=1}^{n} y_i' \frac{\partial f}{\partial y_i'}\right)\bigg|_{x=\xi^-} = \left(f - \sum_{i=1}^{n} y_i' \frac{\partial f}{\partial y_i'}\right)\bigg|_{x=\xi^+} \qquad i = 1, 2, \ldots, n \tag{4.113}$$

There is another kind of discontinuity which can occur whenever the Euler-Lagrange equation has multiple solutions. These special discontinuities usually occur whenever the

Euler-Lagrange differential equation is a differential equation of the first order but not of the first degree. These type of conditions are not considered in this text.

Weierstrass E-function

Weak variations require that the variational curve $Y(x) = y(x) + \epsilon \eta(x)$ behave pretty much like that of the extremum curve $y(x)$. For strong variations additional tests for maximum and minimum values are required and we shall work toward that goal. Consider the functional

$$I = \int_{x_1}^{x_2} f(x, y, y')\, dx \qquad y(x_1) = y_1, \qquad y(x_2) = y_2 \tag{4.114}$$

A necessary condition for an extremal is that y satisfy the Euler-Lagrange equation

$$\frac{\partial f}{\partial y} - \frac{d}{dx}\left(\frac{\partial f}{\partial y'}\right) = 0 \quad \Rightarrow \quad \frac{\partial f}{\partial y} - \frac{\partial^2 f}{\partial y' \partial x} - \frac{\partial^2 f}{\partial y' \partial y} y' - \frac{\partial^2 f}{\partial y'^2} y'' = 0 \tag{4.115}$$

Let C denote an extremal which passes through the points (x_1, y_1) and (x_2, y_2) and then extend the curve C to a near point (x_0, y_0) where $x_0 < x_1$. The situation is illustrated in the figure 4-8. Now consider the one-parameter family of solution curves to the Euler-Lagrange equations which pass through the near point (x_0, y_0) which is also illustrated in the figure 4-8. We denote this family of curves through the point (x_0, y_0) by $y = \phi(x, \gamma)$ where γ is the parameter. Next construct an arbitrary smooth curve $y = Y(x)$ connecting the end points (x_1, y_1) and (x_2, y_2) and denote this curve by C^*. We examine the particular curve from the family $y = \phi(x, \gamma)$ which intersects the curve $y = Y(x)$ at the point (x_3, y_3) as illustrated in the figure 4-8. We denote this special curve by $y = \phi(x, \gamma_3)$.

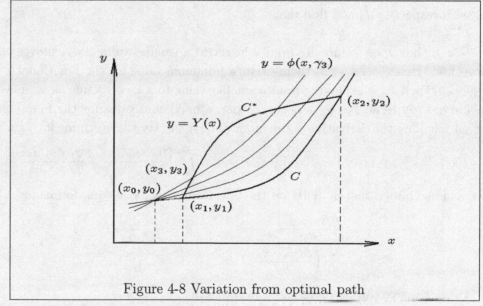

Figure 4-8 Variation from optimal path

We then have the relations

$$y_0 = \phi(x_0, \gamma_3)$$
$$y_3 = \phi(x_3, \gamma_3) \tag{4.116}$$
$$y_3 = Y(x_3)$$

Consider the following integral from the point (x_0, y_0) to the point (x_2, y_2)

$$I(x_3) = \int_{x_0}^{x_3} f(x, \phi(x, \gamma_3), \phi_x(x, \gamma_3))\, dx + \int_{x_3}^{x_2} f(x, Y(x), Y'(x))\, dx \tag{4.117}$$

which is a function of the abscissa x_3. We examine the integral given by equation (4.117) in the limiting cases where x_3 approaches x_1 and x_3 approaches x_2 as the point x_3 moves along the curve $y = Y(x)$. As x_3 tends toward x_1 the equation (4.117) becomes

$$I(x_1) = \int_{x_0}^{x_1} f(x, \phi(x, \gamma_1), \phi_x(x, \gamma_1))\, dx + \int_{x_1}^{x_2} f(x, Y(x), Y'(x))\, dx \tag{4.118}$$

and as x_3 tends toward x_2 along $Y(x)$ the equation (4.117) becomes

$$I(x_2) = \int_{x_0}^{x_1} f(x, \phi(x, \gamma_1), \phi_x(x, \gamma_1))\, dx + \int_{x_1}^{x_2} f(x, \phi(x, \gamma_1), \phi_x(x, \gamma_1))\, dx \tag{4.119}$$

The difference in the integrals given by equations (4.118) and (4.119) is found by subtracting these equations. One finds this difference can be denoted

$$\Delta I = I(x_1) - I(x_2) = \int_{C^*} f(x, Y, Y')\, dx - \int_C f(x, y, y')\, dx \tag{4.120}$$

If the difference $\Delta I > 0$ for all curves $y = Y(x)$, then the functional I given by equation (4.114) corresponds to a minimum value. By a similar argument, if the difference $\Delta I < 0$ for all curves $y = Y(x)$, then the functional I given by equation (4.114) corresponds to a maximum value. That is, if we treat x_3 as a variable quantity and differentiate the equation (4.117) with respect to x_3 and find that

(a) If $\frac{dI}{dx_3} < 0$, then $I(x_3)$ is moving from a larger to a smaller value as x_3 moves along the curve $Y(x)$. Hence, as $x_3 \to x_2$ there exists a minimum value for the functional.

(b) If $\frac{dI}{dx^3} > 0$, then $I(x_3)$ is moving from a smaller value to a larger value as x_3 moves along the curve $Y(x)$. Hence, as $x_3 \to x_2$ there exists a maximum value for the functional.

Let us calculate the derivative of the integral $I(x_3)$ and try to determine its sign. We find

$$\frac{dI}{dx_3} = f(x_3, \phi(x_3, \gamma_3), \phi_x(x_3, \gamma_3)) - f(x_3, Y(x_3), Y'(x_3)) + \int_{x_0}^{x_3} \left[\frac{\partial f}{\partial y} \frac{\partial \phi(x, \gamma_3)}{\partial \gamma_3} + \frac{\partial f}{\partial y'} \frac{\partial \phi_x(x, \gamma_3)}{\partial \gamma_3} \right] \frac{d\gamma_3}{dx_3}\, dx$$

Now we can use integration by parts on the integral in the above equation using

$$u = \frac{\partial f(x, \phi, \phi_x)}{\partial y'} \qquad\qquad dV = \phi_{x\gamma_3} \frac{d\gamma_3}{dx_3}\, dx$$

$$du = \frac{d}{dx}\left(\frac{\partial f(x, \phi, \phi_x)}{\partial y'} \right) dx \qquad\qquad V = \phi_{\gamma_3} \frac{d\gamma_3}{dx_3}$$

so that the derivative can be represented in the form

$$\begin{aligned}
\frac{dI}{dx_3} = {}& f(x_3, y_3, y_3') - f(x_3, Y(x_3), Y'(x_3)) + \left[\frac{\partial f(x, \phi(x, \gamma_3), \phi_x(x, \gamma_3))}{\partial y'} \frac{d\gamma_3}{dx_3} \phi_{\gamma_3}(x, \gamma_3) \right]_{x_0}^{x_3} \\
& + \int_{x_0}^{x_3} \phi_{\gamma_3} \left[\frac{\partial f}{\partial y} - \frac{d}{dx}\left(\frac{\partial f}{\partial y'} \right) \right] \frac{d\gamma_3}{dx_3}\, dx
\end{aligned} \tag{4.121}$$

Note that the equations (4.116) imply that γ_3 is a function of both x_3 and y_3. Differentiating the middle equation of (4.116) with respect to x_3 gives

$$0 = \frac{\partial \phi}{\partial x_3} + \frac{\partial \phi}{\partial \gamma_3}\frac{\partial \gamma_3}{\partial x_3} \quad \Rightarrow \quad \frac{\partial \gamma_3}{\partial x_3} = -\frac{\phi_{x_3}}{\phi_{\gamma_3}} \tag{4.122}$$

and the derivative of the middle equation of (4.116) with respect to y_3 gives

$$1 = \frac{\partial \phi}{\partial \gamma_3}\frac{\partial \gamma_3}{\partial y_3} \quad \Rightarrow \quad \frac{\partial \gamma_3}{\partial y_3} = \frac{1}{\phi_{\gamma_3}} \tag{4.123}$$

Therefore, one can calculate

$$\frac{d\gamma_3}{dx_3} = \frac{\partial \gamma_3}{\partial x_3} + \frac{\partial \gamma_3}{\partial y_3}\frac{dy_3}{dx_3}, \quad \text{where} \quad \frac{dy_3}{dx_3} = Y'(x_3) \tag{4.124}$$

We also find that by differentiating the first equation of (4.116) with respect to γ_3 that

$$\frac{\partial \phi(x_0, \gamma_3)}{\partial \gamma_3} = 0. \tag{4.125}$$

We use the results form equations (4.122), (4.123), (4.124), and (4.125) together with the Euler-Lagrange equation (4.115) to simplify the derivative given by equation (4.121) to the following form

$$\frac{dI}{dx_3} = -\left[f(x_3, Y(x_3), Y'(x_3)) - f(x_3, y_3, y_3') - (Y'(x_3) - y_3')\frac{\partial f(x_3, y_3, y_3')}{\partial y'} \right] \tag{4.126}$$

We know from the equations (4.116) that $y_3 = Y(x_3)$ and denoting the derivative $Y'(x_3)$ by Y_3' the derivative given by equation (4.126) can be represented in the final form

$$\frac{dI}{dx_3} = -E[x_3, y_3, y_3', Y_3'] \tag{4.127}$$

where

$$E[x, y, y', p] = f(x, y, p) - f(x, y, y') - (p - y')\frac{\partial f(x, y, y')}{\partial y'} \tag{4.128}$$

is called the Weierstrass excess function which is often referred to as the E-function. In equation (4.128) note that p represents an arbitrary slope since $y = Y(x)$ is an arbitrary variational arc. The equation (4.128) tells us the following

(a) *A necessary condition for a minimum value to be associated with the functional given by equation (4.114) is for $E \geq 0$.*

(b) *A necessary condition for a maximum value to be associated with the functional given by equation (4.114) is for $E \leq 0$.*

In summary, a necessary and sufficient condition for a strong extremum is

(i) The curve C is an extremal curve satisfying the Euler-Lagrange equation and associated boundary conditions.

(ii) There are no conjugate points to either x_1 or x_2 along C

(iii) The Weierstrass excess function $E(x, y, y', p)$ has a constant sign for all points (x, y) lying sufficiently close to the curve C and arbitrary values for p. The condition $E \geq 0$ corresponds to a minimum value for the functional and the condition $E \leq 0$ corresponds to a maximum value for the functional.

Example 4-8.   Test for weak variation

Consider the problem of finding an extremum for the functional

$$I = \int_{(0,1)}^{(2,0)} y'^2 (y'+1)^2 \, dx$$

Since the integrand does not contain the independent variable x, one can immediately obtain the following first integral of the Euler-Lagrange equation

$$y'^2(y'+1)(3y'+1) = C, \qquad C \text{ is a constant}$$

Let the constant C have the value $C = \beta^2(\beta+1))(3\beta+1)$, then $y' = \beta$ can be integrated to produce $y = \beta x + \alpha$ as one solution of the Euler-Lagrange equation where α and β are constants.

The end conditions $x = 0, y = 1$ and $x = 2, y = 0$ produces the results that $\alpha = 1$ and $\beta = -1/2$ so that $y = -\frac{1}{2}x + 1$ with $y' = -\frac{1}{2}$. The Legendre test gives $F_{y'y'} = 12y'^2 + 12y' + 2 \big|_{y'=-1/2} = -1$ so that the solution $y = -\frac{1}{2}x + 1$ corresponds to a weak maximum.

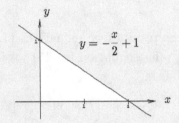

The Jacobi test gives us

$$y_\alpha = 1, \qquad y_\beta = x, \qquad y_\alpha \big|_{x=0} = 1, \qquad y_\beta \big|_{x=0} = 0$$

so that $\begin{vmatrix} x & 1 \\ 0 & 1 \end{vmatrix} = 0$ gives $x = 0$. The value of the given functional is $I = \int_0^2 \left(\frac{-1}{2}\right)^2 \left(\frac{1}{2}\right)^2 dx = \frac{1}{8}$.
Now choose the path of integration

$$y = -4x + 1, \quad 0 < x < 1/8$$

$$y = 1/2, \quad 1/8 < x < 1$$

$$y = -4x + 9/2, \quad 1 < x < 9/8$$

$$y = 0, \quad 9/8 < x < 2$$

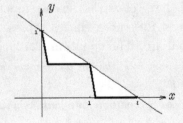

The value of the functional for this path is found to be $I = \int_0^{1/8} 16(9)\,dx + \int_1^{9/8} 16(9)\,dx = 36$ and so the solution $y = -1/2x + 1$ is not a maximum for all paths.

Using the Weierstrass E-function we find that $E = p^2(1+p)^2 - 1/16$. A graph of E vs p illustrates that E is not positive for all values of p so that $y = -\frac{1}{2}x + 1$ does not correspond to a strong maximum.

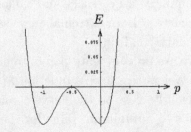

Note that the function E has opposite signs as p varies and therefore a strong extreme does not exist.

∎

In the case of the more general functional

$$I = \int_{x_1}^{x_2} f(x, y_1, y_2, \ldots, y_n, y_1', y_2', \ldots, y_n') \, dx \tag{4.129}$$

one can define the Weierstrass excess function in terms of vectors and gradients. Define the vector quantity $\bar{y} = (y_1, y_2, \ldots, y_n)^T$ and write the functional given by equation (4.129) in the form

$$I = \int_{x_1}^{x_2} f(x, \bar{y}, \bar{y}') \, dx \tag{4.130}$$

The more general form of the Weierstrass excess function can then be written as

$$E(x, \bar{y}, \bar{v}, \bar{w}) = f(x, \bar{y}, \bar{w}) - f(x, \bar{y}, \bar{v}) - \sum_{j=1}^{n} (w_j - v_j) \frac{\partial f(x, \bar{y}, \bar{v})}{\partial v_j} \tag{4.131}$$

where $\bar{y}'$ in the integrand $f(x, \bar{y}, \bar{y}')$ has been replaced by the vectors $\bar{v} = (v_1, v_2, \ldots, v_n)^T$ and $\bar{w} = (w_1, w_2, \ldots, w_n)^T$ in formulating the excess function. One can then say that $\bar{y}$ is an extremal of the functional $I(\bar{y})$ if $\bar{y}$ satisfies the Euler-Lagrange equations and

(i) if $E(x, \bar{y}, \bar{y}', \bar{w}) \geq 0$ for all $x \in [x_1, x_2]$ and all bounded vectors $\bar{w}$, then $I(\bar{y})$ has a strong local minimum.

(ii) if $E(x, \bar{y}, \bar{y}', \bar{w}) \leq 0$ for all $x \in [x_1, x_2]$ and all bounded vectors $\bar{w}$, then $I(\bar{y})$ has a strong local maximum.

Euler-Lagrange equations with constraint conditions

We consider a general variational problem having restrictions imposed by finite equations or differential equations. We employ the method of Lagrange multipliers to solve this type of variational problem. Consider the problem of finding the extremum associated with a functional having the general form

$$I = \int_{t_1}^{t_2} f(t, y_1, y_2, \ldots, y_n, \dot{y}_1, \dot{y}_2, \ldots, \dot{y}_n) \, dt \tag{4.132}$$

where $y_1, y_2, \ldots, y_n$ are required to satisfy the independent constraint conditions

$$\phi_j(t, y_1, y_2, \ldots, y_n, \dot{y}_1, \dot{y}_2, \ldots, \dot{y}_n) = 0, \qquad j = 1, 2, \ldots, N \tag{4.133}$$

where $N < n$. Note that if the constraint conditions ϕ_j are independent of derivatives $\dot{y}_1, \dot{y}_2, \ldots, \dot{y}_n$, then the constraint conditions reduce to a finite set of algebraic equations, rather than a system of differential equations. In order to solve the above variational problem we proceed as before and assume we have a set of functions $y_i(t)$ which produces an extremum and introduce a set of comparison functions $Y_i = Y_i(t) = y_i(t) + \epsilon \xi_i(t)$ for $i = 1, \ldots, n$ where the functions $\xi_i(t)$ are required to satisfy the end conditions $\xi_i(t_1) = \xi_i(t_2) = 0$ for $i = 1, \ldots, n$. Substituting the comparison function into the functional (4.132) gives

$$I = I(\epsilon) = \int_{t_1}^{t_2} f(t, Y_1, Y_2, \ldots, Y_n, \dot{Y}_1, \dot{Y}_2, \ldots, \dot{Y}_n) \, dt \tag{4.134}$$

Now if $I = I(\epsilon)$ has an extremum at $\epsilon = 0$ we require that $\frac{dI}{d\epsilon}\big|_{\epsilon=0} = 0$. One finds that

$$\frac{dI}{d\epsilon} = \int_{t_1}^{t_2} \left\{ \frac{\partial f}{\partial Y_1}\xi_1 + \frac{\partial f}{\partial Y_2}\xi_2 + \cdots + \frac{\partial f}{\partial Y_n}\xi_n + \frac{\partial f}{\partial \dot{Y}_1}\dot{\xi}_1 + \frac{\partial f}{\partial \dot{Y}_2}\dot{\xi}_2 + \cdots + \frac{\partial f}{\partial \dot{Y}_n}\dot{\xi}_n \right\} dt$$

and upon letting $\epsilon \to 0$ we have $(Y_1, Y_2, \ldots, Y_n) \to (y_1, y_2, \ldots, y_n)$ so that

$$\frac{dI}{d\epsilon}\bigg|_{\epsilon=0} = \int_{t_1}^{t_2} \left\{ \frac{\partial f}{\partial y_1}\xi_1 + \frac{\partial f}{\partial y_2}\xi_2 + \cdots + \frac{\partial f}{\partial y_n}\xi_n \right.$$
$$\left. + \frac{\partial f}{\partial \dot{y}_1}\dot{\xi}_1 + \frac{\partial f}{\partial \dot{y}_2}\dot{\xi}_2 + \cdots + \frac{\partial f}{\partial \dot{y}_n}\dot{\xi}_n \right\} dt = 0 \tag{4.135}$$

In the equation (4.135) the functions $\xi_i(t)$ are not independent functions. Note that for $j = 1, \ldots, N < n$ we have

$$\phi_j(\epsilon) = \phi_j(t, Y_1, Y_2, \ldots, Y_n, \dot{Y}_1, \dot{Y}_2, \ldots, \dot{Y}_n) = 0 \tag{4.136}$$

as a set of equations which must be satisfied for all values of ϵ. Differentiating each of the equations (4.136) with respect to ϵ we obtain the relations

$$\frac{d\phi_j}{d\epsilon} = \frac{\partial \phi_j}{\partial Y_1}\xi_1 + \frac{\partial \phi_j}{\partial Y_2}\xi_2 + \cdots + \frac{\partial \phi_j}{\partial Y_n}\xi_n + \frac{\partial \phi_j}{\partial \dot{Y}_1}\dot{\xi}_1 + \frac{\partial \phi_j}{\partial \dot{Y}_2}\dot{\xi}_2 + \cdots + \frac{\partial \phi_j}{\partial \dot{Y}_n}\dot{\xi}_n = 0, \tag{4.137}$$

for $j = 1, 2, \ldots, N$. At $\epsilon = 0$ the above system of equations reduces to

$$\sum_{k=1}^{n} \left(\frac{\partial \phi_j}{\partial y_k}\xi_k + \frac{\partial \phi_j}{\partial \dot{y}_k}\dot{\xi}_k \right) = 0, \qquad j = 1, 2, \ldots, N. \tag{4.138}$$

Multiply the equation (4.138) by $\lambda_j = \lambda_j(t)$ and integrate the result from t_1 to t_2 and then add the results, for each value of j, to the equation (4.135) to obtain

$$\int_{t_1}^{t_2} \left\{ \left[\frac{\partial f}{\partial y_1} + \sum_{j=1}^{N} \lambda_j \frac{\partial \phi_j}{\partial y_1} \right] \xi_1 + \left[\frac{\partial f}{\partial y_2} + \sum_{j=1}^{N} \lambda_j \frac{\partial \phi_j}{\partial y_2} \right] \xi_2 + \cdots + \left[\frac{\partial f}{\partial y_n} + \sum_{j=1}^{N} \lambda_j \frac{\partial \phi_j}{\partial y_n} \right] \xi_n \right.$$
$$\left. + \left[\frac{\partial f}{\partial \dot{y}_1} + \sum_{j=1}^{N} \lambda_j \frac{\partial \phi_j}{\partial \dot{y}_1} \right] \dot{\xi}_1 + \left[\frac{\partial f}{\partial \dot{y}_2} + \sum_{j=1}^{N} \lambda_j \frac{\partial \phi_j}{\partial \dot{y}_2} \right] \dot{\xi}_2 + \cdots + \left[\frac{\partial f}{\partial \dot{y}_n} + \sum_{j=1}^{N} \lambda_j \frac{\partial \phi_j}{\partial \dot{y}_n} \right] \dot{\xi}_n \right\} dt = 0$$

Each term in the last line of the above equation is integrated by parts and the end conditions on $\xi_i(t)$ are used to obtain the requirement

$$\int_{t_1}^{t_2} \left\{ \sum_{i=1}^{n} \left[\frac{\partial H}{\partial y_i} - \frac{d}{dt}\left(\frac{\partial H}{\partial \dot{y}_i} \right) \right] \xi_i(t) \right\} dt = 0 \tag{4.139}$$

where $H = f + \sum_{j=1}^{N} \lambda_j \phi_j$.

In the equation (4.139) the functions $\xi_i(t)$, $i = 1, 2, \ldots, n$ are not arbitrary because of equation (4.138). Observe that the equation (4.138) represents a system of N equations which tells us the functions $\xi_{N+1}(t), \xi_{N+2}(t), \ldots, \xi_n(t)$ can be arbitrary. That is, equation (4.138) implies that the functions $\xi_1(t), \xi(t), \ldots, \xi_N(t)$ are dependent upon the values selected for the $\xi_{N+1}(t), \xi_{N+2}(t), \ldots, \xi_n(t)$ functions which can be arbitrary. Therefore, if we select the Lagrange multipliers $\lambda_1(t), \lambda_2(t), \ldots, \lambda_N(t)$ to be such that

$$\frac{\partial H}{\partial y_i} - \frac{d}{dt}\left(\frac{\partial H}{\partial \dot{y}_i}\right) = 0, \quad \text{for } i = 1, 2, \ldots, N$$

then the first N terms in the summation of equation (4.139) are zero. The equation (4.139) then reduces to

$$\int_{t_1}^{t_2}\left\{\sum_{i=N+1}^{n}\left[\frac{\partial H}{\partial y_i} - \frac{d}{dt}\left(\frac{\partial H}{\partial \dot{y}_i}\right)\right]\xi_i(t)\right\} dt = 0$$

Now the functions $\xi_i(t)$ for $i = N+1, \ldots, n$ are independent and arbitrary and so we can use a form of the basic lemma to conclude that

$$\frac{\partial H}{\partial y_i} - \frac{d}{dt}\left(\frac{\partial H}{\partial \dot{y}_i}\right) = 0, \quad \text{for } i = N+1, \ldots, n$$

Thus, we can conclude that the equation (4.139) is satisfied if we require that the function $H = f + \sum_{j=1}^{N} \lambda_j \phi_j$ satisfies

$$\frac{\partial H}{\partial y_i} - \frac{d}{dt}\left(\frac{\partial H}{\partial \dot{y}_i}\right) = 0 \quad \text{for} \quad i = 1, 2, \ldots, n \tag{4.140}$$

together with the constraint equations

$$\phi_j(t, y_1, y_2, \ldots, y_n, \dot{y}_1, \dot{y}_2, \ldots, \dot{y}_n) = 0, \qquad j = 1, 2, \ldots, N. \tag{4.141}$$

This gives $n + N$ equations in the $n + N$ unknowns $y_1, \ldots, y_n, \lambda_1, \ldots, \lambda_N$ which are functions of the independent variable t. Note that in the study of dynamical systems the Lagrange multipliers λ_i are interpreted as forces that keep a particle on the constraint paths.

Example 4-9. Lagrange multipliers

Find the minimum value associated with the functional $I = \int_0^1 (y'(x))^2\, dx$ subject to the constraint conditions

$$\frac{dw}{dx} + \frac{dy}{dx} - \frac{dz}{dx} = 0 \quad \text{and} \quad \frac{dw}{dx} + 2\frac{dz}{dx} = 0$$

where y, w, z are subject to the end point conditions

$$y(0) = 0, \quad y(1) = 3, \quad w(0) = 0, \quad w(1) = -2, \quad z(0) = 0, \quad z(1) = 1$$

Solution: We use Lagrange multipliers and form

$$H = H(y, w, z, y', w', z') = (y')^2 + \lambda_1(w' + y' - z') + \lambda_2(w' + 2z')$$

and then calculate the derivatives

$$\frac{\partial H}{\partial y} = 0 \qquad\qquad \frac{\partial H}{\partial w} = 0 \qquad\qquad \frac{\partial H}{\partial z} = 0$$

$$\frac{\partial H}{\partial y'} = 2y' + \lambda_1 \qquad \frac{\partial H}{\partial w'} = \lambda_1 + \lambda_2 \qquad \frac{\partial H}{\partial z'} = -\lambda_1 + 2\lambda_2$$

Next form the Euler-Lagrange equations

$$\frac{\partial H}{\partial y} - \frac{d}{dx}\left(\frac{\partial H}{\partial y'}\right) = 0 \qquad \frac{d}{dx}(2y' + \lambda_1) = 0$$

$$\frac{\partial H}{\partial w} - \frac{d}{dx}\left(\frac{\partial H}{\partial w'}\right) = 0 \qquad \frac{d}{dx}(\lambda_1 + \lambda_2) = 0$$

$$\frac{\partial H}{\partial z} - \frac{d}{dx}\left(\frac{\partial H}{\partial z'}\right) = 0 \qquad \frac{d}{dx}(-\lambda_1 + 2\lambda_2) = 0$$

These equations have the first integrals

$$2y' + \lambda_1 = c_1, \qquad \lambda_1 + \lambda_2 = c_2, \qquad -\lambda_1 + 2\lambda_2 = c_3 \tag{4.142}$$

where c_1, c_2, c_3 are constants. The constraint equations can be integrated to obtain the additional equations

$$w + y - z = c_4, \qquad w + 2z = c_5 \tag{4.143}$$

where c_4, c_5 are also constants. From the equations (4.142) one finds that λ_1 and λ_2 are constants. This implies $y' = $ a constant $= (c_1 - \lambda_1)/2 = \alpha$, where α is some new constant. An integration produces the result $y = y(x) = \alpha x + \beta$ where α and β are constants. The boundary conditions $y(0) = 0$ and $y(1) = 3$ allows one to obtain the values $\alpha = 3$ and $\beta = 0$. This gives the solution $y = y(x) = 3x$. The equations (4.143) evaluated at $x = 0$ implies that $c_4 = c_5 = 0$, so that one can write $w = -2z$ and $z = w + y$. One finds that $z = y/3$. Consequently, the desired solutions are

$$y = y(x) = 3x, \qquad z = z(x) = x, \qquad w = w(x) = -2x. \tag{4.144}$$

One finds that $I_{minimum} = \int_0^1 (y'(x))^2 \, dx = 9$.

■

Example 4-10.   Lagrange multipliers

The end points of a string of length ℓ are attached to the fixed points $(a, 0)$ and $(-a, 0)$ as illustrated. Assume that $\ell > 2a$ and find the shape of the string which encloses the maximum area between the string and the x-axis.

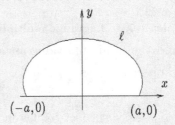

Solution: Let $x = x(t)$ and $y = y(t)$ denote parametric equations defining the shape of the string. The area enclosed by the string can be represented by a special case of Green's theorem (equation 1.47) and written as

$$A = \frac{1}{2}\oint x\,dy - y\,dx = \frac{1}{2}\int_{t_0}^{t_1}\left[x\frac{dy}{dt} - y\frac{dx}{dt}\right]dt$$

where x and y are subject to the constraint condition $\int_{t_0}^{t_1}\sqrt{\dot{x}^2 + \dot{y}^2}\,dt = \ell$. We introduce a Lagrange multiplier λ and form the functional $I = \int_{t_0}^{t_1}\left\{\frac{1}{2}[x\dot{y} - y\dot{x}] + \lambda\sqrt{\dot{x}^2 + \dot{y}^2}\right\}dt$. One can then verify that the Euler-Lagrange equations are given by

$$\dot{y} - \frac{d}{dt}\left[\frac{\lambda\dot{x}}{\sqrt{\dot{x}^2 + \dot{y}^2}}\right] = 0 \qquad \text{and} \qquad \dot{x} + \frac{d}{dt}\left[\frac{\lambda\dot{y}}{\sqrt{\dot{x}^2 + \dot{y}^2}}\right] = 0$$

Integrating these equations there results

$$y - \frac{\lambda\dot{x}}{\sqrt{\dot{x}^2 + \dot{y}^2}} = b \qquad \text{and} \qquad x + \frac{\lambda\dot{y}}{\sqrt{\dot{x}^2 + \dot{y}^2}} = c \tag{4.145}$$

where b and c are constants of integration.

The equations (4.145) imply that

$$(y - b)^2 + (x - c)^2 = \lambda^2$$

which is a circle. The conditions

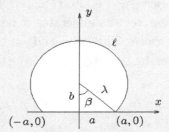

$$x = a, \quad y = 0 \quad \text{requires} \quad b^2 + (a - c)^2 = \lambda^2$$
$$x = -a, \quad y = 0, \quad \text{requires} \quad b^2 + (a + c)^2 = \lambda^2$$

These last two equations imply that $c = 0$ so that the circle is symmetric about the y-axis and has the form

$$(y - b)^2 + x^2 = \lambda^2 \tag{4.146}$$

From the geometry of the circle illustrated one finds

$$\cos\beta = \frac{b}{\lambda}, \qquad \sin\beta = \frac{a}{\lambda}, \qquad b = a\cot\beta$$

so that the equation of the circle can be represented in terms of the angle β as

$$(y - a\cot\beta)^2 + x^2 = a^2\csc^2\beta \tag{4.147}$$

One can also verify that in terms of the angle β the length ℓ of the string is given by $\ell = (\pi - \beta)2a\csc\beta$ and the area enclosed by the string is given by $A = (\pi - \beta)\csc^2\beta + a^2\cot\beta$. ∎

Geodesic curves

A curve on a given surface is called a geodesic curve if it is a curve which produces the shortest distance between two points on the given surface.[†] Recall that a surface can be represented by a set of parametric equations

$$x = x(u, v), \qquad y = y(u, v), \qquad z = z(u, v)$$

with parameters u, v which are called the surface coordinates. A point on the surface can be represented in terms of the u, v surface coordinates by defining the position vector

$$\vec{r} = \vec{r}(u, v) = x(u, v)\,\widehat{e}_1 + y(u, v)\,\widehat{e}_2 + z(u, v)\,\widehat{e}_3$$

[†] A more formal definition is that geodesics are curves of zero geodesic curvature.

where $\widehat{e}_1, \widehat{e}_2, \widehat{e}_3$ are unit base vectors. A surface coordinate system can be constructed by drawing the coordinate curves

$$\vec{r} = \vec{r}(u, v_0) \qquad \text{where } v_0 \text{ is a constant}$$
$$\vec{r} = \vec{r}(u_0, v) \qquad \text{where } u_0 \text{ is a constant}$$

by selecting various values for the constants u_0 and v_0.

The vectors $\frac{\partial \vec{r}}{\partial u}$ and $\frac{\partial \vec{r}}{\partial v}$ evaluated at a point (u_0, v_0) on the surface are tangent vectors to the coordinate curves $\vec{r}(u, v_0)$ and $\vec{r}(u_0, v)$. The cross product of these vectors generates a normal vector $\vec{N}$ to the surface. The magnitude of this normal vector is given by $|\vec{N}| = |\frac{\partial \vec{r}}{\partial u} \times \frac{\partial \vec{r}}{\partial v}|$. Using the vector identity

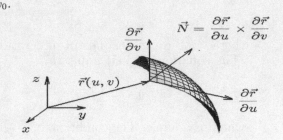

$$(\vec{A} \times \vec{B}) \cdot (\vec{C} \times \vec{D}) = (\vec{A} \cdot \vec{C})(\vec{B} \cdot \vec{D}) - (\vec{A} \cdot \vec{D})(\vec{B} \cdot \vec{C})$$

one can verify that a unit normal vector $\vec{n}$ to the surface is given by

$$\vec{n} = \frac{1}{\pm\sqrt{EG - F^2}} \left(\frac{\partial \vec{r}}{\partial u} \times \frac{\partial \vec{r}}{\partial v} \right)$$

where E, F and G are defined

$$
\begin{aligned}
E = E(u,v) &= \frac{\partial \vec{r}}{\partial u} \cdot \frac{\partial \vec{r}}{\partial u} = \left(\frac{\partial x}{\partial u}\right)^2 + \left(\frac{\partial y}{\partial u}\right)^2 + \left(\frac{\partial z}{\partial u}\right)^2 \\
F = F(u,v) &= \frac{\partial \vec{r}}{\partial u} \cdot \frac{\partial \vec{r}}{\partial v} = \frac{\partial x}{\partial u}\frac{\partial x}{\partial v} + \frac{\partial y}{\partial u}\frac{\partial y}{\partial v} + \frac{\partial z}{\partial u}\frac{\partial z}{\partial v} \\
G = G(u,v) &= \frac{\partial \vec{r}}{\partial v} \cdot \frac{\partial \vec{r}}{\partial v} = \left(\frac{\partial x}{\partial v}\right)^2 + \left(\frac{\partial y}{\partial v}\right)^2 + \left(\frac{\partial x}{\partial v}\right)^2
\end{aligned}
\tag{4.148}
$$

and the $\pm$ sign is to remind you that there are always two normals to a surface.

We desire to find a smooth curve C of shortest length which lies on a surface S and passes through two given fixed points P_1 and P_2 lying on the surface. Such a curve is called a geodesic curve. To calculate this curve we make the following assumptions: (i) the surface S is represented by the parametric equations

$$x = x(u,v), \qquad y = y(u,v), \qquad z = z(u,v) \qquad u,v \in R \tag{4.149}$$

where $x(u,v), y(u,v), z(u,v)$ are well defined continuous functions which are everywhere differentiable and R is some region of the u,v-plane, (ii) the position vector to a variable point on the surface S is given by

$$\vec{r} = \vec{r}(u,v) = x(u,v)\,\widehat{e}_1 + y(u,v)\,\widehat{e}_2 + z(u,v)\,\widehat{e}_3 \qquad u,v \in R \tag{4.150}$$

(iii) Let the fixed points P_1 and P_2 lying on the surface correspond to the parameter values (u_1, v_1) and (u_2, v_2) respectively. (iv) A smooth curve C on the surface S which passes through the fixed points P_1 and P_2 is the image of a smooth curve given by the parametric equations $u = u(t)$ and $v = v(t)$ which passes through the points (u_1, v_1) and (u_2, v_2) in the region R of the u, v-plane. Here we assume there are parametric values t_1 corresponding to the point P_1 and t_2 corresponding to the point P_2. These assumptions allow us to represent the smooth curve C lying on the surface S in the vector form

$$\vec{r} = \vec{r}(t) = x(u(t), v(t))\,\widehat{e}_1 + y(u(t), v(t))\,\widehat{e}_2 + z(u(t), v(t))\,\widehat{e}_3 \qquad t_1 \le t \le t_2 \qquad (4.151)$$

in terms of a single parameter t. An element of arc length ds squared along this curve is then given by $ds^2 = d\vec{r} \cdot d\vec{r}$ so that the length L of the curve C connecting the fixed points P_1 and P_2 is given by

$$L = \int_{t_1}^{t_2} \left[\left(\frac{dx}{dt} \right)^2 + \left(\frac{dy}{dt} \right)^2 + \left(\frac{dz}{dt} \right)^2 \right]^{1/2} dt \qquad (4.152)$$

Here we have

$$d\vec{r} = \frac{\partial \vec{r}}{\partial u}\, du + \frac{\partial \vec{r}}{\partial v}\, dv$$

with element of arc length squared given by

$$ds^2 = d\vec{r} \cdot d\vec{r} = \frac{\partial \vec{r}}{\partial u} \cdot \frac{\partial \vec{r}}{\partial u}\, du^2 + 2\frac{\partial \vec{r}}{\partial u} \cdot \frac{\partial \vec{r}}{\partial v}\, du\, dv + \frac{\partial \vec{r}}{\partial v} \cdot \frac{\partial \vec{r}}{\partial v}\, dv^2 \qquad (4.153)$$

which can also be represented in the form

$$ds^2 = E\, du^2 + 2F\, du dv + G\, dv^2 \qquad (4.154)$$

where E, F, G have been previously defined in the equation (4.148). The length L of the curve C is given by equation (4.152), which can be written in the alternate form

$$L = \int_{t_1}^{t_2} \left[E \left(\frac{du}{dt} \right)^2 + 2F \frac{du}{dt}\frac{dv}{dt} + G \left(\frac{dv}{dt} \right)^2 \right]^{1/2} dt = \int_{t_1}^{t_2} f(t, u(t), v(t), \dot{u}(t), \dot{v}(t))\, dt \qquad (4.155)$$

The geodesic curve through the points P_1 and P_2 lying on the surface S is that curve which minimizes the functional given by equation (4.155). In chapter three we have demonstrated that this functional is minimized when the parametric curve $u = u(t)$ and $v = v(t)$ is selected such that u and v satisfy the Euler-Lagrange equations

$$\begin{aligned} \frac{\partial f}{\partial u} - \frac{d}{dt}\left(\frac{\partial f}{\partial \dot{u}} \right) &= 0 \\ \frac{\partial f}{\partial v} - \frac{d}{dt}\left(\frac{\partial f}{\partial \dot{v}} \right) &= 0 \end{aligned} \qquad (4.156)$$

which are subject to the end point conditions given above.

It is left as an exercise to show that the Euler-Lagrange equations associated with the functional given by equation (4.155) can be represented in the form

$$\begin{aligned} \frac{E_u \dot{u}^2 + 2F_u \dot{u}\dot{v} + G_u \dot{v}^2}{\sqrt{E\dot{u}^2 + 2F\dot{u}\dot{v} + G\dot{v}^2}} - \frac{d}{dt}\left[\frac{2E\dot{u} + 2F\dot{v}}{\sqrt{E\dot{u}^2 + 2F\dot{u}\dot{v} + G\dot{v}^2}} \right] &= 0 \\ \frac{E_v \dot{u}^2 + 2F_v \dot{u}\dot{v} + G_v \dot{v}^2}{\sqrt{E\dot{u}^2 + 2F\dot{u}\dot{v} + G\dot{v}^2}} - \frac{d}{dt}\left[\frac{2F\dot{u} + 2G\dot{v}}{\sqrt{E\dot{u}^2 + 2F\dot{u}\dot{v} + G\dot{v}^2}} \right] &= 0 \end{aligned} \qquad (4.157)$$

where the subscripts denote partial differentiation.

Example 4-11. **Geodesics**

Some examples of well known surfaces that are represented in parametric form are the following.

Sphere of radius r with parameters $u = \theta$ and $v = \phi$

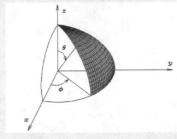

$$x = r \sin\theta \cos\phi$$
$$y = r \sin\theta \sin\phi$$
$$z = r \cos\theta$$

$$0 \le \theta \le \pi, \quad 0 \le \phi \le 2\pi, \quad r = \text{constant}$$

$$\pi/4 \le \phi \le \pi/2, \qquad 0 \le \theta \le \pi/2$$

Cylinder of radius r and height h with parameters $u = \theta$ and $v = z$

$$x = r \cos\theta$$
$$y = r \sin\theta$$
$$z = z$$

$$0 \le \theta \le 2\pi, \quad 0 \le z \le h, \quad r = \text{constant}$$

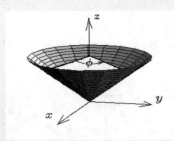

$$\pi/4 \le \theta \le \pi/2, \qquad 0 \le z \le 1$$

Right circular cone with parameters $u = \phi$ and v

$$x = v \sin\alpha \cos\phi$$
$$y = v \sin\alpha \sin\phi$$
$$z = v \cos\alpha$$

$$0 \le \phi \le 2\pi, \quad 0 \le v \le h, \quad \alpha \text{ constant}$$

$$0 \le \phi \le 2\pi, \ 0 \le v \le 1, \ \alpha = \tfrac{\pi}{4}$$

Torus of radius r **with parameters** u **and** v

$$x = (a + r \cos u) \cos v$$

$$y = (a + r \cos u) \sin v$$

$$z = r \sin u$$

$$0 \leq u \leq 2\pi, \quad 0 \leq v \leq 2\pi, \quad a > r$$

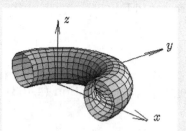

$$0 \leq u \leq 2\pi, \quad 0 \leq v \leq \pi, \quad a = 3, \; r = 1$$

On each of these surfaces one can let $d\vec{r} = \frac{\partial \vec{r}}{\partial u} du + \frac{\partial \vec{r}}{\partial v} dv$ denote a small change in the position vector $\vec{r} = \vec{r}(u, v)$ and calculate the element of arc length squared ds^2 given by

$$ds^2 = d\vec{r} \cdot d\vec{r} = \left(\frac{\partial \vec{r}}{\partial u} du + \frac{\partial \vec{r}}{\partial v} dv \right) \cdot \left(\frac{\partial \vec{r}}{\partial u} du + \frac{\partial \vec{r}}{\partial v} dv \right) = E \, du^2 + 2F \, du dv + G \, dv^2 \qquad (4.158)$$

where

$$E = E(u, v) = \frac{\partial \vec{r}}{\partial u} \cdot \frac{\partial \vec{r}}{\partial u} = \left(\frac{\partial x}{\partial u} \right)^2 + \left(\frac{\partial y}{\partial u} \right)^2 + \left(\frac{\partial z}{\partial u} \right)^2$$

$$F = F(u, v) = \frac{\partial \vec{r}}{\partial u} \cdot \frac{\partial \vec{r}}{\partial v} = \frac{\partial x}{\partial u} \frac{\partial x}{\partial v} + \frac{\partial y}{\partial u} \frac{\partial y}{\partial v} + \frac{\partial z}{\partial u} \frac{\partial z}{\partial v} \qquad (4.159)$$

$$G = G(u, v) = \frac{\partial \vec{r}}{\partial v} \cdot \frac{\partial \vec{r}}{\partial v} = \left(\frac{\partial x}{\partial v} \right)^2 + \left(\frac{\partial y}{\partial v} \right)^2 + \left(\frac{\partial z}{\partial v} \right)^2$$

Alternatively one can use the element of arc length

$$ds = \sqrt{E + 2F \frac{dv}{du} + G \left(\frac{dv}{du} \right)^2} \, du \qquad (4.160)$$

to formulate a functional for distance which is minimized by a function $v = v(u)$.

∎

Example 4-12. Shortest path on surface of sphere

One can verify that for a unit sphere where

$$x = \sin \theta \cos \phi, \qquad y = \sin \theta \sin \phi, \qquad z = \cos \theta \qquad (4.161)$$

$E = 1$, $F = 0$ and $G = \sin^2 \theta$ so that the distance between two points on a unit sphere having the surface coordinates (θ_0, ϕ_0) and (θ_1, ϕ_1) is given by

$$I = \int_{(\theta_0, \phi_0)}^{(\theta_1, \phi_1)} ds = \int_{\theta_0}^{\theta_1} \sqrt{1 + \sin^2 \theta \left(\frac{d\phi}{d\theta} \right)^2} \, d\theta \qquad (4.162)$$

We desire to find a relation between the surface coordinates θ and ϕ of the form $\phi = \phi(\theta)$ which makes the functional given by equation (4.162) a minimum. Here $f = \sqrt{1 + \sin^2 \theta \left(\frac{d\phi}{d\theta} \right)^2}$ and the Euler-Lagrange equation associated with this integrand is given by

$$\frac{\partial f}{\partial \phi} - \frac{d}{d\theta} \left(\frac{\partial f}{\partial \phi'} \right) = 0, \qquad \text{where } \phi' = \frac{d\phi}{d\theta}.$$

Calculating the necessary partial derivatives the Euler-Lagrange equation is found to be represented by

$$\frac{d}{d\theta}\left[\frac{\sin^2\theta\frac{d\phi}{d\theta}}{\sqrt{1+\sin^2\theta\left(\frac{d\phi}{d\theta}\right)^2}}\right]=0. \tag{4.163}$$

Integrating the equation (4.163) and solving for $\frac{d\phi}{d\theta}$ one obtains

$$\frac{d\phi}{d\theta}=\frac{K}{\sin\theta\sqrt{\sin^2\theta-K^2}} \tag{4.164}$$

where K is a constant of integration. In equation (4.164) make the substitutions $\tan\theta=1/\xi$, $\sin\theta=1/\sqrt{1+\xi^2}$ and show $d\theta=\frac{-d\xi}{\xi^2+1}$ and write equation (4.164) in the form

$$d\phi=\frac{-d\xi}{\sqrt{\alpha^2-\xi^2}} \tag{4.165}$$

where $\alpha^2=(1-K^2)/K^2$ is a new constant. Integrating the equation (4.165) produces the result

$$\phi=\cos^{-1}\left(\frac{\xi}{\alpha}\right)+\beta \tag{4.166}$$

where β is another constant of integration. The equation (4.166) has the alternative forms

$$\cos(\phi-\beta)=\frac{\xi}{\alpha}=\frac{1}{\alpha\tan\theta} \tag{4.167}$$
$$\tan\theta\left[\cos\phi\cos\beta+\sin\phi\sin\beta\right]=1/\alpha.$$

One can now use the defining equations given by equation (4.161) and write equation (4.167) in the form

$$x\cos\beta+y\sin\beta=\frac{1}{\alpha}z. \tag{4.168}$$

Note that the equation (4.168) is the equation of a plane which passes through the origin. The constants α and β are selected such that the given points on the sphere also lie on this plane. This plane intersects the sphere in a great circle. Consequently, the shortest arc of a great circle passing through the given points represents the shortest distance between the given points on the surface of the unit sphere.

∎

Isoperimetric problems

Consider the calculus of variation problem to find an extremum of

$$I=\int_{x_0}^{x_1}f(x,y,y')\,dx \tag{4.169}$$

subject to the end conditions $y(x_0)=\alpha$ and $y(x_1)=\beta$ where the class of admissible solution curves must also satisfy the subsidiary condition that the integral

$$J=\int_{x_0}^{x_1}g(x,y,y')\,dx=C=\text{a constant} \tag{4.170}$$

Assume that $y = y(x)$ is an extremum satisfying the end conditions and consider the two-parameter family of comparison functions

$$Y = Y(x) = y(x) + \epsilon_1 \eta_1(x) + \epsilon_2 \eta_2(x) \qquad (4.171)$$

Here we need a two-parameter family of comparison functions because if we used only a one-parameter family, then any change in the functional I would alter the subsidiary condition given by the integral J. We do not want a variation in J to happen because the subsidiary condition must remain constant during any variations. We assume that the functions η_1 and η_2 satisfy the end conditions $\eta_1(x_0) = \eta_1(x_1) = 0$ and $\eta_2(x_0) = \eta_2(x_1) = 0$ and substitute the comparison function Y into both the given functional and the subsidiary condition to obtain

$$I = I(\epsilon_1, \epsilon_2) = \int_{x_0}^{x_1} f(x, Y, Y') \, dx \qquad \text{and} \qquad J = J(\epsilon_1, \epsilon_2) = \int_{x_0}^{x_1} g(x, Y, Y') \, dx = C \qquad (4.172)$$

Note that the second relation $J = J(\epsilon_1, \epsilon_2) = C$ gives us a relationship between the parameters ϵ_1 and ϵ_2 in order for J to maintain a constant value. To solve the extremum problem given by equations (4.172) we use the method of Lagrange multipliers and write

$$F(\epsilon_1, \epsilon_2) = I(\epsilon_1, \epsilon_2) + \lambda \left(J(\epsilon_1, \epsilon_2) - C \right)$$

where λ is a Lagrange multiplier. Here F has an extremum when

$$\frac{\partial F}{\partial \epsilon_1} = 0 \qquad \text{and} \qquad \frac{\partial F}{\partial \epsilon_2} = 0$$

at $\epsilon_1 = 0$ and $\epsilon_2 = 0$. One can verify that

$$\frac{\partial F}{\partial \epsilon_i} = \int_{x_0}^{x_1} \left(\frac{\partial (f + \lambda g)}{\partial Y} \eta_i + \frac{\partial (f + \lambda g)}{\partial Y'} \eta_i' \right) dx \qquad \text{for } i = 1, 2 \qquad (4.173)$$

Then one can write that at an extremum

$$\frac{\partial F}{\partial \epsilon_i} \bigg|_{\epsilon_i = 0} = \int_{x_0}^{x_1} \left(\frac{\partial (f + \lambda g)}{\partial y} \eta_i + \frac{\partial (f + \lambda g)}{\partial y'} \eta_i' \right) dx = 0 \qquad \text{for } i = 1, 2 \qquad (4.174)$$

Use integration by parts on the second term in the integrand of equation (4.174) to show that at an extremum we must have

$$\int_{x_0}^{x_1} \left(\frac{\partial (f + \lambda g)}{\partial y} - \frac{d}{dx} \left(\frac{\partial (f + \lambda g)}{\partial y'} \right) \right) \eta_i(x) \, dx = 0 \qquad \text{for } i = 1, 2 \qquad (4.175)$$

Applying the basic lemma to either of the equations (4.175) produces the Euler-Lagrange necessary condition for an extremum

$$\frac{\partial (f + \lambda g)}{\partial y} - \frac{d}{dx} \left(\frac{\partial (f + \lambda g)}{\partial y'} \right) = 0 \qquad (4.176)$$

The solution to equation (4.176) can be expressed in the parametric form $y = y(x, c_1, c_2, \lambda)$ which uses a notation which indicates that the solution depends upon constants of integration

182

c_1, c_2 as well as the Lagrange multiplier λ. This gives three unknowns c_1, c_2, λ to be determined from the end point conditions and the subsidiary condition.

Generalization

The previous discussions can be extended to functionals of the form

$$I = I(y_1, y_2, \ldots, y_n) = \int_{x_0}^{x_1} f(t, y_1, y_2, \ldots, y_n, y_1', y_2', \ldots, y_n') \, dx$$

subject to the boundary conditions $y_i(x_0) = \alpha_i$ and $y_i(x_1) = \beta_i$ for $i = 1, \ldots, n$. An isoperimetric problem results if we impose the $N < n$ constraint or subsidiary conditions

$$J_k = \int_{x_0}^{x_1} g_k(t, y_1, y_2, \ldots, y_n, y_1', y_2', \ldots, y_n') \, dx \qquad k = 1, \ldots, N \tag{4.177}$$

This problem can also be solved by the use of Lagrange multipliers λ_k, for $k = 1, \ldots, N$. One can verify that a necessary condition for an extremum is for

$$\frac{\partial\left(f + \sum_{k=1}^N \lambda_k g_k\right)}{\partial y_i} - \frac{d}{dx}\left(\frac{\partial\left(f + \sum_{k=1}^N \lambda_k g_k\right)}{\partial y_i'}\right) = 0, \quad \text{for } i = 1, 2, \ldots, n \tag{4.178}$$

This set of equations has solutions containing $2n$ arbitrary constants and N-parameters $\lambda_1, \ldots, \lambda_N$ which are determined from the given boundary conditions and N subsidiary conditions.

Example 4-13.   Isoperimetric problem

The ancient problem of Queen Dido to find the closed curve of given perimeter which encloses the maximum area was well known to the Greeks. The circle is the curve with given perimeter that encloses the maximum area. This problem can be formulated as follows. Let $x = x(t)$ and $y = y(t)$ denote the parametric equations of a closed curve in the x, y-plane. If the curve is closed, then there exists parameter values t_1 and $t_2 \neq t_1$ such that

$$x(t_1) = x(t_2) = x_0 \quad \text{and} \quad y(t_1) = y(t_2) = y_0. \tag{4.179}$$

We use the area formula given by equation (1.18) which states that the area enclosed by the above parametric curve is given by the line integral

$$I = \int_{t_1}^{t_2} f(t, x, y, x', y') \, dt = \frac{1}{2} \int_{t_1}^{t_2} \left(x\frac{dy}{dt} - y\frac{dx}{dt}\right) dt \tag{4.180}$$

and the total length of the closed curve is given by

$$J = \int_{t_1}^{t_2} g(t, x, y, x', y') \, dt = \int_{t_1}^{t_2} ds = \int_{t_1}^{t_2} \sqrt{\left(\frac{dx}{dt}\right)^2 + \left(\frac{dy}{dt}\right)^2} \, dt \tag{4.181}$$

where J is to have a given constant value, say L. We employ Lagrange multipliers and calculate the associated integrand for the above problem. Let

$$F = F(x, y, x', y') = f + \lambda g = \frac{1}{2}(xy' - yx') + \lambda\left[(x')^2 + (y')^2\right]^{1/2} \tag{4.182}$$

from which one can calculate the partial derivatives

$$\frac{\partial F}{\partial x} = \frac{1}{2}y' \qquad\qquad \frac{\partial F}{\partial y} = -\frac{1}{2}x$$

$$\frac{\partial F}{\partial x'} = -\frac{1}{2}y + \frac{\lambda x'}{\sqrt{(x')^2 + (y')^2}} \qquad \frac{\partial F}{\partial y'} = \frac{1}{2}x + \frac{\lambda y'}{\sqrt{(x')^2 + (y')^2}} \qquad (4.183)$$

The Euler-Lagrange equations (4.178) become

$$\frac{\partial F}{\partial x} - \frac{d}{dx}\left(\frac{\partial F}{\partial x'}\right) = 0 \quad\Rightarrow\quad \frac{dy}{dt} - \lambda\frac{d}{dt}\left(\frac{x'}{\sqrt{(x')^2 + (y')^2}}\right) = 0$$

$$\frac{\partial F}{\partial y} - \frac{d}{dt}\left(\frac{\partial F}{\partial y'}\right) = 0 \quad\Rightarrow\quad \frac{dx}{dt} + \lambda\frac{d}{dt}\left(\frac{y'}{\sqrt{(x')^2 + (y')^2}}\right) = 0 \qquad (4.184)$$

Note that the above equations simplify if one selects the parameter t to be the arc length s. In this special case $\left(\frac{dx}{ds}\right)^2 + \left(\frac{dy}{ds}\right)^2 = 1$ so that the equations (4.184) simplify to

$$\frac{dy}{ds} - \lambda\frac{d^2x}{ds^2} = 0 \qquad \text{and} \qquad \frac{dx}{ds} + \lambda\frac{d^2y}{ds^2} = 0 \qquad (4.185)$$

One can integrate the equations (4.185) to obtain

$$y - \lambda\frac{dx}{ds} = c_1 \qquad \text{and} \qquad x + \lambda\frac{dy}{ds} = c_2 \qquad (4.186)$$

where c_1 and c_2 are constants of integration. The equations (4.186) imply that

$$\left(\frac{dx}{ds}\right)^2 + \left(\frac{dy}{ds}\right)^2 = \left(\frac{y - c_1}{\lambda}\right)^2 + \left(\frac{x - c_2}{\lambda}\right)^2 = 1. \qquad (4.187)$$

Eliminating x from the equations (4.186) we find that y must satisfy the second order differential equation

$$\lambda^2\frac{d^2y}{ds^2} + y = c_1. \qquad (4.188)$$

This differential equation has the general solution

$$y = c + \alpha\cos(s/\lambda) + \beta\sin(s/\lambda) \qquad (4.189)$$

with α, β constants. One then finds that

$$x = -\lambda\frac{dy}{ds} + c_2 = c_2 + \alpha\sin(s/\lambda) - \beta\cos(s/\lambda) \qquad (4.190)$$

The equations (4.189) and (4.190) can be written in the alternate form

$$y = c_1 + R\cos(s/\lambda - \theta)$$
$$x = c_2 + R\sin(s/\lambda - \theta) \qquad (4.191)$$

where $R = \sqrt{\alpha^2 + \beta^2}$ is a new constant and $\tan\theta = \beta/\alpha$ where θ is called a phase angle. The equations (4.191) describe the equation of a circle with center (c_2, c_1) and radius R. The perimeter of this circle is $L = 2\pi R$ and so $\lambda = R = L/2\pi$.

∎

Modifying natural boundary conditions

Consider the variational problem to find an extremum associated with the functional

$$I = \int_a^b \left[\frac{1}{2}(y')^2 - y\psi(x) \right] dx \tag{4.192}$$

where $\psi(x)$ is a given function of x which is well defined over the interval (a, b). The first variation of I is given by

$$\delta I = \int_a^b [y'\delta y' - \delta y\psi(x)] \, dx. \tag{4.193}$$

By using integration by parts on the first integral of equation (4.193), there results

$$\delta I = y'\delta y \Big|_{x=a}^{x=b} - \int_a^b \left[\frac{d}{dx}(y') + \psi(x) \right] \delta y \, dx \tag{4.194}$$

At an extremum we require that $\delta I = 0$. This is accomplished if we require y to satisfy the Euler-Lagrange equation

$$\frac{d^2 y}{dx^2} + \psi(x) = 0 \tag{4.195}$$

together with either fixed end point conditions ($\delta y = 0$) or natural boundary conditions (transversality conditions) where $\frac{dy}{dx} = 0$ at the end points. We consider the problem with the natural boundary conditions to solve equation (4.195) subject to the natural boundary conditions

$$\frac{dy}{dx} \Big|_{x=a} = 0, \quad \text{and} \quad \frac{dy}{dx} \Big|_{x=b} = 0. \tag{4.196}$$

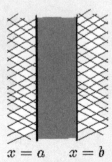

$x = a \qquad x = b$

The equations (4.195) and (4.196) have the physical interpretation of representing the mathematical model for heat conduction in a solid slab of material. The quantity $y = y(x)$ represents temperature within the slab between the values $x = a$ and $x = b$ and $\psi(x)$ represents the rate of internal heat generation per unit volume within the slab. The natural boundary conditions imply that no heat flows across the boundaries at $x = a$ and $x = b$ as these boundaries are insulated.

An integration of the equation (4.195) gives $\frac{dy}{dx} = -\int_a^x \psi(x) \, dx$ and this equation must satisfy $\frac{dy}{dx} = 0$ at $x = a$. Note that the no heat flow condition $\frac{dy}{dx} \Big|_{x=b} = 0$ requires the condition $0 = \int_a^b \psi(x) \, dx$. This places restrictions on the rate of heat generation within the slab of material and one must have an equality of heat sources and sinks within the slab to maintain this condition. The boundary conditions impose this special type of situation to exist.

One can construct other types of natural boundary conditions by modifying the functional given by equation (4.192). The general procedure is to add additional terms to the statement of the variational problem. The added terms then modify the natural boundary conditions. For example, consider the problem to find an extremum for the functional

$$I = \int_a^b \left[\frac{1}{2}(y')^2 - y\psi(x) \right] dx + \frac{\alpha}{2} y^2 \Big|_{x=a} \tag{4.197}$$

where α is a constant representing a heat loss coefficient. The calculation of the first variation in I, originally given by equation (4.194), is modified by the additional term and one finds

$$\delta I = y'\delta y \Big|_{x=a}^{x=b} - \int_a^b \left[\frac{d}{dx}(y') + \psi(x) \right] \delta y \, dx + \alpha y \delta y \Big|_{x=a} \tag{4.198}$$

For an extremum we require y to satisfy the Euler-Lagrange equation

$$\frac{d^2 y}{dx^2} + \psi(x) = 0$$

subject to the natural boundary conditions

$$-\frac{dy}{dx} + \alpha y = 0 \quad \text{at } x = a \quad \text{and} \quad \frac{dy}{dx} = 0 \quad \text{at } x = b.$$

Note that by adding an additional term to the functional we have modified the boundary condition at $x = a$ to give a heat loss boundary condition.

The above discussions for one-dimensional problems can be generalized to two- and three-dimensional problems. Consider the variational problem

$$I = \iiint_V \left[\nabla u \cdot \nabla u - 2\psi(x, y, z) \, u \right] d\tau \tag{4.199}$$

where V represents a volume, $d\tau$ is a volume element and $u = u(x, y, z)$ represents a temperature. Here

$$\nabla u = \operatorname{grad} u = \frac{\partial u}{\partial x} \hat{\mathbf{e}}_1 + \frac{\partial u}{\partial y} \hat{\mathbf{e}}_2 + \frac{\partial u}{\partial z} \hat{\mathbf{e}}_3 \tag{4.200}$$

with

$$\nabla u \cdot \nabla u = \left(\frac{\partial u}{\partial x} \right)^2 + \left(\frac{\partial u}{\partial y} \right)^2 + \left(\frac{\partial u}{\partial z} \right)^2 \tag{4.201}$$

and so the equation (4.199) can be written in the form

$$I = \iiint_V f(x, y, z, u, u_x, u_y, u_z) \, d\tau \tag{4.202}$$

where

$$f = f(x, y, z, u, u_x, u_y, u_z) = \left(\frac{\partial u}{\partial x} \right)^2 + \left(\frac{\partial u}{\partial y} \right)^2 + \left(\frac{\partial u}{\partial z} \right)^2 - 2\psi(x, y, z) \, u \tag{4.203}$$

In equation (4.202) assume that u is the function which makes I an extremum and then replace u by the set of comparison functions

$$U = U(x, y, z) = u(x, y, z) + \epsilon \eta(x, y, z)$$

to obtain

$$I = I(\epsilon) = \iiint_V f(x, y, z, U, U_x, U_y, U_z) \, d\tau$$

with

$$I'(\epsilon) = \iiint_V \left(\frac{\partial f}{\partial U} \eta + \frac{\partial f}{\partial U_x} \eta_x + \frac{\partial f}{\partial U_y} \eta_y + \frac{\partial f}{\partial U_z} \eta_z \right) d\tau.$$

At an extremum we want

$$I'(0) = \iiint_V \left(\frac{\partial f}{\partial u} \eta + \frac{\partial f}{\partial u_x} \eta_x + \frac{\partial f}{\partial u_y} \eta_y + \frac{\partial f}{\partial u_z} \eta_z \right) d\tau = 0. \tag{4.204}$$

To simplify the equation (4.204) we employ the Gauss divergence theorem

$$\iiint \operatorname{div} \vec{A} \, d\tau = \oiint_S \vec{A} \cdot \hat{n} \, d\sigma \tag{4.205}$$

where $\vec{A} = P\,\hat{e}_1 + Q\,\hat{e}_2 + R\,\hat{e}_3$, is a continuous vector field over a closed volume V, with unit exterior normal to the surface S which bounds the volume V given by $\hat{n} = n_1\,\hat{e}_1 + n_2\,\hat{e}_2 + n_3\,\hat{e}_3$ and $d\sigma$ is an element of surface area. We let

$$P = \eta \frac{\partial f}{\partial u_x}, \qquad Q = \eta \frac{\partial f}{\partial u_y}, \qquad R = \eta \frac{\partial f}{\partial u_z} \tag{4.206}$$

and write the Gauss divergence theorem in the form

$$\iiint_V \left[\eta \frac{\partial}{\partial x} \left(\frac{\partial f}{\partial u_x} \right) + \eta \frac{\partial}{\partial y} \left(\frac{\partial f}{\partial u_y} \right) + \eta \frac{\partial}{\partial z} \left(\frac{\partial f}{\partial u_z} \right) \right.$$
$$\left. + \frac{\partial f}{\partial u_x} \eta_x + \frac{\partial f}{\partial u_y} \eta_y + \frac{\partial f}{\partial u_z} \eta_z \right] d\tau = \oiint_S \eta \left[\frac{\partial f}{\partial u_x} n_1 + \frac{\partial f}{\partial u_y} n_2 + \frac{\partial f}{\partial u_z} n_3 \right] d\sigma \tag{4.207}$$

Rearrange of the terms in the equation (4.207) enables one to simplify the equation (4.204) to the form

$$I'(0) = \iiint_V \left[\frac{\partial f}{\partial u} - \frac{\partial}{\partial x} \left(\frac{\partial f}{\partial u_x} \right) - \frac{\partial}{\partial y} \left(\frac{\partial f}{\partial u_y} \right) - \frac{\partial}{\partial z} \left(\frac{\partial f}{\partial u_z} \right) \right] \eta \, d\tau$$
$$+ \oiint_S \eta \left[\frac{\partial f}{\partial u_x} n_1 + \frac{\partial f}{\partial u_y} n_2 + \frac{\partial f}{\partial u_z} n_3 \right] d\sigma = 0 \tag{4.208}$$

We require that the Euler-Lagrange equation

$$\frac{\partial f}{\partial u} - \frac{\partial}{\partial x} \left(\frac{\partial f}{\partial u_x} \right) - \frac{\partial}{\partial y} \left(\frac{\partial f}{\partial u_y} \right) - \frac{\partial}{\partial z} \left(\frac{\partial f}{\partial u_z} \right) = 0 \tag{4.209}$$

be satisfied everywhere within the volume V and on the surface S of V we require that the following natural boundary condition be satisfied

$$\left. \frac{\partial f}{\partial u_x} n_1 + \frac{\partial f}{\partial u_y} n_2 + \frac{\partial f}{\partial u_z} n_3 \right|_{(x,y,z) \in S = \partial V} = 0. \tag{4.210}$$

Differentiating the equation (4.203) we find

$$f_u = -2\psi(x, y, z), \qquad f_{u_x} = 2u_x, \qquad f_{u_y} = 2u_y, \qquad f_{u_z} = 2u_z$$

so that the Euler-Lagrange equation (4.209) becomes

$$\nabla^2 u + \psi(x, y, z) = 0$$

which is Poisson's equation. This is the steady state heat equation with a source term. The boundary condition, from equation (4.210), requires that $\nabla u \cdot \hat{n} = 0$ which is the requirement that there is no heat loss across the boundary surface or the condition that the boundary surface is insulated.

"Physical concepts are free creations of the human mind, and are not, however it may seem, uniquely determined by the external world. In our endeavor to understand reality we are somewhat like a man trying to understand the mechanism of a closed watch. He sees the face and the moving hands, even hears its ticking, but he has no way of opening the case. If he is ingenious he may form some picture of a mechanism which could be responsible for all the things he observes, but he may never be quite sure his picture is the only one which could explain his observations. He will never be able to compare his picture with the real mechanism and he cannot even imagine the possibility of the meaning of such a comparison."

Albert Einstein, (1879-1955)

Exercises Chapter 4

▶ **1.**   Find the extremal for the functional $I = \int_0^{\pi/2} \left[(y'')^2 - y^2 + x^2 \right] dx$ subject to the end point conditions $y(0) = 1$, $y'(0) = 0$, $y(\pi/2) = 0$, $y'(\pi/2) = -1$.

▶ **2.**   Find the shortest distance between two surface points (θ_1, z_1) and (θ_2, z_2) on the surface of a cylinder of radius r.

▶ **3.**   Find the general form for the curve of shortest length between two surface points (v_1, θ_1) and (v_2, θ_2) on a right circular cone. Hint: Use spherical coordinates.

▶ **4.**   Find the general form for the curve of shortest length between two surface points (u_1, v_1) and (u_2, v_2) on a torus with parameters r and a fixed and $a > r$. (See Example 4-11)

▶ **5.**   Find the extremum for the functional $I = \int_1^2 \left[y' + x^2 (y')^2 \right] dx$ with boundary conditions $y(1) = 1$ and $y(2) = 2$. What are the natural boundary conditions for this problem?

▶ **6.**   Find the extremum for the functional $I = \int_0^{\pi} \left[(y')^2 + 2y \sin x \right] dx$ with boundary conditions $y(0) = 0$ and $y(\pi) = 0$. What are the natural boundary conditions for this problem?

▶ **7.**   Find the curve joining the points $(0,0)$ and $(1,0)$ for which the functional

$$I = \int_0^1 (y'')^2 \, dx$$

is a minimum and y is subject to the conditions $y'(0) = a$ and $y'(1) = b$, where a, b are given constants. What are the natural boundary conditions for his problem?

▶ **8.** Derive the Euler-Lagrange equation for the functional

$$I = \iiint_V f(x, y, z, w, w_x, w_y, w_z)\, dxdydz$$

where V is a given volume. Find the natural boundary conditions for this problem.
Hint: Use the divergence theorem

$$\iiint_V \left(\frac{\partial P}{\partial x} + \frac{\partial Q}{\partial y} + \frac{\partial R}{\partial z}\right) dxdydz = \iint_S (Pn_1 + Qn_2 + Rn_3)\, d\sigma$$

where S is the surface representing the boundary of V, $d\sigma$ is an element of surface area, $\hat{n} = n_1\hat{\mathbf{e}}_1 + n_2\hat{\mathbf{e}}_2 + n_3\hat{\mathbf{e}}_3$ is a unit outward normal to the boundary surface S. Let $P = \eta\frac{\partial f}{\partial w_x}$, $Q = \eta\frac{\partial f}{\partial w_y}$, $R = \eta\frac{\partial f}{\partial w_z}$ where $\eta = \eta(x, y, z) = 0$ for (x, y, z) on the surface S.

▶ **9.** Determine the Euler-Lagrange equation associated with the functional

$$I = I(u) = \iint_R F(x, y, z, z_x, z_y, z_{xx}, z_{xy}, z_{yy})\, dxdy$$

Show that the boundary condition for this problem can be written
$\left(Q\hat{\mathbf{e}}_1 - P\hat{\mathbf{e}}_2\right)\cdot\hat{\mathbf{n}}\Big|_{x,y\in\partial R} = 0$ where $Q = \left[\frac{\partial F}{\partial z_x} - \frac{\partial}{\partial x}\left(\frac{\partial F}{\partial z_{xx}}\right)\right]\eta + \frac{\partial F}{\partial z_{xx}}\eta_x + \frac{\partial F}{\partial z_{xy}}\eta_y$
and $P = \left[-\frac{\partial F}{\partial z_y} + \frac{\partial}{\partial y}\left(\frac{\partial F}{\partial z_{yy}}\right) + \frac{\partial}{\partial x}\left(\frac{\partial F}{\partial z_{xy}}\right)\right]\eta - \frac{\partial F}{\partial z_{yy}}\eta_y$

▶ **10.** Find the functions $x = x(t)$ and $y = y(t)$, such that the functional

$$I = \int_0^L (x^2 + y^2)\, dt \qquad\qquad 10(a)$$

is an extremum such that x and y are related by the differential equation

$$\frac{dx}{dt} = y - x, \qquad\qquad 10(b)$$

and subject to the boundary conditions $x(0) = 1$, $y(L) = 1$.
(a) Solve by substitution of y from equation 10(b) into equation 10(a).
(b) Solve by using a Lagrange multiplier λ which is a function of position so that $\lambda = \lambda(t)$. Note from the Euler-Lagrange equations that $\lambda = 0$ when $t = L$.

▶ **11.** Find the function $y(x)$ which produces an extremal for the functional
$I = \int_0^\pi \left(y^2 - 2y\sin x + (y')^2\right) dx$ satisfying the end point conditions $y(0) = 0$ and $y(\pi) = \sinh\pi$.

▶ **12.**

Consider the extreme value of the functional $I = \int_{x_1}^{x_2} f(x, y, v)\, dx$ subject to the constraint condition $v = \frac{dy}{dx}$. Solve this problem using a Lagrange multiplier $\lambda = \lambda(x)$ using variations with respect to both y and v and derive the Euler-Lagrange equation for the functional $I = \int_{x_1}^{x_2} f(x, y, y')\, dx$

▶ **13.** Let $x = x(u, v), y = y(u, v)$ and $z = z(u, v)$ define a smooth closed surface S for $u, v \in R_{uv}$.

(a) Show the unit normal to the surface S is given by

$$\hat{\mathbf{n}} = \frac{1}{\pm\sqrt{EG - F^2}} \begin{vmatrix} \hat{\mathbf{e}}_1 & \hat{\mathbf{e}}_2 & \hat{\mathbf{e}}_3 \\ \frac{\partial x}{\partial u} & \frac{\partial y}{\partial u} & \frac{\partial z}{\partial u} \\ \frac{\partial x}{\partial v} & \frac{\partial y}{\partial v} & \frac{\partial z}{\partial v} \end{vmatrix}$$

$$\hat{\mathbf{n}} = \frac{1}{\pm\sqrt{EG - F^2}} \left[(y_u z_v - y_v z_u)\hat{\mathbf{e}}_1 + (z_u x_v - z_v x_u)\hat{\mathbf{e}}_2 + (x_u y_v - x_v y_u)\hat{\mathbf{e}}_3 \right]$$

where E, F and G are defined by equation (4.148). Note that at a point on a closed surface S there are always two normal vectors, an inward normal and an outward normal. In most applications, the outward normal is selected. If the surface is not closed, then you must select one of the normals.

(b) Show the surface area is given by $S = \iint_{R_{uv}} \sqrt{EG - F^2}\, du\, dv$

Hint: How is the element of surface area related to the vector $d\vec{r}$ lying in the tangent plane to a point on the surface?

(c) Use the results from problem 41, chapter 1, to show the volume enclosed by the smooth closed surface can be represented

$$V = \frac{1}{3} \iint_{R_{uv}} \left[x(y_u z_v - y_v z_u) + y(z_u x_v - z_v x_u) + z(x_u y_v - x_v y_u) \right] du\, dv$$

(d) Write an outline on how you would go about finding the system of differential equations defining the smooth closed surface which encloses the maximum volume subject to the constraint that the surface area has a fixed constant value S_0. Hint: See [f8] and the isoperimetric derivation on pages 180-181.

▶ **14.** Apply the Lagrange multiplier rule to the problem of finding $w = w(x)$ such that the functional $I = \int_0^{\pi/2} w^2(x)\, dx$ is to be a minimum, where w is subject to the constraints

$$\frac{dw}{dx} + y - (y - z)^2 y = 0, \qquad \frac{dy}{dx} - w = 0$$

with w, y, z subject to the boundary conditions

$$y(0) = 0, \quad z(0) = 0, \quad w(0) = 1, \quad y(\pi/2) = 1, \quad z(\pi/2) = 1, \quad w(\pi/2) = 0.$$

▶ **15.** A special case associated with the functional given by equation (4.80) is

$$I = I(y) = \int_{x_1}^{x_2} p(x, y)\sqrt{1 + (y')^2}\, dx, \qquad p(x, y) \text{ is a known function.}$$

For this functional find the transversality condition if the point (x_1, y_1) is to lie on a given curve $y = g(x)$ and the point (x_2, y_2) is to lie on another curve $y = h(x)$. In this special case, show the transversality condition reduces to an orthogonality condition.

▶ **16.** Find the Euler-Lagrange equation together with essential and natural boundary conditions associated with the given functional

$$I = I(y) = \int_{x_1}^{x_2} f(x, y, y')\, dx + g(y)\Big|_{x=x_1}$$

▶ **17.** Find the Euler-Lagrange equation together with essential and natural boundary conditions associated with the given functional

$$I = \int_{x_1}^{x_2} f(x, y_1, y_2, y_1', y_2')\, dx + g(y_1, y_2)\Big|_{x=x_2}$$

▶ **18.** Consider the extreme value of the functional

$$I = \int_{x_1}^{x_2} f\left(x, y(x), y'(x), \int_{x_1}^{x_2} g(s, y(s), y'(s))\, ds\right) dx$$

and how to re-write the problem using Lagrange multipliers.

(a) Let $v = \int_{x_1}^{x_2} g(x, y, y')\, dx$ and show that one can introduce a Lagrange multiplier λ and write the above problem in the form

$$I = I(y, v) = \int_{x_1}^{x_2} \left[f(x, y, y', v) + \lambda\left(g(x, y, y') - \frac{v}{x_2 - x_1}\right) \right] dx$$

(b) Find the Euler-Lagrange equations associated with this new functional.

▶ **19.** Find the Euler-Lagrange equations and natural boundary conditions associated with the functional $I = \int_a^b f(x, y, y')\, dx + g(x, y)\Big|_{x=b} - h(x, y)\Big|_{x=a}$

▶ **20.** Find the Euler-Lagrange equations and natural boundary conditions associated with the functional $\quad I = \int_a^b f(x, y, y', y'')\, dx + g(x, y, y')\Big|_{x=b} - h(x, y, y')\Big|_{x=a}$

▶ **21.** Find the Euler-Lagrange equations and natural boundary conditions associated with the functional $I = \iint_S f(x, y, w, w_x, w_y)\, d\sigma + \oint_C \left[g(x, y, w)\frac{dx}{ds} - h(x, y, w)\frac{dy}{ds} \right] ds$ where C is a boundary curve on a given surface S and boundary conditions are specified along the curve C.

▶ **22.** Find the Euler-Lagrange equations and natural boundary conditions associated with the functional

$$I = \iiint_V f(x, y, z, w, w_x, w_y, w_z)\, d\tau + \oiint_S \vec{G} \cdot \hat{n}\, d\sigma$$

where the vector $\vec{G}$ is defined $\vec{G} = g_1(x, y, z, w)\,\widehat{e}_1 + g_2(x, y, z, w)\,\widehat{e}_2 + g_3(x, y, z, w)\,\widehat{e}_3$ and represents a continuous vector field for $x, y, z \in V$ and $\hat{n} = n_1\,\widehat{e}_1 + n_2\,\widehat{e}_2 + n_3\,\widehat{e}_3$ is a unit exterior normal vector to the surface S which encloses the volume V.

▶ **23.** Find the Euler-Lagrange equation and boundary conditions associated with the variational problem to find $u = u(x, y, z)$ such that

$$I = \iiint_V [\nabla u \cdot \nabla u - 2\psi(x, y, z)\, u]\, d\tau + \oiint_S [2uf(x, y, z) + u^2 g(x, y, z)]\, d\sigma$$

is an extremum. Give a physical interpretation to your results.

▶ **24.** Find extreme values for $I = \int_{(x_1, y_1)}^{(6,8)} \sqrt{1 + y'^2}\, dx$ where (x_1, y_1) is restricted to lie on the circle $x^2 + y^2 = 25$.

▶ **25.**

Find the Euler-Lagrange equations and natural boundary conditions associated with finding the function $u = u(x, y)$, where $x, y \in R$, such that

$$I = \int_{x_1}^{x_2} \int_{y_1}^{y_2} F(x, y, u, u_x, u_y, u_{xx}, u_{xy}, u_{yy})\, dx dy$$

is an extremum. The region R being the square region illustrated.

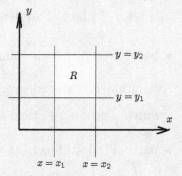

▶ **26.** Find the curve $y = y(x)$ for which the functional

$$I = \int_0^{x_1} \frac{1}{y}\sqrt{1 + (y')^2}\, dx, \qquad y(0) = 0$$

can have an extremum if the point (x_1, y_1) is to lie on the curve $y = x - \alpha$ where $\alpha > 0$ is a constant.

▶ **27.** Find the shortest distance from the point $(3, 0)$ to the parabola $y = x^2$ by finding the minimum value of $I = \int_{x_1}^3 \sqrt{1 + (y')^2}\, dx$ subject to the constraint that the point (x_1, y_1) is on the given parabola.

▶ **28.** Find the shortest distance between the curve $\psi(x) = x$ and the curve $\phi(x) = \sqrt{x - 1}$ by finding the minimum value of $I = \int_a^b \sqrt{1 + (y')^2}\, dx$ where a and b are variable end points on the given curves.

▶ **29.** Consider a smooth curve $y(x)$ which makes the functional $I = I(y) = \int_{x_1}^{x_2} f(x, y, y')\, dx$ have an extremum. Show that if the end points of the curve $y(x)$ are required to lie on two different given curves having the implicit forms $\phi_1(x, y) = 0$ and $\phi_2(x, y) = 0$, then the transversality conditions can be written in the form

$$\left(f - y'\frac{\partial f}{\partial y'}\right)\frac{\partial \phi_1}{\partial y} - \frac{\partial f}{\partial y'}\frac{\partial \phi_1}{\partial x}\bigg|_{(x_1, y_1)} = 0 \quad \text{and} \quad \left(f - y'\frac{\partial f}{\partial y'}\right)\frac{\partial \phi_2}{\partial y} - \frac{\partial f}{\partial y'}\frac{\partial \phi_2}{\partial x}\bigg|_{(x_2, y_2)} = 0$$

▶ **30.** Determine the Euler-Lagrange equation associated with the functional

$$I = I(u) = \iint_R \left[\alpha(x,y) \left(\frac{\partial u}{\partial x} \right)^2 + \beta(x,y) \left(\frac{\partial u}{\partial y} \right)^2 + \gamma(x,y)u^2 + 2f(x,y)u \right] dxdy$$

where u is subject to the boundary conditions that $u(x,y) = g(x,y)$ for $x, y \in C = \partial R$. Assume α, β, γ and f are well defined continuous functions with derivatives which are also continuous over the region R.

▶ **31.** Find the curve in polar coordinates (r, θ) where $I = \int_{(1,0)}^{(\sqrt{2}, \pi/6)} r \, ds$ is a minimum. Here ds is an element of arc length in polar coordinates.

▶ **32.** Determine a minimum value for the functional $\int_{(0,1)}^{(1,y_2)} \left(\frac{dy}{dx} \right)^2 dx$ where y_2 is subject to variation and the constraint condition $\int_0^1 \frac{y}{y_2} dx = -1$ is to be satisfied. Hint: If $I = \int_0^1 f \, dx$ with $f = (y')^2 + \lambda \frac{y}{y_2}$, then $\delta I = \int_0^1 \left(\frac{\partial f}{\partial x} \delta x + \frac{\partial f}{\partial y} \delta y + \frac{\partial f}{\partial y'} \delta y' + \frac{\partial f}{\partial y_2} \delta y_2 \right) dx$ since y_2 is allowed to vary.

▶ **33.** Find all functions $y = y(x)$ which produce a stationary value for the functional given by $\int_0^\ell \left(\frac{dy}{dx} \right)^2 dx$ and satisfy the conditions $y(0) = 0$ and $y(\ell) = 0$ while subject to the constraint condition that $\int_0^\ell y^2 \, dx = c^2 \ell / 2$ where $c > 0$ and $\ell > 0$ are constants.

▶ **34.**
Consider the problem to find $y = y(x)$ which maximizes the integral $\int_0^2 y \, dx$ subject to the constraint condition $\int_0^2 \sqrt{1 + y'^2} \, dx = \ell$. where $2 < \ell$ Give a physical interpretation of this problem. Discuss the cases where discontinuities are allowed. Consider $\ell < \pi$, $\ell = \pi$ and $\ell > \pi$ i.e. $\ell = \ell_1 + \ell_2 + \ell_3$ (see figures).

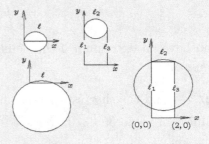

▶ **35.** Determine a minimum value for the functional $\int_{(0,1)}^{(1,y_2)} \frac{y'^2}{x^3} dx$ where y_2 is subject to variation and the constraint condition $\int_0^1 \frac{y}{y_2} dx = 1/3$ is to be satisfied. Hint: If $I = \int_0^1 f \, dx$ with $f = \frac{y'^2}{x^3} + \lambda \frac{y}{y_2}$, then $\delta I = \int_0^1 \left(\frac{\partial f}{\partial x} \delta x + \frac{\partial f}{\partial y} \delta y + \frac{\partial f}{\partial y'} \delta y' + \frac{\partial f}{\partial y_2} \delta y_2 \right) dx$ since y_2 is allowed to vary.

▶ **36.**
(a) Find the minimum value of $I = \int_{x_1}^{x_2} \sqrt{1 + y'^2} \, dx$ subject to the constraint condition that (x_1, y_1) lies on the curve $y = g(x) = x^2$ and (x_2, y_2) lies on the curve $y = h(x) = 1 - (x - 4)^2$. Give a physical interpretation to this problem. Sketch the curves.
(b) Show the square of the distance between the points $(x_1, g(x_1))$ and $(x_2, h(x_2))$ is given by $F(x_1, x_2) = \ell^2 = (x_2 - x_1)^2 + \left[1 - (x_2 - 4)^2 - x_1^2 \right]^2$ and find the minimum value for F.

▶ **37.** Find the curve $y = y(x)$ such that $I = \int_{(0,0)}^{(x,y)} y^2(1 + y'^2)\, dx$ has a stationary value, the upper end point being required to lie on the line $y = 1$.

▶ **38.** Minimize $\int_{(x_1,y_1)}^{(x_2,y_2)} \sqrt{1 + y'^2}\, dx$ subject to the constraint that (x_1, y_1) lie on the curve $y = g(x) = 4 + (x - 3)^2$ and the point (x_2, y_2) lie on the curve $y = h(x) = -2x - 1$. Give a physical interpretation to this problem and find $y = y(x)$ and (x_2, y_2) and (x_1, y_1).

▶ **39.**
(a) Consider the integral $I = \int_{(1,1)}^{(2,3)} ds = \int_{(1,1)}^{(2,3)} \sqrt{1 + y'^2}\, dx$ where ds is an element of arc length.
   (i) Obtain the Euler-Lagrange differential equation.
   (ii) Solve the Euler-Lagrange equation to obtain solution family $y = y(x, \alpha, \beta)$
   (iii) Apply the end condition requirements and show $y = y(x) = 2x - 1$
(b) Show the Legendre condition $f_{y'y'} > 0$ is satisfied.
(c) Show the Weierstrass excess function is $E(x, y, y', p) = \sqrt{1 + p^2} - \sqrt{5} - \dfrac{2}{\sqrt{5}}(p - 2)$
   (i) Expand $\sqrt{1 + p^2}$ in a Taylors series about $p = 2$ and verify that

$$\sqrt{1 + p^2} = \sqrt{5} + \frac{2}{\sqrt{5}}(p - 2) + R_2, \quad \text{where} \quad R_2 = \frac{1}{(1 + \xi^2)^{3/2}} \frac{(p - 2)^2}{2!}, \quad 2 < \xi < p$$

   (ii) Show $E > 0$
(d) Show there are no conjugate points between 1 and 2
(e) Determine if the solution y corresponds to a maximum or minimum value for I.

▶ **40.** Find the Euler-Lagrange equation and solve if $I = \int_0^1 (\dot{y}^2 + \dot{x}^2 - 4t\dot{x} - 4x)\, dt$ is to have a stationary value subject to the constraint conditions $x(0) = y(0) = 0$, $x(1) = y(1) = 1$ and $\int_0^1 (\dot{y}^2 - t\dot{y} - \dot{x}^2)\, dt = 2$.

▶ **41.** Find the extremal for the functional $I = \int_0^1 \left[xy' + (y')^2 \right] dx$ subject to the end point conditions $y(0) = 1$, $y(1) = 0$.

▶ **42.** Show that the change of variables $x = x(u, v)$, $y = y(u, v)$ converts the integral $\iint_{R_{xy}} dx dy$ to the form $\iint_{R_{uv}} \sqrt{EG - F^2}\, du dv = \iint_{R_{uv}} \left| \frac{\partial(x,y)}{\partial(u,v)} \right| du dv$ where R_{uv} is a region determined by the curvilinear coordinates of the transformation and the region R_{xy}.

▶ **43.** Write a discussion of how you would derive the Euler-Lagrange equations which describe the extremal arc for the isoperimetric problem that $I = \int_{(x_0,y_0)}^{(x_1,y_1)} F(x, y, y')\, dx$ be an extremum subject to the constraints

$$J_1 = \int_{(x_0,y_0)}^{(x_1,y_1)} G_1(x, y, y')\, dx = C_1 \quad \text{and} \quad J_2 = \int_{(x_0,y_0)}^{(x_1,y_1)} G_2(x, y, y')\, dx = C_2$$

where C_1 and C_2 are given constants.

▶ **44.**

A surface called a helicoid is defined by the parametric equations $x = r\cos\theta$, $y = r\sin\theta$, $z = \alpha\theta$ where α is a constant, $0 \le r \le r_0$, and $0 \le \theta \le 2\pi$. Find the equation which defines a geodesic curve on a helicoid.

▶ **45.**

Consider a general surface of revolution defined by

$$x = r\cos\theta, \qquad y = r\sin\theta, \qquad z = f(r)$$

where $f(r) \in \mathcal{C}^1$, $0 \le \theta \le 2\pi$ and $0 \le r \le r_0$. Find the differential equation which defines the geodesic on this surface.

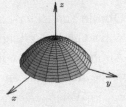

▶ **46.**

Consider the cone of revolution defined by the parametric equations $x = r\cos\theta$, $y = r\sin\theta$, $z = a - r$ where $a > 0$ is a constant, $0 \le \theta \le 2\pi$ and $0 \le r \le a$. Find the equation which defines a geodesic curve between two points on this cone.

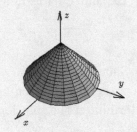

▶ **47.** Let $P_0(x_0, y_0, z_0)$ and $P_1(x_1, y_1, z_1)$ denote two different points lying on a surface defined by an equation of the form $g(x, y, z) = 0$

(a) Show that the shortest distance between the two given points on the surface is given by $I = \int_{x_0}^{x_1} \sqrt{1 + (y')^2 + (z')^2}\, dx$ subject to the constraint that $g(x, y, z) = 0$.

(b) Find the system of Euler-Lagrange equations for determining the functions $y = y(x)$ and $z = z(x)$ such that the shortest path can be described by $\vec{r} = x\,\widehat{e}_1 + y(x)\,\widehat{e}_2 + z(x)\,\widehat{e}_3$ for $x_1 \le x \le x_2$.

▶ **48.**

(a) Obtain the Euler-Lagrange equation for producing a stationary value associated with the functional $I = \int_{(0,1)}^{(1,6)} \frac{1 + y'^2}{y}\, dx$

(b) Solve the resulting Euler-Lagrange equation and show $y = \frac{1}{4\alpha^2}(x + \beta)^2 - \alpha^2$ is a two-parameter family of solutions.

(c) From the result in part (b) obtain two linearly independent solutions of the Jacobi differential equation.

(d) Test your result to determine if a maximum or minimum arc exists.

Chapter 5
Applications of the Variational Calculus

In this chapter we present selected applications from the fields of science and engineering where the calculus of variations is employed to obtain an extremal associated with a derived functional. The resulting Euler-Lagrange equations are either ordinary differential equations or partial differential equations with boundary conditions. We begin with a presentation of some of the more classical applications where the calculus of variations is utilized.

The brachistrochrone problem

In 1696 John Bernoulli formulated the following problem. Imagine a bead sliding smoothly without friction on a thin wire. Given two points P_0 and P_1 on the wire, find the shape of the wire such that the bead moves under the action of gravity from point P_0 to point P_1 in the shortest time. This is called the brachistrochrone problem. The word brachistrochrone comes form the Greek *brachistos* for shortest and the Greek *chrones* for time. Assume the points $P_0 = (x_0, y_0)$ and $P_1 = (x_1, y_1)$ are points in a vertical plane not on the same vertical line with $x_1 > x_0$ and $y_1 > y_0$. By constructing a coordinate system with the y-axis pointing downward the situation can be described by the illustration in figure 5-1.

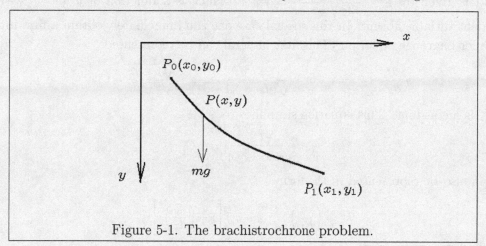

Figure 5-1. The brachistrochrone problem.

Let $y = y(x)$ denote the curve representing the shape of the thin wire which connects the points P_0 and P_1 and let V_0 denote the initial velocity of the bead of mass m at the point P_0 and let V denote the velocity of the bead at a general point $P(x, y)$ on the curve $y = y(x)$. Employ the work-energy theorem from physics which states that in the absence of frictional forces the work done by forces acting on a particle must equal the change in the kinetic energy of the particle. The work done is force times distance and kinetic energy is $T = \frac{1}{2}mV^2$ where m is the mass and V is the velocity.

The work done represents the change in potential[‡] energy in moving the bead from point y_0 to y and is given by $mg(y-y_0)$. Here m is the mass of the particle, and g is the acceleration of gravity. The change in kinetic energy of the particle is given by $\frac{1}{2}m(V^2 - V_0^2)$, where V is the final velocity and V_0 is the initial velocity. The work-energy theorem requires

$$mg(y - y_0) = \frac{1}{2}m(V^2 - V_0^2) \tag{5.1}$$

and solving for the velocity squared we find that

$$V^2 = V_0^2 + 2gy - 2gy_0.$$

Without loss of generality we can assume that the particle starts from rest so that $V_0 = 0$ and consequently the particle velocity $V = \frac{ds}{dt}$ is represented by the change in distance s with time t, and so one can write

$$\frac{ds}{dt} = V = \sqrt{2g(y - y_0)}. \tag{5.2}$$

The arc length along the extremal curve $y = y(x)$ is obtained by integrating $ds = \sqrt{1 + y'^2}\, dx$ so that the total time T taken to move from P_0 to P_1 can then be expressed

$$T = \int_{x_0}^{x_1} \frac{ds}{V} = \frac{1}{\sqrt{2g}} \int_{x_0}^{x_1} \frac{\sqrt{1 + y'^2}}{\sqrt{y - y_0}}\, dx. \tag{5.3}$$

We are to find the curve $y = y(x)$ such that the total time T is a minimum. Here the integrand is given by $f = \frac{1}{\sqrt{2g}}\sqrt{\frac{1 + y'^2}{y - y_0}} = f(y, y')$ which is a function of y and y' with the independent variable absent. In this special case one can immediately obtain a first integral of the Euler-Lagrange equation. This first integral can be represented

$$y'\frac{\partial f}{\partial y'} - f = C_1 \quad \text{or} \quad \frac{y'^2}{\sqrt{(y - y_0)(1 + y'^2)}} - \frac{\sqrt{1 + y'^2}}{\sqrt{y - y_0}} = C_1. \tag{5.4}$$

where C_1 is a constant. This equation simplifies to

$$\frac{-1}{[(y - y_0)(1 + y'^2)]^{1/2}} = C_1 \tag{5.5}$$

which can also be represented in the form

$$k^2 = \left(\frac{-1}{C_1}\right)^2 = (y - y_0)\left[1 + \left(\frac{dy}{dx}\right)^2\right] \tag{5.6}$$

[‡] If $\vec{F}$ is a force acting on a particle, then one can define the potential energy function at a point P by $V(P) = -\int_\alpha^P \vec{F} \cdot d\vec{r}$ where α is some convenient reference point. The difference in potential between two points P_1 and P_2 is given by

$$V(P_2) - V(P_1) = -\int_\alpha^{P_2} \vec{F} \cdot d\vec{r} + \int_\alpha^{P_1} \vec{F} \cdot d\vec{r} = -\int_{P_1}^{P_2} \vec{F} \cdot d\vec{r}$$

and represents the negative of the work done in moving from P_1 to P_2. For a conservative force system $\oint_C \vec{F} \cdot d\vec{r} = 0$ where C is a simple closed path. The choice of zero potential energy is arbitrary because we are only concerned with differences in potential energy.

where k^2 is some new constant. This form is easier to integrate because one can separate the variables in equation (5.6) to obtain

$$\frac{\sqrt{y - y_0}\, dy}{\sqrt{k^2 - (y - y_0)}} = dx. \tag{5.7}$$

Make the substitution $y = y_0 + k^2 \sin^2 \theta$ with $dy = 2k^2 \sin \theta \cos \theta \, d\theta$ and verify that equation (5.7) simplifies to

$$dx = 2k^2 \sin^2 \theta \, d\theta. \tag{5.8}$$

This equation can now be integrated to obtain

$$x = x_0 + k^2 \left(\theta - \frac{1}{2} \sin 2\theta \right).$$

This gives the solution of the Euler-Lagrange equation as the parametric equations

$$x = x_0 + \frac{1}{2} k^2 \left(2\theta - \sin 2\theta \right), \qquad y = y_0 + \frac{1}{2} k^2 \left(1 - \cos 2\theta \right) \tag{5.9}$$

which represents a family of cycloids. Here the constants k and x_0 can be selected such that the curve passes through a given point (x_1, y_1). Note that the point y_0 is not arbitrary since by equation (5.2) we know that when $y = y_1$ we must have $V = V_1$ so that y_0 must satisfy $V_1 = \sqrt{2g(y_1 - y_0)}$ or $y_0 = y_1 - V_1^2/2g$.

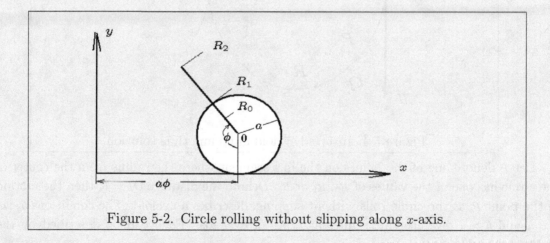

Figure 5-2. Circle rolling without slipping along x-axis.

There are many types of cycloids. Consider a circle of radius $R = a$ which rolls without slipping along the x-axis. Let $0R_0R_1R_2$ denote a line segment extended from the center of the circle, with $0 < R_0 < R_1 < R_2$ and $R_1 = a$, as illustrated in the figure 5-2.

198

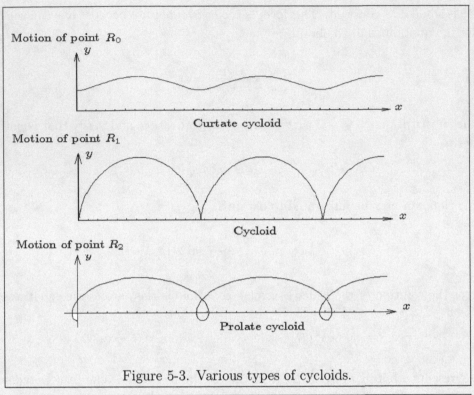

Figure 5-3. Various types of cycloids.

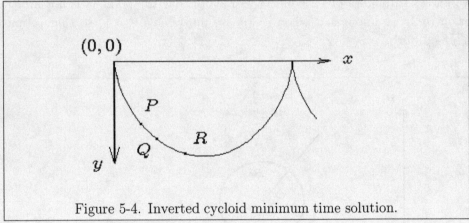

Figure 5-4. Inverted cycloid minimum time solution.

Let P denote one of the points on the line segment where the radius from the center of the circle has one of the values of R_0, R_1 or R_2. Denote the distance $OP = b$, then the motion of the point P, as the circle rolls without slipping, describes a cycloid. The conditions $b < a$, $b = a$ and $b > a$ produce three different types of cycloids. Each can be described by the parametric equations

$$x = a\phi - b\sin\phi, \qquad y = a - b\cos\phi \qquad (5.10)$$

where ϕ is a parameter and a is the radius of the circle. The case $b > a$ gives a prolate cycloid. The case $b = a$ gives a cycloid and the case $b < a$ gives a curtate cycloid. These cases are illustrated in the figure 5-3. Note that the inverted cycloid, figure 5-4, is the shortest time problem solution. Additional properties of the brachistrochrone curve can be found in the exercises for chapter five.

The hanging chain, rope or cable

Consider a flexible chain, rope or cable of length L hanging between the two points (ℓ, h) and $(-\ell, h)$, with $h > 0$ as illustrated in the figure 5-5. One can use telephone poles and hanging telephone lines as an example of this problem. Assume the length L satisfies the inequality $L > 2\ell$.

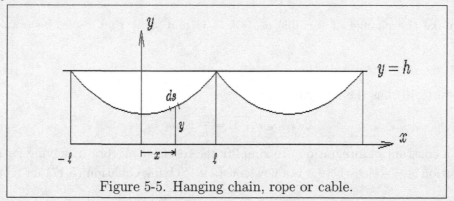

Figure 5-5. Hanging chain, rope or cable.

Consider a flexible cable whose center line gives a curve which defines the shape of the cable. This curve is assumed to lie in the xy-plane and in addition it is assumed that the cable is in a stable equilibrium configuration so that the potential energy associated with the cable is a minimum. Let ϱ denote the mass density per unit length of the cable and assume this lineal density is a constant. Consider an element of mass dm located at a general point (x, y) on the hanging cable. This element of mass has the potential energy $dV = (dm)gh = (\varrho ds)gy$ where $ds = \sqrt{1 + (y')^2}\, dx$ is a element of arc length along the cable. The potential energy of the cable is then obtained by integration over the length of the cable to obtain

$$V = \int_{-\ell}^{\ell} dV = \varrho g \int_{-\ell}^{\ell} y\sqrt{1 + (y')^2}\, dx, \qquad y' = \frac{dy}{dx}. \tag{5.11}$$

Here we must find the curve $y = y(x)$ such that the integral given by equation (5.11) is a minimum subject to the constraint conditions (i) the length of the cable between (ℓ, h) and $(-\ell, h)$ is a given length L so that $L = \int_{-\ell}^{\ell} \sqrt{1 + (y')^2}\, dx$ is a constant and (ii) the curve must satisfy the boundary conditions $y(-\ell) = h$ and $y(\ell) = h$. This is an isoperimetric problem in the calculus of variations. Introduce a Lagrange multiplier λ and form the integrand

$$f = f(x, y, y') = \varrho g y \sqrt{1 + (y')^2} + \lambda \sqrt{1 + (y')^2} \tag{5.12}$$

The Euler-Lagrange equation associated with this integrand is readily verified to have the representation

$$\frac{d}{dx}\left(\frac{\partial f}{\partial y'}\right) - \frac{\partial f}{\partial y} = 0, \qquad \Rightarrow \qquad \frac{d}{dx}\left[\frac{\varrho g y y' + \lambda y'}{\sqrt{1 + (y')^2}}\right] - \varrho g \sqrt{1 + (y')^2} = 0 \tag{5.13}$$

Note that the independent variable x is absent and consequently the equation (3.25) can be employed to immediately obtain the first integral $y'\frac{\partial f}{\partial y'} - f = \alpha =$ a constant. This first integral can be represented in the form

$$(\varrho g y + \lambda)\left[\frac{(y')^2}{\sqrt{1+(y')^2}} - \sqrt{1+(y')^2}\right] = \alpha \tag{5.14}$$

Solve the equation (5.14) for $(y')^2$ and verify that

$$(y')^2 = \frac{(\varrho g y + \lambda)^2}{\alpha^2} - 1 \tag{5.15}$$

and then make the change of variable $\varrho g y + \lambda = \alpha z$ and show that z can be obtained by integrating

$$\frac{dz}{\sqrt{z^2-1}} = \frac{\varrho g}{\alpha}\, dx \tag{5.16}$$

The equation (5.16) has the integral

$$\ln(z + \sqrt{z^2-1}) = \frac{\varrho g}{\alpha}\, x + b_1 \tag{5.17}$$

where b_1 is a constant of integration. To simplify the resulting algebra in solving for z, make the substitution $b_1 = -\frac{\varrho g}{\alpha}\beta$ where β is a new constant. Solving equation (5.17) for z produces the result

$$z = \frac{\varrho g y + \lambda}{\alpha} = \cosh\left[\frac{\varrho g(x - \beta)}{\alpha}\right] \tag{5.18}$$

and therefore the curve defining the shape of the cable is given by

$$y = y(x) = -\frac{\lambda}{\varrho g} + \frac{\alpha}{\varrho g}\cosh\left[\frac{\varrho g(x - \beta)}{\alpha}\right] \tag{5.19}$$

This gives the shape of the hanging rope, chain or cable as a hyperbolic cosine function. The Latin word for chain is *catena* and so the shape of a hanging chain, rope or cable is called a catenary.

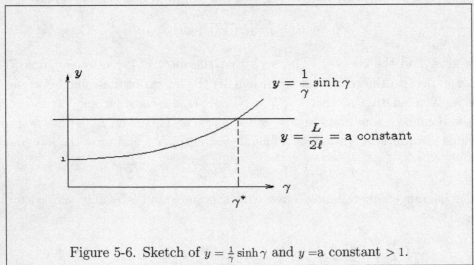

Figure 5-6. Sketch of $y = \frac{1}{\gamma}\sinh\gamma$ and $y =$ a constant > 1.

Note the symmetry of the problem and select the constant $\beta = 0$ in equation (5.19) so that the boundary conditions $y(-\ell) = y(\ell) = h$ are satisfied by an appropriate selection for the constant λ once the constant α is known. The solution given by equation (5.19) must satisfy the length restriction that

$$L = \int_{-\ell}^{\ell} \sqrt{1 + (y')^2}\, dx = \int_{-\ell}^{\ell} \cosh\left(\frac{\varrho g x}{\alpha}\right) dx = \frac{2\alpha}{\varrho g} \sinh\left(\frac{\varrho g \ell}{\alpha}\right) \tag{5.20}$$

Here the fixed length L satisfies $L > 2\ell$ so that $L/2\ell > 1$ and so equation (5.20) will always have a solution for the constant α which can be determined from equation (5.20) using numerical methods as follows. Let $\gamma = \frac{\varrho g \ell}{\alpha}$ and write equation (5.20) in the form

$$\frac{L}{2\ell} = \frac{1}{\gamma} \sinh \gamma$$

A sketch of the curves $y = \frac{1}{\gamma} \sinh \gamma$ and $y = \frac{L}{2\ell}$ are illustrated in the figure 5-6 The problem is now reduced to finding the value γ^* where the two curves intersect. This is a root finding exercise in numerical methods.

Soap film problem

Consider a curve $y = y(x)$ passing through two given points (x_1, y_1) and (x_2, y_2) which is rotated about the x-axis as illustrated in the accompanying figure. One can formulate the variational problem to find the curve $y = y(x)$ such that the surface of revolution has a minimum surface area.

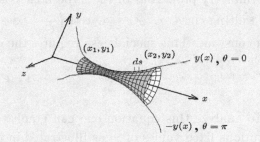

A ribbon element of surface area constructed on the surface is given by

$$dS = 2\pi y\, ds = 2\pi y \sqrt{1 + (y')^2}\, dx$$

where ds is an element of arc length along the original curve $y = y(x)$. One can then formulate the variational problem to find $y = y(x)$ such that

$$S = \int_{x_1}^{x_2} 2\pi y \sqrt{1 + (y')^2}\, dx$$

is a minimum. Here we have

$$f = f(x, y, y') = 2\pi y \sqrt{1 + (y')^2}$$

which is an integrand independent of x and consequently the Euler-Lagrange equation will have the first integral given by equation (3.25)

$$y' \frac{\partial f}{\partial y'} - f = \alpha \quad \text{or} \quad \frac{y(y')^2}{\sqrt{1 + (y')^2}} - y\sqrt{1 + (y')^2} = \alpha$$

where α is a constant. This equation can be simplified to the form

$$\frac{y}{\sqrt{1+(y')^2}} = C_1$$

where $C_1 = -\alpha$ is a new constant. This equation can be further simplified to the form

$$\frac{dx}{dy} = \frac{C_1}{\sqrt{y^2 - C_1^2}}$$

which can be integrated to produce the result

$$y = y(x) = C_1 \cosh\left(\frac{x - C_2}{C_1}\right)$$

where C_2 is a constant. The constants C_1 and C_2 are found from the requirement that the curve $y = y(x)$ passes through the given points (x_1, y_1) and (x_2, y_2). The surface generated by the curve $y = y(x)$ is a catenary of revolution which is called a catenoid.

Let us consider a special case of the above problem where we can make use of the symmetry properties of the hyperbolic cosine function. We set $C_2 = 0$ and require that the resulting curve $y = y(x) = C_1 \cosh\left(\frac{x}{C_1}\right)$ pass through the points $(-x_1, 1)$ and $(x_1, 1)$ where x_1 is a constant. This special case requires the constant C_1 to be selected to satisfy the equation

$$1 = C_1 \cosh\left(\frac{x_1}{C_1}\right). \tag{5.21}$$

To analyze this equation one can graph the functions $y = 1$ and $y = C_1 \cosh\left(\frac{x_1}{C_1}\right)$ vs C_1 for various fixed values of x_1 as illustrated in the figure 5-7.

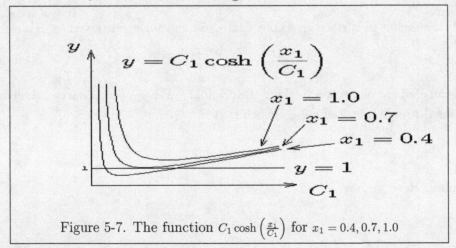

Figure 5-7. The function $C_1 \cosh\left(\frac{x_1}{C_1}\right)$ for $x_1 = 0.4, 0.7, 1.0$

The figure 5-7 illustrates the following cases. Case 1: There are values of x_1 for which equation (5.21) has no solutions for C_1. Case 2: There are values of x_1 for which one solution exists for C_1. Case 3: There are values of x_1 for which two solutions can exist for C_1. To understand what is happening consider a wire frame having two circles of radius unity where the circles lie in parallel planes separated by a distance $2x_1$. When the circles are dipped into a soap solution and removed a surface film is formed representing the minimal surface. If x_1 is small, then two possible values for C_1 can exist which gives rise to two different soap films. One of the soap films will correspond to the minimum surface area. The reference G.A. Bliss has given an examination of the two catenaries of revolution that produce the two soap films. The upper catenary is a flat curve for $y(x)$ and the lower catenary gives a deep curve for $y(x)$. The flat curve corresponds to an absolute minimum for the surface area while the lower curve does not produce a minimum value. As the end circles move further apart the quantity x_1 reaches a critical point where equation (5.21) has only one solution. Increasing x_1 beyond this critical point causes the soap film to break and no solution exists for equation (5.21). When this occurs, there is formed a circular film over each circle on the ends of the wire frame . This is known as the Goldschmidt discontinuous solution. For additional experiments with soap films the reference Courant and Robbins is suggested.

Hamilton's principle and Euler-Lagrange equations

Consider a system of N particles and denote by $\vec{r}_i = \vec{r}_i(t)$ the position vector of the ith particle of mass m_i. Newton's law of motion for the ith particle states that the time rate of change of linear momentum must equal the force applied to the ith particle. Assume that all the particles have a constant mass and write Newton's law of motion for the ith particle

$$m_i \ddot{\vec{r}}_i = \vec{F}_i \qquad \text{or} \qquad m_i \ddot{\vec{r}}_i - \vec{F}_i = \vec{0} \tag{5.22}$$

where $\vec{F}_i$ denotes the total force acting on the ith particle. Let $\delta \vec{r}_i$ denote the variation in the position of the ith particle and take the dot product of the ith equation (5.22) with the $\delta \vec{r}_i$ and then sum over all particles i to obtain

$$\sum_{i=1}^{N} \left(m_i \ddot{\vec{r}}_i - \vec{F}_i \right) \cdot \delta \vec{r}_i = \vec{0} \qquad \text{or} \qquad \sum_{i=1}^{N} m_i \ddot{\vec{r}}_i \cdot \delta \vec{r}_i - \sum_{i=1}^{N} \vec{F}_i \cdot \delta \vec{r}_i = \vec{0} \tag{5.23}$$

The second term on the left-hand side of equation (5.23) represents a summation of terms where each term represents a force times a distance or work done. The total work done δW is a summation of the forces $\vec{F}_i$ due to the displacements $\delta \vec{r}_i$. One can write

$$\delta W = \sum_{i=1}^{N} \vec{F}_i \cdot \delta \vec{r}_i \tag{5.24}$$

Here the force $\vec{F}$ has units of *Newtons* [N], $\delta \vec{r}$ has units of *meters* [m] and the work W has units of *joules* [J]. Here 1 *joule* = 1 *newton · meter*. To simplify the left-hand side of equation (5.23) we employ the identity

$$\sum_{i=1}^{N} \left[m_i \frac{d}{dt} \left(\dot{\vec{r}}_i \cdot \delta \vec{r}_i \right) - \delta \left(\frac{m_i}{2} \dot{\vec{r}}_i \cdot \dot{\vec{r}}_i \right) \right] = \sum_{i=1}^{N} m_i \ddot{\vec{r}}_i \cdot \delta \vec{r}_i \tag{5.25}$$

The kinetic energy T of the system of particles is defined as a summation of one-half the mass times velocity squared for each particle. This gives the kinetic energy

$$T = \sum_{i=1}^{N} \frac{m_i}{2} \dot{\vec{r}}_i \cdot \dot{\vec{r}}_i = \sum_{i=1}^{N} \frac{m_i}{2} v_i^2 \tag{5.26}$$

and so by using the identity (5.25) the left-hand side of equation (5.23) can be expressed

$$\sum_{i=1}^{N} m_i \ddot{\vec{r}}_i \cdot \delta \vec{r}_i = \sum_{i=1}^{N} m_i \frac{d}{dt}(\dot{\vec{r}}_i \cdot \delta \vec{r}_i) - \delta T \tag{5.27}$$

where δT is the variation in kinetic energy. Substitute the equations (5.27) and (5.24) into the equation (5.23) and express the result in the form

$$\delta T + \delta W = \sum_{i=1}^{N} m_i \frac{d}{dt}\left(\dot{\vec{r}}_i \cdot \delta \vec{r}_i\right) \tag{5.28}$$

Assume that the variations in path satisfy the conditions $\delta \vec{r}_i(t_1) = 0$ and $\delta \vec{r}_i(t_2) = 0$ and integrate the equation (5.28) from t_1 to t_2 to obtain

$$\int_{t_1}^{t_2} (\delta T + \delta W)\, dt = \sum_{i=1}^{N} m_i (\dot{\vec{r}}_i \cdot \delta \vec{r}_i) \Big|_{t_1}^{t_2} = 0 \tag{5.29}$$

If there exists a potential function V such that $\delta W = -\delta V$, then the equation (5.29) can be written in the form

$$\int_{t_1}^{t_2} (\delta T - \delta V)\, dt = 0. \tag{5.30}$$

Define the quantity $L = T - V$ as the Lagrangian function associated with the motion of the system of particles, where T is the kinetic energy of the system of particles and V is the potential energy of the system of particles. Hamilton's principle can then be expressed

$$I = \int_{t_1}^{t_2} L\, dt \quad \text{is an extremum, so that} \quad \delta I = \delta \int_{t_1}^{t_2} L\, dt = 0.$$

The integral

$$I = \int_{t_1}^{t_2} L\, dt \tag{5.31}$$

is called a Hamilton integral and Hamilton's principle is that the path of motion of a particle or system of particles is such that the Hamilton integral is an extremum, usually a minimum. The equations of motion of the particles are then along the paths defined by the system of Euler-Lagrange equations. If the Lagrangian is a function of position coordinates $q_1, q_2, \ldots, q_N$ and velocities $\dot{q}_1, \dot{q}_2, \ldots, \dot{q}_N$, called generalized coordinates and their time derivatives, then the Lagrangian can be written $L = L(q_1, q_2, \ldots, q_n, \dot{q}_1, \dot{q}_2, \ldots, \dot{q}_n)$ and the equations of motion of the system are represented by the Euler-Lagrange equations

$$\frac{\partial L}{\partial q_i} - \frac{d}{dt} \frac{\partial L}{\partial \dot{q}_i} = 0 \quad \text{for} \quad i = 1, 2, \ldots, N \tag{5.32}$$

Example 5-1. Equations of motion

Let $\vec{r}$ $\vec{r}(t) = x(t)\,\hat{\mathbf{e}}_1 + y(t)\,\hat{\mathbf{e}}_2 + z(t)\,\hat{\mathbf{e}}_3$ denote the position vector of a single particle of constant mass m which is moving in a force field $\vec{F} = \vec{F} = F_x\,\hat{\mathbf{e}}_1 + F_y\,\hat{\mathbf{e}}_2 + F_z\,\hat{\mathbf{e}}_3$. Newton's second law describing the particle motion is written

$$m\ddot{\vec{r}} - \vec{F} = \vec{0} \quad \text{resulting in the scalar equations} \quad \begin{cases} m\ddot{x} = F_x \\[4pt] m\ddot{y} = F_y \\[4pt] m\ddot{z} = F_z \end{cases}$$

Multiply the above vector equation by $\delta\vec{r}\,dt$ and then integrate from t_1 to t_2 to define the integral

$$\delta I = \int_{t_1}^{t_2} \left(m\ddot{\vec{r}} \cdot \delta\vec{r} - \vec{F} \cdot \delta\vec{r} \right) dt = 0$$

Use integration by parts to integrate the first term and obtain

$$\delta I = m\dot{\vec{r}} \cdot \delta\vec{r}\,\Big|_{t_1}^{t_2} - \int_{t_1}^{t_2} \left(m\dot{\vec{r}} \cdot \delta\dot{\vec{r}} + \vec{F} \cdot \delta\vec{r} \right) dt = 0$$

Assume that the variation at the end points is zero so that the first term in the above equation is zero. Using the relation $\delta(\dot{\vec{r}})^2 = 2\dot{\vec{r}} \cdot \delta\dot{\vec{r}}$ the above relation can be written in the form

$$\delta I = \int_{t_1}^{t_2} \left[\delta\left(\frac{m}{2}\dot{\vec{r}}^{\,2} \right) + \vec{F} \cdot \delta\vec{r} \right] dt = \int_{t_1}^{t_2} (\delta T + \delta W)\,dt = 0$$

$$\text{where} \quad I = \int_{t_1}^{t_2} (T + W)\,dt \tag{5.33}$$

with T denoting the kinetic energy and $\delta W = \vec{F} \cdot \delta\vec{r}$ the variation of the work done in moving the particle in the force field.

If the vector field $\vec{F}$ is irrotational, then $\nabla \times \vec{F} = \operatorname{curl}\vec{F} = \vec{0}$ and the work done in moving the particle through the force field is independent of path taken. Such a force is called conservative and is derivable from a scalar function V called a force potential. The force is obtained by the gradient operation $\vec{F} = -\operatorname{grad}V = -\nabla V$. The work done in moving from the point P_1 where $t = t_1$ to the point P_2 where $t = t_2$ is given by

$$W_{12} = \int_{t_1}^{t_2} \vec{F} \cdot \delta\vec{r} = \int_{P_1}^{P_2} -\operatorname{grad}V \cdot \delta\vec{r} = V(P_1) - V(P_2) \tag{5.34}$$

which shows that for a conservative force field the work done is independent of the path taken by the particle in moving from P_1 to P_2. The change in potential energy of the particle is defined as minus the work done W_{12} by the conservative force field in moving from point P_1 to the point P_2. One can then write

$$W_{12} = \int_{t_1}^{t_2} \vec{F} \cdot dr = \int_{t_1}^{t_2} m\ddot{\vec{r}} \cdot d\vec{r} = \int_{t_1}^{t_2} m\ddot{\vec{r}} \cdot \frac{d\vec{r}}{dt}\,dt = \int_{t_1}^{t_2} \frac{m}{2}\frac{d}{dt}\left(\dot{\vec{r}} \cdot \dot{\vec{r}} \right) dt = \frac{m}{2}v_2^2 - \frac{m}{2}v_1^2 = T_2 - T_1 \tag{5.35}$$

where $v = \left|\frac{d\vec{r}}{dt}\right|$ denotes the magnitude of the particle velocity. This result shows that the work done equals the change in kinetic energy of the particle.

Equating the equations (5.34) and (5.35) gives the result

$$V(P_1) - V(P_2) = T_2 - T_1 \qquad \text{or} \qquad T_1 + V(P_1) = T_2 + V(P_2) \tag{5.36}$$

which gives the principle of conservation of energy for a particle. That is, if the force acting on a particle is conservative and derivable from a potential function V, then the total energy $T + V$ of the particle is conserved.

Here $\delta W = -\delta V$ so that the equation (5.33) becomes

$$I = \int_{t_1}^{t_2} (T - V)\, dt = \int_{t_1}^{t_2} L\, dt, \qquad L = T - V \tag{5.37}$$

with $\delta I = 0$ The quantity $L = T - V$ is called the Lagrangian with kinetic energy T and potential energy V. The motion of the particle is such that the integral of the Lagrangian is a minimum with the requirement $\delta I = 0$. This is known as Hamilton's principle. The kinetic energy associated with a particle of mass m at position (x, y, z) is given by

$$T = \frac{1}{2} m(\dot{x}^2 + \dot{y}^2 + \dot{z}^2) \tag{5.38}$$

and the potential energy V of the particle is such that

$$F_x\,\widehat{\mathbf{e}}_1 + F_y\,\widehat{\mathbf{e}}_2 + F_z\,\widehat{\mathbf{e}}_3 = -\operatorname{grad} V = -\frac{\partial V}{\partial x}\,\widehat{\mathbf{e}}_1 - \frac{\partial V}{\partial y}\,\widehat{\mathbf{e}}_2 - \frac{\partial V}{\partial z}\,\widehat{\mathbf{e}}_3 \tag{5.39}$$

The Lagrangian can then be written

$$L = L(x, y, z, \dot{x}, \dot{y}, \dot{z}) = T - V = \frac{1}{2} m(\dot{x}^2 + \dot{y}^2 + \dot{z}^2) - V(x, y, z) \tag{5.40}$$

and the variation $\delta I = 0$ requires that the following Euler-Lagrange equations be satisfied

$$
\begin{aligned}
\frac{\partial L}{\partial x} - \frac{d}{dt}\left(\frac{\partial L}{\partial \dot{x}}\right) &= 0 &\Rightarrow&\quad m\ddot{x} + \frac{\partial V}{\partial x} = 0 &\Rightarrow&\quad m\ddot{x} - F_x = 0 \\
\frac{\partial L}{\partial y} - \frac{d}{dt}\left(\frac{\partial L}{\partial \dot{y}}\right) &= 0 &\Rightarrow&\quad m\ddot{y} + \frac{\partial V}{\partial y} = 0 &\Rightarrow&\quad m\ddot{y} - F_y = 0 \\
\frac{\partial L}{\partial z} - \frac{d}{dt}\left(\frac{\partial L}{\partial \dot{z}}\right) &= 0 &\Rightarrow&\quad m\ddot{z} + \frac{\partial V}{\partial z} = 0 &\Rightarrow&\quad m\ddot{z} - F_z = 0
\end{aligned}
\tag{5.41}
$$

which represent the scalar form of the equations of motion.

Generalization

Consider a system of p particles which are acted upon by a conservative force system which are subject to given geometric constraints. Let the ith particle have position (x_i, y_i, z_i) and let there exist a potential function

$$V = V(x_1, y_1, z_1, \ldots, x_i, y_i, z_i, \ldots, x_p, y_p, z_p) \tag{5.42}$$

so that the force acting on the ith particle is given by

$$\vec{F}^{(i)} = -\nabla_{(i)}V, \qquad \text{where} \qquad \nabla_{(i)} = \frac{\partial V}{\partial x_i} + \frac{\partial V}{\partial y_i} + \frac{\partial V}{\partial z_i}$$

Then each particle can be labeled with a position, kinetic energy and force as follows.

particle	position	kinetic energy	Force derivable from potential
1	(x_1, y_1, z_1)	$\frac{1}{2}m_1(\dot{x}_1^2 + \dot{y}_1^2 + \dot{z}_1^2)$	$\vec{F}^{(1)} = -\nabla_{(1)}V$
$\vdots$	$\vdots$	$\vdots$	$\vdots$
i	(x_i, y_i, z_i)	$\frac{1}{2}m_i(\dot{x}_i^2 + \dot{y}_i^2 + \dot{z}_i^2)$	$\vec{F}^{(i)} = -\nabla_{(i)}V$
$\vdots$	$\vdots$	$\vdots$	$\vdots$
p	(x_p, y_p, z_p)	$\frac{1}{2}m_p(\dot{x}_p^2 + \dot{y}_p^2 + \dot{z}_p^2)$	$\vec{F}^{(p)} = -\nabla_{(p)}V$

Here the positions of the particles, kinetic energies and forces are described by 3p position coordinates. The total kinetic energy of the system of particles is given by

$$T = \frac{1}{2}\sum_{j=1}^{p} m_j(\dot{x}_j^2 + \dot{y}_j^2 + \dot{z}_j^2) \tag{5.43}$$

If the particles are subject to independent constraint conditions of the form

$$\phi_j(x_1, y_1, z_1, \ldots, x_p, y_p, z_p) \quad \text{for } j = 1, 2, \ldots, k \tag{5.44}$$

where k is less than 3p, then these constraints reduce the number of independent coordinates describing the position of the particles. The number of independent coordinates needed to describe the system of particles is reduced to $N = 3p - k$ where k is the number of constraint equations. Here the constraint equations can be used to eliminate k variables from the original system of $3p$ variables.

It is convenient to choose any set of $N = 3p - k$ independent coordinates to describe the position of the particles. The coordinates selected may or may not have a physical meaning associated with the system of particles. The N independent coordinates selected are called generalized coordinates. If $q_1, q_2, \ldots, q_N$ are selected as generalized coordinates, then these coordinates are used to replace the constraint equations (5.44). One introduces a set of $3p$ equations related to the position coordinates by a set of transformation equations of the form

$$\begin{aligned} x_i &= x_i(q_1, q_2, \ldots, q_N) \\ y_i &= y_i(q_1, q_2, \ldots, q_N) \\ z_i &= z_i(q_1, q_2, \ldots, q_N) \end{aligned} \tag{5.45}$$

for $i = 1, \ldots, p$. It is customary to interpret the generalized coordinates as representing parameters so that the equations (5.45) can be viewed as a set of parametric equations describing the coordinates of the particles. Note that a set of parametric equations of the form (5.45) is not unique. Differentiating the position coordinates with respect to time produces the equations

$$\dot{x}_i = \sum_{j=1}^{N} \frac{\partial x_i}{\partial q_j}\dot{q}_j, \qquad \dot{y}_i = \sum_{j=1}^{N} \frac{\partial y_i}{\partial q_j}\dot{q}_j, \qquad \dot{z}_i = \sum_{j=1}^{N} \frac{\partial z_i}{\partial q_j}\dot{q}_j, \tag{5.46}$$

Substitute the transformation equations (5.45) and derivatives given by equation (5.46) into the kinetic energy equation (5.43). The potential energy equation (5.42) allows the construction of the Lagrangian function in terms of the generalized coordinates having the form

$$L = T - V = L(q_1, q_2, \ldots, q_N, \dot{q}_1, \dot{q}_2, \ldots, \dot{q}_N) \tag{5.47}$$

where

$$T = \frac{1}{2} \sum_{j=1}^{p} m_j(\dot{x}_j^2 + \dot{y}_j^2 + \dot{z}_j^2) = \frac{1}{2} \sum_{j=1}^{p} m_j (\dot{q}_i)^2 \left[\sum_{i=1}^{N} \left\{ \left(\frac{\partial x_j}{\partial q_i} \right)^2 + \left(\frac{\partial y_j}{\partial q_i} \right)^2 + \left(\frac{\partial z_j}{\partial q_i} \right)^2 \right\} \right] \tag{5.48}$$

is the kinetic energy which is a homogeneous function of degree two in the generalized velocity components $\dot{q}_1, \dot{q}_2, \ldots, \dot{q}_N$.

The equations of motion of the system of particles can then be determined from Hamilton's principle and the Euler-Lagrange equations

$$\frac{\partial L}{\partial q_j} - \frac{d}{dt} \left(\frac{\partial L}{\partial \dot{q}_j} \right) = 0, \qquad \text{for } j = 1, 2, \ldots, N. \tag{5.49}$$

The more general form of the equation (5.49) is given by

$$\frac{d}{dt} \left(\frac{\partial T}{\partial \dot{q}_j} \right) - \frac{\partial T}{\partial q_j} = Q_j \tag{5.50}$$

where Q_j represents a generalized force associated with the generalized coordinate q_j. Here Q_i for $i = 1, \ldots, N$ can be obtained from the principle of virtual work. If

$$\delta W = Q_1 \delta q_1 + Q_2 \delta q_2 + \cdots Q_N \delta q_N$$

represents the work done by the generalized forces, then δW is represented

$$\delta W = \sum_{j=1}^{N} \vec{F}_j \cdot \delta \vec{r}_j = \sum_{j=1}^{N} \vec{F}_j \cdot \sum_{i=1}^{N} \frac{\partial \vec{r}_j}{\partial q_i} \delta q_i = \sum_{i=1}^{N} Q_i \delta q_i$$

which implies that $Q_i = \sum_{j=1}^{N} \vec{F}_j \cdot \frac{\partial \vec{r}_j}{\partial q_i}$ is the generalized force associated with the generalized coordinate q_i.

If the Lagrangian L does not contain the independent time variable t explicitly, employ the results from equations (3.61) and (3.62) to obtain a first integral of the system of equations (5.49). This first integral can be represented

$$\sum_{j=1}^{N} \dot{q}_j \frac{\partial L}{\partial \dot{q}_j} - L = E \tag{5.51}$$

where E is a constant. Also note that the kinetic energy T is a homogeneous function of degree 2 in the generalized velocities $\dot{q}_1, \dot{q}_2, \ldots, \dot{q}_N$. A generalization of Euler's theorem, equation (1.123), allows us to write

$$\sum_{j=1}^{N} \dot{q}_j \frac{\partial T}{\partial \dot{q}_j} = 2T \tag{5.52}$$

The potential energy function V is independent of the generalized velocities so that

$$\frac{\partial T}{\partial \dot{q}_j} = \frac{\partial L}{\partial \dot{q}_j} \quad \text{and} \quad Q_j = -\frac{\partial V}{\partial q_j} \tag{5.53}$$

Substituting the results from the equations (5.52) and (5.53) into the first integral given by equation (5.51) produces the result

$$2T - L = 2T - (T - V) = E \quad \text{or} \quad T + V = E. \tag{5.54}$$

This states that for a conservative system the sum of the kinetic energy and potential energy is a constant. The constant E is the total energy of the system. The value of the total energy E can be determined by substituting the initial values for the generalized positions q_j and generalized velocities $\dot{q}_j$ into equation (5.54).

Nonconservative systems

The Euler-Lagrange equations of motion having the form

$$\frac{d}{dt}\left(\frac{\partial T}{\partial \dot{q}_j}\right) - \frac{\partial T}{\partial q_j} = Q_j \tag{5.55}$$

do not require that the generalized forces Q_j be derivable from a potential function. If they are derivable from a potential function, then the system is called a conservative system, otherwise it is called nonconservative. One type of nonconservative system is characterized by dissipative forces such as frictional forces and viscous damping forces which are sometimes modeled as being proportional to the velocities. If the dissipative forces Q_i, for $i = 1, \ldots, n$, are proportional to the velocities, then

$$Q_i = -c_{i1}\dot{q}_1 - c_{i2}\dot{q}_2 - \cdots - c_{in}\dot{q}_n = -\sum_{j=1}^{n} c_{ij}\dot{q}_j \tag{5.56}$$

where c_{ij} for $j = 1, \ldots, n$ are constants. The total work done by these dissipative forces per unit of time can be expressed

$$Work = W = Q_1\dot{q}_1 + Q_2\dot{q}_2 + \cdots + Q_n\dot{q}_n = -\sum_{i=1}^{n}\sum_{j=1}^{n} c_{ij}\dot{q}_i\dot{q}_j \tag{5.57}$$

When nonconservative forces of this type are present, a dissipation function D can be introduced. The dissipation function D is defined by

$$D = -\frac{1}{2}\text{total work done by forces} = \frac{1}{2}\sum_{i=1}^{n}\sum_{j=1}^{n} c_{ij}\dot{q}_i\dot{q}_j \tag{5.58}$$

where the coefficients c_{ij} represent constants. Note that the dissipative forces can be derived from the derivative

$$Q_i = -\frac{\partial D}{\partial \dot{q}_i} \tag{5.59}$$

The introduction of a dissipative function D associated with forces proportional to the velocities allows the modification of the Euler-Lagrange equations of motion (5.55) to include both conservative and nonconservative forces. The resulting Euler-Lagrange equations can be written

$$\frac{d}{dt}\left(\frac{\partial T}{\partial \dot{q}_i}\right) - \frac{\partial T}{\partial q_i} - Q_j + \frac{\partial D}{\partial \dot{q}_i} = 0 \quad \text{or} \quad \frac{d}{dt}\left(\frac{\partial T}{\partial \dot{q}_i}\right) - \frac{\partial T}{\partial q_i} + \frac{\partial V}{\partial q_i} + \frac{\partial D}{\partial \dot{q}_i} = 0 \quad (5.60)$$

Lagrangian for spring-mass system

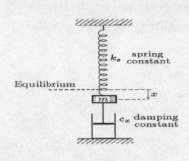

Consider the spring-mass system illustrated and assume the spring restoring force F_s is proportional to the spring displacement x. The element of work done in moving the spring a distance dx is $dW = -F_s\,dx = -k_s x\,dx$ where k_s is the spring constant. An integration gives $W = -\int_0^x k_s x\,dx = -\frac{1}{2}k_s x^2$ as the work done. The potential function from which the spring force is derivable is therefore $V = -W = \frac{1}{2}k_s x^2$ where k_s is the linear spring constant. The kinetic energy of the vibrating mass is given by $T = \frac{1}{2}m\dot{x}^2$. In order to represent the damping force we introduce the dissipation function $D = \frac{1}{2}c_x\dot{x}^2$, where c_x is a constant.

Here we have

$$\frac{\partial T}{\partial \dot{x}} = m\dot{x}, \quad \frac{\partial V}{\partial x} = k_s x, \quad \frac{\partial D}{\partial \dot{x}} = c_x\dot{x}$$

so that the modified Euler-Lagrange equation of motion is given by

$$\frac{d}{dt}(m\dot{x}) + c_x\dot{x} + k_s x = 0 \quad \text{or} \quad m\frac{d^2x}{dt^2} + c_x\frac{dx}{dt} + k_s x = 0.$$

Hamilton's equations of motion

Consider a system which can be described by n generalized coordinates $q_1, q_2, \ldots, q_n$, with Lagrangian having the form

$$L = L(q_1, q_2, \ldots, q_n, \dot{q}_1, \dot{q}_2, \ldots, \dot{q}_n, t). \quad (5.61)$$

Define the generalized momentum p_i associated with the coordinate q_i by the relation

$$p_i = \frac{\partial L}{\partial \dot{q}_i}, \quad \text{for} \quad i = 1, 2, \ldots, n. \quad (5.62)$$

We shall assume that the potential energy function does not contain velocity terms and write

$$p_i = \frac{\partial L}{\partial \dot{q}_i} = \frac{\partial T}{\partial \dot{q}_i}. \quad (5.63)$$

Note that from the Lagrange equations of motion there results

$$\frac{d}{dt}\left(\frac{\partial L}{\partial \dot{q}_i}\right) - \frac{\partial L}{\partial q_i} = 0 \quad \text{or} \quad \frac{dp_i}{dt} = \dot{p}_i = \frac{\partial L}{\partial q_i} \quad (5.64)$$

for $i = 1, 2, \ldots, n$. The general form for the kinetic energy T is

$$T = \frac{1}{2} \sum_{i=1}^{n} \sum_{j=1}^{n} a_{ij} \dot{q}_i \dot{q}_j \tag{5.65}$$

where the $a_{ij} = a_{ji}$ coefficients are functions of the coordinates only or possibly a function of coordinates and time. This form is called a quadratic form involving the generalized velocities. Differentiating equation (5.65) one obtains the generalized momentum

$$p_i = \frac{\partial T}{\partial \dot{q}_i} = \sum_{i=1}^{n} a_{ij} \dot{q}_i \quad \text{for} \quad i = 1, 2, \ldots, n \tag{5.66}$$

which is a linear form in the generalized velocities $\dot{q}_i$, $i = 1, 2 \ldots, n$. That is, it is possible to solve the system of equations (5.66) and write $\dot{q}_i$ in terms of the generalized momentum p_i. The general form for the equations of motion given by the equation

$$\frac{d}{dt}\left(\frac{\partial T}{\partial \dot{q}_j}\right) - \frac{\partial T}{\partial q_j} = Q_j \tag{5.67}$$

can then be expressed in the form

$$\frac{dp_j}{dt} = Q_j + \frac{\partial T}{\partial q_j}, \quad \text{for } j = 1, 2, \ldots, n \tag{5.68}$$

Here the term $\frac{\partial T}{\partial q_j}$ is termed an inertia force because it depends upon the coordinate system selected to represent the problem.

The quantity

$$H = H(q_1, q_2, \ldots, q_n, p_1, p_2, \ldots, p_n, t) = \sum_{i=1}^{n} p_i \dot{q}_i - L \tag{5.69}$$

is called the Hamiltonian function. The total differential of H is given by

$$\begin{aligned}
dH &= \sum_{i=1}^{n} \frac{\partial H}{\partial q_i} dq_i + \sum_{i=1}^{n} \frac{\partial H}{\partial p_i} dp_i + \frac{\partial H}{\partial t} dt \\
dH &= \sum_{i=1}^{n} p_i d\dot{q}_i + \sum_{i=1}^{n} \dot{q}_i dp_i - \sum_{i=1}^{n} \frac{\partial L}{\partial q_i} dq_i - \sum_{i=1}^{n} \frac{\partial L}{\partial \dot{q}_i} d\dot{q}_i - \frac{\partial L}{\partial t}
\end{aligned} \tag{5.70}$$

The first and fourth terms of the equation (5.70) add to zero so that there results

$$dH = \sum_{i=1}^{n} \dot{q}_i dp_i - \sum_{i=1}^{n} \frac{\partial L}{\partial q_i} dq_i - \frac{\partial L}{\partial t}. \tag{5.71}$$

Equating the first line of equation (5.70) with equation (5.71) and employing the equation (5.64) we produce the Hamiltonian equations of motion

$$\frac{\partial H}{\partial p_i} = \dot{q}_i, \qquad \frac{\partial H}{\partial q_i} = -\frac{\partial L}{\partial q_i} = -\dot{p}_i, \qquad \frac{\partial H}{\partial t} = -\frac{\partial L}{\partial t}. \tag{5.72}$$

In the special case were the Lagrangian is independent of the time t and has the form

$$L = T - V = L(q_1, q_2, \ldots, q_n, \dot{q}_1, \dot{q}_2, \ldots, \dot{q}_n),$$

then the Hamiltonian equations of motion reduce to

$$\frac{\partial H}{\partial p_i} = \dot{q}_i, \qquad \frac{\partial H}{\partial q_i} = -\frac{\partial L}{\partial q_i} = -\dot{p}_i, \qquad \frac{\partial H}{\partial t} = \sum_i \left[\frac{\partial H}{\partial p_i} \frac{dp_i}{dt} + \frac{\partial H}{\partial q_i} \frac{dq_i}{dt} \right] = \sum_i \left[\dot{q}_i \dot{p}_i - \dot{p}_i \dot{q}_i \right] = 0 \qquad (5.73)$$

and the Hamiltonian H is a constant. Another way to show the Hamiltonian is a constant is to note that the Lagrangian equations of motion are written

$$\frac{d}{dt}\left(\frac{\partial L}{\partial \dot{q}_i} \right) - \frac{\partial L}{\partial q_i} = 0$$

for $i = 1, \ldots, n$. Consequently,

$$\begin{aligned}
\frac{dH}{dt} &= \frac{d}{dt}\left(\sum_{j=1}^{n} p_j \dot{q}_j - L \right) = \frac{d}{dt}\left(\sum_{j=1}^{n} \dot{q}_j \frac{\partial L}{\partial \dot{q}_j} - L \right) \\
&= \sum_{j=1}^{n} \dot{q}_j \frac{d}{dt}\left(\frac{\partial L}{\partial \dot{q}_j} \right) + \sum_{j=1}^{n} \ddot{q}_j \frac{\partial L}{\partial \dot{q}_j} - \sum_{j=1}^{n} \frac{\partial L}{\partial q_j} \dot{q}_j - \sum_{j=1}^{n} \frac{\partial L}{\partial \dot{q}_j} \ddot{q}_j \\
&= \sum_{j=1}^{n} \dot{q}_j \left[\frac{d}{dt}\left(\frac{\partial L}{\partial \dot{q}_j} \right) - \frac{\partial L}{\partial q_j} \right] = 0
\end{aligned}$$

Thus, the derivative of H is zero which implies H is a constant.

Alternatively, make use of the symmetry of the coefficients a_{ij} and differentiate the equation (5.65) to obtain

$$\frac{\partial T}{\partial \dot{q}_k} = \frac{1}{2} \sum_{i=1}^{n} \sum_{j=1}^{n} \left(a_{ij} \dot{q}_i \frac{\partial \dot{q}_j}{\partial \dot{q}_k} + a_{ij} \frac{\partial \dot{q}_i}{\partial \dot{q}_k} q_j \right) = \frac{1}{2} \sum_{i=1}^{n} \sum_{j=1}^{n} \left(a_{ij} \dot{q}_i \delta_{jk} + a_{ij} \delta_{ik} \dot{q}_j \right)$$

$$\frac{\partial T}{\partial \dot{q}_k} = \frac{1}{2} \left(\sum_{i=1}^{n} a_{ik} \dot{q}_i + \sum_{j=1}^{n} a_{kj} \dot{q}_j \right) = \sum_{j=1}^{n} a_{jk} \dot{q}_j$$

We use the result from equation (5.65) to write the equation (5.69) in the following form

$$H = \sum_{i=1}^{n} p_i \dot{q}_i - L = \sum_{i=1}^{n} \frac{\partial T}{\partial \dot{q}_i} \dot{q}_i - (T - V) = \sum_{i=1}^{n} \sum_{j=1}^{n} a_{ij} \dot{q}_i \dot{q}_j - (T - V)$$

$$H = 2T - (T - V) = T + V = \text{a constant} = E$$

That is, the Hamiltonian is the sum of the kinetic and potential energies and is a constant for scleronomous systems.

Inverse square law of attraction

Consider a particle with mass m at a distance r from the origin which is attracted toward the origin by an inverse square law force given by $\frac{m\gamma}{r^2}$ where γ is a constant. Using polar coordinates this force is derivable from the potential function $V = V(r) = -\frac{m\gamma}{r}$. The kinetic energy of the particle can be represented in rectangular or polar coordinates as

$$T = \frac{1}{2}m\left(\dot{x}^2 + \dot{y}^2\right) \qquad \text{or} \qquad T = \frac{1}{2}m\left(\dot{r}^2 + r^2\dot{\theta}^2\right) \tag{5.74}$$

Consequently, the Lagrangian can be written in polar coordinates (r, θ) as

$$L = T - V = \frac{1}{2}m\left(\dot{r}^2 + r^2\dot{\theta}^2\right) + \frac{m\gamma}{r} \tag{5.75}$$

and this equation can be differentiated to obtain

$$\frac{\partial L}{\partial \dot{r}} = p_r = m\dot{r} \qquad \text{and} \qquad \frac{\partial L}{\partial \dot{\theta}} = p_\theta = mr^2\dot{\theta} \tag{5.76}$$

Here the generalized coordinates are $q_1 = r$, $q_2 = \theta$ with generalized momentum $p_1 = p_r$ and $p_2 = p_\theta$. The Hamiltonian can be written

$$H = \sum_{k=1}^{2} p_k \dot{q}_k - L = p_r\dot{r} + p_\theta\dot{\theta} - \frac{1}{2}m\left(\dot{r}^2 + r^2\dot{\theta}^2\right) - \frac{m\gamma}{r} \tag{5.77}$$

Note that H must be a function of q_k, p_k for $k = 1, 2$. We employ the equations (5.76) and make the substitutions $\dot{r} = p_r/m$ and $\dot{\theta} = p_\theta/mr^2$ to express the Hamiltonian in the form

$$H = \frac{1}{2}\frac{p_r^2}{m} + \frac{1}{2}\frac{p_\theta^2}{mr^2} - \frac{m\gamma}{r} \tag{5.78}$$

One can then calculate the Hamiltonian equations of motion

$$\frac{\partial H}{\partial r} = -\dot{p}_r = \frac{m\gamma}{r^2} - \frac{p_\theta^2}{mr^3} \qquad \text{and} \qquad \frac{\partial H}{\partial \theta} = -\dot{p}_\theta = 0 \tag{5.79}$$

This last equation implies that $p_\theta = mr^2\dot{\theta}$ is a constant. For convenience we select this constant to be $m\beta$ where β is a constant. One can then write

$$p_\theta = mr^2\dot{\theta} = m\beta \tag{5.80}$$

The total energy of the system is written

$$T + V = E \qquad \text{or} \qquad \frac{1}{2}m\left(\dot{r}^2 + r^2\dot{\theta}^2\right) - \frac{m\gamma}{r} = E \tag{5.81}$$

We solve for $\dot{\theta}$ using equation (5.80) and substitute the result into equation (5.81) and then solve for $\dot{r}^2$ to obtain

$$\dot{r}^2 = 2\left(\mathcal{E} + \frac{\gamma}{r} - \frac{\beta^2}{2r^2}\right) \tag{5.82}$$

where $\mathcal{E} = E/m$ represents total energy per unit mass. The right-hand side of equation (5.82) must satisfy $\mathcal{E} + \frac{\gamma}{r} - \frac{\beta^2}{2r^2} \geq 0$ in order that $\dot{r}$ be real. By analyzing the right-hand side of equation (5.82) one can better understand the type of motion described by the inverse square law force of attraction. This can be accomplished by representing the right-hand side

214

of equation (5.82) as the difference of two curves. One curve is $y_1 = \mathcal{E} = constant$ and the other curve is $y_2 = \frac{\beta^2}{2r^2} - \frac{\gamma}{r}$. In figure 5-8 we plot a graph of the function $y_2 = \frac{\beta^2}{2r^2} - \frac{\gamma}{r}$ vs r for selected values of $\beta > 0$ and $\gamma > 0$ and find the general shape illustrated. Also in the figure 5-8 we sketch the straight lines $y_1 = \mathcal{E}_1 > 0$, $y_1 = \mathcal{E}_2 = 0$, $y_1 = \mathcal{E}_3 < 0$ and $y_1 = \mathcal{E}_4 < 0$.

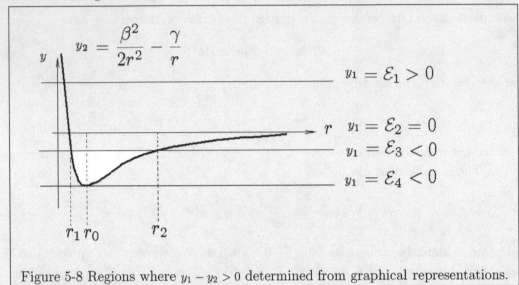

Figure 5-8 Regions where $y_1 - y_2 > 0$ determined from graphical representations.

We can then interpret the right-hand side of equation (5.82) as determining the region in figure 5-8 where the difference in curve heights $y_1 - y_2$ is positive. We consider the following cases.

(i) $y_1 = \mathcal{E}_1 > 0$ intersects the curve y_2 at some value $r_{min} < r_1$ so that $y_1 - y_2 > 0$ for $r > r_{min}$. Here the particle moves from a large value of r to r_{min} and then r increases again.

(ii) $y_1 = \mathcal{E}_2 = 0$ intersects the curve y_2 at some value near r_{min} and again the particle motion moves toward this minimum value and then away from it.

(iii) $y_1 = \mathcal{E}_3 < 0$ intersects the y_2 curve at two points labeled r_1 and r_2 and $y_1 - y_2 > 0$ in the region $r_1 \leq r \leq r_2$. The particle oscillates between these positions having a maximum velocity at the position $r = r_0$ where $y_1 - y_2$ has its maximum value.

(iv) $y_1 = \mathcal{E}_4 < 0$ intersects the y_2 curve where $r = r_0$. This is a special case where r is a constant and so the motion of the particle is circular.

Let us analyze the case (iii) in more detail. Note that when the particle is at one of the positions r_1 or r_2, then $\dot{r} = 0$. In this case

$$\mathcal{E} - \frac{\beta^2}{2r^2} + \frac{\gamma}{r} = 0, \qquad \text{or} \qquad 2\mathcal{E}r^2 + 2\gamma r - \beta^2 = 0 \qquad (5.83)$$

for $\mathcal{E} = \mathcal{E}_3$. This implies the values r_1 and r_2 are determined by the roots of the equation (5.83) which are given by

$$r_1 = \frac{-2\gamma + \sqrt{4\gamma^2 + 8\mathcal{E}\beta^2}}{4\mathcal{E}}, \qquad r_2 = \frac{-2\gamma - \sqrt{4\gamma^2 + 8\mathcal{E}\beta^2}}{4\mathcal{E}} \qquad (5.84)$$

Note that equation (5.83) can also be written in the form

$$r^2 + \frac{\gamma}{\mathcal{E}}r - \frac{\beta^2}{2\mathcal{E}} = (r - r_1)(r - r_2) = r^2 - (r_1 + r_2)r + r_1 r_2 = 0$$

so that one can write $r_1 + r_2 = -\frac{\gamma}{\mathcal{E}}$ and $r_1 r_2 = -\frac{\beta^2}{2\mathcal{E}}$ all evaluated at $\mathcal{E} = \mathcal{E}_3 < 0$.

We solve equation (5.82) for $\dot{r}$ and then separate the variables and integrate to obtain

$$\int_{r_1}^{r} \frac{dr}{\sqrt{2\left(\mathcal{E} + \frac{\gamma}{r} - \frac{\beta^2}{2r^2}\right)}} = \int_{t_1}^{t} dt = t - t_1 \tag{5.85}$$

Once we know r as a function of time t, we can find θ as a function of time t. One can separate the variables in the equation (5.80) and integrate to obtain

$$\int_{t_1}^{t} \frac{\beta}{r^2} dt = \int_{\theta_1}^{\theta} d\theta = \theta - \theta_1 \tag{5.86}$$

In order to find r as a function of θ we must employ chain rule differentiation which requires that

$$\frac{dr}{dt} = \frac{dr}{d\theta}\frac{d\theta}{dt} = \frac{\beta}{r^2}\frac{dr}{d\theta}$$

so that r and θ can be related by the differential equation

$$\frac{dr}{d\theta} = \frac{r^2}{\beta}\sqrt{2\left(\mathcal{E} + \frac{\gamma}{r} - \frac{\beta^2}{2r^2}\right)} \tag{5.87}$$

Separate the variables in equation (5.87) and then integrate to obtain

$$\int_{r_1}^{r} \frac{\beta\, dr}{r^2\sqrt{2\left(\mathcal{E} + \frac{\gamma}{r} - \frac{\beta^2}{2r^2}\right)}} = \int_{\theta_1}^{\theta} d\theta = \theta - \theta_1 \tag{5.88}$$

The substitution $u = 1/r$ with $du = -dr/r^2$ converts the integral given by equation (5.88) to the more standard form

$$\int_{u_1}^{u} \frac{\beta\, du}{\sqrt{2\mathcal{E} + 2\gamma u - \beta^2 u^2}} = \theta_1 - \theta \tag{5.89}$$

which can be evaluated by using a good table of integrals. From a table of integrals one can find the integral

$$\int \frac{dx}{\sqrt{a \pm 2bx - cx^2}} = \frac{1}{\sqrt{c}}\sin^{-1}\left(\frac{cx \mp b}{\sqrt{b^2 + ac}}\right) \tag{5.90}$$

where a, b and c are constants. Comparing the integrals (5.89) and (5.90) we set $a = 2\mathcal{E}$, $b = \gamma$, and $c = \beta^2$, then integrate equation (5.89) to obtain the result

$$\left[\sin^{-1}\left(\frac{\beta^2 u - \gamma}{\sqrt{\gamma^2 + 2\mathcal{E}\beta^2}}\right)\right]_{u_1}^{u} = \theta_1 - \theta \tag{5.91}$$

Suppose we select $\theta_1 = 0$ when $r = r_1$. At this value of r we have $\dot{r} = 0$ and from equation (5.84) we find that

$$\frac{1}{r_1} = u_1 = \frac{\gamma}{\beta^2} + \frac{\sqrt{\gamma^2 + 2\mathcal{E}\beta^2}}{\beta^2}$$

which implies $\frac{\beta^2 u_1 - \gamma}{\sqrt{\gamma^2 + 2\mathcal{E}\beta^2}} = 1$ so that $\sin^{-1}(1) = \frac{\pi}{2}$. The solution given by equation (5.91) then can be written in any of the forms

$$\sin^{-1}\left(\frac{\beta^2 u - \gamma}{\sqrt{\gamma^2 + 2\mathcal{E}\beta^2}}\right) - \frac{\pi}{2} = -\theta$$

$$\beta^2 u - \gamma = \sqrt{\gamma^2 + 2\mathcal{E}\beta^2} \sin\left(\frac{\pi}{2} - \theta\right) \tag{5.92}$$

$$\frac{1}{r} = u = \frac{\gamma}{\beta^2} + \frac{\sqrt{\gamma^2 + 2\epsilon\beta^2}}{\beta^2} \cos\theta$$

Recall from calculus that the equation of a conic section is given by

$$r = \frac{\epsilon p}{1 + \epsilon \cos\theta} \quad \text{or} \quad \frac{1}{r} = \frac{1}{\epsilon p} + \frac{1}{p}\cos\theta \tag{5.93}$$

where ϵ is the eccentricity and p is a constant. We compare this last equation with our solution for the particle motion in polar coordinates. Here

$$\frac{1}{\epsilon p} = \frac{\gamma}{\beta^2} \quad \text{and} \quad \frac{1}{p} = \frac{\sqrt{\gamma^2 + 2\mathcal{E}\beta^2}}{\beta^2} \quad \text{which implies} \quad \epsilon = \frac{\beta^2}{\gamma p} = \sqrt{1 + 2\mathcal{E}\beta^2}/\gamma^2 \tag{5.94}$$

For $\epsilon = 1$ $(\mathcal{E} = 0)$, the path described by the conic section is a parabola. For $\epsilon > 1$ $(\mathcal{E} > 0)$, the path described by a conic section is a hyperbola. For $\epsilon = 0$, $(\mathcal{E} = \mathcal{E}_4)$ the particle moves in a circle. For $\epsilon < 1$, $(\mathcal{E} < 0)$, the path described by the conic section is an ellipse. We are interested in the case where $\mathcal{E} = \mathcal{E}_3 < 0$ so that from equation (5.94) we find that $\epsilon < 1$ so that the particle moves in an elliptical path about the origin.

Spring-mass system

Two masses connected by linear springs are given the displacements x_1 and x_2 from their equilibrium positions and then released. Neglect frictional forces and find the equations of motion of the spring-mass system.

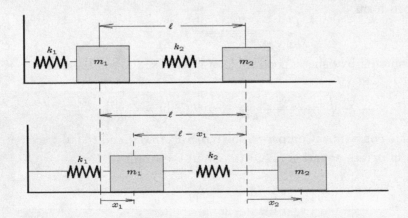

Assume the initial displacements x_1, x_2 and initial velocities $\dot{x}_1, \dot{x}_2$ are specified at time $t = 0$. Further assume that k_1 and k_2 are known linear spring constants.

Solution: Here we let $q_1 = x_1$ and $q_2 = x_2$ denote the generalized coordinates representing the respective distances as measured from the equilibrium positions of the masses m_1 and m_2. The kinetic energy of the system can be written

$$T = \frac{1}{2}m_1\dot{x}_1^2 + \frac{1}{2}m_2\dot{x}_2^2 \tag{5.95}$$

The springs are linear so that the spring force is proportional to the displacements. Using k_1 and k_2 to denote the spring constants, the element of work done on the mass m_1 is given by

$$\delta W_1 = -F_{s_1}\,\delta x_1 = -(k_1 x_1)\,\delta x_1 \quad \text{or} \quad W_1 = -\frac{k_1}{2}x_1^2. \tag{5.96}$$

Similarly, the element of work done on the mass m_2 is given by

$$\delta W_2 = -F_{s_2}\,\delta x_2 = -k_2(x_2 - x_1)\,\delta x_2, \quad \text{or} \quad W_2 = -\frac{k_2}{2}(x_2 - x_1)^2 \tag{5.97}$$

where $x_2 - x_1$ is the relative displacement of the two masses. The potential energy V is the negative of the work done which gives

$$V = -(W_1 + W_2) = \frac{k_1}{2}x_1^2 + \frac{k_2}{2}(x_2 - x_1)^2 \tag{5.98}$$

This gives the Lagrangian

$$L = T - V = \frac{1}{2}m_1\dot{x}_1^2 + \frac{1}{2}m_2\dot{x}_2^2 - \frac{k_1}{2}x_1^2 - \frac{k_2}{2}(x_2 - x_1)^2 \tag{5.99}$$

The Lagrangian equations of motion are then

$$\begin{aligned}
\frac{\partial L}{\partial x_1} - \frac{d}{dt}\left(\frac{\partial L}{\partial \dot{x}_1}\right) &= 0, \quad \Rightarrow \quad m_1\ddot{x}_1 = -k_1 x_1 + k_2(x_2 - x_1) \\
\frac{\partial L}{\partial x_2} - \frac{d}{dt}\left(\frac{\partial L}{\partial \dot{x}_2}\right) &= 0, \quad \Rightarrow \quad m_2\ddot{x}_2 = -k_2(x_2 - x_1)
\end{aligned} \tag{5.100}$$

These equations of motion are a set of coupled second order linear ordinary differential equations subject to the given initial conditions. These equations can be solved by standard methods such as Laplace transforms or matrix methods.

If we add an external force $f(t)$ and damping forces which is proportional to the velocity we obtain the spring-mass system illustrated

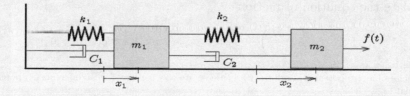

One must add a generalized force $f(t)$ and the Rayleigh dissipation function

$$D = \frac{C_1}{2}\dot{x}_1^2 + \frac{C_2}{2}(\dot{x}_2 - \dot{x}_1)^2$$

to the previous equations. It is left as an exercise to show that in this more general case the equations of motion are given by

$$m_1\ddot{x}_1 = -k_1 x_1 + k_2(x_2 - x_1) - (C_1 + C_2)\dot{x}_1 + C_2\dot{x}_2$$
$$m_2\ddot{x}_2 = -k_2(x_2 - x_1) + C_2(\dot{x}_1 - \dot{x}_2) + f(t)$$

(5.101)

Pendulum systems

We illustrated the modeling of several pendulum systems and present the material by way of examples. We begin with the simple pendulum.

Example 5-2.   Simple pendulum

A simple pendulum is defined by a thin wire (treated as a massless rod) attached to a large mass m as illustrated. Calculate the Lagrangian and determine the equation which describes the motion of the simple pendulum.

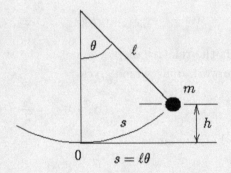

Solution:

Let θ denote the generalized coordinate, then the kinetic energy of the mass is given by

$$T = \frac{1}{2}mv^2 = \frac{1}{2}m\left(\frac{ds}{dt}\right)^2 = \frac{1}{2}m\ell^2\dot{\theta}^2$$

and the potential energy is given by the product of the weight $w = mg$ times its height h above the zero reference point[‡]

$$V = mgh = mg\ell(1 - \cos\theta)$$

where g is the acceleration of gravity. The Lagrangian is therefore

$$L = T - V = \frac{1}{2}m\ell^2\dot{\theta}^2 - mg\ell(1 - \cos\theta).$$

We calculate the equation of motion

$$\frac{d}{dt}\left(\frac{\partial L}{\partial\dot{\theta}}\right) - \frac{\partial L}{\partial\theta} = 0 \quad \Rightarrow \quad m\ell^2\ddot{\theta} + mgl\sin\theta = 0$$

[‡] The choice of a zero potential is arbitrary because we are only concerned with differences in potential energy from some reference point.

which simplifies to

$$\ddot{\theta} + \frac{y}{\ell} \sin\theta = 0. \qquad (5.102)$$

The second order ordinary nonlinear differential equation (5.102) is often times replaced by a system of first order differential equations by introducing a new variable $y = \frac{d\theta}{dt}$, then the equation (5.102) can be replaced by the first order system

$$\frac{d\theta}{dt} = y$$
$$\frac{dy}{dt} = -\frac{g}{\ell} \sin\theta \qquad (5.103)$$

A parametric plot of $y = \dot{\theta}$ vs θ is called a phase plane plot.

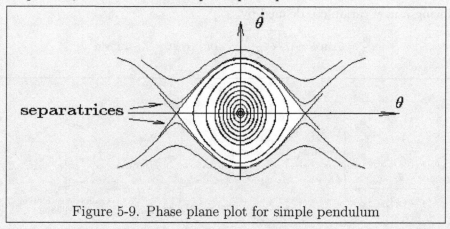

Figure 5-9. Phase plane plot for simple pendulum

By multiplying the equation (5.102) by $\dot{\theta}$ and then integrating, produces the first integral

$$E = \dot{\theta}^2 - 2\frac{g}{\ell} \cos\theta \qquad (5.104)$$

where E is a constant of integration. The phase plane plot for the level curves $E = constant$ associated with equation (5.104) are illustrated in the figure 5-9. If E is small then the phase plane plot gives closed curves representing small oscillations or periodic solutions with center at a stable equilibrium position about the point where $\theta = 0, \dot{\theta} = 0$. If E is too large, then the rotatory curves above and below the closed curves result. These curves depict a situation where the pendulum never reaches a point where $\dot{\theta} = 0$ which means the pendulum has too much energy and flips over continuously. The curves which separate the flipping over, rotatory or spinning motions of the pendulum from the oscillatory motions are called separatrices. The saddle points where $\dot{\theta} = 0$ and $\theta = \pm\pi$ represent unstable equilibrium positions where the pendulum becomes inverted. We investigate the cases of periodic and rotary motion of the pendulum as illustrated in the figure 5-9.

Let us derive an analytical representation of the solution representing periodic or oscillatory solutions. Let $\theta = \theta_M$ denote the maximum displacement of the pendulum from its equilibrium position at $\theta = 0$. At this maximum angular displacement we will have $\frac{d\theta}{dt} = 0$

when $\theta = \theta_M$ and when these values are substituted into the equation (5.104) the constant E becomes $E = -2\frac{g}{\ell}\cos\theta_M$. This value for E then simplifies the equation (5.104) to the form

$$\frac{d\theta}{dt} = \sqrt{\frac{2g}{\ell}}\sqrt{\cos\theta - \cos\theta_M} \tag{5.105}$$

or upon using the trigonometric identity $\cos\theta - \cos\theta_M = 2\left(\sin^2\frac{\theta_M}{2} - \sin^2\frac{\theta}{2}\right)$ there results the alternative expression

$$\frac{d\theta}{dt} = 2\sqrt{\frac{g}{\ell}}\sqrt{\sin^2\frac{\theta_M}{2} - \sin^2\frac{\theta}{2}} \tag{5.106}$$

This last equation is simplified by introducing the constant k defined by

$$k = \sin\frac{\theta_M}{2} = \sqrt{\frac{1 - \cos\theta_M}{2}}. \tag{5.107}$$

and introducing a new variable ϕ defined by

$$\sin\frac{\theta}{2} = k\sin\phi = \sqrt{\frac{1 - \cos\theta}{2}}, \quad \text{or} \quad \cos\theta = 1 - 2k^2\sin^2\phi. \tag{5.108}$$

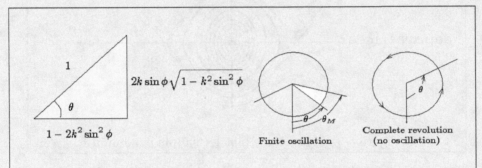

Figure 5-10. Transformation triangle and types of pendulum motion.

The transformation equations (5.107) and (5.108) reduces the differential equation (5.105) or (5.106), after some algebra and simplification, to the much simpler form

$$\sqrt{\frac{g}{\ell}}dt = \frac{d\phi}{\sqrt{1 - k^2\sin^2\phi}} \tag{5.109}$$

Now integrate the equation (5.109) to obtain

$$\int_0^t \sqrt{\frac{g}{\ell}}dt = \int_0^\phi \frac{d\phi}{\sqrt{1 - k^2\sin^2\phi}} = u = t\sqrt{\frac{g}{\ell}} \tag{5.110}$$

This gives the result

$$\operatorname{sn}(u, k) = \operatorname{sn}\left(t\sqrt{\frac{g}{\ell}}, k\right) = \sin\phi \tag{5.111}$$

where sn denotes a Jacobian elliptic function. Note that equation (5.108) requires

$$2k^2\sin^2\phi = 1 - \cos\theta$$

$$k\sin\phi = \sqrt{\frac{1 - \cos\theta}{2}} = \sin\frac{\theta}{2} \tag{5.112}$$

so that $\qquad \sin\phi = \dfrac{1}{k}\sin\dfrac{\theta}{2}.$

By combining the equations (5.112) and (5.111) the solution for θ can be represented in terms of the Jacobian elliptic function sn as

$$\theta = 2\sin^{-1}\left[k\,\text{sn}\left(t\sqrt{\frac{g}{\ell}},k\right)\right], \qquad \text{where} \quad k = \sin\frac{\theta_M}{2}. \tag{5.113}$$

Note that when $\theta = \theta_M$, then $\phi = \frac{\pi}{2}$ and $\dot\theta = 0$ and the time T taken to reach this point is found from

$$T = \sqrt{\frac{\ell}{g}}\int_0^{\pi/2}\frac{d\phi}{\sqrt{1-k^2\sin^2\phi}} = \sqrt{\frac{\ell}{g}}F(\frac{\pi}{2},k). \tag{5.114}$$

This time represents one quarter of the total time to complete an oscillation.

The rotary motion can be described by assigning a value to E in equation (5.104) so that $\dot\theta$ always remains greater than zero. To make the algebra easier we select the special value $E = \frac{2g}{\ell}(\frac{h^*}{\ell}-1) > 0$ where h^* is a constant. This value for E reduces the equation (5.104) to the form

$$\dot\theta^2 = \frac{2g}{\ell}\left[\frac{h^*}{\ell}-(1-\cos\theta)\right]$$

$$\dot\theta^2 = \frac{2g}{\ell}\left[\frac{h^*}{\ell}-2\sin^2\frac{\theta}{2}\right] \tag{5.115}$$

$$\dot\theta^2 = \frac{2gh^*}{\ell^2}\left[1-\frac{2\ell}{h^*}\sin^2\frac{\theta}{2}\right].$$

Note that when $\theta = \pi$ we want $\dot\theta^2 = \frac{2gh^*}{\ell^2}\left(1-\frac{2\ell}{h^*}\right) > 0$, so that we must require that the constant h^* satisfy the inequality $h^* > 2\ell$. The equation (5.115) simplifies to the differential equation

$$\frac{d\theta}{dt} = \frac{\sqrt{2gh^*}}{\ell}\sqrt{1-k^2\sin^2\frac{\theta}{2}}, \qquad \text{where} \quad k^2 = \frac{2\ell}{h^*} < 1. \tag{5.116}$$

The transformation $\theta = 2\phi$ further simplifies the differential equation (5.116) to the form

$$\frac{d\phi}{\sqrt{1-k^2\sin^2\phi}} = \frac{\sqrt{2gh^*}}{2\ell}dt \tag{5.117}$$

Integrating the equation (5.117) we find

$$u = F(\phi,k) = \int_0^\phi \frac{d\phi}{\sqrt{1-k^2\sin^2\phi}} = \frac{\sqrt{2gh^*}}{2\ell}t, \qquad k^2 = \frac{2\ell}{h^*} < 1 \tag{5.118}$$

where $F(\phi,k)$ is an elliptic integral of the first kind. At the top of its rotary motion there results one-half of a complete revolution. In this situation $\theta = \pi$ and $\phi = \frac{\pi}{2}$. Therefore, the time T taken to complete a one-half revolution is obtained from

$$\frac{\sqrt{2gh^*}}{2\ell}T = F(\pi/2,k), \qquad \text{or} \qquad T = \frac{2\ell}{\sqrt{2gh^*}}F(\pi/2,k). \tag{5.119}$$

The equation (5.118) can be modified to represent the angle θ since

$$\phi = \text{am}\,u = \text{am}\left(\frac{\sqrt{2gh^*}t}{2\ell}\right) = \frac{\theta}{2} \tag{5.120}$$

so that

$$\sin\phi = \text{sn}\,(u, k), \qquad \text{or} \qquad \sin\frac{\theta}{2} = \text{sn}\left(\frac{\sqrt{2gh^*}\,t}{2\ell}, k\right) \tag{5.121}$$

giving the result

$$\theta = \theta(t) = 2\sin^{-1}\left[\text{sn}\left(\frac{\sqrt{2gh^*}\,t}{2\ell}, k\right)\right], \qquad k^2 = \frac{2\ell}{h^*} < 1. \tag{5.122}$$

Many texts greatly simplify the problem and restrict motions to small oscillations in θ so that the approximation $\sin\theta \approx \theta$ can be employed to reduce the equation (5.102) to the form

$$\frac{d^2\theta}{dt^2} + \omega^2\theta = 0, \qquad \text{where} \quad \omega^2 = \frac{g}{\ell}. \tag{5.123}$$

The equation (5.123) is now a linear ordinary differential equation with the general solution

$$\theta = C_1\sin\omega t + C_2\cos\omega t = A\sin(\omega t + \psi) \tag{5.124}$$

where C_1, C_2 are arbitrary constants, $A = \sqrt{C_1^2 + C_2^2}$, and $\psi = \arctan(C_2/C_1)$. The solution given by equation (5.124) is valid for small values of θ.

$\blacksquare$

Example 5-3.   Spherical pendulum

A spherical pendulum consists of a large mass m suspended at the end of a thin rod or wire of length ℓ. Neglect the weight of the rod and use the Lagrangian equations to find the equations of motion of the mass in terms of the spherical coordinates θ, ϕ illustrated in the figure 5-11.

Solution: The position of the mass m is given by

$$x = \ell\sin\theta\cos\phi, \qquad y = \ell\sin\theta\sin\phi, \quad z = \ell(1 - \cos\theta) \tag{5.125}$$

Treating θ and ϕ as functions of time t the equations (5.125) have the time derivatives

$$\begin{aligned}
\dot{x} &= \ell\sin\theta\sin\phi\,\dot{\phi} + \ell\cos\theta\cos\phi\,\dot{\theta} \\
\dot{y} &= \ell\sin\theta\cos\phi\,\dot{\phi} + \ell\cos\theta\sin\phi\,\dot{\theta} \\
\dot{z} &= \ell\sin\theta\,\dot{\theta}
\end{aligned} \tag{5.126}$$

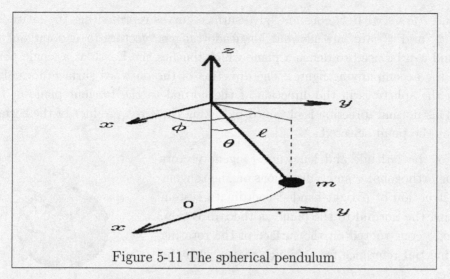

Figure 5-11 The spherical pendulum

The kinetic energy can then be represented as

$$T = \frac{1}{2}m(\dot{x}^2 + \dot{y}^2 + \dot{z}^2) = \frac{1}{2}m\ell^2\left(\sin^2\theta\,\dot{\phi}^2 + \dot{\theta}^2\right) \tag{5.127}$$

and the potential energy is, as in the previous example, given by

$$V = mgh = mg\ell(1 - \cos\theta) \tag{5.128}$$

so that there results the Lagrangian

$$L = T - V = \frac{1}{2}m\ell^2\left(\dot{\theta}^2 + \sin^2\theta\,\dot{\phi}^2\right) - mg\ell(1 - \cos\theta). \tag{5.129}$$

This gives the Lagrangian equations of motion

$$\frac{\partial L}{\partial \theta} - \frac{d}{dt}\left(\frac{\partial L}{\partial \dot{\theta}}\right) = 0, \quad \Rightarrow \quad \frac{d}{dt}\left(m\ell^2\dot{\theta}\right) - m\ell^2\sin\theta\cos\theta\dot{\phi}^2 + mg\ell\sin\theta = 0$$
$$\frac{\partial L}{\partial \phi} - \frac{d}{dt}\left(\frac{\partial L}{\partial \dot{\phi}}\right) = 0, \quad \Rightarrow \quad \frac{d}{dt}\left(m\ell^2\sin^2\theta\dot{\phi}\right) = 0 \tag{5.130}$$

Example 5-4.   The Foucault Pendulum

The Foucault pendulum is a modification of the above problem which considers additional forces acting on a spherical pendulum positioned on the surface of the Earth. Let the Earth be considered as a sphere where a point on the surface of the sphere of radius R is given by the position vector

$$\vec{r}(u, v) = R\cos\mu\cos\lambda\,\hat{e}_1 + R\sin\mu\cos\lambda\,\hat{e}_2 + R\sin\lambda\,\hat{e}_3, \qquad 0 \le \mu \le 2\pi, \quad -\pi/2 \le \lambda \le \pi/2 \tag{5.131}$$

where the parameters μ and λ are associated with the longitude and latitude coordinates on the surface of the Earth. That is, the curves $\vec{r}(\mu_0, \lambda)$ with μ_0 constant, give the longitude surface curves and $\vec{r}(\mu, \lambda_0)$ with λ_0 constant, give surface curves representing the latitudes. The derivatives $\frac{\partial \vec{r}}{\partial \mu}$ and $\frac{\partial \vec{r}}{\partial \lambda}$ are latitude and longitude tangent vectors to a point on the Earths surface and can be used to define a plane which touches the Earth at a single point as illustrated in the accompanying figure. The direction of the outward surface normal at a fixed point on the sphere is in the direction of the normal to the tangent plane at the point selected. This normal direction is obtained by taking the cross product of the surface tangent vectors at the point selected.

The directions of the latitude and longitude tangent vectors are used to define orthogonal x and y directions on the tangent plane. The construction of a right-handed coordinate system results by selecting the normal to the plane as the z direction. The x, y and z axes constructed on the surface of the rotating Earth is a non inertial reference frame which moves with the Earth as it rotates.

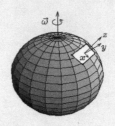

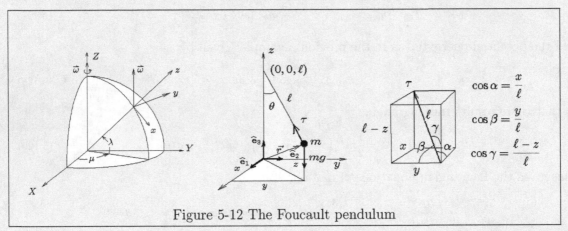

Figure 5-12 The Foucault pendulum

We use the local x, y, z-axes described above to construct a pendulum with wire of length ℓ attached to a mass m supported at a point on the z axis. We translate axes slightly in the z direction so that the pendulum does not hit the surface of the sphere. We also neglect the frictional forces and torques at the point of pendulum support on the z axis which we label $(0, 0, \ell)$. Let $\vec{r} = x\,\widehat{e}_1 + y\,\widehat{e}_2 + z\,\widehat{e}_3$ denote the position of the mass m and assume that the pendulum motion in the z direction is very small in comparison to the length ℓ and that θ undergoes small changes as the pendulum oscillates. If ℓ is large and θ changes are small, then the motion of the mass m occurs approximately in the x, y-tangent plane. The angular velocity of the Earth's rotation is assumed to be a constant small quantity (it has a magnitude of $7.3\,(10^{-5})\,rad/sec$) and so products of the angular velocity ω will be neglected. The forces acting on the pendulum are the tension τ in the attached wire and the force of gravity g acting on the mass m illustrated in the figure 5-12.

We use the Euler-Lagrange equations (5.55), with $(q_1, q_2, q_3) = (x, y, z)$, to derive the equations of motion of the Foucault pendulum. The velocity of the mass m in a moving reference frame is calculated as follows. The position of the mass m is given by $\vec{r} = x\,\hat{e}_1 + y\,\hat{e}_2 + z\,\hat{e}_3$ with derivative

$$\frac{d\vec{r}}{dt} = \dot{x}\,\hat{e}_1 + \dot{y}\,\hat{e}_2 + \dot{z}\,\hat{e}_3 + x\frac{d\,\hat{e}_1}{dt} + y\frac{d\,\hat{e}_2}{dt} + z\frac{d\,\hat{e}_3}{dt}$$

where the change of the unit vectors $\hat{e}_1$, $\hat{e}_2$, $\hat{e}_3$ with respect to time is given by

$$\frac{d\,\hat{e}_1}{dt} = \vec{\omega} \times \hat{e}_1, \qquad \frac{d\,\hat{e}_2}{dt} = \vec{\omega} \times \hat{e}_2, \qquad \frac{d\,\hat{e}_3}{dt} = \vec{\omega} \times \hat{e}_3$$

where $\vec{\omega}$ is the angular velocity of the Earth. This gives $\frac{d\vec{r}}{dt} = \vec{v} + \vec{\omega} \times \vec{r}$ so that the kinetic energy of the mass m can be calculated from the relation

$$T = \frac{1}{2}m\frac{d\vec{r}}{dt} \cdot \frac{d\vec{r}}{dt} = \frac{1}{2}m(\vec{v} + \vec{\omega} \times \vec{r}) \cdot (\vec{v} + \vec{\omega} \times \vec{r}) \tag{5.132}$$

where $\vec{v} = \dot{x}\,\hat{e}_1 + \dot{y}\,\hat{e}_2 + \dot{z}\,\hat{e}_3$. The angular velocity components with respect to the x, y, z-axes is given by $\vec{\omega} = \omega_x\,\hat{e}_1 + \omega_y\,\hat{e}_2 + \omega_z\,\hat{e}_3$. The components of the angular velocity vector are calculated

$$\omega_x = -\omega\cos\lambda, \qquad \omega_y = 0, \qquad \omega_z = \omega\sin\lambda \tag{5.133}$$

where $\omega = |\vec{\omega}|$. These components are obtained by projecting $\vec{\omega}$ onto the x, y and z axes respectively.

Expanding the equation (5.132) and neglecting the product term $(\vec{\omega} \times \vec{r}) \cdot (\vec{\omega} \times \vec{r})$ since it is small, (i.e. $|\vec{\omega}| = \omega$ is small), one obtains the following expression for the kinetic energy

$$T = \frac{1}{2}m\left[\dot{x}^2 + \dot{y}^2 + \dot{z}^2 + 2\dot{x}\left(\omega_y z - \omega_z y\right) + 2\dot{y}\left(\omega_z x - \omega_x z\right) + 2\dot{z}\left(\omega_x y - \omega_y x\right)\right] \tag{5.134}$$

An examination of the figure 5-12 one finds the components of the external force acting on the mass m are

$$Q_x = -\frac{x}{\ell}\tau, \qquad Q_y = -\frac{y}{\ell}\tau, \qquad Q_z = -mg + \frac{(\ell - z)}{\ell}\tau \tag{5.135}$$

where τ is the tension in the wire, m is the mass, g is the acceleration of gravity and ℓ is the length of the wire. The external force can thus be represented

$$\vec{Q} = -mg\,\hat{e}_3 - \frac{x}{\ell}\tau\,\hat{e}_1 - \frac{y}{\ell}\tau\,\hat{e}_2 + \frac{(\ell - z)}{\ell}\tau\,\hat{e}_3.$$

We employ the equations (5.55) to construct the equations describing the motion of the mass m and obtain the system of differential equations

$$\begin{aligned}
\ddot{x} - 2\omega\dot{y}\sin\lambda + \left(\frac{\tau}{\ell m}\right)x &= 0 \\
\ddot{y} + 2\omega\dot{z}\cos\lambda + 2\omega\dot{x}\sin\lambda + \left(\frac{\tau}{\ell m}\right)y &= 0 \\
\ddot{z} - 2\omega\dot{y}\cos\lambda + g - \frac{\ell - z}{\ell m}\tau &= 0
\end{aligned} \tag{5.136}$$

The resulting motion of the pendulum mass is a back and forth planar oscillation where the plane of oscillation slowly rotates clockwise in the northern hemisphere. The resulting motion is due to the axial rotation of the Earth.

The Foucault pendulum experiment was first constructed by Jean-Bernard-Leon Foucault (1819-1868) who constructed a pendulum of length 67 meters with a mass of 28 kilograms and then suspended it from the dome of the Pantheon in Paris. Similar Foucault pendulum demonstrations can be found in many museums and public buildings throughout the world.

∎

Navier equations

Let $\vec{u} = u_1(x,y)\,\widehat{e}_1 + u_2(x,y)\,\widehat{e}_2$ denote a displacement field within the material of a thin plate. If the thin plate is in equilibrium, the theory of elasticity requires that the potential energy associated with deformations of the thin plate be a minimum. The potential energy of the thin plate is given by the integral

$$I = \iint_R \left\{ \frac{1}{2}u\left[\left(\frac{\partial u_1}{\partial y}\right)^2 + \left(\frac{\partial u_2}{\partial x}\right)^2 \right] + \left(\frac{1}{2}\lambda + \mu\right)\left[\left(\frac{\partial u_1}{\partial x}\right)^2 + \left(\frac{\partial u_2}{\partial y}\right)^2 \right] + (\lambda + \mu)\frac{\partial u_1}{\partial x}\frac{\partial u_2}{\partial y} \right\} dx\,dy$$

where λ and μ are called the Lame's constants and R is the region of integration. One can verify that the resulting Euler-Lagrange equations have the form

$$\begin{aligned}
(\lambda + 2\mu)\frac{\partial^2 u_1}{\partial x^2} + \mu\frac{\partial^2 u_1}{\partial y^2} + (\lambda + \mu)\frac{\partial^2 u_2}{\partial x \partial y} &= 0 \\
(\lambda + 2\mu)\frac{\partial^2 u_2}{\partial y^2} + \mu\frac{\partial^2 u_2}{\partial x^2} + (\lambda + \mu)\frac{\partial^2 u_1}{\partial x \partial y} &= 0
\end{aligned}$$

(5.137)

These equations can be written in the vector form

$$\mu \nabla^2 \vec{u} + (\lambda + \mu)\nabla(\nabla \cdot \vec{u}) = \vec{0}$$

(5.138)

which is know in the theory of elasticity as the two-dimensional Navier's equation. This is a vector equation that the displacement vector must satisfy inside a thin elastic material in the absence of body forces. One must add to the Navier equations boundary conditions, either specified displacements at the boundary or prescribed stress at the boundary of the material.

Example 5-5.   Navier equations

In the theory of elasticity it is customary to define a point (x,y,z) in an isotropic elastic solid using the subscript notation $(x,y,z) = (x_1, x_2, x_3)$. An elastic body (B) is a three-dimensional isotropic material with boundary surface (S) which is subjected to body forces $\vec{F}_B$ and surface traction forces $\vec{F}_S$ When the elastic body is subjected to internal and external forces, which can change with time t, each point (x_1, x_2, x_3) within the elastic isotropic body experiences a displacement vector

$$\vec{u} = u_1(t, x_1, x_2, x_3)\,\widehat{e}_1 + u_2(t, x_1, x_2, x_3)\,\widehat{e}_2 + u_3(t, x_1, x_2, x_3)\,\widehat{e}_3$$

The elastic strain tensor ϵ_{ij}, $i,j = 1,2,3$, is related to the derivatives of the displacement vector u_i, $i = 1,2,3$ by the relations

$$\epsilon_{ij} = \frac{1}{2}\left(\frac{\partial u_i}{\partial x_j} + \frac{\partial u_j}{\partial x_i}\right) \qquad \text{for } i,j = 1,2,3 \tag{5.139}$$

For a linear elastic material the stresses σ_{ij}, $i,j = 1,2,3$ at a point are related to the strains by the relations

$$\sigma_{ij} = \lambda \delta_{ij} \sum_{m=1}^{3} \epsilon_{mm} + 2\mu\epsilon_{ij} \tag{5.140}$$

where λ and μ are known as the Lame's constants and δ_{ij} is the Kronecker delta which satisfies

$$\delta_{ij} = \begin{cases} 1, & \text{if } i = j \\ 0, & \text{if } i \neq j \end{cases} \tag{5.141}$$

The kinetic energy associated with a point within an elastic material is given by

$$T = \frac{1}{2}\iiint_B \rho \vec{u}_t \cdot \vec{u}_t \, d\tau = \frac{1}{2}\rho \iiint_B \left(\frac{\partial u}{\partial t}\right)^2 d\tau \tag{5.142}$$

where ρ is the density of the material, assumed constant, and $d\tau$ is an element of volume within the material. The potential energy at a point within the elastic body is given by

$$V = \frac{1}{2}\iiint_B \sum_{i=1}^{2} \sum_{j=1}^{3} \sigma_{ij}\epsilon_{ij} \, d\tau \tag{5.143}$$

The Navier equations of elasticity can be derived by finding the extremum associated with the functional

$$I = \int_{t_0}^{t_1} (T - V) \, dt = \int_{t_0}^{t_1} \iiint_B L \, d\tau dt \tag{5.144}$$

where the integrand L of equation (5.144) is obtained from equations (5.142) and (5.143). We assume that the displacements u_i for $i = 1,2,3$ are specified on the boundary surface (S) so that $\delta u_i \big|_{(x_1,x_2,x_3)\in S} = 0$ for $i = 1,2,3$. At a stationary value we require that $\delta I = 0$. In expanded form the Lagrangian integrand L is represented

$$\begin{aligned} L = &\frac{1}{2}\rho \sum_{k=1}^{3} \left(\frac{\partial u_k}{\partial t}\right)^2 \\ &- \frac{1}{4}\left\{\left[\lambda \sum_{m=1}^{3} \frac{\partial u_m}{\partial x_m} + 2\mu \frac{\partial u_1}{\partial x_1}\right]\left(2\frac{\partial u_1}{\partial x_1}\right) + \mu\left(\frac{\partial u_1}{\partial x_2} + \frac{\partial u_2}{\partial x_1}\right)^2 + \mu\left(\frac{\partial u_1}{\partial x_3} + \frac{\partial u_3}{\partial x_1}\right)^2\right\} \\ &- \frac{1}{4}\left\{\mu\left(\frac{\partial u_2}{\partial x_1} + \frac{\partial u_1}{\partial x_2}\right)^2 + \left[\lambda \sum_{m=1}^{3} \frac{\partial u_m}{\partial x_m} + 2\mu \frac{\partial u_2}{\partial x_2}\right]\left(2\frac{\partial u_2}{\partial x_2}\right) + \mu\left(\frac{\partial u_2}{\partial x_3} + \frac{\partial u_3}{\partial x_2}\right)^2\right\} \\ &- \frac{1}{4}\left\{\mu\left(\frac{\partial u_3}{\partial x_1} + \frac{\partial u_1}{\partial x_3}\right)^2 + \mu\left(\frac{\partial u_3}{\partial x_2} + \frac{\partial u_2}{\partial x_3}\right)^2 + \left[\lambda \sum_{m=1}^{3} \frac{\partial u_m}{\partial x_m} + 2\mu \frac{\partial u_3}{\partial x_3}\right]\left(2\frac{\partial u_3}{\partial x_3}\right)\right\} \end{aligned} \tag{5.145}$$

Note that this integrand is such that the functional given by equation (5.144) has the form of the previously studied functional [f11] from chapter 3. This relation can be discerned by

228

making the substitution $(x_1, x_2, x_3) = (x, y, z)$ and $(u_1, u_2, u_3) = (u, v, w)$. Using the results from [f11] of chapter 3 the Euler-Lagrange equations are found to have the form

$$\rho\frac{\partial^2 u_1}{\partial t^2} = \mu\left(\frac{\partial^2 u_1}{\partial x_1^2} + \frac{\partial^2 u_1}{\partial x_2^2} + \frac{\partial^2 u_1}{\partial x_3^2}\right) + (\lambda + \mu)\frac{\partial}{\partial x_1}\left(\frac{\partial u_1}{\partial x_1} + \frac{\partial u_2}{\partial x_2} + \frac{\partial u_3}{\partial x_3}\right)$$

$$\rho\frac{\partial^2 u_2}{\partial t^2} = \mu\left(\frac{\partial^2 u_2}{\partial x_1^2} + \frac{\partial^2 u_2}{\partial x_2^2} + \frac{\partial^2 u_2}{\partial x_3^2}\right) + (\lambda + \mu)\frac{\partial}{\partial x_2}\left(\frac{\partial u_1}{\partial x_1} + \frac{\partial u_2}{\partial x_2} + \frac{\partial u_3}{\partial x_3}\right)$$

$$\rho\frac{\partial^2 u_3}{\partial t^2} = \mu\left(\frac{\partial^2 u_3}{\partial x_1^2} + \frac{\partial^2 u_3}{\partial x_2^2} + \frac{\partial^2 u_3}{\partial x_3^2}\right) + (\lambda + \mu)\frac{\partial}{\partial x_3}\left(\frac{\partial u_1}{\partial x_1} + \frac{\partial u_2}{\partial x_2} + \frac{\partial u_3}{\partial x_3}\right)$$

These are the Navier equations from elasticity which can be written in the more compact vector form

$$\rho\frac{\partial^2 \vec{u}}{\partial t^2} = \mu\nabla^2\vec{u} + (\lambda + \mu)\nabla\left(\nabla \cdot \vec{u}\right) \tag{5.146}$$

Elastic beam theory

Consider a beam of length L in the x-direction with a constant and symmetric cross-section as illustrated in the figure. Let $y = y(x)$ denote the deflection of the neutral axis of the beam and let $q = q(x)$ denote a loading per unit length. The Bernoulli-Euler law from beam theory states that the bending moment M_z is given by

$$M_z = \frac{EI_{zz}}{R} = EI_{zz}\frac{y''}{[1 + (y')^2]^{3/2}} \tag{5.147}$$

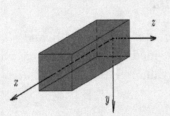

where E is Young's modulus of elasticity, I_{zz} is the moment of inertia of the cross section and R is the radius of curvature of the neutral axis of the beam. Assume that the deflection is small with y' small so that we can use the approximation $M_z = EI_{zz}y''$. For a symmetric cross section the bending stress σ_{xx} is given by $\sigma_{xx} = -\frac{M_z z}{I_{zz}}$.

The work done by a force can be represented as the area under a force-displacement curve. For a linear stress-strain relation the stress is proportional to strain and so one can write $\sigma_{xx} = E\epsilon_{xx}$. This relation is a straight line on a force-displacement diagram as illustrated. The area under this curve is given by

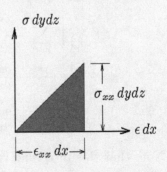

$$\frac{1}{2}\sigma_{xx}\epsilon_{xx}\,d\tau = \frac{\sigma_{xx}^2}{2E}d\tau = \frac{M_z^2 z^2}{2EI_{zz}^2}d\tau = \frac{E}{2}(y'')^2 z^2\,d\tau$$

where $d\tau = dzdydx$ is an element of volume.

The quantity $\sigma_{xx}^2/2E$ denoting the strain energy per unit volume within the beam. The total strain energy or potential energy is obtained by integrating over the beam volume to obtain

$$V_1 = \iiint \frac{E}{2}(y'')^2 z^2 \, dz dy dx = \int_0^L \frac{EI_{zz}}{2}(y'')^2 \, dx \qquad (5.148)$$

where $I_{zz} = \iint z^2 \, dy dz$ is the moment of inertial of the cross section.

The work done by the external force per unit length is force times displacement and is given by $dw = q(x)y(x) \, dx$ and integrating over the length of the beam one obtains the potential energy

$$V_2 = -\int_0^L q(x)y \, dx. \qquad (5.149)$$

Here V_2 is the negative of the work done and represents the potential energy due to the external loading on the beam. The potential energy can then be represented by the integral

$$I = V_1 + V_2 = \int_0^L \left[\frac{EI_{zz}}{2}(y'')^2 - q(x)y \right] \, dx \qquad (5.150)$$

Applying the principle of minimum potential energy we want to find the displacement $y = y(x)$ which minimizes this integral. This gives the Euler-Lagrange equation plus a boundary condition equation.

In equation (5.150) we replace $y = y(x)$ by $Y(x) = y(x) + \epsilon\eta(x)$ to obtain

$$I = I(\epsilon) = \int_0^L \left[\frac{EI_{zz}}{2}(y'' + \epsilon\eta'')^2 - q(x)(y(x) + \epsilon\eta(x)) \right] \, dx$$

with derivative

$$\frac{dI}{d\epsilon} = \int_0^L \left[EI_{zz}(y'' + \epsilon\eta'')\eta'' - q(x)\eta(x) \right] \, dx \qquad (5.151)$$

Integrate the first part of equation (5.151) using integration by parts twice and verify that there results the equation

$$\begin{aligned}
\frac{dI}{d\epsilon} = & \left[EI_{zz}(y'' + \epsilon\eta'')\eta' \right]_0^L - \left[EI_{zz}\frac{d}{dx}(y'' + \epsilon\eta'')\eta \right]_0^L \\
& + \int_0^L \left[EI_{zz}\frac{d^2}{dx^2}(y'' + \epsilon\eta'') - q(x) \right] \eta \, dx
\end{aligned} \qquad (5.152)$$

We require a stationary value at $\epsilon = 0$ so that

$$\delta I = \frac{dI}{d\epsilon}\bigg|_{\epsilon=0} = EI_{zz}\left[y''(x)\eta'(x) - y'''(x)\eta(x) \right]_0^L + \int_0^L \left[EI_{zz}\frac{d^2}{dx^2}(y'') - q(x) \right] \eta dx = 0 \qquad (5.153)$$

The integrand in the last term of equation (5.153) is zero because we recognize it as the Euler-Lagrange equation

$$EI_{zz}y^{iv}(x) - q(x) = 0 \qquad (5.154)$$

associated with the functional given by equation (5.150). This reduces the equation (5.153) to the form

$$\delta I = EI_{zz}\left[y''(L)\eta'(L) - y'''(L)\eta(L) - y''(0)\eta'(0) + y'''(0)\eta(0) \right] = 0 \qquad (5.155)$$

The general solution to the Euler-Lagrange equation (5.154) contains four arbitrary constants, hence we need four boundary conditions in order to determine these constants. We consider the following cases.

Case 1: Beam with clamped ends

Because the ends are fixed we require that

$$y(0) = 0 \qquad y(L) = 0$$
$$y'(0) = 0 \qquad y'(L) = 0$$

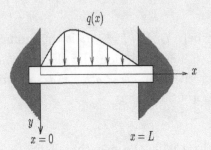

This implies $\delta y = 0$ and $\delta y' = 0$ at $x = 0$ and $x = L$. That is, $\eta(0) = \eta(L) = 0$ and $\eta'(0) = \eta'(L) = 0$ so that equation (5.155) is satisfied

These are imposed boundary conditions and so there are no natural boundary conditions.

Case 2: Simply supported beam

Here we have zero displacements at the points of support so that

$$y(0) = 0 \quad \text{and} \quad y(L) = 0$$

These conditions imply that $\delta y = 0$ at $x = 0$ and $x = L$. That is, $\eta(0) = 0$ and $\eta(L) = 0$ so that equation (5.155) reduces to the form

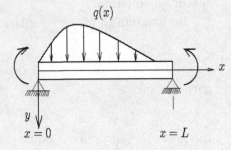

$$\delta I = EI_{zz}\Big[y''(L)\eta'(L) - y''(0)\eta'(0)\Big] = 0$$

which requires that we impose the natural boundary conditions $y''(0) = 0$ and $y''(L) = 0$ so that equation (5.155) is now satisfied.

Case 3: Beam with one end clamped and one end free

Assume that the beam is clamped at $x = 0$. This requires the boundary conditions

$$y(0) = 0 \quad \text{and} \quad y'(0) = 0$$

which implies $\delta y = 0$ and $\delta y' = 0$ at $x = 0$.

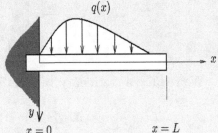

That is, $\eta(0) = \eta'(0) = 0$ which reduces the boundary equation (5.155) to the form

$$\delta I = EI_{zz}\Big[y''(L)\eta'(L) - y'''(L)\eta(L)\Big] = 0$$

Therefore, we require that the additional boundary conditions $y''(L) = 0$ and $y'''(L) = 0$ be satisfied. These are the natural boundary conditions.

"Mathematicians have tried in vain to this day to discover some order in the sequence of prime numbers, and we have reason to believe that it is a mystery into which the human mind will never penetrate. "

Leonhard Euler (1707-1783)

Exercises Chapter 5

▶ **1.** Consider the brachistrochrone problem with the y-axis in the upward direction as illustrated in the accompanying figure.

(a) Show by the principle of conservation of energy that

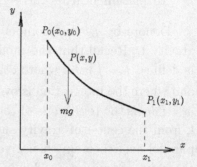

$$\frac{1}{2}mV^2 = mg(y_0 - y)$$

where m is the mass of the particle, V is the particle velocity and g is the acceleration of gravity.

(b) Show that

$$dt = \frac{ds}{V} = \frac{\sqrt{1 + (y')^2}}{\sqrt{2g(y_0 - y)}}\, dx$$

(c) Show the time required to move from point P_0 to point P_1 is

$$T = \int_{x_0}^{x_1} \sqrt{\frac{1 + (y')^2}{2g(y_0 - y)}}\, dx$$

(d) Obtain a first integral of the Euler-Lagrange equation associated with this functional and show that it can be written in the form

$$\frac{C_1}{1 + (y')^2} = y_0 - y$$

where C_1 is a constant.

(e) Make the substitution $y' = \tan\theta$ and show $y = y_0 - C_1\cos^2\theta$

(f) Show that a parametric solution of the Euler-Lagrange equation is

$$x = \frac{1}{2}C_1(\sin 2\theta + 2\theta) + C_2 \qquad \text{and} \qquad y = y_0 - \frac{1}{2}C_1(1 + \cos 2\theta)$$

where C_1 and C_2 are constants.

▶ **2.**

Consider a ball of mass m which falls under the action of gravity.

(a) Show the Lagrangian is given by $L = \frac{1}{2}m\dot{y}^2 + mgy$ where y is the distance fallen.

(b) Find the equation of motion from the Lagrangian.

(c) Show the Hamiltonian is given by $H = \frac{1}{2m}p^2 - mgy$

(d) Find the equations of motion from the Hamiltonian.

▶ **3.**

The compound pendulum is defined by a rigid body of arbitrary shape which is free to make oscillations about an axis through a point of support which we label point S in the accompanying figure. The point G denotes the center of gravity of the compound pendulum with $q_1 = \theta$ the generalized displacement from the vertical.

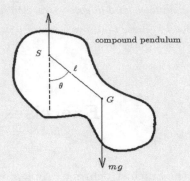
compound pendulum

Denote by I_G the moment of inertial of the compound pendulum about the center of gravity G. Recall that the moment of inertia about the axis perpendicular to the point G is defined $I = \int r^2 dm$ where dm is an element of mass a radial distance r from the point G and where the integration is over the region defining the compound pendulum. The quantity $m = \int dm$ is the total mass of the compound pendulum. If a point mass m is placed a distance k from the center of gravity such that $I = mk^2$, then k is called the radius of gyration. The parallel axis theorem then gives the moment of inertia about the axis of rotation through the point of support S. This moment of inertia is

$$I_S = I_G + m\ell^2 = m(k^2 + \ell^2)$$

where ℓ is the distance from the point S to the point G.

(a) Show the rotational kinetic energy with respect to the point S is given by $T = \frac{1}{2}I_S\dot{\theta}^2$

(b) Show the potential energy is given by $V = mg\ell(1 - \cos\theta)$

(c) Show the Lagrange equation of motion is given by

$$I_S\ddot{\theta} + mg\ell\sin\theta = 0 \qquad (3a)$$

(d) Solve the equation (3a) for I_S, m, g, ℓ constants.

▶ **4.** Show that if there exists a function $U = U(q_i, \dot{q}_i)$ such that the generalized forces can be derived from the relation

$$Q_i = -\frac{\partial U}{\partial q_i} + \frac{d}{dt}\left(\frac{\partial U}{\partial \dot{q}_i}\right)$$

then the equations of motion can be obtained from the Lagrangian equations

$$\frac{d}{dt}\left(\frac{\partial L}{\partial \dot{q}_i}\right) - \frac{\partial L}{\partial q_i} = 0$$

where $L = T - U$.

▶ **5.** For the spring-mass system illustrated in the figure 5-13(b) derive the equation of motion for the displacement x from the equilibrium position. State all assumptions made. (a) By summing forces and using Newton's second law. (b) By introducing a kinetic energy, potential energy and dissipation function. (c) Note the electrical-mechanical-torsional analogs given in the accompanying table. Derive these equations also.

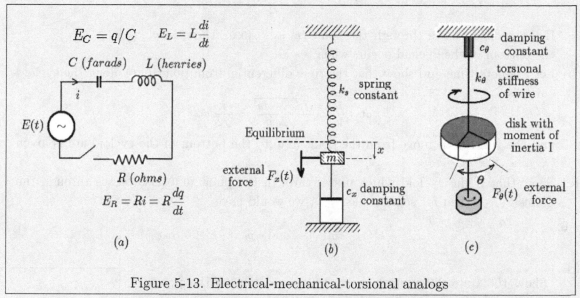

Figure 5-13. Electrical-mechanical-torsional analogs

Table: Electrical-Mechanical-Torsional Analogs								

Electrical circuit $L\dfrac{d^2q}{dt^2} + R\dfrac{dq}{dt} + \dfrac{1}{C}q = E(t)$

Spring-mass system $m\dfrac{d^2x}{dt^2} + c_x\dfrac{dx}{dt} + k_s x = F_x(t)$

Torsional vibration $I\dfrac{d^2\theta}{dt^2} + c_\theta\dfrac{d\theta}{dt} + k_\theta\,\theta = F_\theta(t)$

Electrical circuit			Spring-mass system			Torsional vibration		
Quantity	Symbol	Units	Quantity	Symbol	Units	Quantity	Symbol	Units
inductance	L	henry	mass	m	Kg	moment of inertia	I	$\text{Kg}\cdot\text{m}^2$
charge	q	coulomb	displacement	x	m	angular displacement	θ	rad
resistance	R	ohm	damping coefficient	c_x	$\dfrac{\text{N}}{\text{(m/s)}}$	torsional damping	c_θ	$\dfrac{\text{N}\cdot\text{m}}{\text{rad/s}}$
capacitance	C	farad	spring constant	k_s	$\dfrac{\text{N}}{\text{m}}$	torsional stiffness	k_θ	$\dfrac{\text{N}\cdot\text{m}}{\text{rad}}$
current	$i = \dfrac{dq}{dt}$	ampere	velocity	$v = \dfrac{dx}{dt}$	$\dfrac{\text{m}}{\text{s}}$	angular velocity	$\omega = \dfrac{d\theta}{dt}$	$\dfrac{\text{rad}}{\text{s}}$
applied voltage	$E(t)$	volt	external force	$F_x(t)$	N	external torque	$F_\theta(t)$	$\text{N}\cdot\text{m}$

▶ **6.**

(a)

In the equations (5.9) make the substitution $\phi = 2\theta$ and show the equation of the cycloid can be written

$$x = x_0 + \frac{1}{2}k^2(\phi - \sin\phi), \quad \text{and} \quad y = y_0 + \frac{1}{2}k^2(1 - \cos\phi)$$

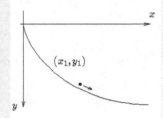

If the cycloid passes through the point $(x_0, y_0) = (0, 0)$, then show the bottom of the cycloid occurs when $\phi = \pi$.

(b) Let t denote time and show that the time differential from point $(0,0)$ along the cycloid curve is

$$dt = \frac{ds}{V} = \frac{\sqrt{dx^2 + dy^2}}{\sqrt{2gy}} = \frac{k}{\sqrt{2g}}d\phi$$

and the time T_0 to move from the point $(0,0)$ to the bottom of the cycloid arc is given by $T_0 = \frac{k}{\sqrt{2g}}\pi$

(c) Show that if the cycloid which passes through the point $(0,0)$ also passes through the point (x_1, y_1), then for some value $\phi = \phi_1$ we would have

$$x_1 = \frac{1}{2}k^2(\phi_1 - \sin\phi_1), \quad \text{and} \quad y_1 = \frac{1}{2}k^2(1 - \cos\phi_1)$$

(d)

Show the time differential along the cycloid joining the point (x_1, y_1) to the lowest point on the cycloid is given by

$$dt = \frac{ds}{V} = \frac{\sqrt{dx^2 + dy^2}}{\sqrt{2g(y - y_1)}} = \frac{k}{\sqrt{2g}}\sqrt{\frac{1 - \cos\phi}{\cos\phi_1 - \cos\phi}}\, d\phi$$

(e) Show the total time T_1 to move from the point (x_1, y_1) to the lowest point on the cycloid arc is given by

$$T_1 = \int_{\phi_1}^{\pi} \frac{k}{\sqrt{2g}}\sqrt{\frac{1 - \cos\phi}{\cos\phi_1 - \cos\phi}}\, d\phi = \frac{k}{\sqrt{2g}}\int_{\phi_1}^{\pi} \frac{\sin\frac{1}{2}\phi\, d\phi}{\sqrt{\cos^2\left(\frac{\phi_1}{2}\right) - \cos^2\left(\frac{\phi}{2}\right)}}$$

Hint: Make use of the trigonometric identities $\sin\frac{\theta}{2} = \sqrt{\frac{1-\cos\theta}{2}}$ and $\cos\frac{\theta}{2} = \sqrt{\frac{1+\cos\theta}{2}}$

(f) In part (e) make the substitutions

$$u = \frac{\cos\frac{\phi}{2}}{\cos\frac{\phi_1}{2}} \quad \text{and} \quad du = \frac{-\sin\frac{\phi}{2}\, d\phi}{2\cos\frac{\phi_1}{2}}$$

to show

$$T_1 = -2\frac{k}{\sqrt{2g}}\int_1^0 \frac{du}{\sqrt{1 - u^2}} = \frac{k}{\sqrt{2g}}\pi$$

(g) Compare your answers in parts (b) and (f) and then look up the word tautochrone which comes from the Greek *tauto* meaning "same" and *chrone* meaning "time".

► 7.

Assume a thin rigid rod, which is weightless, connects two masses and is supported by the spring system illustrated. Assume each spring is a linear spring with spring constant k. Let q_1 and q_2 denote the vertical displacements of the masses from the equilibrium position and assume that the system is constructed so that the masses move in the vertical direction. Assume the system is given an initial displacement from the equilibrium position and released. Find the equations governing the motion of the spring-mass system.

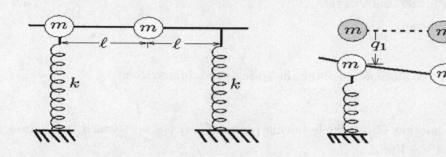

(a) Show that $x = 2q_2 - q_1$

(b) Show that the system kinetic energy is given by $T = \dfrac{m}{2}(\dot{q}_1^2 + \dot{q}_2^2)$

(c) Show the system potential energy can be written $V = \dfrac{k}{2}q_1^2 + \dfrac{k}{2}(2q_2 - q_1)^2$

(d) Form the Lagrangian L.

(e) Show the differential equations describing the motions $q_1 = q_1(t)$ and $q_2 = q_2(t)$ are given by

$$m\ddot{q}_1 + 2k(q_1 - q_2) = 0 \qquad \text{and} \qquad m\ddot{q}_2 + 2k(2q_2 - q_1) = 0$$

► 8.

Consider the linear spring-mass system illustrated.

(a) Show the Lagrangian is given by $L = \frac{1}{2}m\dot{x}^2 - \frac{1}{2}k_s x^2$

(b) Find the equation of motion from the Lagrangian.

(c) Show the Hamiltonian is given by $H = \frac{1}{2m}p^2 + \frac{1}{2}k_s x^2$

(d) Find the equation of motion from the Hamiltonian.

► 9.

(a) Construct the kinetic energy, potential energy and Lagrangian for the simple pendulum given in the example 6-4. Use the displacement angle θ as the generalized coordinate.

(b) Find the equation of motion of the simple pendulum system.

(c) Write the Hamiltonian for this system.

(d) Determine the equations of motion from Hamilton's equations.

▶ **10.**

For the spring-pendulum system illustrated the point of support 0 can slide without friction along the bar AA. Linear springs, with spring constant k, apply a restoring force to the point of support. Show that the kinetic energy T and potential energy V of the system are given by

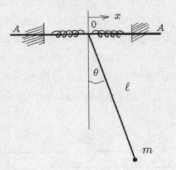

$$T = \frac{m}{2}\left(\dot{x}^2 + 2\dot{x}\dot{\theta}\ell\cos\theta + \ell^2\dot{\theta}^2\right)$$
$$V = mg\ell(1 - \cos\theta) + kx^2$$

Find the equations of motion describing the spring-pendulum system.

▶ **11.** Consider the motion of a particle having position $(x(t), y(t), z(t))$ which moves in a potential field $V = V(x, y, z)$.

(a) Show the kinetic energy of the particle is given by $T = \frac{1}{2}\left[\dot{x}^2 + \dot{y}^2 + \dot{z}^2\right]$

(b) Calculate the Lagrangian $L = T - V$

(c) Show that the trajectory of the particle is governed by the Newtonian equations of motion

$$m\frac{d^2x}{dt^2} = -\frac{\partial V}{\partial x}, \qquad m\frac{d^2y}{dt^2} = -\frac{\partial V}{\partial y}, \qquad m\frac{d^2z}{dt^2} = -\frac{\partial V}{\partial z}$$

(d) Show that by introducing the notation $y_1 = x$, $y_2 = y$ and $y_3 = z$, the equations of motion can be expressed in the more compact form

$$m\frac{d^2y_i}{dt^2} = -\frac{\partial V}{\partial y_i}, \qquad \text{for} \quad i = 1, 2, 3.$$

▶ **12.** Consider the problem of finding an extreme value for the functional
$$I = \int_{x_1}^{x_2} \left(\frac{d^2y}{dx^2}\right)^2 dx = \int_{x_1}^{x_2} f(x, y, y', y'')\, dx.$$
Show that the end points must satisfy the variational equation

$$f\,\delta x + \left[\frac{\partial f}{\partial y'} - \frac{d}{dx}\left(\frac{\partial f}{\partial y''}\right)\right](\delta y - y'\,\delta x) + \frac{\partial^2 f}{\partial y''^2}(\delta y' - y''\,\delta x) = 0.$$

Determine these end point conditions for the integrand $f = (y'')^2$.

▶ **13.**

(a) For the simply supported beam illustrated having a constant external loading q_0 lbs per unit length find the displacement $y = y(x)$ of the centerline for $0 \le x \le L$.

(b) Find the maximum deflection of the beam.

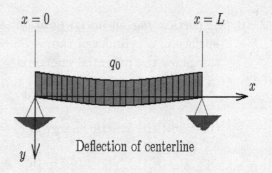

Deflection of centerline

▶ **14.**

(a) For the cantilever beam illustrated having a constant external loading q_0 lbs per unit length find the displacement $y = y(x)$ of the centerline for $0 \le x \le L$.

(b) Find the maximum deflection of the beam.

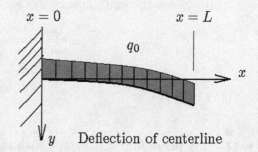

Deflection of centerline

▶ **15.** **The Foucault pendulum** (computer problem)

(a) The assumption that ℓ is large and θ changes are small allows one to assume that $\ddot{z}$ and $\dot{z}$ are both small and can be neglected in the system of equations (5.136). Show that these assumptions give the equations of motion

$$\ddot{x} + \alpha^2 x = 2\omega \dot{y} \sin \lambda, \qquad \ddot{y} + \alpha^2 y = -2\omega \dot{x} \sin \lambda$$

where $\alpha^2 = \tau/\ell m$, with ω and λ constants.

(b) Let $2\omega \sin \lambda = 0.0001$, this requires that $\lambda \approx 43°$ latitude and let $\alpha^2 = \tau/\ell m = 4(10)^{-6}$ and numerically solve the equations in part (a) subject to the initial conditions $x(0) = 1$, $\dot{x}(0) = 0$, $y(0) = 0$ and $\dot{y}(0) = 0$. You should obtain results similar to the following figures

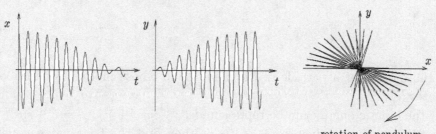

rotation of pendulum

▶ **16.**
A particle has an initial position $(R\sin\theta, R\cos\theta, z)$ on the surface of a cylinder of radius R.

(a) Show the potential energy is

$$V = mgR\sin\theta + constant$$

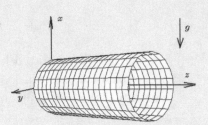

(b) Show the kinetic energy is

$$T = \frac{m}{2}\left[R^2\dot\theta^2 + \dot z^2\right]$$

(c) Show the equations of motion are given by

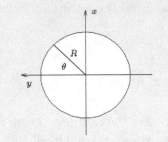

$$mR\frac{d^2\theta}{dt^2} = -mg\cos\theta$$
$$m\frac{d^2z}{dt^2} = 0$$

▶ **17.** A pendulum of length ℓ with mass m is attached to the rim of a wheel of radius r_0. The wheel rotates with constant angular velocity ω. Let θ denote the angular displacement of the pendulum from the vertical. Find the equation describing the change of θ with time t.

(a) Show the position of the mass m is given by
$$x = r_0\cos\omega t + \ell\sin\theta, \qquad y = r_0\sin\omega t - \ell\cos\theta$$

(b) Show the kinetic energy of the mass m is given by
$$T = \frac{m}{2}\left[r_0^2\omega^2 + \ell^2\dot\theta^2 + 2r_0\ell\omega\dot\theta\sin(\theta - \omega t)\right]$$

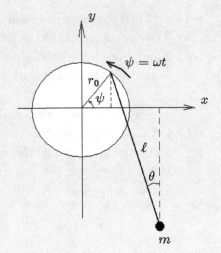

(c) Show the potential energy is
$$V = mg\ell(1 - \cos\theta) + mgr_0\sin\omega t + constant$$

(d) Show the differential equation describing the motion of the angle θ is given by
$$\ddot\theta + \frac{g}{\ell}\sin\theta - \omega^2\frac{r_0}{\ell}\cos(\theta - \omega t) = 0$$

(e) What happens if $\omega = 0$?

▶ **18.**
Consider the motion of a particle in the plane $z = 0$ whose motion is described using polar coordinates (r, θ).

(a) Show that the kinetic energy can be represented $T = \frac{1}{2}m(\dot x^2 + \dot y^2) = \frac{1}{2}m(\dot r^2 + r^2\dot\theta^2)$

(b) If the potential energy is given by $V = V(r, \theta)$, then find the Lagrangian equations of motion for the particle.

▶ **19.**

In the study of statistical mechanics on finds that the energy distribution of a system of particles is described by a probability distribution function $f(E)$. If the system of particles is in equilibrium, then the probability function $f(E)$ must be selected to maximize the integral $I = -\int_0^\infty f(E) \ln f(E)\, dE$ subject to the constraint conditions

$$\int_0^\infty f(E)\, dE = 1 \qquad \text{and} \qquad \int_0^\infty E f(E)\, dE = E_0 = \text{a constant}$$

Introduce Lagrange multipliers λ_1 and λ_2 and determine the probability function $f(E)$.

▶ **20.**

Two particles of equal mass m are connected by an inextensible string which passes through a hole in a smooth (friction free) horizontal table. The first particle rests on the table and the second particle is suspended vertically as illustrated in the accompanying figure. Assume that the particle on the table is caused to describe a circular path about the hole so that the second particle is suspended in equilibrium.

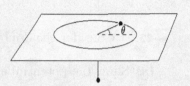

(a) Assume the particle on the table has an angular velocity $\dot\theta = \omega = \sqrt{g/a}$ where a is the radius of the circular path and g is the acceleration of gravity. If at time $t = 0$ the suspended mass is pulled down a short distance x and then released, while the first mass on the table continues to rotate, show that the Lagrangian is given by

$$L = m\left(\dot{x}^2 + \frac{1}{2}(a-x)^2 \dot\theta^2 + gx \right).$$

(b) Obtain the equations describing the motion of the suspended mass.

(c) Show the equation of motion in the θ-direction can be integrated to obtain

$$(a-x)^2 \dot\theta = a\sqrt{ag}.$$

(d) Show that the equation of motion in the x-direction can be written in the form
$$2\ddot{x} + \left[\frac{1}{(1-x/a)^3} - 1 \right] g = 0.$$

(e) Show that if $x \ll a$, then

$$\frac{1}{(1-x/a)^3} = 1 + 3\frac{x}{a} + 6\left(\frac{x}{a}\right)^2 + \cdots$$

is approximately $1 + 3x/a$

(f) Use the approximation in part (e) to show the suspended mass oscillates with period given by $2\pi\sqrt{2a/3g}$

240

▶ **21.**
Calculate the kinetic energy T in a spherical coordinate system (r, θ, ϕ). Here the generalized coordinates are given by $(q_1, q_2, q_3) = (r, \theta, \phi)$ with r denoting the distance from the origin to a general point (x, y, z), with $0 \leq \theta \leq \pi$ and $0 \leq \phi \leq 2\pi$. The relations between the coordinates (x, y, z) and (r, θ, ϕ) are given by the transformation equations

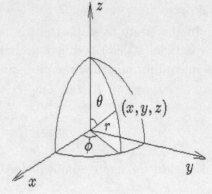

$$x = r \sin \theta \cos \phi, \quad y = r \sin \theta \sin \phi, \quad z = r \cos \theta$$

Assume that the potential energy is given by a function $V = V(\rho, \theta, \phi)$. Construct the Lagrangian and find the Euler-Lagrange equations of motion in spherical coordinates.

▶ **22.** Consider the double pendulum system illustrated.

(a) Show the potential energy is given by

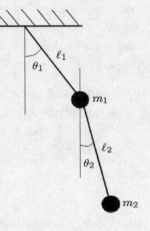

$$V = m_1 g \ell_1 (1 - \cos \theta_1) + m_2 g \left[\ell_1 (1 - \cos \theta_1) + \ell_2 (1 - \cos \theta_2) \right]$$

(b) Show the kinetic energy can be represented

$$T = \frac{1}{2} m_1 \ell_1^2 \dot{\theta}_1^2 + \frac{1}{2} m_2 \left[\ell_1^2 \dot{\theta}_1^2 + \ell_2^2 \dot{\theta}_2^2 + 2 \ell_1 \ell_2 \dot{\theta}_1 \dot{\theta}_2 \cos(\theta_1 - \theta_2) \right]$$

(c) Show the Euler-Lagrange equations of motion are given by

$$\ell_1 (m_1 + m_2) \ddot{\theta}_1 + m_2 \ell_2 \ddot{\theta}_2 \cos(\theta_1 - \theta_2) + m_2 \ell_2 (\dot{\theta}_2)^2 \sin(\theta_1 - \theta_2) +$$
$$\ell_2 \ddot{\theta}_2 + \ell_1 \ddot{\theta}_1 \cos(\theta_1 - \theta_2) - \ell_1 (\dot{\theta}_1)^2 \sin(\theta_1 - \theta_2) + g \sin \theta_2 = 0$$

▶ **23.** Consider the functional $I = I(y) = \int_0^1 \left[\frac{1}{2} (y')^2 + f(x) y \right] dx + \alpha y(0) - \beta y(1)$ where α and β are constants.

(a) Show $\delta I = \left[y'(1) - \beta \right] \delta y(1) - \left[y'(0) - \alpha \right] \delta y(0) - \int_0^1 \left[y'' - f(x) \right] \delta y \, dx$

(b) Find conditions for a stationary value if

  (i) The essential boundary conditions $y(0) = y_0$ and $y(1) = y_1$ are specified.

  (ii) The boundary conditions $y(0)$ and $y(1)$ are not specified.

  (iii) The boundary condition $y(0) = y_0$ is specified but $y(1)$ is not specified.

  (iv) The boundary condition $y(1) = y_1$ is specified but $y(0)$ is not specified.

  (v) What are the natural and essential boundary conditions for this problem?

▶ **24.** Consider the problem of finding a stationary value of the functional $I = \int_{(x_0,y_0)}^{(x_1,y_1)} f(x, y, y')\, dx$ where $y = y(x)$ is a curve connecting the initial point (x_0, y_0) to a point (x_1, y_1) which lies on a given curve $g(x, y) = 0$.

(a) Show that if the desired curve is represented in the parametric form $x = x(t)$, $y = y(t)$, $t_0 \leq t \leq t_1$ satisfying $x(t_0) = x_0$, $y(t_0) = y_0$ and $x(t_1) = x_1$, $y(t_1) = y_1$, then the functional can be written in the form $I = \int_{t_0}^{t_1} f\left(x, y, \frac{\dot{y}}{\dot{x}}\right) \dot{x}\, dt$ with the subsidiary condition that $g(x(t_1), y(t_1)) = 0$.

(b) Introduce a Lagrange multiplier λ and construct $I^* = \int_{t_0}^{t_1} f\left(x, y, \frac{\dot{y}}{\dot{x}}\right) \dot{x}\, dt + \lambda g(x(t_1), y(t_1))$ and show that the first variation of I can be written

$$\delta I^* = \int_{t_0}^{t_1} \left[f\delta\dot{x} + \dot{x}\left(\frac{\partial f}{\partial x}\delta x + \frac{\partial f}{\partial y}\delta y + \frac{\partial f}{\partial \dot{x}}\delta\dot{x} + \frac{\partial f}{\partial \dot{y}}\delta\dot{y} \right) \right] dt + \lambda \left[\frac{\partial g}{\partial x}\delta x + \frac{\partial g}{\partial y}\delta y \right]_{t=t_1} + \delta\lambda\, g(x(t_1), y(t_1))$$

(c) Use integration by parts and show a stationary value exists if

$$\dot{x}\frac{\partial f}{\partial x} - \frac{d}{dt}\left(f + \dot{x}\frac{\partial f}{\partial \dot{x}} \right) = 0, \qquad \dot{x}\frac{\partial f}{\partial y} - \frac{d}{dt}\left(f + \dot{x}\frac{\partial f}{\partial \dot{y}} \right) = 0,$$

$$g(x(t_1), y(t_1)) = 0, \qquad \dot{x}\frac{\partial f}{\partial \dot{y}} + \lambda\frac{\partial g}{\partial y}\bigg|_{t=t_1} = 0, \qquad f + \dot{x}\frac{\partial f}{\partial \dot{x}} + \lambda\frac{\partial g}{\partial x}\bigg|_{t=t_1} = 0$$

where $\delta x = \delta y = 0$ at $t = t_0$.

(d) Eliminate λ and derive the transversality condition $\left(f + \dot{x}\frac{\partial f}{\partial \dot{x}} \right)\frac{\partial g}{\partial y} - \dot{x}\frac{\partial f}{\partial \dot{y}}\frac{\partial g}{\partial x} = 0$

▶ **25.** The potential energy function associated with the vertical loading P of a uniform cantilever beam is given by

$$\Pi = \int_0^L \left[\frac{EI}{2}\left(\frac{d^2 y}{dx^2} \right)^2 - \frac{P}{2}\left(\frac{dy}{dx} \right)^2 \right] dx$$

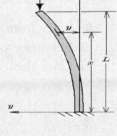

(a) Show that

$$\delta\Pi = \left[EIy''\delta y' - (EIy''' + Py')\delta y \right]_0^L + \int_0^L \left(EIy'''' + Py'' \right) \delta y\, dx$$

(b) Find the Euler-Lagrange equation and boundary conditions required for a stationary value of the potential energy to exist.

▶ **26.** Consider a simply supported beam with uniformly distributed load q and constant cross section. It can be shown that by considering only strain energy due to bending the potential function for this system is given by $\Pi = \int_0^L \left[\frac{EI}{2}\left(\frac{d^2 w}{dx^2} \right)^2 - qw \right] dx$

(a) Show that $\delta\Pi = [EIw'' - EIw'''\delta w]_0^l - \int_0^L [EIw'''' - q]\,\delta w\, dx$

(b) Find the Euler-Lagrange equation and boundary conditions required for a stationary value of he potential energy to exist.

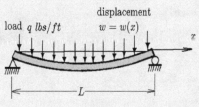

load q lbs/ft displacement $w = w(x)$

242

▶ **27.**

(a) Show that the functional

$$I = I(z) = \iint_R \left[\left(\frac{\partial z}{\partial x} \right)^2 + \left(\frac{\partial z}{\partial y} \right)^2 \right] dx\, dy$$

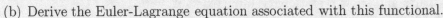

remains invariant under a rotation of axes given by

$$\overline{x} = x\cos\theta - y\sin\theta, \quad \overline{y} = x\sin\theta + y\cos\theta, \quad \overline{z} = z$$

Hint: Show that

$$I(\overline{z}) = \int_{\overline{R}} \left[\left(\frac{\partial \overline{z}}{\partial \overline{x}} \right)^2 + \left(\frac{\partial \overline{z}}{\partial \overline{y}} \right)^2 \right] d\overline{x}\, d\overline{y} = \iint_R \left[\left(\frac{\partial z}{\partial x} \right)^2 + \left(\frac{\partial z}{\partial y} \right)^2 \right] \frac{\partial(\overline{x}, \overline{y})}{\partial(x, y)} \, dx\, d,$$

and then calculate the Jacobian of the transformation.

(b) Derive the Euler-Lagrange equation associated with this functional.

(c) What boundary conditions can be assigned to the Euler-Lagrange equation?

▶ **28.** Consider a ray of light which travels in a medium of variable index of refraction given by $\mu = \mu(x, y, z)$. The light ray travels from one fixed point to another within the medium along a path which minimizes the integral $I = \int_{s_0}^{s_1} \mu\, ds$ where ds is an element of arc length. The equations which describe the path of the ray of light must be such that

$$\delta I = \delta \int_{s_0}^{s_1} \mu\, ds = \int_{s_0}^{s_1} (\delta\mu\, ds + \mu\, \delta ds) = 0.$$

(a) Show from the element of arc length squared that $2ds\, \delta\, ds = 2dx\, \delta\, dx + 2dy\, \delta\, dy + 2dz\, \delta\, dx$

(b) Show the variation in the refractive index can be represented $\delta\mu = \frac{\partial\mu}{\partial x}\,\delta x + \frac{\partial\mu}{\partial y}\,\delta y + \frac{\partial\mu}{\partial z}\,\delta z$

(c) Show that

$$\delta I = \int_{s_0}^{s_1} \left(\frac{\partial\mu}{\partial x}\,\delta x + \frac{\partial\mu}{\partial y}\,\delta y + \frac{\partial\mu}{\partial z}\,\delta z \right) ds + \mu \left(\frac{dx}{ds}\, d\delta x + \frac{dy}{dx}\, d\delta y + \frac{dz}{ds}\, d\delta z \right) = 0$$

(d) Use integration by parts to show

$$\delta I = \left[\mu\frac{dx}{dx}\delta x + \mu\frac{dy}{ds}\delta y + \mu\frac{dz}{ds}\delta z \right]_{s_0}^{s_1} +$$
$$\int_{s_0}^{s_1} \left[\left\{ \frac{\partial\mu}{\partial x} - \frac{d}{ds}\left(\mu\frac{dx}{ds} \right) \right\}\delta x + \left\{ \frac{\partial\mu}{\partial y} - \frac{d}{ds}\left(\mu\frac{dy}{ds} \right) \right\}\delta y + \left\{ \frac{\partial\mu}{\partial z} - \frac{d}{ds}\left(\mu\frac{dz}{ds} \right) \right\}\delta z \right] ds = 0$$

(e) Show that the light ray must follow the path described by the system of differential equations

$$\frac{\partial\mu}{\partial x} - \frac{d}{ds}\left(\mu\frac{dx}{ds} \right) = 0, \quad \frac{\partial\mu}{\partial y} - \frac{d}{ds}\left(\mu\frac{dy}{ds} \right) = 0, \quad \frac{\partial\mu}{\partial z} - \frac{d}{ds}\left(\mu\frac{dz}{ds} \right) = 0$$

Chapter 6
Additional Applications

Applications of the variational calculus can be found in such diverse areas as chemistry, biology, physics, engineering and mechanics. Many of these applications require special knowledge of a subject area in order to formulate a variational problem. In this chapter we consider a few such applications which do not require excessive background material for their development. We also consider numerical methods and approximation techniques for obtaining extreme values associated with functions and functionals.

The vibrating string

A string is placed under tension between given fixed points. The zero displacement state of the string is called the equilibrium position of the string. The string is given an initial displacement from its equilibrium position and then released from rest. We desire to develop the equation describing the motion of the string. We introduce symbols, use some basic physics and use Hamilton's principle to construct the equation of motion.

Let x denote distance with the points $x = 0$ and $x = L$ the fixed end points of the string. Further, let $u = u(x,t)$ [m] denote the string displacement of a general point x, [m] for $0 \le x \le L$, at a time t [sec], then at a general point x at time t the quantities $\frac{\partial u}{\partial x}$ and $\frac{\partial u}{\partial t}$ denote respectively the slope and velocity of a general point on the string.

We neglect damping forces and assume that the slope of the string $\frac{\partial u(x,t)}{\partial x}$ at any point x is small such that $|\frac{\partial u}{\partial x}| << 1$. To employ Hamilton's principle we must calculate the kinetic energy and potential energy for the string. To accomplish this let ϱ [kg/m] denote the lineal mass density of the string so that $\varrho\,dx$ denotes the mass of a string element of length dx. The kinetic energy of the string can then be expressed as a summation of one-half the mass times velocity squared for all elements along the string. This gives the kinetic energy

$$T = \frac{1}{2} \int_0^L \varrho \left[\frac{\partial u(x,t)}{\partial t} \right]^2 dx \qquad \left[\frac{kg}{m} \right] \left[\frac{m}{sec} \right]^2 [m] = [joule] \tag{6.1}$$

The potential energy of the string is written as $V = V_1 + V_2$ where V_1 is the work done in distorting the string from its equilibrium position together with V_2 denoting the work done by the stretching that occurs at a boundary point. Assume that each element of the string experiences a constant stretching force of τ [nt]. The work done in distorting a string element of length dx in its equilibrium position to a length $ds = \sqrt{1 + \left(\frac{\partial u}{\partial x}\right)^2}\,dx$ at some general time t is given by a force times a distance or

$$dV_1 = \tau(ds - dx) = \tau \left(\sqrt{1 + \left(\frac{\partial u}{\partial x} \right)^2} - 1 \right) dx \tag{6.2}$$

244

We can expand the square root term using a binomial expansion to obtain

$$\left[1+\left(\frac{\partial u}{\partial x}\right)^2\right]^{1/2} = 1 + \frac{1}{2}\left(\frac{\partial u}{\partial x}\right)^2 - \frac{1}{8}\left(\frac{\partial u}{\partial x}\right)^4 + \frac{1}{16}\left(\frac{\partial u}{\partial x}\right)^6 - \cdots$$

and if $|\frac{\partial u}{\partial x}| << 1$ we can neglect higher order terms in this expansion. The equation (6.2) can then be expressed in the simplified form

$$dV_1 = \frac{1}{2}\tau\left(\frac{\partial u}{\partial x}\right)^2 dx \tag{6.3}$$

so that by summing over all values of x we obtain the first part of the potential

$$V_1 = \int_0^L \frac{1}{2}\tau\left(\frac{\partial u}{\partial x}\right)^2 dx \qquad [nt]\,[m] = [joule] \tag{6.4}$$

There is also work done in stretching the ends of the string. The ends of the string can be thought of as springs with spring constant κ_1 at $x = 0$ and spring constant κ_2 at $x = L$. The springs are treated as linear springs so that the spring force is proportional to the displacement. If the displacement of a boundary point is denoted by ξ, then the element of work done by these end point forces is given by force $(\kappa\xi)$ times displacement $(d\xi)$. The work done at the boundaries can therefore be expressed

$$\int_0^{u(0,t)} \kappa_1\xi\,d\xi = \frac{1}{2}\kappa_1 u^2(0,t) \qquad \text{and} \qquad \int_0^{u(L,t)} \kappa_2\xi\,d\xi = \frac{1}{2}u^2(L,t) \tag{6.5}$$

These integrals produce the potential function

$$V_2 = \frac{1}{2}\kappa_1 u^2(0,t) + \frac{1}{2}\kappa_2 u^2(L,t) \tag{6.6}$$

associated with the displacement of the boundaries. The Lagrangian is then found to be

$$L = T - V_1 - V_2 = \int_0^L \left[\frac{1}{2}\varrho\left(\frac{\partial u}{\partial t}\right)^2 - \tau\left(\frac{\partial u}{\partial x}\right)^2\right] dx - \frac{1}{2}\kappa_1 u^2(0,t) - \frac{1}{2}\kappa_2 u^2(L,t) \tag{6.7}$$

Hamilton's principle requires that the integral

$$I = \int_{t_1}^{t_2} L\,dt$$
$$I = \int_{t_1}^{t_2}\int_0^L \left[\frac{1}{2}\varrho\left(\frac{\partial u}{\partial t}\right)^2 - \tau\left(\frac{\partial u}{\partial x}\right)^2\right] dxdt - \frac{1}{2}\kappa_1\int_{t_1}^{t_2} u^2(0,t)\,dt - \frac{1}{2}\kappa_2\int_{t_1}^{t_2} u^2(L,t)\,dt \tag{6.8}$$

be a minimum for arbitrary times t_1 and $t_2 > t_1$.

Assume that $u(x,t)$ is the function that minimizes the Hamiltonian integral given by equation (6.8) and then consider a comparison function $U = U(x,t) = u(x,t) + \epsilon\eta(x,t)$ where $\eta(x,t_1) = 0$ and $\eta(x,t_2) = 0$ for all x satisfying $0 \le x \le L$. That is, we assume there is no variation at the beginning time t_1 and ending time t_2. The equation (6.8) then becomes the functional

$$I = I(\epsilon) = \int_{t_1}^{t_2}\int_0^L \left[\frac{1}{2}\varrho\left(\frac{\partial u}{\partial t} + \epsilon\frac{\partial\eta}{\partial t}\right)^2 - \tau\left(\frac{\partial u}{\partial x} + \epsilon\frac{\partial\eta}{\partial x}\right)^2\right] dxdt$$
$$-\frac{1}{2}\kappa_1\int_{t_1}^{t_2}\left(u(0,t) + \epsilon\eta(0,t)\right)^2 dt - \frac{1}{2}\kappa_2\int_{t_1}^{t_2}\left(u(L,t) + \epsilon\eta(L,t)\right)^2 dt \tag{6.9}$$

which has a minimum at $\epsilon = 0$. A necessary condition for a minimum is that $\frac{dI}{d\epsilon}\big|_{\epsilon=0} = 0$. We differentiate the equation (6.9) with respect to ϵ and set the result equal to zero. The resulting equation can be written in the form

$$\frac{dI}{d\epsilon}\Big|_{\epsilon=0} = \int_{t_1}^{t_2} \int_0^L [\varrho u_t \eta_t - \tau u_x \eta_x]\, dx dt - \kappa_1 \int_{t_1}^{t_2} u(0,t)\eta(0,t)\, dt - \kappa_2 \int_{t_1}^{t_2} u(L,t)\eta(L,t)\, dt = 0 \qquad (6.10)$$

where the subscripts denote partial derivatives. Observe that for ϱ and τ constant one can calculate the derivatives

$$\frac{\partial}{\partial x}\left[-\tau u_x(x,t)\eta(x,t)\right] = -\tau u_x(x,t)\eta_x(x,t) - \tau u_{xx}(x,t)\eta(x,t)$$
$$\frac{\partial}{\partial t}\left[\varrho u_t(x,t)\eta(x,t)\right] = \varrho u_t(x,t)\eta_t(x,t) + \varrho u_{tt}(x,t)\eta(x,t)$$

$\qquad (6.11)$

which can be substituted into the equation (6.10) to obtain

$$\frac{dI}{d\epsilon}\Big|_{\epsilon=0} = \int_{t_1}^{t_2} \int_0^L [-\varrho u_{tt} + \tau u_{xx}]\, \eta\, dx dt - \kappa_1 \int_{t_1}^{t_2} u(0,t)\eta(0,t)\, dt - \kappa_2 \int_{t_1}^{t_2} u(L,t)\eta(L,t)\, dt$$
$$+ \int_{t_1}^{t_2} \int_0^L \frac{\partial}{\partial x}\left[-\tau u_x(x,t)\eta(x,t)\right] dx dt + \int_{t_1}^{t_2} \frac{\partial}{\partial t}\left[\varrho u_t(x,t)\eta(x,t)\right] dx dt = 0$$

$\qquad (6.12)$

One can integrate the last integral in equation (6.12) with respect to t and show the result is zero. This is because of our original assumption the η is zero at t_1 and t_2. The second to last integral in equation (6.12) can be integrated with respect to x to obtain

$$\int_{t_1}^{t_2} \left[-\tau u_x(x,t)\eta(x,t)\right]_0^L\, dt = \int_{t_1}^{t_2} \tau\left[u_x(0,t)\eta(0,t) - u_x(L,t)\eta(L,t)\right] dt \qquad (6.13)$$

The equation (6.12) can then be written in the form

$$\frac{dI}{d\epsilon}\Big|_{\epsilon=0} = \int_{t_1}^{t_2} \int_0^L [-\varrho u_{tt} + \tau u_{xx}]\, \eta\, dx dt$$
$$- \int_{t_1}^{t_2} \left[\kappa_1 u(0,t) - \tau u_x(0,t)\right]\eta(0,t)\, dt$$
$$- \int_{t_1}^{t_2} \left[\kappa_2 u(L,t) + \tau u_x(L,t)\right]\eta(L,t)\, dt = 0$$

$\qquad (6.14)$

The equation (6.14) can now be assigned boundary conditions and analyzed.

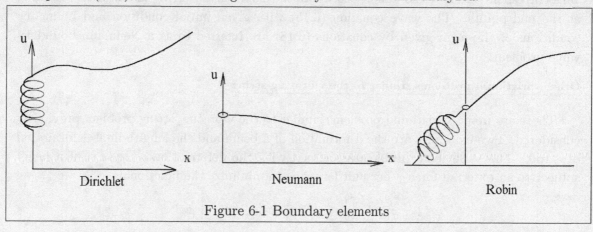

Figure 6-1 Boundary elements

(i) **Dirichlet boundary value problem**

Assume the ϱ and τ are constants. If $\eta(0,t) = 0$ and $\eta(L,t) = 0$ with $\eta(x,t)$ otherwise arbitrary we obtain the Euler-Lagrange equation

$$u_{tt}(x,t) = \alpha^2 u_{xx}(x,t), \qquad 0 \leq x \leq L, \quad t > 0 \text{ where } \alpha^2 = \tau/\varrho \tag{6.15}$$

which is known as the wave equation. The wave equation is subject to the initial condition that $u(x,0) = f(x)$ describes the initial shape of the string and the boundary conditions $u(0,t)$ and $u(L,t)$ are specified. The resulting boundary value problem is referred to as a Dirichlet boundary value problem.

(ii) **Robin boundary value problem**

If $\eta(0,t)$ and $\eta(L,t)$ are arbitrary, then in addition to the Euler-Lagrange equation (6.15) we require the natural boundary conditions

$$\begin{aligned}
\tau u_x(0,t) - \kappa_1 u(0,t) &= 0 \\
\tau u_x(L,t) + \kappa_2 u(L,t) &= 0
\end{aligned} \tag{6.16}$$

The normal vectors at the end of string are $\hat{n} = -\hat{\mathbf{e}}_1$ at $x = 0$ and $\hat{n} = +\hat{\mathbf{e}}_1$ at $x = L$ so that in terms of normal derivatives the above boundary conditions can be written in the form

$$\begin{aligned}
\frac{\partial u(0,t)}{\partial n} + h_1 u(0,t) &= 0, \qquad h_1 = \kappa_1/\tau \\
\frac{\partial u(L,t)}{\partial n} + h_2 u(L,t) &= 0, \qquad h_2 = \kappa_2/\tau
\end{aligned} \tag{6.17}$$

The wave equation (6.15) with given initial condition and boundary conditions of the form given by equations (6.17) is referred to as a Robin boundary value problem.

(iii) **Neumann boundary value problem**

In the special case where $\kappa_1 = 0$ and $\kappa_2 = 0$ the boundary conditions given by equation (6.17) reduce to the form

$$\frac{\partial u(0,t)}{\partial x} = 0 \quad \text{and} \quad \frac{\partial u(L,t)}{\partial x} = 0 \tag{6.18}$$

This corresponds to free end conditions where the slopes of the string maintain a zero slope at the end points. The wave equation (6.15) with given initial condition and boundary conditions of the form given by equations (6.18) are referred to as a Neumann boundary value problem.

Other variational problems similar to the vibrating string

There are many variational problems similar to the vibrating string problem previously considered. Two examples are the deformation of a beam and the longitudinal deformation of a rod. The variational problem associated with the deformation $u = u(x,t)$ of a beam subject to an external force f per unit length is to minimize the functional

$$I = \int_{t_1}^{t_2} \int_0^L \left[\frac{1}{2}\rho \left(u_t\right)^2 - \frac{1}{2}EI \left(u_{xx}\right)^2 + fu \right] dxdt$$

where E is Young's modulus of elasticity and I is the moment of inertia of the beam cross section. The variational problem associated with the deformation of $u = u(x,t)$ of a rod subject to an external force per unit length is to minimize the functional

$$I = \int_{t_1}^{t_2} \int_0^L \left[\frac{1}{2} \rho \left(u_t \right)^2 - \frac{1}{2} EA \left(u_x \right)^2 + fu \right] \, dxdt$$

where A is the cross sectional area of the rod with f the external force per unit length.

The vibrating membrane

The equations of motion for a vibrating membrane is developed in a manner which is very similar to our previous development for the equations of motion of the vibrating string. Let $u = u(x,y,t)$ denote the displacement of a thin elastic membrane from its equilibrium position $u = 0$. The thin membrane extends over a region R which is bounded by a simple closed curve $C = \partial R$. The boundary curve C is assumed to be fixed. Let $\varrho = \varrho(x,y)$ $[kg/m^2]$ denote the mass per unit area of the membrane. The kinetic energy per unit area of the membrane is given by $\frac{1}{2} \varrho \left(\frac{\partial u}{\partial t} \right)^2$ $[kg/m^2][m/sec]^2 = [Nt\,m]/m^2$. This is summed over the area R to obtain the total kinetic energy

$$T = \frac{1}{2} \iint_R \varrho \left(\frac{\partial u}{\partial t} \right)^2 \, dxdy \qquad [Nt \cdot m] = [joule] \tag{6.19}$$

The potential energy is composed of two parts. The first part represents the work needed to move the membrane from its equilibrium position $u = 0$ to some shape defined by $u = u(x,y,t)$ and the second part represents the work done in stretching the boundary from its fixed position. For the first part of the potential energy assume the membrane experiences a constant surface tension per unit of length which is given by τ $[Nt/m]$. The work done in moving an element of area $dA = dxdy$ in the equilibrium position to an element of area dS on the surface is given by

$$dV_1 = \tau \left(dS - dA \right) \quad [Nt/m][m^2] = [Nt \cdot m] \tag{6.20}$$

Recall that if $\vec{r} = x\,\widehat{\mathbf{e}}_1 + y\,\widehat{\mathbf{e}}_2 + u\,\widehat{\mathbf{e}}_3$ is a position vector to a point on the surface, then the differential of $\vec{r}$ is given by $d\vec{r} = \frac{\partial \vec{r}}{\partial x} \, dx + \frac{\partial \vec{r}}{\partial y} \, dy$ which represents a small change on the surface. The components $\frac{\partial \vec{r}}{\partial x} \, dx$ and $\frac{\partial \vec{r}}{\partial y} \, dy$ of this vector are the sides of an element of surface area dS. This element of area is given by

$$dS = \left| \frac{\partial \vec{r}}{\partial x} \, dx \times \frac{\partial \vec{r}}{\partial y} \, dy \right| = \sqrt{ \left(\frac{\partial \vec{r}}{\partial x} \cdot \frac{\partial \vec{r}}{\partial x} \right) \left(\frac{\partial \vec{r}}{\partial y} \cdot \frac{\partial \vec{r}}{\partial y} \right) - \left(\frac{\partial \vec{r}}{\partial x} \cdot \frac{\partial \vec{r}}{\partial y} \right)^2 } \, dxdy = \sqrt{ 1 + \left(\frac{\partial u}{\partial x} \right)^2 + \left(\frac{\partial u}{\partial y} \right)^2 } \, dxdy \tag{6.21}$$

The element of work done in moving dA to dS can therefore be expressed in the form

$$dV_1 = \tau \left(\sqrt{ 1 + u_x^2 + u_y^2 } - 1 \right) dxdy \tag{6.22}$$

We make the assumption that the slopes $\left| \frac{\partial u}{\partial x} \right|$ and $\left| \frac{\partial u}{\partial y} \right|$ are small so that we can make the approximation

$$\sqrt{ 1 + u_x^2 + u_y^2 } = 1 + \frac{1}{2}(u_x^2 + u_y^2) - \frac{1}{4}(u_x^2 + u_y^2)^2 + \cdots \tag{6.23}$$

If we neglect terms $(u_x^2 + u_y^2)^m$ for $m \geq 2$ in equation (6.23), then the potential energy of moving an element of area can be represented in the form

$$dV_1 = \frac{1}{2}\tau \left(u_x^2 + u_y^2\right) dxdy \qquad (6.24)$$

Summing over the region R gives the potential energy function

$$V_1 = \frac{1}{2}\iint_R \tau \left(u_x^2 + u_y^2\right) dxdy \qquad [Nt \cdot m] \qquad (6.25)$$

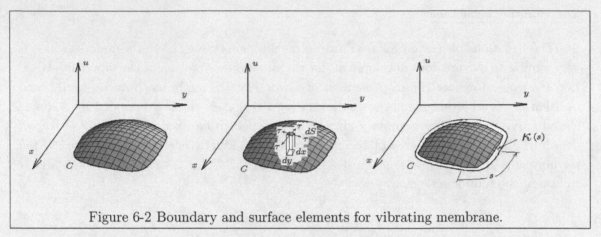

Figure 6-2 Boundary and surface elements for vibrating membrane.

The stretching of the membrane along the boundary is treated in a manner similar to that of the stretched string. Assume that each point of the boundary has associated with it a spring constant $\kappa(s)$ which varies with arc length s measured from some reference point on the boundary. The spring force is assumed to be linear and consequently the total work done in stretching the boundary can be written as a sum of the forces around the bounding curve C. Following the example of the vibrating string we construct the potential function

$$V_2 = \frac{1}{2}\int_C \kappa(s)u^2(s,t)\,ds$$

where s is an element of arc length and $u(s,t)$ denotes the displacement of a boundary point from its equilibrium position.

The Lagrangian can then be written $L = T - V_1 - V_2$ and the Hamiltonian can be expressed as

$$I = \int_{t_1}^{t_2} L\,dt = \int_{t_1}^{t_2}\iint_R \left(\frac{1}{2}\varrho u_t^2 - \frac{1}{2}\tau\left[u_x^2 + u_y^2\right]\right) dxdydt - \frac{1}{2}\int_{t_1}^{t_2}\int_c \kappa(s)u^2(s,t)\,dt \qquad (6.26)$$

Hamilton's principle requires that the Hamiltonian be minimized. To find the function $u = u(x,y,t)$ which minimizes the integral given by equation (6.26) we assume that $u(x,y,t)$ minimizes I and then consider the family of comparison functions given by the variation $U = U(x,y,t) = u(x,y,t) + \epsilon\eta(x,y,t)$. Here we assume that the conditions $\eta(x,y,t_1) = 0$ and $\eta(x,y,t_2) = 0$ are satisfied at the end points of the arbitrary time interval. Substituting the comparison function into the integral (6.26) gives

$$I = I(\epsilon) = \int_{t_1}^{t_2}\iint_R \left(\frac{1}{2}\varrho U_t^2 - \frac{1}{2}\tau\left[U_x^2 + U_y^2\right]\right) dxdydt - \frac{1}{2}\int_{t_1}^{t_2}\int_c \kappa(s)U^2(s,t)\,dt \qquad (6.27)$$

which is to have a minimum at $\epsilon = 0$. A necessary condition for a minimum is that $\frac{dI}{d\epsilon}\Big|_{\epsilon=0} = 0$. Differentiate the equation (6.27) with respect to ϵ to obtain the equation

$$\frac{dI}{d\epsilon}\Big|_{\epsilon=0} = \int_{t_1}^{t_2} \iint_R \{\varrho u_t \eta_t - \tau[u_x \eta_x + u_y \eta_y]\}\,dxdydt - \int_{t_1}^{t_2} \kappa(s)u(s,t)\eta\,ds = 0 \tag{6.28}$$

One can employ the derivatives

$$\frac{\partial}{\partial t}(u_t \eta) = u_t \eta_t + u_{tt}\eta\ , \qquad \frac{\partial}{\partial x}(u_x \eta) = u_x \eta_x + u_{xx}\eta\ , \qquad \frac{\partial}{\partial y}(u_y \eta) = u_y \eta_y + u_{yy}\eta$$

to write equation (6.28) in the form

$$\frac{dI}{d\epsilon}\Big|_{\epsilon=0} = \int_{t_1}^{t_2} \iint_R [-\varrho u_{tt} + \tau(u_{xx} + u_{yy})]\,\eta\,dxdydt - \int_{t_1}^{t_2} \kappa(s)u(s,t)\,\eta\,ds$$
$$- \int_{t_1}^{t_2} \iint_R \tau\left[\frac{\partial}{\partial x}(u_x \eta) + \frac{\partial}{\partial y}(u_y \eta)\right]dxdydt + \int_{t_1}^{t_2} \frac{\partial}{\partial t}(u_t \eta)\,dxdydt = 0 \tag{6.29}$$

The last integral in equation (6.29) can be integrated with respect to time t with the result being zero because of the assumption that $\eta(x,y,t_1) = 0$ and $\eta(x,y,t_2) = 0$ at the end points of the arbitrary time interval. The two-dimensional form of the Green's theorem can then be employed to replace the second to last integral in equation (6.29) in terms of a line integral. For τ constant, one can verify that

$$\iint_R \left[\frac{\partial}{\partial x}(u_x \eta) + \frac{\partial}{\partial y}(u_y \eta)\right]dxdy = \int_C (u_x \eta\,dy - u_y \eta\,dx) = \int_C \frac{\partial u}{\partial n}\eta ds \tag{6.30}$$

where

$$\frac{\partial u}{\partial n} = \text{grad}\,u \cdot \hat{n} = \left(\frac{\partial u}{\partial x}\,\hat{e}_1 + \frac{\partial u}{\partial y}\,\hat{e}_2\right) \cdot \left(\frac{dy}{ds}\,\hat{e}_1 - \frac{dx}{ds}\,\hat{e}_2\right) \tag{6.31}$$

That is, if $\vec{r} = x\,\hat{e}_1 + y\,\hat{e}_2$ defines the boundary curve C for the membrane, then the unit normal to the boundary is given by $\hat{n} = \frac{dy}{ds}\,\hat{e}_1 - \frac{dx}{ds}\,\hat{e}_2$.

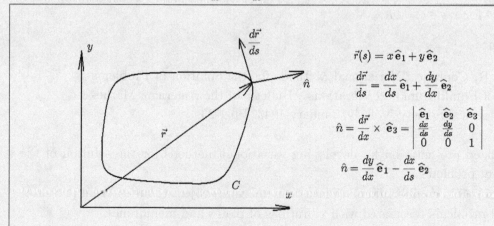

Figure 6-3 Boundary curve C with unit tangent and normal vectors to boundary.

The equation (6.29) can then be simplified to the form

$$\frac{dI}{d\epsilon}\bigg|_{\epsilon=0} = \int_{t_1}^{t_2} \iint_R \left[-\varrho u_{tt} + \tau\left(u_{xx} + u_{yy}\right)\right] \eta\, dxdydt - \int_{t_1}^{t_2} \int_C \left(\kappa u + \tau\frac{\partial u}{\partial n}\right)\eta\, dsdt = 0 \qquad (6.32)$$

which can now be analyzed.

(i) **Fixed boundary conditions**

If $\eta(x,y,t) = 0$ for all $x,y \in C$ and all times t, then the last term in equation (6.32) vanishes. We can then employ a form of the basic lemma to obtain the Euler-Lagrange equation

$$\varrho u_{tt} = \tau\left(u_{xx} + u_{yy}\right) \qquad \text{or} \qquad u_{tt} = \alpha^2\left(u_{xx} + u_{yy}\right), \qquad \alpha^2 = \tau/\varrho \qquad (6.33)$$

This equation is called the two-dimensional wave equation. It is to be solved subject to the initial condition $u(x,y,0) = f(x,y)$ which defines the initial shape of the membrane. Here $\eta = 0$ along the boundary and therefore the boundary condition is assumed to be specified, say $u(x,y,t) = G(x,y,t)$ for $x,y \in C$.

(ii) **Elastic boundary condition**

If η is arbitrary, then the natural boundary conditions from the boundary term occurring in equation (6.32) requires that

$$\kappa(s)u(s,t) + \tau\frac{\partial u}{\partial n} = 0 \qquad \text{for} \quad s \in C = \partial R \qquad (6.34)$$

This type of boundary condition is called an elastic boundary condition.

(iii) **Free and fixed boundary condition**

If $\kappa(s) = 0$, then a free boundary condition is said to exist. This requires that $\frac{\partial u}{\partial n} = 0$ for $x,y \in C = \partial R$. The other extreme is for $\kappa(s)$ to increase without bound. Dividing equation (6.34) by κ and then letting κ increase without bound produces the fixed boundary condition that $u(x,y,t) = 0$ for $x,y \in C = \partial R$.

The reference

R. Courant, "Variational Methods for the Solution of Problems of Equilibrium and Vibrations", Bulletin of the American Mathematical Society, Vol. 49, January 1943, Pp 1-23.

gives a generalized presentation for developing variational methods for the solution of the following type of problems.

(i) Problems of stable equilibrium of a plate or membrane subject to an external pressure

(ii) Variational problems associated with vibrations of plates and membranes.

The reader should have some basic background knowledge from the theory of elasticity before consulting the above reference.

Generalized brachistrochrone problem

A generalization of the brachistrochrone problem is to develop the equations of motion of a particle which is acted upon by a conservative force system. Find the path of motion such that the particle moves from point P_1 to a point P_2 in the shortest time. Another form of this problem is to find the path or curve from one surface to another surface such that the particle moves in the shortest time. All solution paths or curves which represent solutions to the above type of problems are called brachistrochrones. Let $PE = m\phi(x, y, z)$ denote the potential energy of the force system and let $KE = \frac{1}{2}mv^2$ denote the kinetic energy of the particle where m is the mass of the particle, v denotes the velocity of the particle and ϕ is the potential energy per unit mass. The force per unit mass acting on the particle is

$$\vec{F} = -\text{grad}\,\phi = -\frac{\partial \phi}{\partial x}\,\widehat{\mathbf{e}}_1 - \frac{\partial \phi}{\partial y}\,\widehat{\mathbf{e}}_2 - \frac{\partial \phi}{\partial z}\,\widehat{\mathbf{e}}_3$$

and the components represent the rate of decrease of the potential energy because of the minus sign. The total energy E of the particle is the sum of the kinetic energy and potential energy and can be written

$$E = KE + PE = \frac{1}{2}mv^2 + m\phi = \text{Constant}$$

Holonomic and nonholonomic systems

Dynamical systems are classified as scleronomic if there are constraint equations independent of time t of the form $f_i(q^1, q^2, \ldots, q^n) = 0$ where q^i, $i = 1, 2, \ldots, n$ are generalize coordinates. The system is classified as rheonomic if the time variable t is needed for the representation of the constraint conditions so that one would have constraint equations of the form $f_i(q^1, q^2, \ldots, q^n, t) = 0$. If there exists constraint equations of the form $f_j(q^1, \ldots, q^n, \dot{q}^1, \ldots, \dot{q}^n, t) = 0$ which are all integrable or constraint equations exist of the form $f_j(q^1, q^2, \ldots, q_n, t) = 0$ for $j = 1, 2, \ldots, m$, then a definite relation can be found between the generalized coordinates and so the system is termed a holonomic system. If at least one constraint equation is not integrable or a definite relationship is not specified by any of the constraint equations, then the system is said to be nonholonomic.

An example of a holonomic constraint would be that of a particle constrained to move on a given surface $G(x, y, z) = 0$. Under such conditions one of the coordinate unknowns, either x, y or z can be eliminated from the problem. That is, it is theoretically possible to solve the equation $G(x, y, z) = 0$, for say the variable z as a function of x and y. Constraints which can not be integrated to find a definite relation between the variables or where inequalities are involve, then these type of constraints are called nonholonomic constraints. As an example of such a constraint consider the motion of gas particles moving inside a cylinder container of radius r_0. The radial position r of a gas molecule is subject to the constraint condition $r \le r_0$. Another example of a nonholonomic constraint is that of a particle released on the top of a sphere. Here the particle will roll on the surface of a sphere, due to the gravitational force acting on the particle, and then it eventually falls off the sphere. If r_0 is the radius of the sphere, then the constraint condition can be expressed in the form $r^2 - r_0^2 \geq 0$, where r is the radial position of the particle from the center of the coordinate system defining the sphere.

Example 6-1. Rolling cylinders

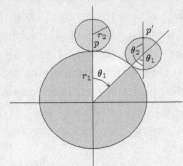

Consider a cylinder of radius r_2 and mass m which rolls without slipping on the surface of another fixed cylinder of radius $r_1 > r_2$. At some point the outer cylinder falls off the inner cylinder. Find the equations of motion of the outer cylinder. Find the angle θ_1^* at which the cylinders separate.

Solution: Introduce as generalized coordinates the angle θ_1 measured clockwise and the angle θ_2 measured counterclockwise from the vertical lines through the center of the cylinders together with the generalized coordinate r representing the distance of the center of the smaller cylinder from the origin of the larger cylinder. The angles θ_1 and θ_2 are illustrated in the above sketch of the problem which illustrates the point p moving to the point p' as θ_1 changes. The constraint condition that the smaller cylinder rolls without slipping can be represented by the constraint equation that

$$s = \text{arc length} = r_1\theta_1 = r_2(\theta_2 - \theta_1) \tag{6.35}$$

which states that the arc lengths swept out by the point p must be the same on both cylinders. The further constraint

$$r = r_1 + r_2 = \text{a constant} \tag{6.36}$$

states that the smaller cylinder remains on the larger cylinder as it rolls. These constraint conditions can be represented

$$\begin{aligned} G_1 &= (r_1 + r_2)\theta_1 - r_2\theta_2 = 0 \\ G_2 &= \quad r - (r_1 + r_2) = 0 \end{aligned} \tag{6.37}$$

The kinetic energy of the smaller cylinder can be broken up into a translational component involving the change in arc length $s = r\theta_1$ and change in radial distance r together with a rotational component $\frac{1}{2}I\omega^2$ involving the moment of inertial $I = \frac{1}{2}mr_2^2$ of the smaller cylinder and angular velocity $\omega = \dot{\theta}_2$. The kinetic energy T can then be written as

$$T = \underbrace{\frac{1}{2}m\left(\dot{r}^2 + r^2\dot{\theta}_1^2\right)}_{\text{translational}} + \underbrace{\frac{1}{2}\left[\frac{1}{2}mr_2^2\right]\dot{\theta}_2^2}_{\text{rotational}} \tag{6.38}$$

The potential energy can be represented

$$V = mgh + \text{constant} = mgr\cos\theta_1 + \text{constant} \tag{6.39}$$

The Lagrangian $L = T - V$ can then be formed. Using the method of Lagrange multipliers it is customary to modify the integrand of a functional to involve the constraint conditions. There are two constraint conditions and so we introduce two Lagrange multipliers λ_1 and λ_2 and form the integrand

$$F = F(r, \theta_1, \theta_2, \dot{r}, \dot{\theta}_1, \dot{\theta}_2) = L + \lambda_1 G_1 + \lambda_2 G_2 \tag{6.40}$$

and require that the $\int_{t_1}^{t_2} F\, dt$ be an extremum. This is a modification of Hamilton's principle using Lagrange multipliers to bring the constraint conditions into the problem. The integrand F can therefore be represented in the form

$$F = \frac{1}{2}m(\dot{r}^2 + r^2\dot{\theta}_1^2) + \frac{1}{4}mr_2^2\dot{\theta}_1^2 - mgr\cos\theta_1$$
$$+ \lambda_1[(r_1 + r_2)\theta_1 - r_2\theta_2] + \lambda_2[r - (r_1 + r_2)]. \tag{6.41}$$

Here the Lagrange multipliers λ_1 and λ_2 are given the physical interpretation of representing forces needed to maintain the constraint conditions. The Lagrange multiplier λ_1 is called the no-slip force and the Lagrange multiplier λ_2 is called the restraining force which prevents the rolling cylinder from leaving the larger cylinder. It is left as an exercise to verify that the Euler-Lagrange equations of motion which describes the rolling of the smaller cylinder upon the larger cylinder can be represented

$$\frac{\partial F}{\partial r} - \frac{d}{dt}\left(\frac{\partial F}{\partial \dot{r}}\right) = 0 \quad\Rightarrow\quad m\ddot{r} - mr\dot{\theta}_1^2 + mg\cos\theta_1 - \lambda_2 = 0 \tag{6.42}$$

$$\frac{\partial F}{\partial \theta_1} - \frac{d}{dt}\left(\frac{\partial F}{\partial \dot{\theta}_1}\right) = 0 \quad\Rightarrow\quad mr^2\ddot{\theta}_1 + 2mr\dot{r}\dot{\theta}_1 - mgr\sin\theta_1 - \lambda_1(r_1 + r_2) = 0 \tag{6.43}$$

$$\frac{\partial F}{\partial \theta_2} - \frac{d}{dt}\left(\frac{\partial F}{\partial \dot{\theta}_2}\right) = 0 \quad\Rightarrow\quad \frac{1}{2}mr_2^2\ddot{\theta}_2 + r_2\lambda_1 = 0 \tag{6.44}$$

The generalized coordinate r represents the distance of the center of the smaller cylinder from the origin of the larger cylinder. This distance remains constant while the smaller cylinder rolls on the surface of the larger cylinder. Consequently, both $\dot{r}$ and $\ddot{r}$ are zero. This reduces the equation (6.42) to the form

$$\lambda_2 = -mr\dot{\theta}_1^2 + mg\cos\theta_1. \tag{6.45}$$

The equations (6.37) show that

$$\theta_2 = \left(\frac{r_1 + r_2}{r_2}\right)\theta_1 \quad\text{so that}\quad \ddot{\theta}_2 = \left(\frac{r_1 + r_2}{r_2}\right)\ddot{\theta}_1 \tag{6.46}$$

which simplifies the equation (6.44) to the form

$$\lambda_1 = -\frac{1}{2}mr_2\ddot{\theta}_2 = -\frac{1}{2}mr_2\left(\frac{r_1 + r_2}{r_2}\right)\ddot{\theta}_1 = -\frac{1}{2}m(r_1 + r_2)\ddot{\theta}_1 \tag{6.47}$$

During the time interval when $r = r_1 + r_2$ remains constant, the equation (6.43) simplifies to the form

$$\lambda_1 = -mg\sin\theta_1 + mr\ddot{\theta}_1. \tag{6.48}$$

Equate the equations (6.48) and (6.47) to obtain

$$\frac{3}{2}r\ddot{\theta}_1 - g\sin\theta_1 = 0. \tag{6.49}$$

Now multiply equation (6.49) by $\dot{\theta}_1$ and integrate to find that θ_1 must satisfy

$$\frac{3}{4}r\dot{\theta}_1^2 + g\cos\theta_1 = E = \text{a constant} \tag{6.50}$$

Assume initial conditions for θ_1 that at time $t = 0$, we have $\theta_1 = 0$ and $\dot{\theta}_1 = 0$. This gives the constant $E = g$ and consequently the equation (6.50) can be written in the form

$$\dot{\theta}_1^2 = \frac{4g(1 - \cos\theta_1)}{3r} \tag{6.51}$$

At the point where the smaller cylinder rolls off of the larger cylinder the work-energy principle from physics requires that the work done equal the change in kinetic energy or

$$mgr(1 - \cos\theta_1) = \frac{1}{2}mr^2\dot{\theta}_1^2 + \frac{1}{4}mr_2^2\dot{\theta}_2^2. \tag{6.52}$$

Here θ_2 is related to θ_1 by the equation (6.46) and so the work-energy relation, given by equation (6.52) can also be written in the form

$$g(1 - \cos\theta_1) = \frac{1}{2}r\dot{\theta}_1^2 + \frac{1}{4}r\dot{\theta}_1^2 = \frac{3}{4}r\dot{\theta}_1^2$$

which is the same result that we obtained in equation (6.51).

The smaller cylinder rolls on the larger cylinder up to a point where separation occurs. At separation we require that $\lambda_2 = 0$. Setting $\lambda_2 = 0$ in equation (6.45) requires that at separation

$$\dot{\theta}_1^2 = \frac{g}{r}\cos\theta_1 \tag{6.53}$$

Now equate the equations (6.53) and (6.51) to obtain

$$\dot{\theta}_1^2 = \frac{g}{r}\cos\theta_1 = \frac{4g}{3}\frac{(1 - \cos\theta_1)}{r}$$

with $r = r_1 + r_2$ constant at point of separation. This equation can now be solved for the separation angle θ_1. Performing the necessary algebra one finds that at separation

$$\theta_1 = \cos^{-1}(4/7) = 0.962551 \text{ radians} \approx 55.15 \text{ degrees}.$$

■

Eigenvalues and eigenfunctions

A linear second order differential equation having the form

$$L(y) = \frac{d}{dx}\left[p(x)\frac{dy}{dx}\right] - q(x)y = -\lambda w(x)y, \qquad a \le x \le b \tag{6.54}$$

is called a Sturm-Liouville differential equation. In equation (6.54) $L(\ )$ is a linear differential operator, λ is a parameter and the coefficients $p(x)$, $q(x)$ and $w(x)$ are real-valued continuous and bounded functions of x over the interval $a \le x \le b$. In addition $p(x)$ is assumed to be continuous and differentiable with the requirement that $p(x) > 0$ and $w(x) > 0$ for $a \le x \le b$. The Sturm-Liouville equation (6.54) is called regular whenever $p(x)$ and $w(x)$ are well defined and different from zero for $a \le x \le b$. In the cases where $p(x)$ or $q(x)$ become undefined over the interval $a \le x \le b$, then the Sturm-Liouville equation is said to be singular.

A Sturm-Liouville system is the differential equation (6.54) subject to end point conditions. One form for the boundary conditions is to require that the solutions of the Sturm-Liouville equation satisfy the end conditions

$$c_1 y'(a) + c_2 y(a) = 0$$
$$c_3 y'(b) + c_4 y(b) = 0$$

(6.55)

In the boundary conditions (6.55) the coefficients c_1, c_2, c_3, c_4 are real constants not all zero. The special cases where $c_1 = c_2 = 0$ or $c_3 = c_4 = 0$ lead to the trivial solution and are not considered. If boundary conditions are of the form $y(a) = y(b)$ and $y'(a) = y'(b)$, then they are called periodic end point conditions.

Definitions

Values of λ for which a Sturm-Liouville system has a nonzero solution are called eigenvalues and the corresponding nonzero solution is called an eigenfunction. The set of all eigenvalues of a regular Sturm-Liouville system is called the spectrum of the system.

Sturm-Liouville systems have the following properties which are stated without proofs. For proofs of these properties the reader should see the reference Birkhoff and Rota.

Properties of a Sturm-Liouville system

1. A regular Sturm-Liouville system has an infinite sequence of real discrete eigenvalues which can be ordered

$$\lambda_1 < \lambda_2 < \lambda_3 < \cdots < \lambda_n < \cdots$$

where λ_n increases without bound as n increases.

2. Corresponding to each eigenvalue λ_n there is a solution $y_n(x) = y(x; \lambda_n)$ called an eigenfunction.

3. The set of eigenfunctions $\{y_n(x)\}$, for $n = 1, 2, 3, \ldots$ are orthogonal over the interval (a, b) with respect to the weight function $w(x)$ and satisfy the inner product relation

$$(y_n, y_m) = \int_a^b w(x) y_n(x) y_m(x)\, dx = \begin{cases} 0, & \text{if } n \ne m \\ \| y_m \|^2, & \text{if } n = m \end{cases}$$

where the special inner product

$$(y_n, y_n) = \| y_n \|^2 = \int_a^b w(x) y_n^2(x)\, dx, \quad n = 1, 2, 3, \ldots$$

is called a norm squared.

4. If the eigenfunctions are scaled such that $\| y_n \|^2 = 1$, for $n = 1, 2, 3, \ldots$, then the eigenfunctions are called orthonormal.

5. The nth eigenfunction $y_n(x)$ has $(n - 1)$ zeros in the interval $a \leq x \leq b$.

Alternate form for Sturm-Liouville system

Consider the problem to represent the Sturm-Liouville system

$$L(y) = \frac{d}{dx}\left[p(x)\frac{dy}{dx} \right] - q(x)y = -\lambda w(x)y, \qquad a \leq x \leq b$$
$$y(a) = 0 \qquad y(b) = 0$$

(6.56)

in a variational form. Toward this goal we first consider the variational problem to find an extremum for the functional

$$I = \int_a^b \left\{ p(x)\left(y'\right)^2 + q(x)y^2 \right\} dx + \alpha_1 y^2 \bigg|_{x=a} + \alpha_2 y^2 \bigg|_{x=b}$$

(6.57)

where α_1, α_2 are constants. The solution $y = y(x)$ is to be subject to the constraint condition

$$\int_a^b w(x)y^2(x)\, dx = 1.$$

(6.58)

To solve this variational problem we introduce a Lagrange multiplier $-\lambda$ and formulate the associated functional

$$H = \int_a^b \left\{ p(x)\left(y'\right)^2 + q(x)y^2 - \lambda w(x)y^2 \right\} dx + \alpha_1 y^2 \bigg|_{x=a} + \alpha_2 y^2 \bigg|_{x=b}$$

(6.59)

The function $y = y(x)$ which makes the functional (6.59) an extremum must be such that the first variation δH is zero. This requires that

$$\delta H = \int_a^b \left\{ 2p(x)y'\delta y' + 2q(x)y\delta y - 2\lambda w(x)y\delta y \right\} dx + 2\alpha_1 y\delta y \bigg|_{x=a} + 2\alpha_2 y\delta y \bigg|_{x=b} = 0$$

(6.60)

Now use integration by parts on the first term under the integral in equation (6.60) to obtain

$$\delta H = \int_a^b 2\left\{ -\frac{d}{dx}\left(p(x)y'\right) + [q(x)y - \lambda w(x)y] \right\} \delta y\, dx + 2p(x)y'\delta y \bigg|_a^b + 2\alpha_1 y\delta y \bigg|_{x=a} + 2\alpha_2 y\delta y \bigg|_{x=b} = 0 \quad (6.61)$$

The fundamental lemma gives us the Euler-Lagrange equation

$$\frac{d}{dx}\left(p(x)y'\right) - q(x)y + \lambda w(x)y = 0$$

(6.62)

with the essential boundary conditions

$$y(a) = 0, \quad \text{and} \quad y(b) = 0$$

(6.63)

and natural boundary conditions

$$p(x)y'(x) + \alpha_2 y(x) \bigg|_{x=b} = 0, \quad \text{and} \quad p(x)y'(x) - \alpha_1 y(x) \bigg|_{x=a} = 0$$

(6.64)

Observe that the differential equation (6.62) and boundary condition (6.63) can be replaced by the equivalent variational problem given by equations (6.57) and (6.58).

Also note that if we multiply equation (6.62) by $y\,dx$ and then integrate this equation from a to b, one obtains

$$\int_a^b \frac{d}{dx}\left(p(x)y'\right)y\,dx - \int_a^b q(x)y^2\,dx = -\lambda\int_a^b w(x)y^2\,dx \tag{6.65}$$

Now perform integration by parts on the first integral in equation (6.65) to obtain

$$p(x)yy'\big]_a^b - \int_a^b p(x)y'^2\,dx - \int_a^b q(x)y^2\,dx = -\lambda\int_a^b w(x)y^2\,dx \tag{6.66}$$

The essential boundary conditions and the constraint condition simplifies the equations (6.66) and (6.57) to the form

$$I = \int_a^b \left\{p(x)(y')^2 + q(x)y^2\right\}dx = \lambda \tag{6.67}$$

The variational formulation given by the equations (6.57) and (6.58) are often times used in place of the Sturm-Liouville system (6.56) in order to generate approximate solutions. We will investigate approximate solutions later in this chapter.

Schrödinger equation

We begin with the De Broglie hypothesis (1924) that there is a dual wave-particle character associated with atomic particles such as electrons, protons, atoms and even molecules. The energy E (ergs) assigned to a particle of radiant energy or photon is given by

$$E = h\nu \tag{6.68}$$

where h is Planck's constant $(h = 6.6252\,(10)^{-27}\text{ erg sec})$ and ν (sec^{-1}) is the frequency of the radiation. The frequency and wavelength λ (cm) of the radiation determine the wave velocity v (cm/sec) from the relation

$$v = \lambda\nu \tag{6.69}$$

The momentum p of the particle having velocity v can be represented in either of the forms

$$p = mv = m\lambda\nu = \frac{h}{\lambda} \qquad (erg\,sec/cm) \tag{6.70}$$

Recall that wave propagation in three-dimensions is represented by the wave equation

$$\nabla^2 U = \frac{\partial^2 U}{\partial x^2} + \frac{\partial^2 U}{\partial y^2} + \frac{\partial^2 U}{\partial z^2} = \frac{1}{v^2}\frac{\partial^2 U}{\partial t^2} \tag{6.71}$$

where U is called the wave function, t is time (sec), v (cm/sec) is the wave velocity and x, y, z have units of length (cm). Typical examples of the wave equation occur when U represents a component of the electric vector $\vec{\mathcal{E}}$ or magnetic vector $\vec{\mathcal{M}}$ propagating through space as represented by Maxwell's equations. Other examples of where the wave equation occurs are modeling of a vibrating string, a vibrating drum head and spherical sound waves.

The wave equation can be represented in various forms. In the case of a single particle of mass m with velocity v which is treated as a conservative system, the kinetic energy T of the particle can be written in terms of the difference of the total energy E and potential energy V using the relation

$$T = \frac{1}{2}mv^2 = E - V \tag{6.72}$$

This is because $T + V = E$ for a conservative system. Consequently, one can solve for the velocity v in equation (6.72) and express the particle momentum in the form

$$p = mv = \sqrt{2m(E - V)} \tag{6.73}$$

The equations (6.69), (6.70), and (6.73) can then be employed to express the wave velocity in the form

$$v = \lambda \nu = \frac{\lambda}{h}(h\nu) = \frac{E}{p} = \frac{h\nu}{p} = \frac{h\nu}{\sqrt{2m(E - V)}} \tag{6.74}$$

We substitute this velocity into the wave equation (6.71) to obtain

$$\nabla^2 U = \frac{\partial^2 U}{\partial x^2} + \frac{\partial^2 U}{\partial y^2} + \frac{\partial^2 U}{\partial z^2} = \frac{2m(E - V)}{h^2 \nu^2} \frac{\partial^2 U}{\partial t^2} \tag{6.75}$$

Assume a solution to equation (6.75) having the form

$$U = u\,e^{2\pi\nu t}, \qquad u = u(x, y, z) \tag{6.76}$$

where u is a function of position only. Substitute this assumed solution into the wave equation (6.75) and simplify the results to obtain

$$\nabla^2 u = \frac{\partial^2 u}{\partial x^2} + \frac{\partial^2 u}{\partial y^2} + \frac{\partial^2 u}{\partial z^2} = -\frac{8\pi^2 m}{h^2}(E - V)\,u \tag{6.77}$$

which is one form of Schrödinger's wave equation for a single particle. Solutions u of equation (6.77) for a given potential function $V = V(x, y, z)$ are called wave functions.

Observe that the solution u of the Schrödinger equation (6.77) can be viewed as a function which makes the volume integral

$$I = \iiint_V \left[\left(\frac{\partial u}{\partial x}\right)^2 + \left(\frac{\partial u}{\partial y}\right)^2 + \left(\frac{\partial u}{\partial z}\right)^2 + \frac{8\pi^2 m}{h^2}(V - E)\,u^2 \right] dx\,dy\,dz \tag{6.78}$$

an extremum. The solutions u of Schrödinger's equation represent probabilities and so are selected to satisfy the normalization condition

$$\iiint_V u\,u^*\,dx\,dy\,dz = 1 \tag{6.79}$$

where u^* represents the conjugate of u.

Solutions of the Schrödinger equation for the hydrogen atom

The hydrogen atom consists of a nucleus with a single electron. The nucleus and electron rotate about the center of mass of the atom and we consider the motion of the electron relative to this point. Based upon the center of mass coordinates we introduce spherical coordinates (r, θ, ϕ) as illustrated in the figure 6-4.

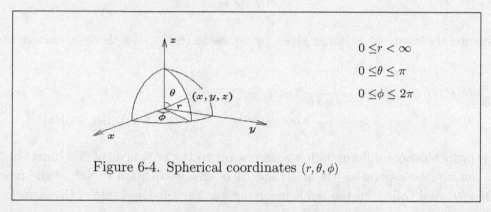

$$0 \leq r < \infty$$
$$0 \leq \theta \leq \pi$$
$$0 \leq \phi \leq 2\pi$$

Figure 6-4. Spherical coordinates (r, θ, ϕ)

We can generalize our calculations of the Schrödinger equation by considering an atom or ion which has only one electron. In this case the potential energy is treated as that of a point charge associated with a spherically symmetric potential and is determined from Coulomb's law as

$$V = V(r) = -\frac{Ze^2}{4\pi\epsilon_0 \, r} \tag{6.80}$$

where $+Ze$ is the charge of the nucleus, with $-e$ the charge of the electron and r is the distance between the nucleus and the electron. The constant $\epsilon_0 = 8.9\,(10^{-12})\,\left(C^2/N\,m^2\right)$ is a proportionality constant called the permitivity of free space. Here Z denotes the number of charged nucleons in the atom with $Z = 1$ representing the hydrogen atom. We express the Laplacian operator $\nabla^2 u$ in spherical coordinates (r, θ, ϕ) and write the Schrödinger equation (6.77) in the spherical coordinate form

$$\nabla^2 u = \frac{1}{r^2}\frac{\partial}{\partial r}\left(r^2\frac{\partial u}{\partial r}\right) + \frac{1}{r^2\sin\theta}\frac{\partial}{\partial\theta}\left(\sin\theta\frac{\partial u}{\partial\theta}\right) + \frac{1}{r^2\sin^2\theta}\frac{\partial^2 u}{\partial\phi^2} = -\frac{2m_e}{\hbar^2}(E - V(r))u \tag{6.81}$$

where $\hbar = h/2\pi$ and m_e is the mass of the electron. To solve this equation we employ the method of separation of variables and assume a solution having the form

$$u = u(r, \theta, \phi) = F(r)G(\theta)H(\phi) \tag{6.82}$$

which, in spherical coordinates, is required to satisfy the normalization condition

$$\int_0^\infty \int_0^\pi \int_0^{2\pi} u\,u^*r^2\sin\theta\,d\phi d\theta\,dr = 1 \tag{6.83}$$

Later we will find that F and G are real functions and H is a complex function. This implies that F, G and H must be selected to satisfy the normalization conditions

$$
\begin{aligned}
\int_0^\infty F^2(r)r^2\,dr &= 1 \\
\int_0^\pi G^2(\theta)\sin\theta\,d\theta &= 1 \\
\int_0^{2\pi} H(\phi)H^*(\phi)\,d\phi &= 1
\end{aligned}
\tag{6.84}
$$

We substitute the assumed solution given by equation (6.82) into the equation (6.81) to obtain

$$
\frac{1}{r^2}\frac{\partial}{\partial r}\left(r^2 F'(r)\right)G(\theta)H(\phi) + \frac{1}{r^2\sin\theta}\frac{\partial}{\partial\theta}\left(\sin\theta G'(\theta)\right)F(r)H(\phi)
$$
$$
+ \frac{1}{r^2\sin^2\theta}F(r)G(\theta)H''(\phi) = -\frac{2m_e}{\hbar^2}(E-V(r))F(r)G(\theta)H(\phi)
\tag{6.85}
$$

where the primes denote differentiation with respect to the argument of the function. The variables can now be separated one at a time. If we divide equation (6.85) by the product $F(r)G(\theta)H(\phi)/r^2$ and collect all the terms involving r on the left-hand side of the equation we obtain

$$
\frac{1}{F(r)}\frac{d}{dr}\left(r^2 F'(r)\right) + \frac{2m_e r^2}{\hbar^2}(E-V(r)) = \frac{-1}{\sin^2\theta}\frac{H''(\phi)}{H(\phi)} - \frac{1}{G(\theta)\sin\theta}\frac{d}{d\theta}\left(\sin\theta G'(\theta)\right)
\tag{6.86}
$$

The variables r,θ,ϕ are independent variables so that the left and right-hand sides of equation (6.86) must equal a constant λ_1 which is called a separation constant. One then obtains the equations

$$
\frac{1}{F(r)}\frac{d}{dr}\left(r^2 F'(r)\right) + \frac{2m_e r^2}{\hbar^2}(E-V(r)) = \lambda_1
\tag{6.87}
$$

$$
\frac{-1}{\sin^2\theta}\frac{H''(\phi)}{H(\phi)} - \frac{1}{G(\theta)\sin\theta}\frac{d}{d\theta}\left(\sin\theta G'(\theta)\right) = \lambda_1
\tag{6.88}
$$

We next separate the variables in equation (6.88) to obtain

$$
\frac{H''(\phi)}{H(\phi)} = \frac{-\sin\theta}{G(\theta)}\frac{d}{d\theta}\left(\sin\theta G'(\theta)\right) - \lambda_1\sin^2\theta = \lambda_2
\tag{6.89}
$$

where λ_2 is another separation constant. This produces the additional two equations

$$
H''(\phi) - \lambda_2 H(\phi) = 0
\tag{6.90}
$$

$$
\sin\theta\frac{d}{d\theta}\left(\sin\theta G'(\theta)\right) + \left(\lambda_2 + \lambda_1\sin^2\theta\right)G(\theta) = 0
\tag{6.91}
$$

For convenience we select the separation constants λ_1 and λ_2 to have the special forms

$$
\lambda_1 = \ell(\ell+1) \qquad \text{and} \qquad \lambda_2 = -m^2
\tag{6.92}
$$

where m and ℓ are for now just some new constants. In summary, the method of separation of variables produces the three differential equations (6.87), (6.90) and (6.91) which we reproduce with the new separation constants in the form

$$\frac{d^2H}{d\phi^2} + m^2H = 0 \tag{6.93}$$

$$\frac{1}{\sin\theta}\frac{d}{d\theta}\left(\sin\theta\frac{dG}{d\theta}\right) + \left(\ell(\ell+1) - \frac{m^2}{\sin^2\theta}\right)G = 0 \tag{6.94}$$

$$\frac{1}{r^2}\frac{d}{dr}\left(r^2\frac{dF}{dr}\right) - \frac{\ell(\ell+1)}{r^2}F + \frac{2m_e}{\hbar^2}(E - V(r))F = 0 \tag{6.95}$$

The above equations are subject to periodic boundary conditions. We shall find that the equation (6.93) possesses nonzero solutions only for certain values of the constant m, the equation (6.94) possesses nonzero solutions only for certain values of the constant ℓ and the equation (6.95) has nonzero solutions only for certain values of the energy E. These special values correspond to discrete stationary states associated with the Hydrogen atom.

The solution of equation (6.93) subject to periodic boundary conditions $H(\phi) = H(\phi+2\pi)$ requires that m be an integer having any of the values $m = 0, \pm1, \pm2, \pm3, \ldots$. The constant m is called the magnetic quantum number. The equation (6.93) is that of a harmonic oscillator and the special solutions are represented in the complex form

$$H = H_m(\phi) = \alpha e^{im\phi} \tag{6.96}$$

where α is a constant which is selected to satisfy the normalization condition specified by the equation (6.84). This requires that

$$\alpha^2 \int_0^{2\pi} e^{im\phi}e^{-im\phi}\,d\phi = 2\pi\alpha^2 = 1.$$

We solve for the constant α and find $\alpha = 1/\sqrt{2\pi}$. This gives the normalized periodic solutions

$$H_m(\phi) = \frac{1}{\sqrt{2\pi}}e^{im\phi}, \qquad m = 0, \pm1, \pm2, \ldots \tag{6.97}$$

In equation (6.94), which defines $G = G(\theta)$, one can make the substitution $x = \cos\theta$ and verify that the equation (6.94) reduces to the form

$$\frac{d}{dx}\left[(1 - x^2)\frac{dG}{dx}\right] + \left[\ell(\ell+1) - \frac{m^2}{1 - x^2}\right]G = 0, \qquad -1 \le x \le 1 \tag{6.98}$$

The equation (6.98) can be found in advanced differential equations textbooks where it defines the associated Legendre functions $P_\ell^m(x)$ and $Q_\ell^m(x)$ of the first and second kind. Only the associated Legendre function $P_\ell^m(x)$ is a bounded function over the interval $(-1, 1)$. The associated Legendre function of the first kind is defined

$$P_\ell^m(x) = (1 - x^2)^{m/2}\frac{d^m}{dx^m}P_\ell(x) \tag{6.99}$$

where ℓ and m are integers and $P_\ell(x)$ is the Legendre function of the first kind of order ℓ defined by

$$P_\ell(x) = \frac{1}{2^\ell \ell!} \frac{d^\ell}{dx^\ell} \left[(x^2 - 1)^\ell \right]. \tag{6.100}$$

Here ℓ can have any of the values

$$\ell = m, m+1, m+2, \ldots \tag{6.101}$$

where m is a positive integer. The integers ℓ are call azimuthal quantum numbers. Some examples of associated Legendre functions of the first kind are given by

$$\begin{aligned}
P_1^1(x) &= \sqrt{1 - x^2} & P_3^1(x) &= \frac{3}{2}(5x^2 - 1)\sqrt{1 - x^2} \\
P_2^1(x) &= 3x\sqrt{1 - x^2} & P_3^2(x) &= 15x(1 - x^2) \\
P_2^2(x) &= 3(1 - x^2) & P_3^3(x) &= 15(1 - x^2)^{3/2}
\end{aligned} \tag{6.102}$$

and $P_\ell^m = 0$ for $m > \ell$.

The solution to equation (6.94) can therefore be represented in the form

$$G = G_{\ell m} = \beta P_\ell^m(\cos\theta) \tag{6.103}$$

where β is a constant. It is known that the associated Legendre functions given by equation (6.99) satisfies the inner product relation

$$\| P_\ell^m \|^2 = (P_\ell^m, P_\ell^m) = \int_{-1}^{1} P_\ell^m(x) P_\ell^m(x)\, dx = \frac{2(\ell + m)!}{(2\ell + 1)(\ell - m)!} \tag{6.104}$$

Also note the normalization condition given by equation (6.84) requires that the constant β be selected such that

$$\beta^2 \int_0^\pi P_{\ell m}(\cos\theta) P_{\ell m}(\cos\theta)\, \sin\theta\, d\theta = 1. \tag{6.105}$$

These two conditions can be related if one observes that the substitution $x = \cos\theta$ reduces the integral given by equation (6.105) to the integral given in equation (6.104). Consequently, one can show

$$\beta = \left[\frac{(2\ell + 1)(\ell - m)!}{2(\ell + m)!} \right]^{1/2} \tag{6.106}$$

where m is a positive integer.

The differential equation (6.95) which defines $F = F(r)$ can be expanded and written in the form

$$\frac{d^2 F}{dr^2} + \frac{2}{r}\frac{dF}{dr} - \frac{\ell(\ell+1)}{r^2}F + \frac{2m_e}{\hbar^2}\left(E + \frac{Ze^2}{4\pi\epsilon_0\, r} \right) F = 0 \tag{6.107}$$

where we have substituted for $V = V(r)$ from equation (6.80). This equation, with the proper substitutions, can also be reduced to a standard form which can be found in most advanced differential equation textbooks. Toward this objective we make the following substitutions in equation (6.107). For negative values for E we let

$$\gamma^2 = -\frac{2m_e E}{\hbar^2}, \qquad \lambda = \frac{m_e Z e^2}{4\pi\epsilon_0 \gamma \hbar^2}, \qquad \rho = 2\gamma r \tag{6.108}$$

with

$$\frac{dF}{dr} = \frac{dF}{d\rho} 2\gamma \qquad \text{and} \qquad \frac{d^2F}{dr^2} = \frac{d^2F}{d\rho^2} 4\gamma^2 \tag{6.109}$$

so that equation (6.107) takes on the form

$$\frac{d^2F}{d\rho^2} + \frac{2}{\rho}\frac{dF}{d\rho} - \frac{\ell(\ell+1)}{\rho^2}F - \frac{1}{4}F + \frac{\lambda}{\rho}F = 0 \tag{6.110}$$

We now make one more transformation of this equation to get it into a standard form which you may or may not recognize. Make the change of variable

$$F = e^{-\rho/2}\rho^\ell Y \tag{6.111}$$

and show that equation (6.110) can be reduced to the form

$$\rho\frac{d^2Y}{d\rho^2} + \{2(\ell+1) - \rho\}\frac{dY}{d\rho} + \{\lambda - (\ell+1)\}Y = 0 \tag{6.112}$$

The differential equation

$$\rho\frac{d^2Y}{d\rho^2} + (M+1-\rho)\frac{dY}{d\rho} + (N-M)Y = 0 \tag{6.113}$$

is known as the Laguerre associated differential equation and polynomial solutions for non-negative integers M and N are given by the associated Laguerre polynomials defined by

$$Y = Y(\rho) = L_N^M(\rho) = \frac{d^M}{d\rho^M}L_N(\rho) \tag{6.114}$$

where $L_N(\rho)$ are the Laguerre polynomials defined by

$$L_N(\rho) = e^\rho \frac{d^N}{d\rho^N}\left(\rho^N e^{-\rho}\right) \tag{6.115}$$

Some examples of associated Laguerre polynomials are

$$\begin{array}{ll}
L_1^1(x) = -1 & L_3^1(x) = -3x^2 + 18x - 18 \\
L_2^1(x) = 2x - 4 & L_3^2(x) = -6x + 18 \\
L_2^2(x) = 2 & L_3^3(x) = -6
\end{array} \tag{6.116}$$

and $L_N^M(x) = 0$ if $M > N$.

Comparing the equation (6.113) with the equation (6.112) we find

$$2(\ell+1) - \rho = M + 1 - \rho \qquad \text{and} \qquad \lambda - (\ell+1) = N - M$$

which implies

$$N - \lambda + \ell \qquad \text{and} \qquad M = 2\ell + 1 \tag{6.117}$$

Because N and M must be positive integers we must require that λ be a positive integer. Therefore, we define

$$n = \lambda = \frac{m_e Z e^2}{4\pi\epsilon_0 \gamma \hbar^2} \tag{6.118}$$

as an integer which is called the total quantum number. We can now solve equation (6.118) for γ to obtain

$$\gamma = \frac{m_e Z e^2}{4\pi\epsilon_0 n \hbar^2} \quad \text{which implies} \quad \gamma^2 = -\frac{2m_e E}{\hbar^2} = \frac{m_e^2 Z^2 e^4}{16\pi^2 \epsilon_0^2 n^2 \hbar^4} \tag{6.119}$$

so that the energy states for the hydrogen atom can be expressed in terms of the total quantum number n as

$$E = E_n = -\frac{m_e Z^2 e^4}{32\pi^2 \epsilon_0^2 n^2 \hbar^2} = -(2.1787\,(10)^{-18}\ J)\,(Z^2/n^2) \tag{6.120}$$

From this relation one can see that as the quantum number n increases, the electron moves to a higher energy and the bonding force to the nucleus is reduced. We can now write the solution to the differential equation (6.112) in the form

$$Y = Y(\rho) = c L_{n+\ell}^{2\ell+1}(\rho) \quad \text{where} \quad \rho = 2\gamma r = \frac{2m_e Z e^2}{4\pi\epsilon_0 n \hbar^2} r = \frac{2Z}{na_0} r \tag{6.121}$$

where $a_0 = \dfrac{4\pi\epsilon_0 \hbar^2}{m_e e^2} = 52.92\,(pm)$ is a constant known as the smallest radius for an orbit in a hydrogen atom, called the Bohr radius, and c is a normalization constant selected as follows. It is known that the associated Laguerre polynomials $L_N^M(x)$ are orthogonal over the interval $(0,\infty)$ with respect to the weight function $x^M e^{-x}$ and that it satisfies the integral relation

$$\int_0^\infty e^{-x} x^M L_N^M(x) L_N^M(x)\, dx = \frac{(2N - M + 1)\,(N!)^3}{(N - M)!} \tag{6.122}$$

The normalization condition given by equation (6.84) requires that we select the constant c such that

$$c^2 \int_0^\infty e^{-\rho} \rho^{2\ell} \left[L_{n+\ell}^{2\ell+1}(\rho) \right]^2 \frac{\rho^2}{\left[\frac{2Z}{na_0} \right]^3}\, d\rho = 1 \tag{6.123}$$

Using the result from equation (6.122) the integral (6.123) can be evaluated and consequently one can show that the normalization constant is given by

$$c = \left[\left(\frac{2Z}{na_0} \right)^3 \frac{(n - \ell - 1)!}{2n \{(n + \ell)!\}^3} \right]^{1/2} \tag{6.124}$$

One can now verify that the solution to the original differential equation (6.107) can be written in the form

$$F(r) = F_{n\ell}(r) = \left[\left(\frac{2Z}{na_0} \right)^3 \frac{(n - \ell - 1)!}{2n \{(n + \ell)!\}^3} \right]^{1/2} e^{-\rho/2} \rho^\ell L_{n+\ell}^{2\ell+1}(\rho) \quad \text{where} \quad \rho = \frac{2Z}{na_0} r \tag{6.125}$$

In summary, the wave functions depend upon the quantum numbers

$$m = 0, \pm 1, \pm 2, \pm 3, \dots$$
$$\ell = |m|, |m| + 1, |m| + 2, \dots$$
$$n = \ell + 1, \ell + 2, \ell + 3, \dots$$

which are independent and usually expressed in the form

total or principal quantum number $\quad n = 1, 2, 3, \ldots$

azimuthal quantum number $\quad \ell = 0, 1, 2, 3, \ldots, n-1 \qquad$ (6.126)

magnetic quantum number $\quad m = -\ell, -\ell+1, \ldots, -1, 0, 1, , \ldots, \ell-1, \ell$

Note that some chemistry textbook write the magnetic quantum number as m_ℓ. The quantum numbers n, ℓ and m are used as subscripts to represent the various wave functions in the form

$$u_{n\ell m}(r, \theta, \phi) = F_{n\ell}(r) G_{\ell m}(\theta) H_m(\phi) \qquad (6.127)$$

where

$$F_{n\ell}(r) = \left[\left(\frac{2Z}{na_0} \right)^3 \frac{(n-\ell-1)!}{2n\{(n+\ell)!\}^3} \right]^{1/2} e^{-\rho/2} \rho^\ell L_{n+\ell}^{2\ell+1}(\rho), \quad \text{with} \quad \rho = \frac{2Z}{na_0} r$$

$$G_{\ell m}(\theta) = \left[\frac{(2\ell+1)(\ell-|m|)!}{2(\ell+|m|)!} \right]^{1/2} P_\ell^{|m|}(\cos\theta) \qquad (6.128)$$

$$H_m(\phi) = \frac{1}{\sqrt{2\pi}} e^{im\phi}$$

If real solutions are desired one usual writes the normalized solutions for H in the form

$$H_0(\phi) = \frac{1}{\sqrt{2\pi}}, \qquad H_{|m|\cos\phi} = \frac{1}{\sqrt{\pi}} \cos|m|\phi, \qquad H_{|m|\sin\phi} = \frac{1}{\sqrt{\pi}} \sin|m|\phi \qquad (6.129)$$

Also note that in order that the functions $F_{n\ell}(r)$ be positive for small values of r the function $F_{n\ell}(r)$ is sometimes replaced by $-F_{n\ell}(r)$. Allowed wave functions for the hydrogen atom are called orbitals. These orbitals describe an electron density in space having a characteristic energy and characteristic shape.

The principal quantum number n determines the energy of the electron which is given by equation (6.120). The principal quantum number can have any of the values $n = 1, 2, 3, \ldots$ but can never be zero. Electrons in an atom which have the same principal quantum number n are said to be at the same level or shell. Sublevels or subshells are governed by the azimuthal quantum number ℓ which is sometimes referred to as the angular momentum quantum number or orbital quantum number. The quantum number ℓ determines sublevels related to the quantum number n by the relation $\ell = 0, 1, 2, 3, \ldots, n-1$, which says there are n sublevels. The quantum number ℓ also determines the shape of the orbital. The magnetic quantum number m sometimes called the magnetic orbital quantum number and often written as m_ℓ determines the orientation of the electron probability cloud. For a given value of ℓ the quantum number m_ℓ can have any integer values between $-\ell$ and ℓ, including the value zero and one can write $m_\ell = \ell, \ell-1, \ell-2, \ldots, 1, 0, -1, \ldots, -\ell$ which represents $2\ell+1$ different orientations. There is a fourth quantum number m_s called a magnetic spin quantum number. The quantum number m_s can have one of two possible values $+1/2$ or $-1/2$ and is not related to the quantum numbers n, ℓ, m_ℓ. The notation $\uparrow = +1/2$ and $\downarrow = -1/2$ is often used to denote the quantum number m_s. In the year 1925 Wolfgang Pauli (1900-1958) (Noble prize in physics in the year 1945) formulated the following exclusion principle.

Pauli exclusion principle
The Pauli exclusion principle states that no two electrons in an atom can have the same set of quantum numbers $\{n, \ell, m_\ell, m_s\}$

One can construct a table of allowed quantum states for electrons in an atom. As an example, the table 6.1 illustrates how one could construct such a table using the principal quantum numbers $n = 1, 2, 3$.

Table 6.1 Allowed Quantum States for Electron in Atoms													
quantum level **n**	1	2			3								
sublevel ℓ	0	0	1			0	1			2				
orbital $\mathbf{m}_\ell$	0	0	1	0	-1	0	1	0	-1	2	1	0	-1	-2
magnetic spin $\mathbf{m}_s$	↑↓	↑↓	↑↓	↑↓	↑↓	↑↓	↑↓	↑↓	↑↓	↑↓	↑↓	↑↓	↑↓	↑↓

The number of electrons in any given level can be determined by examining the number of allowed electrons in a sublevel ℓ associated with a given principal quantum number n. The table 6.2 gives such a representation.

Table 6.2 Population of Electronic Levels in Atoms						
principle quantum number **n**	Number of allowed electrons in sublevel ℓ					Total number in level or shell
	0	1	2	3	4	
1	2					2
2	2	6				8
3	2	6	10			18
4	2	6	10	14		32
5	2	6	10	14	18	50

In chemistry the quantum number ℓ is replaced by terminology developed by spectroscopists for the sharp (s), principal (p), diffuse (d) and fundamental (f) series observed in atomic spectra. Chemists replace $\ell = 0$ by the name s-orbitals, $\ell = 1$ by the name p-orbitals, $\ell = 2$ by the name d-orbitals, $\ell = 3$ by the name f-orbitals, $\ell = 4$ by the name g-orbitals, $\ell = 5$ by the name h-orbitals, etc.

The tables 6.3, 6.4 and 6.5 give representations for the functions H_m, $G_{\ell m}$ and $F_{n\ell}$ which are the components used to define the wave functions associated with a given set of quantum numbers.

Table 6.3 The $H_m(\phi)$ component of the wave function	
$m = 0$	$H_0(\phi) = \frac{1}{\sqrt{2\pi}}$
$m = \pm 1$	$H_m(\phi) = \frac{1}{\sqrt{2\pi}} e^{im\phi}$
$m = \pm 2$	$H_m(\phi) = \frac{1}{\sqrt{2\pi}} e^{im\phi}$
$\vdots$	$\vdots$

Table 6.4 The $G_{\ell m}(\theta)$ component of the wave function		
$\ell = 0, \ m = 0$	$G_{00}(\theta) = \frac{\sqrt{2}}{2}$	s-orbitals
$\ell = 1, \ m = 0$ $\ell = 1, \ m = \pm 1$	$G_{10}(\theta) = \frac{\sqrt{6}}{2}\cos\theta$ $G_{1\,m}(\theta) = \frac{\sqrt{3}}{2}\cos\theta$	p-orbitals
$\ell = 2, \ m = 0$ $\ell = 2, \ m = \pm 1$ $\ell = 2, \ m = \pm 2$	$G_{20}(\theta) = \frac{\sqrt{10}}{4}\left(3\cos^2\theta - 1\right)$ $G_{2\,m}(\theta) = \frac{\sqrt{15}}{2}\sin\theta\cos\theta$ $G_{2\,m}(\theta) = \frac{\sqrt{15}}{4}\sin^2\theta$	d-orbitals
$\ell = 3, \ m = 0$ $\ell = 3, \ m = \pm 1$ $\ell = 3, \ m = \pm 2$ $\ell = 3, \ m = \pm 3$	$G_{30}(\theta) = \frac{3\sqrt{14}}{4}\left(\frac{5}{3}\cos^3\theta - \cos\theta\right)$ $G_{3\,m}(\theta) = \frac{\sqrt{42}}{8}\sin\theta\left(5\cos^2\theta - 1\right)$ $G_{3\,m}(\theta) = \frac{\sqrt{105}}{4}\sin^2\theta\cos\theta$ $G_{3\,m}(\theta) = \frac{\sqrt{70}}{8}\sin^3\theta$	f-orbitals
$\ell = 4, \ m = 0$ $\ell = 4, \ m = \pm 1$ $\ell = 4, \ m = \pm 2$ $\ell = 4, \ m = \pm 3$ $\ell = 4, \ m = \pm 4$	$G_{40}(\theta) = \frac{9\sqrt{2}}{16}\left(\frac{35}{3}\cos^4\theta - 10\cos^2\theta + 1\right)$ $G_{4\,m}(\theta) = \frac{9\sqrt{10}}{8}\sin\theta\left(\frac{7}{3}\cos^3\theta - \cos\theta\right)$ $G_{4\,m}(\theta) = \frac{3\sqrt{5}}{8}\sin^2\theta\left(7\cos^2\theta - 1\right)$ $G_{4\,m}(\theta) = \frac{3\sqrt{70}}{8}\sin^3\theta\cos\theta$ $G_{4\,m}(\theta) = \frac{3\sqrt{35}}{16}\sin^4\theta$	g-orbitals
$\ell = 5, \ m = 0$ $\ell = 5, \ m = \pm 1$ $\ell = 5, \ m = \pm 2$ $\ell = 5, \ m = \pm 3$ $\ell = 5, \ m = \pm 4$ $\ell = 5, \ m = \pm 5$	$G_{50}(\theta) = \frac{15\sqrt{22}}{16}\left(\frac{21}{5}\cos^5\theta - \frac{14}{3}\cos^3\theta + \cos\theta\right)$ $G_{5\,m}(\theta) = \frac{\sqrt{165}}{16}\sin\theta\left(21\cos^4\theta - 14\cos^2\theta + 1\right)$ $G_{5\,m}(\theta) = \frac{\sqrt{1155}}{8}\sin^2\theta\left(3\cos^3\theta - \cos\theta\right)$ $G_{5\,m}(\theta) = \frac{\sqrt{770}}{32}\sin^3\theta\left(9\cos^2\theta - 1\right)$ $G_{5\,m}(\theta) = \frac{3\sqrt{385}}{16}\sin^4\theta\cos\theta$ $G_{5\,m}(\theta) = \frac{3\sqrt{154}}{32}\sin^5\theta$	h-orbitals

Note that the spherical harmonic function $Y_n^m(\theta, \phi)$ which is defined

$$Y_n^m(\theta, \phi) = (-1)^m \sqrt{\frac{2n+1}{4\pi}\frac{(n-m)!}{(n+m)!}}\, P_n^m(\cos\theta)\, e^{im\phi} \tag{6.130}$$

can be used to give graphical representations of the two-dimensional angular part of the hydrogen atom orbitals. The hydrogen atom wave function product $G_{\ell m}(\theta)H_m(\phi)$ can be replaced by the spherical harmonic function. Don't worry about the $(-1)^m$ term as we will be taking absolute values. Note that the hydrogen wave function solutions to the Schrödinger equation can be represented in the alternate form

$$u_{n\ell m} = F_{n\ell}(r)Y_\ell^m(\theta, \phi) \tag{6.131}$$

where n is the principal quantum number, ℓ is the angular momentum quantum number and m is the magnetic quantum number.

Table 6.5 The $F_{n\ell}(r)$ component of the wave function with $\rho = \frac{2Z}{na_0} r$		
$n = 1,\ \ell = 0$	1s	$F_{10}(r) = \left(\frac{Z}{a_0}\right)^{3/2} 2 e^{\rho/2}$
$n = 2,\ \ell = 0$	2s	$F_{20}(r) = \left(\frac{Z}{a_0}\right)^{3/2} \frac{1}{2\sqrt{2}} (2 - \rho) e^{-\rho/2}$
$n = 2,\ \ell = 1$	2p	$F_{21}(r) = \left(\frac{Z}{a_0}\right)^{3/2} \frac{1}{2\sqrt{6}} \rho e^{-\rho/2}$
$n = 3,\ \ell = 0$	3s	$F_{30}(r) = \left(\frac{Z}{a_0}\right)^{3/2} \frac{1}{9\sqrt{3}} (6 - 6\rho + \rho^2) e^{-\rho/2}$
$n = 3,\ \ell = 1$	3p	$F_{31}(r) = \left(\frac{Z}{a_0}\right)^{3/2} \frac{1}{9\sqrt{6}} (4 - \rho)\rho e^{-\rho/2}$
$n = 3,\ \ell = 2$	3d	$F_{32}(r) = \left(\frac{Z}{a_0}\right)^{3/2} \frac{1}{9\sqrt{30}} \rho^2 e^{-\rho/2}$
$n = 4,\ \ell = 0$	4s	$F_{40}(r) = \left(\frac{Z}{a_0}\right)^{3/2} \frac{1}{96} (24 - 36\rho + 12\rho^2 - \rho^3) e^{-\rho/2}$
$n = 4,\ \ell = 1$	4p	$F_{41}(r) = \left(\frac{Z}{a_0}\right)^{3/2} \frac{1}{32\sqrt{15}} (20 - 10\rho + \rho^2)\rho e^{-\rho/2}$
$n = 4,\ \ell = 2$	4d	$F_{42}(r) = \left(\frac{Z}{a_0}\right)^{3/2} \frac{1}{96\sqrt{5}} (6 - \rho)\rho^2 e^{-\rho/2}$
$n = 4,\ \ell = 3$	4f	$F_{43}(r) = \left(\frac{Z}{a_0}\right)^{3/2} \frac{1}{96\sqrt{35}} \rho^3 e^{-\rho/2}$
$n = 5,\ \ell = 0$	5s	$F_{50}(r) = \left(\frac{Z}{a_0}\right)^{3/2} \frac{1}{300\sqrt{5}} (120 - 240\rho + 120\rho^2 - 20\rho^3 + \rho^4) e^{-\rho/2}$
$n = 5,\ \ell = 1$	5p	$F_{51}(r) = \left(\frac{Z}{a_0}\right)^{3/2} \frac{1}{150\sqrt{30}} (120 - 90\rho + 18\rho^2 - \rho^3)\rho e^{-\rho/2}$
$n = 5,\ \ell = 2$	5d	$F_{52}(r) = \left(\frac{Z}{a_0}\right)^{3/2} \frac{1}{150\sqrt{70}} (42 - 14\rho + \rho^2)\rho^2 e^{-\rho/2}$
$n = 5,\ \ell = 3$	5f	$F_{53}(r) = \left(\frac{Z}{a_0}\right)^{3/2} \frac{1}{300\sqrt{70}} (8 - \rho)\rho^3 e^{-\rho/2}$
$n = 5,\ \ell = 4$	5g	$F_{54}(r) = \left(\frac{Z}{a_0}\right)^{3/2} \frac{1}{900\sqrt{70}} \rho^4 e^{-\rho/2}$

One can generate surface plots illustrating the angular dependence associated with the hydrogen atom orbitals by plotting the parametric surface defined by

$$\vec{r} = \vec{r}(\theta, \phi) = x(\theta, \phi)\,\widehat{\mathbf{e}}_1 + y(\theta, \phi)\,\widehat{\mathbf{e}}_2 + z(\theta, \phi)\,\widehat{\mathbf{e}}_3 \qquad (6.132)$$

where

$$x = x(\theta, \phi) = |\,Y_n^m(\theta, \phi)\,|\,\sin\theta\cos\phi, \quad y = y(\theta, \phi) = |\,Y_n^m(\theta, \phi)\,|\,\sin\theta\sin\phi, \quad z = z(\theta, \phi) = |\,Y_n^m(\theta, \phi)\,|\,\cos\theta$$

for $0 \leq \theta \leq \pi$ and $0 \leq \phi \leq 2\pi$.

Note that real functions can be obtained by adding and subtracting certain combinations of the spherical harmonic functions and then taking the absolute value. For example,

$$|\,Y_n^{-m}(\theta, \phi) - Y_n^m(\theta, \phi)\,| \quad \text{and} \quad |\,Y_n^{-m}(\theta, \phi) + Y_n^m(\theta, \phi)\,| \qquad (6.133)$$

These functions can be scaled and plotted to give images of the angular dependence associated with the wave functions for the hydrogen atom. This is equivalent to taking the real

(Re) and imaginary (Im) parts associated with the spherical harmonic function. Selected images of the hydrogen orbital angular dependence are given in the figures 6-5.

The function $|u_{n\ell m}(r, \theta, \phi)|^2$ represents the probability of finding an electron in an element of volume $d\tau$ at a position (r, θ, ϕ). The radial variation of the wave function is illustrated in the figure 6-6. The figure 6-6(b) represents the shape of regions where there is a 90 per cent chance that the electron will occupy that region. These figures represent charge cloud distributions associated with the 1s, 2s and 3s orbitals. Additional three dimensional graphic representation of the wave functions can be found on the internet.

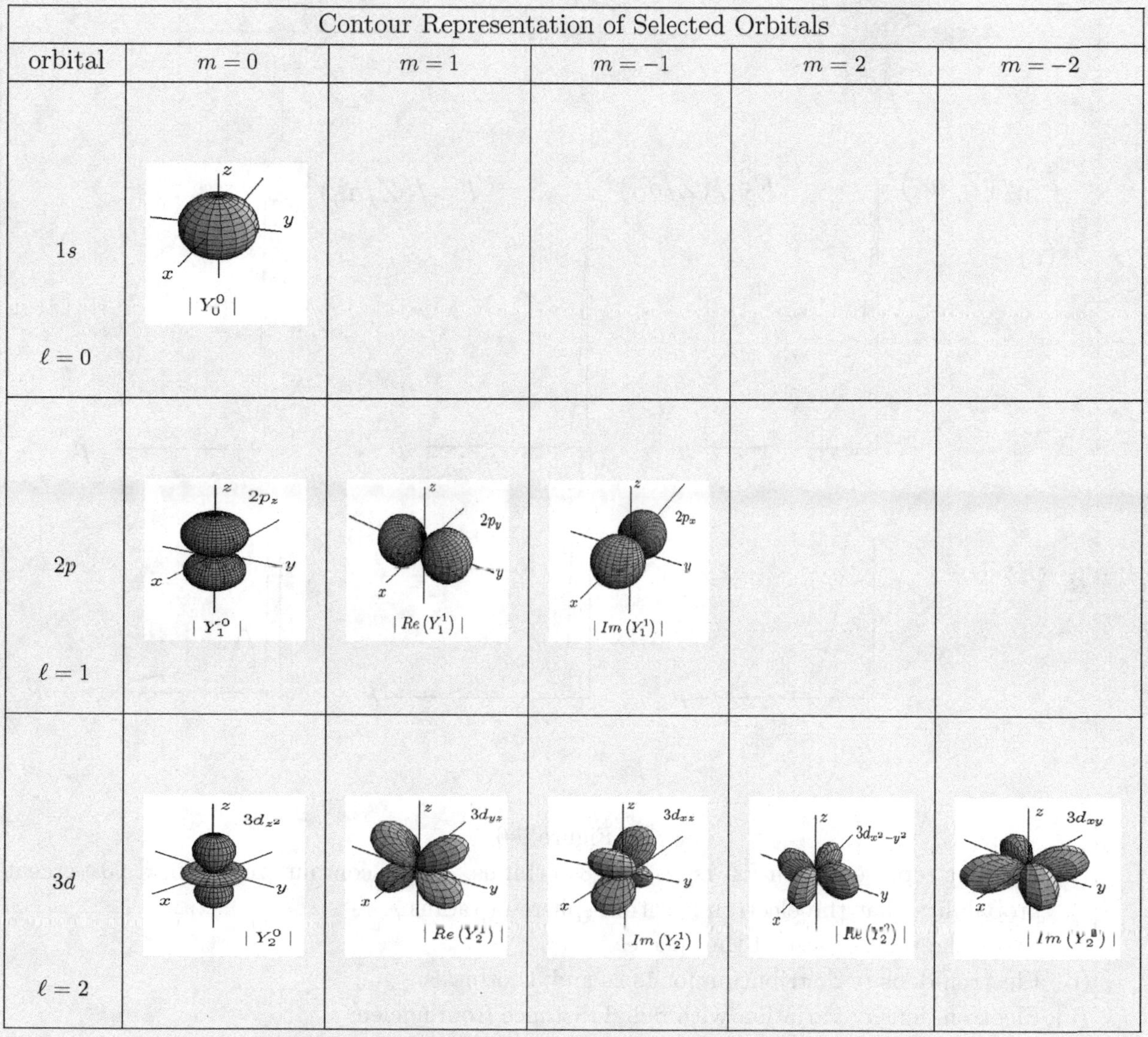

Contour Representation of Selected Orbitals					
orbital	$m = 0$	$m = 1$	$m = -1$	$m = 2$	$m = -2$
$1s$ $\ell = 0$	$\lvert Y_0^0 \rvert$				
$2p$ $\ell = 1$	$\lvert Y_1^0 \rvert$	$\lvert Re\,(Y_1^1) \rvert$	$\lvert Im\,(Y_1^1) \rvert$		
$3d$ $\ell = 2$	$\lvert Y_2^0 \rvert$	$\lvert Re\,(Y_2^1) \rvert$	$\lvert Im\,(Y_2^1) \rvert$	$\lvert Re\,(Y_2^2) \rvert$	$\lvert Im\,(Y_2^2) \rvert$

Figure 6-5 Selected Contour Representations

270

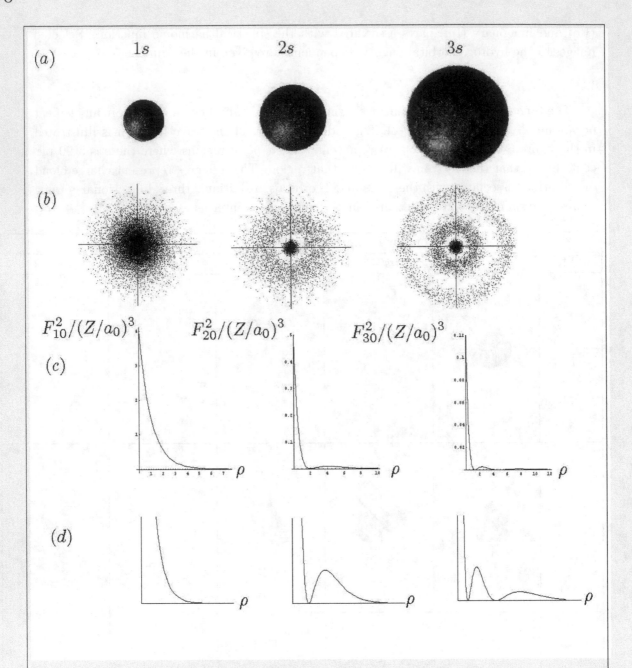

Figure 6-6.

(a) Contour representation for 1s, 2s and 3s orbitals. These contours represent a 90 percent probability that the electron is within sphere of radius ρ.
   (Note the relative size of the spheres.)

(b) Electron-density distribution for 1s, 2s and 3s orbitals.

(c) Electron-density variation with radial distance from nucleus.
   (Note points where $F_{i0}(\rho) = 0$, $i = 1, 2, 3$ are called nodes.)

(d) Same graphs as (c) with different scaling of axes.

The charge cloud representation for a single electron around an atom can be used for representation of charge clouds associated with several electrons around an atom. The principal of superposition of charge clouds associated with electrons occupying different orbitals is how these more complicated representations are obtained. The electron configurations together with the geometric properties of the charge clouds are important topics in studying forces holding atoms together in compound substances. Consequently, the above quantum mechanical concepts are preliminary background material needed to study the subject of chemical bonding found in most chemistry books.

Maxwell's equations

The Maxwell's equations can be written

$$\nabla \cdot \vec{B} = 0 \tag{6.134}$$

$$\nabla \times \vec{E} + \frac{\partial \vec{B}}{\partial t} = 0, \qquad \text{(Faraday's law)} \tag{6.135}$$

$$\nabla \cdot \vec{E} = \rho/\epsilon_0, \qquad \text{(Gauss's law)} \tag{6.136}$$

$$\nabla \times \vec{B} - \epsilon_0 \mu_0 \frac{\partial \vec{E}}{\partial t} = \mu_0 \vec{J}, \qquad \text{(Ampere's law)} \tag{6.137}$$

where ρ is the charge density with units $[coulomb/m^3]$ which is related to the current density $\vec{J}$ by $\vec{J} = \rho \vec{V} = J_1 \hat{e}_1 + J_2 \hat{e}_2 + J_3 \hat{e}_3$ where $\vec{V}$, with units $[m/sec]$, is the velocity of the charge density. In Maxwell's equations $\vec{E}$ denotes the electric force field having units $[Newton/coulomb]$, $\vec{B}$ is the magnet induction field with units $[Weber/m^2]$, ϵ_0 is the permitivity of free space with units $[farads/m]$ and μ_0 is the permeability of free space with units $[henry/m]$.

The first two Maxwell's equations, equations (6.134) and (6.135) define the electric field $\vec{E}$ and magnetic induction field $\vec{B}$. These vector fields can be represented in terms of a scalar potential ϕ and vector potential $\vec{A}$ by the relations

$$\vec{E} = -\nabla\phi - \frac{\partial \vec{A}}{\partial t} \tag{6.138}$$

$$\vec{B} = \nabla \times \vec{A} \tag{6.139}$$

Note that one can employ the vector identities

$$\nabla \times \nabla\phi = 0 \quad \text{and} \quad \nabla \cdot \left(\nabla \times \vec{A}\right) = 0 \tag{6.140}$$

so that by taking the divergence of both sides of equation (6.139) one obtains the equation (6.134) and taking the curl of both sides of equation (6.138) one obtains the equation (6.135).

The Maxwell's equations can be derived from the Lagrangian per unit volume

$$L = \frac{1}{2}\left(\epsilon_0 E^2 - \frac{1}{\mu_0}B^2\right) - \rho\phi + \vec{J} \cdot \vec{A} \tag{6.141}$$

which involves the vector potential $\vec{A} = A_1 \hat{e}_1 + A_2 \hat{e}_2 + A_3 \hat{e}_3$ and scalar potential ϕ, where $E^2 = \vec{E} \cdot \vec{E}$ and $B^2 = \vec{B} \cdot \vec{B}$. The equation (6.138) tells us that the E^2 term in the Lagrangian can be treated as a function of the derivatives ϕ_x, ϕ_y, ϕ_z. If we write the Lagrangian given by equation (6.141) as an integrand of the form $L = L(\phi, \phi_x, \phi_y, \phi_z)$, then we can use the results from [f9] of chapter three to obtain the Euler-Lagrange equation

$$\frac{\partial L}{\partial \phi} - \frac{\partial}{\partial x}\left(\frac{\partial L}{\partial \phi_x}\right) - \frac{\partial}{\partial y}\left(\frac{\partial L}{\partial \phi_y}\right) - \frac{\partial}{\partial z}\left(\frac{\partial L}{\partial \phi_z}\right) = 0 \qquad (6.142)$$

Differentiate the Lagrangian L given by equation (6.141) to obtain the derivatives

$$\frac{\partial L}{\partial \phi} = -\rho$$

$$\frac{\partial L}{\partial \phi_x} = \epsilon_0 \vec{E} \cdot \frac{\partial \vec{E}}{\partial \phi_x} = \epsilon_0 \vec{E} \cdot (-\widehat{\mathbf{e}}_1) = -\epsilon_0 E_1$$

$$\frac{\partial L}{\partial \phi_y} = \epsilon_0 \vec{E} \cdot \frac{\partial \vec{E}}{\partial \phi_y} = \epsilon_0 \vec{E} \cdot (-\widehat{\mathbf{e}}_2) = -\epsilon_0 E_2$$

$$\frac{\partial L}{\partial \phi_z} = \epsilon_0 \vec{E} \cdot \frac{\partial \vec{E}}{\partial \phi_z} = \epsilon_0 \vec{E} \cdot (-\widehat{\mathbf{e}}_3) = -\epsilon_0 E_3$$

so that the Euler-Lagrange equation (6.142) becomes

$$-\rho + \epsilon_0 \left(\frac{\partial E_1}{\partial x} + \frac{\partial E_2}{\partial y} + \frac{\partial E_3}{\partial z}\right) = 0 \qquad \text{or} \qquad \nabla \cdot \vec{E} = \rho/\epsilon_0 \qquad (6.143)$$

which is the Gauss law given by equation (6.136).

The vector components A_1, A_2, A_3 are functions of t, x, y, z and so we use the integrand [f11] from chapter 3 with u, v, w replaced by A_1, A_2, A_3 to obtain the three Euler-Lagrange equations

$$\frac{\partial L}{\partial A_i} - \frac{\partial}{\partial x}\left(\frac{\partial L}{\partial A_{ix}}\right) - \frac{\partial}{\partial y}\left(\frac{\partial L}{\partial A_{iy}}\right) - \frac{\partial}{\partial z}\left(\frac{\partial L}{\partial A_{iz}}\right) - \frac{\partial}{\partial t}\left(\frac{\partial L}{\partial A_{it}}\right) = 0, \qquad i = 1, 2, 3 \qquad (6.144)$$

We substitute

$$\vec{B} = \nabla \times \vec{A} = \begin{vmatrix} \widehat{\mathbf{e}}_1 & \widehat{\mathbf{e}}_2 & \widehat{\mathbf{e}}_3 \\ \frac{\partial}{\partial x} & \frac{\partial}{\partial y} & \frac{\partial}{\partial z} \\ A_1 & A_2 & A_3 \end{vmatrix}$$

$$\vec{B} = (A_{3y} - A_{2z})\,\widehat{\mathbf{e}}_1 + (A_{1x} - A_{3x})\,\widehat{\mathbf{e}}_2 + (A_{2x} - A_{1y})\,\widehat{\mathbf{e}}_3 \qquad (6.145)$$

into the Lagrangian and calculate the derivatives

$$\frac{\partial L}{\partial A_1} = \vec{J} \cdot \widehat{\mathbf{e}}_1 = J_1$$

$$\frac{\partial L}{\partial A_{1x}} = 0$$

$$\frac{\partial L}{\partial A_{1y}} = -\frac{1}{\mu_0}\vec{B} \cdot \frac{\partial \vec{B}}{\partial A_{1y}} = \frac{1}{\mu_0}B_3$$

$$\frac{\partial L}{\partial A_{1z}} = -\frac{1}{\mu_0}\vec{B} \cdot \frac{\partial \vec{B}}{\partial A_{1z}} = -\frac{1}{\mu_0}B_2$$

$$\frac{\partial L}{\partial A_{1t}} = \epsilon_0 \vec{E} \cdot \frac{\partial \vec{E}}{\partial A_{1t}} = -\epsilon_0 E_1$$

The Euler-Lagrange equation (6.144) then becomes

$$\text{for } i = 1 \qquad \mu_0 J_1 = \left(\frac{\partial B_3}{\partial y} - \frac{\partial B_2}{\partial z}\right) - \epsilon_0 \mu_0 \frac{\partial E_1}{\partial t} \tag{6.146}$$

It is left as an exercise to show the remaining Euler-Lagrange equations (6.144) are

$$\text{for } i = 2 \qquad \mu_0 J_2 = \left(\frac{\partial B_1}{\partial z} - \frac{\partial B_3}{\partial x}\right) - \epsilon_0 \mu_0 \frac{\partial E_2}{\partial t}$$
$$\text{for } i = 3 \qquad \mu_0 J_3 = \left(\frac{\partial B_2}{\partial x} - \frac{\partial B_1}{\partial y}\right) - \epsilon_0 \mu_0 \frac{\partial E_3}{\partial t} \tag{6.147}$$

Observe that the Euler-Lagrange equations (6.146) and (6.147) are the components of Ampere's law given by equation (6.137).

Numerical Methods

In the previous chapters we developed techniques for determining the maximum or minimum values associated with a function or a functional. For example, we have shown how to construct the Euler-Lagrange equations which is a necessary condition to be satisfied if a functional is to have an extremal. You have observed that many of the resulting Euler-Lagrange equations are nonlinear and difficult to solve. Whenever there occurs a maximum-minimum problem that is just too difficult to solve in a closed form, then in such cases one can always employ approximation techniques and/or numerical methods to solve the resulting equations. In this section we present selected approximation methods and numerical methods which are based upon optimization concepts.

Search techniques

A standard nonlinear programming problem with equality constraints can be expressed in the form

$$\text{minimize} \qquad f = f(x_1, x_2, x_3, \ldots, x_n)$$
$$\text{subject to the constraints} \qquad g_1(x_1, x_2, \ldots, x_n) = 0$$
$$g_2(x_1, x_2, \ldots, x_n) = 0$$
$$\vdots$$
$$g_m(x_1, x_2, \ldots, x_n) = 0 \tag{6.148}$$

where there must be fewer constraints than variables so that $m < n$. Here f is called the objective function to be optimized and the variables $x_1, x_2, \ldots, x_n$ are independent variables which are related by the constraint functions.

The above problem can be reformulated in terms of penalty weights and written in the form

$$\text{minimize} \qquad F = f(x_1, x_2, \ldots, x_n) + \sum_{i=1}^{m} p_i g_i^2(x_1, x_2, \ldots, x_n) \tag{6.149}$$

where each $p_i > 0$ are constants called penalty weights. Here the terms $p_i g_i^2$, for $i = 1, 2, \ldots, m$, act like penalty functions to the function to be minimized. If the functions g_i are not near zero for all values of the index i, then a penalty is added to the function to be minimized.

Consider the problem of finding a local minimum associated with a real valued function ϕ in n-dimensional space. We write the function in the form $\phi = \phi(\vec{x})$ where $\vec{x} = (x_1, x_2, \ldots, x_n)^T$ and let

$$g(\vec{x}) = \left[\frac{\partial \phi}{\partial x_1}, \frac{\partial \phi}{\partial x_2}, \ldots, \frac{\partial \phi}{\partial x_n} \right]^T \tag{6.150}$$

denote the gradient function of ϕ. The problem of determining a point where a local minimum occurs requires one to solve the system of equations

$$g(\vec{x}) = \vec{0} \tag{6.151}$$

which in general is a system of nonlinear equations. If $\vec{x}^{(k)}$ denotes a fixed point in the n-dimensional space, then the function $g(\vec{x})$ can be expanded in a Taylor series about this point. The Taylor series expansion in n-dimensions has the form

$$g(\vec{x}) = g(\vec{x}^{(k)}) + g'(\vec{x}^{(k)})(\vec{x} - \vec{x}^{(k)}) + \mathcal{O}\left(\parallel \vec{x} - \vec{x}^{(k)} \parallel^2 \right) \tag{6.152}$$

where

$$g'(\vec{x}^{(k)}) = J(\vec{x}^{(k)}) = \left[\frac{\partial^2 \phi}{\partial x_i \partial x_j} \right]_{\vec{x} = \vec{x}^{(k)}} \tag{6.153}$$

is the $n \times n$ Jacobian matrix sometimes referred to as the Hessian of ϕ and $\parallel \parallel$ denotes some convenient norm. The search for point $\vec{x}^{(k+1)}$, where $g(\vec{x}^{(k+1)}) = \vec{0}$ produces the equation of first approximation

$$g(\vec{x}^{(k+1)}) = g(\vec{x}^{(k)}) + J(\vec{x}^{(k)})(\vec{x}^{(k+1)} - \vec{x}^{(k)}) = \vec{0} \tag{6.154}$$

which represents a linear system of equations to be solved. If the Jacobian matrix is nonsingular, then $\vec{x}^{(k+1)}$ can be solved for. In many cases the system of equations (6.154) represents a large number of linear equations to be solved and so the equation (6.154) is usually written in the iterative form

$$\vec{x}^{(k+1)} = \vec{x}^{(k)} - J^{-1}(\vec{x}^{(k)}) g(\vec{x}^{(k)}) \tag{6.155}$$

where J^{-1} is the inverse of the Jacobian matrix. The iterative form given by equation (6.155) is known as the Newton-Raphson iterative method. There exists a variety of search methods for a local minimum which are based upon modifications of the Newton-Raphson iterative method. Many of these modified search techniques have the following steps in common.

(i) Determine a search direction $\vec{d}^{(k)}$ at the k-th step of the iteration.

(ii Construct the line $\vec{x}^{(k+1)} = \vec{x}^{(k)} \pm \lambda \vec{d}^{(k)}$ where λ is a scalar.

(iii) Vary λ and try to find a point along the line where $\phi(\vec{x}^{(k+1)}) < \phi(\vec{x}^{(k)})$

(iv) Keep repeating this process until some terminating condition is achieved.

There are many computer search techniques that have been constructed to solve or approximately solve the above type of problem. Some of the more common names associated with these computer search techniques are: Random search, which requires a random number generator, random walk method, Fibonacci search, gradient search methods, the Newton-Raphson search method, the Fletcher and Reeves method, the Fletcher-Powell gradient search technique, Stewart-Davidon-Fletcher-Powell technique which is a modification

275

of the Fletcher-Powell method, the method of steepest ascent, pattern search techniques such as Hooke-Jeeves which is a direction and pattern movement combination in search of an optimum value, quadratic interpolation methods, cubic interpolation methods . Many of these search techniques are readily available in certain computer library packages. Some search techniques can use up a lot of computer time and wind up not finding a minimum or maximum value.

Scaling

One of the difficulties that is sometimes associated with using these computer search techniques is that occasionally one of the variables $x_1, x_2, x_3, \ldots, x_n$ tends to become quite large without changing the value of the objective function. In such cases the computer search usually terminates with an error indicating that an optimum value was not achieved. To avoid this kind of a problem in using a computer search one can reformulate the problem and place bounds on the search variables x_i for $i = 1, 2, \ldots, n$. In many cases an understanding of the physical nature of the problem allows one to define lower bounds α_i and upper bounds β_i to be associated with each search variable x_i for $i = 1, 2, \ldots, n$. One can then introduce auxiliary search variables ξ_i where $-\infty < \xi_i < \infty$, and scaling variables s_i which are related to the original variables x_i by the relations

$$x_i = \left(\frac{\beta_i + \alpha_i}{2}\right) + \left(\frac{\beta_i - \alpha_i}{2}\right)\tanh(s_i\xi_i) \tag{6.156}$$

for $i = 1, 2, \ldots, n$. Then for each value of $i = 1, 2, \ldots, n$, a graph of the true variable x_i vs the dummy variable ξ_i will appear as in the figure 6-7. Observe that as the dummy variable ξ_i varies from $-\infty$ to $+\infty$ the variable x_i satisfies the inequalities $\alpha_i \le x_i \le \beta_i$ for $i = 1, 2, \ldots, n$. Also note that the scaling variables s_i control the steepness of the change from the lower limit α_i to the upper limit β_i. The scaling variables s_i are a way of preventing sudden changes to occur within the objective function and influencing the computer algorithm in an adverse way. The transformations given by the equation (6.156) provide a way of placing bounds upon each of the search variables so that they do not move too far from the search area of interest and thereby preventing termination of many search routines.

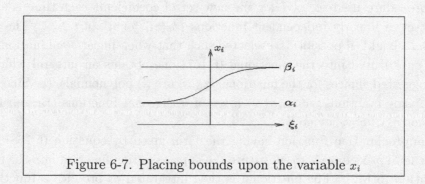

Figure 6-7. Placing bounds upon the variable x_i

Many of these search techniques are developed for application in engineering, industrial, economic, and military applications where one must decide how to use various resources in an efficient manor. These type of problems are more fully discussed in operational research textbooks and mathematical programming textbooks where mathematics is used to develop theories for the allocation of resources in some optimal fashion.

Rayleigh-Ritz method for functionals

The Rayleigh-Ritz method is an approximation method for obtaining solutions to various variational problems. Consider the construction of approximate solutions to variational problems involving functions of a single variable. We consider the problem to find a function $y = y(x)$ which is an extremum for the functional

$$I = \int_a^b f(x, y, y') \, dx \tag{6.157}$$

subject to given boundary conditions. The Rayleigh-Ritz method requires one to generate a sequence of approximate solutions $\{y_n(x)\}$, for $n = 1, 2, 3, \ldots$, to the variational problem given by equation (6.157). Each approximate solution is required to produce an extremum to a closely related problem associated with the equation (6.157). The nth approximate solution has the form

$$y = y(x) \approx y_n(x) = \phi_0(x) + c_1\phi_1(x) + c_2\phi_2(x) + \cdots + c_n\phi_n(x) \tag{6.158}$$

where $c_1, c_2, \ldots, c_n$ are constant coefficients to be determined and $\phi_0(x), \phi_1(x), \phi_2(x), \ldots, \phi_n(x)$ are a set of linearly independent functions selected such that $\phi_0(x)$ satisfies the boundary conditions and the functions $\phi_i(x)$, for $i = 1, \ldots, n$ satisfy zero boundary conditions. In this way the boundary conditions are satisfied for all choices of the coefficients $c_i, i = 1, 2, \ldots, n$ The $(n+1)$ approximate solution has the form

$$y = y(x) \approx y_{n+1}(x) = \phi_0(x) + \bar{c}_1\phi_1(x) + \bar{c}_2\phi_2(x) + \cdots + \bar{c}_n\phi_n(x) + \bar{c}_{n+1}\phi_{n+1}(x) \tag{6.159}$$

where the $\bar{c}_i$, for $i = 1, \ldots, n+1$, are new constants to be determined. In general, a new set of coefficients must be solved for each time a new approximate solution is generated. The general procedure used to solve for the new set of coefficients each time is as follows. First select a set of linearly independent functions $\{\phi_n(x)\}$ for $n = 0, 1, 2, \ldots$. The functions $\phi_n(x)$, $n = 1, 2, \ldots$ should, if possible, be selected such that when linear combinations of these functions are substituted into the functional (6.157) one obtains an integral which can be integrated. Suggested choices for the functions $\{\phi_n(x)\}$ are (i) polynomials, (ii) Sine functions $(\sin nx)$, (iii) Cosine functions $(\cos nx)$, (iv) a set of orthogonal functions (Bessel, Legendre, Hermite, Laguerre, Chebyshev,etc).

Once an approximation function having the form given by equation (6.158) has been selected the general procedure is to substitute the approximation function into the functional given by equation (6.157). The functional is then integrated to produce a function which is dependent upon the coefficients $c_1, c_2, \ldots, c_n$. That is, the equation (6.157) reduces to the form

$$I = I(c_1, c_2, \ldots, c_n) \tag{6.160}$$

after the approximation (6.158) is substituted into equation (6.157) and the result is integrated. The constants c_i for $i = 1, \ldots, n$ are selected to have the values which makes I have a maximum or minimum value. In this way the problem has been reduced to just an ordinary maximum-minimum problem. One must solve for the critical points determined from the system of equations

$$\frac{\partial I}{\partial c_i} = 0, \quad \text{for} \quad i = 1, \ldots, n \tag{6.161}$$

The equations (6.161) represent a system of n-equations in the n-unknowns $c_1, c_2, \ldots, c_n$. Solve this system of equations and substitute the constants into equation (6.158) to obtain the desired approximate solution.

Example 6-2.   Rayleigh-Ritz method

Apply the Rayleigh-Ritz method to obtain an approximate solution for the extremum associated with the functional

$$I = \int_0^1 \left[(y')^2 - y^2 - 2xy \right] dx, \qquad y(0) = y(1) = 0 \tag{6.162}$$

Solution: We use the polynomial functions $\phi_n(x) - x(1-x)^n$ for $n = 1, 2, \ldots$, because each of these functions satisfies the given boundary conditions. We let $\phi_0 = 0$ which certainly satisfies the boundary conditions for all choices of the constant c_i, $i = 1, 2, \ldots$. As a first approximation assume

$$y_1(x) = c_1 x(1-x) \quad \text{with derivative} \quad y_1'(x) = c_1(1 - 2x).$$

This approximation is substituted into the functional given by equation (6.162) to obtain

$$I = I(c_1) = \int_0^1 \left[c_1^2(1 - 2x)^2 - c_1^2 x^2(1-x)^2 - 2xc_1 x(1-x) \right] dx. \tag{6.163}$$

The equation (6.163) is now integrated to obtain

$$I = I(c_1) = \frac{3}{10}c_1^2 - \frac{1}{6}c_1.$$

A critical point for I occurs when

$$\frac{\partial I}{\partial c_1} = \frac{6}{10}c_1 - \frac{1}{6} = 0 \qquad \text{or} \qquad c_1 = \frac{5}{18}.$$

This gives the first approximation $y_1(x) = \frac{5}{18}x(1-x)$ which produces a minimum value for I.

A second approximation having the form $y_2(x) = \alpha x(1-x) + \beta x(1-x)^2$ is constructed, where α and β are new constants to be determined. This approximation is substituted into the functional given by equation (6.162) to obtain

$$I = I(\alpha, \beta) = \int_0^1 \left\{ [\alpha(1-2x) + \beta(1 - 4x + 3x^2)]^2 - [\alpha x(1-x) + \beta x(1-x)^2]^2 - 2x[\alpha x(1-x) + \beta x(1-x)^2] \right\} dx$$

This result is integrated giving the equation

$$I = I(\alpha, \beta) = -\frac{\alpha}{6} + \frac{3}{10}\alpha^2 - \frac{\beta}{15} + \frac{3}{10}\alpha\beta + \frac{13}{105}\beta^2 \qquad (6.164)$$

The function I has critical points where

$$\frac{\partial I}{\partial \alpha} = -\frac{1}{6} + \frac{6}{10}\alpha + \frac{3}{10}\beta = 0$$
$$\frac{\partial I}{\partial \beta} = -\frac{1}{15} + \frac{3}{10}\alpha + \frac{26}{105}\beta = 0$$

One can solve this system of equations and verify that the solution is given by

$$\alpha = \frac{134}{369}, \qquad \text{and} \qquad \beta = -\frac{7}{41}$$

This gives the second approximation

$$y_2(x) = \frac{134}{369}x(1-x) - \frac{7}{41}x(1-x)^2 \qquad (6.165)$$

One can verify that the Euler-Lagrange equation associated with the given functional is

$$y'' + y = -x, \qquad y(0) = y(1) = 0$$

with exact solution given by

$$y_{exact} = \frac{\sin x}{\sin(1)} - x.$$

One can define the difference

$$\text{Error} = y_{exact} - y_2(x)$$

and then plot sketches of $y_1(x)$, $y_2(x)$, y_{exact} and the Error term, as a function of x, to obtain the accompanying figures.

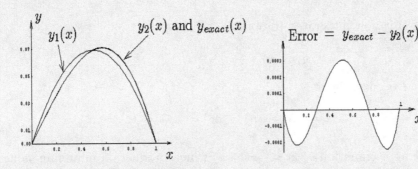

Note that the second approximation is quite good. The second approximation usually turns out to be good in most applications (but not all) of the Rayleigh-Ritz method.

The Rayleigh-Ritz method can also be applied to obtain approximate solutions for functions which minimize or maximize functionals involving more than one variable. Consider a functional of the form

$$I = I(u) = \iint_R f\left(x, y, u, \frac{\partial u}{\partial x}, \frac{\partial u}{\partial y}\right) dxdy, \qquad u = u(x,y), \quad \text{for } x, y \in R \qquad (6.166)$$

which is subject to the boundary condition

$$u = g(x,y), \quad \text{for } x, y \in C = \partial(R).$$

The approximation procedure is similar to the one-dimensional case previously considered. For the nth approximation one assumes a solution of the form

$$u(x,y) = u_n(x,y) = \phi_0(x,y) + c_1\phi_1(x,y) + c_2\phi_2(x,y) + \cdots + c_n\phi_n(x,y) \qquad (6.167)$$

where the c_i, $i = 1, 2, \ldots, n$ are constants to be determined. The $(n+1)$st approximation is

$$u_{n+1}(x,y) = \phi_0(x,y) + \bar{c}_1\phi_1(x,y) + \bar{c}_2\phi_2(x,y) + \cdots + \bar{c}_{n+1}\phi_{n+1}(x,y) \qquad (6.168)$$

where the $\bar{c}_i$, $i = 1, 2, \ldots, n+1$ are new constants to be determined. In all approximations it is customary to select $\phi_0(x,y)$ such that $\phi_0(x,y) = g(x,y)$ for $x, y \in C = \partial(R)$, and then select the remaining functions $\phi_i(x,y)$ for $i = 1, \ldots, n$ such that $\phi_i(x,y) = 0$ for $x, y \in C = \partial(R)$.

The equation (6.109) is just one form that can be used to construct an approximate solution. Alternatively, one can select a sequence of functions having the form

$$\begin{aligned} u_1 &= u_1(x, y, c_1) \\ u_2 &= u_2(x, y, c_1, c_2) \\ u_3 &= u_3(x, y, c_1, c_2, c_3) \\ &\vdots \\ u_n &= u_n(x, y, c_1, c_2, c_3, \ldots, c_n) \end{aligned} \qquad (6.169)$$

where each function satisfies the given boundary condition. When the nth approximate solution is substituted into the functional given by equation (6.166) and the integrations are performed, and one obtains a function

$$I = I(c_1, c_2, \ldots, c_n) \qquad (6.170)$$

which depends upon the n-constants. At an extreme value or stationary value for I the constants c_i, $i = 1, \ldots, n$ must be selected such that

$$\frac{\partial I}{\partial c_i} = 0, \qquad \text{for } i = 1, 2, \ldots, n \qquad (6.171)$$

This gives a system of n-equations in the n-unknowns $c_1, c_2, \ldots, c_n$ which must be solved.

Galerkin method

In order to develop a numerical technique for the construction of approximate solutions to boundary value problems, B.G. Galerkin (1871-1945), a Russian mathematician/Engineer, made use of the following concepts.

(a) A linear operator satisfies the properties

(i) $L(0) = 0$

(ii) $L(cu) = cL(u)$ for all constants c and functions u (6.172)

(iii) $L(u + v) = L(u) + L(v)$ for all functions u and v

(b) The inner product of two real functions $f(x)$ and $g(x)$ over an interval $a \leq x \leq b$ with respect to a weight function $w(x) > 0$ is defined

$$(f, g) = \int_a^b w(x)f(x)g(x)\, dx \tag{6.173}$$

(c) Two real functions $f(x)$ and $g(x)$ are said to be orthogonal over an interval $a \leq x \leq b$ with respect to a weight function $w(x)$ if the inner product is zero. That is, $f(x)$ and $g(x)$ are called orthogonal with respect to the weight function $w(x)$ if

$$(f, g) = \int_a^b w(x)f(x)g(x)\, dx = 0.$$

Here it is assumed that the functions $f(x)$ and $g(x)$ are nonzero, bounded and are such that the inner product integrals exist.

(d) The inner product of a function with itself is called a norm squared and is written

$$\| f \|^2 = (f, f) = \int_a^b w(x)f^2(x)\, dx$$

Here the norm squared is nonzero unless $f(x)$ is identically zero.

The above concepts associated with real functions of a single variable can also be extended to apply to real functions of several variables. For example,

(e) The inner product of two real functions $F(x, y)$ and $G(x, y)$ over a region R of the x, y-plane with respect to a weight function $W(x, y) > 0$ is defined

$$(F, G) = \iint_R W(x, y)F(x, y)G(x, y)\, dxdy \tag{6.174}$$

(f) Two real functions $F(x, y)$ and $G(x, y)$ are said to be orthogonal over a region R of the x, y-plane with respect to a weight function $W(x, y)$ if the inner product is zero. That is, $F(x, y)$ and $G(x, y)$ are called orthogonal with respect to a weight function $W(x, y)$ if

$$(F, G) = \iint_R W(x, y)F(x, y)G(x, y)\, dxdy = 0.$$

(g) The inner product of a function with itself is called a norm squared and written

$$\| F \|^2 = (F, F) = \iint_R W(x, y) F^2(x, y)\, dx\, dy$$

The Galerkin technique can be applied to boundary value problems involving either linear ordinary differential equations or linear partial differential equations. The technique also can be applied to the associated variational problem directly, if such a problem exits.

Consider an extremum of the functional

$$I = I(u) = \int_a^b \left[p(x) \left(\frac{du}{dx} \right)^2 + q(x) u^2 + 2 f(x) u \right] dx \tag{6.175}$$

where u is subject to the boundary conditions $u(a) = u_a$ and $u(b) = u_b$. The corresponding Euler-Lagrange equation is given by

$$L(u) = \frac{d}{dx} \left(p(x) \frac{du}{dx} \right) - q(x) u = f(x), \qquad a \leq x \leq b \tag{6.176}$$

subject to the boundary conditions

$$u(a) = u_a \qquad \text{and} \qquad u(b) = u_b \tag{6.177}$$

The Galerkin method assumes an approximate solution to this problem exists of the form

$$u = u(x) \approx \phi_0(x) + \sum_{j=1}^n c_j \phi_j(x), \quad c_1, c_2, \ldots, c_n \text{ are constants,} \tag{6.178}$$

where the functions $\phi_i(x)$ for $i = 0, 1, \ldots, n$ are linearly independent functions and $\phi_0(x)$ is selected to satisfy the given nonhomogeneous boundary conditions while the remaining functions $\phi_j(x)$ for $j = 1, \ldots, n$ are to satisfy homogeneous boundary conditions $\phi_j(a) = 0$ and $\phi_j(b) = 0$ for $j = 1, \ldots, n$. The operator L is a linear operator and consequently when the assumed solution (6.178) is substituted into the differential equation (6.176) the constants c_i, for $i = 1, \ldots, n$ are to be selected such that

$$L(u) = L(\phi_0) + L \left(\sum_{j=1}^n c_j \phi_j \right) = f \tag{6.179}$$

$$\text{or} \qquad L(u) = L(\phi_0) + \sum_{j=1}^n c_j L(\phi_j) = f$$

To solve for the n-constants $c_1, c_2, \ldots, c_n$ one selects a set of weight functions $\psi_1, \psi_2, \ldots, \psi_n$ and then forms the inner products

$$(\psi_i, L(\phi_0)) + \sum_{j=1}^n c_j (\psi_i, L(\phi_j)) = (\psi_i, f) \tag{6.180}$$

for $i = 1, 2, \ldots, n$. This gives a system of n-equations in n-unknowns $c_1, c_2, \ldots, c_n$ to be solved. The weight functions $\psi_i(x)$ are usually selected such that the resulting inner product integrals can be easily performed and the resulting system of equations has a unique solution. In many cases the functions $\phi_i(x)$ and $\psi_i(x)$ are selected as orthogonal functions. In this case many of the inner product integrals are zero. In many applications, the weight functions $\psi_i(x)$ are selected to be the same as the original independent set of functions $\phi_i(x)$. In this case the resulting system of equations has the form

$$
\begin{aligned}
(\phi_1, L(\phi_0)) + c_1(\phi_1, L(\phi_1)) + c_2(\phi_1, L(\phi_2)) + \cdots + c_n(\phi_1, L(\phi_n)) &= (\phi_1, f) \\
(\phi_2, L(\phi_0)) + c_1(\phi_2, L(\phi_1)) + c_2(\phi_2, L(\phi_2)) + \cdots + c_n(\phi_2, L(\phi_n)) &= (\phi_2, f) \\
\vdots \qquad\qquad &= \quad \vdots \\
(\phi_n, L(\phi_0)) + c_1(\phi_n, L(\phi_1)) + c_2(\phi_n, L(\phi_2)) + \cdots + c_n(\phi_n, L(\phi_n)) &= (\phi_n, f)
\end{aligned}
\tag{6.181}
$$

In the case where $L(\ )$ is a linear partial differential operator we consider the boundary-value problem to solve

$$
L(u) = f(x,y), \qquad x, y \in R
\tag{6.182}
$$

subject to the boundary condition that $u(x,y)\Big|_{(x,y)\in C=\partial R} = g(x,y)$. The Galerkin method assumes a solution of the form

$$
u_n(x,y) = \phi_0(x,y) + \sum_{i=1}^{n} c_i \phi_i(x,y)
\tag{6.183}
$$

where $\phi_i(x,y)$, for $i = 1, 2, \ldots, n$, are functions which satisfy the homogeneous boundary conditions $\phi_i(x,y) = 0$ for $(x,y) \in C = \partial R$ and $\phi_0(x,y) = g(x,y)$ for $(x,y) \in C = \partial R$ and c_i for $i = 1, 2, \ldots, n$ are constants to be determined. In general, the assumed solution given by equation (6.183) does not satisfy the partial differential equation (6.182). When the assumed solution (6.183) is substituted into the given equation (6.182) there results an equation of the form

$$
L(u_n) - f(x,y) = R_n(x,y)
\tag{6.184}
$$

where $R_n(x,y)$ is called a residual or error. Galerkin assumed that the series

$$
u(x,y) = \phi_0(x,y) + \sum_{i=1}^{\infty} c_i \phi_i(x,y)
$$

is a complete series and thought of $u_n(x,y)$, for $n = 1, 2, 3, \ldots$ as a sequence of partial sums which approximates the solution $u(x,y)$. Galerkin imposed the condition that the residual or error should be orthogonal, over the domain R, to each of the independent functions $\phi_k(x,y)$ for $k = 1, \ldots, n$. This inner product orthogonality condition is written

$$
(R_n, \phi_k) = \iint_R [L(u_n) - f(x,y)] \phi_k(x,y)\, dx dy = 0, \quad \text{for } k = 1, 2, \ldots, n
\tag{6.185}
$$

This represents n linear equations in the n unknowns $c_1, c_2, \ldots, c_n$ to be solved. Note that one must solve for a new set of coefficients c_i for $i = 1, \ldots, n$ each time the value of n is increased.

Rayleigh-Ritz, Galerkin and Collocation Methods for differential equations.

We present the least squares Rayleigh-Ritz, Galerkin and collocation methods by way of examples. These are numerical methods for constructing approximate solutions to boundary value problems in ordinary differential equations. We examine only second order ordinary differential having the general form

$$f(x, y, y', y'') = 0, \qquad a \le x \le b \tag{6.186}$$

which are subject to boundary conditions $y(a) = y_a$ and $y(b) = y_b$. From these examples one can get the general idea of how these techniques can be applied to higher ordered differential equations. The general procedure for each method is to select a set of independent functions $\phi_k(x)$ for $k = 0, 1, 2, \ldots, n$ and assume a solution to the given differential equation (6.186) having the form

$$y(x) = \phi_0(x) + c_1 \phi_1(x) + c_2 \phi_2(x) + \cdots + c_n \phi_n(x) \tag{6.187}$$

where $c_1, c_2, \ldots, c_n$ are constants. The function $\phi_0(x)$ is usually selected to satisfy the given end point conditions and the remaining functions $\phi_i(x)$, $i = 1, 2, \ldots, n$ are then selected to be zero at the end points $x = a$ and $x = b$. The assumed solution given by equation (6.187) is substituted into the given differential equation (6.186). Unless you are extremely lucky in your initial guess, the result of substituting the assumed solution into the differential equation will give a result called a residual which is different from zero. In general, one obtains a residual

$$R(x) = f(x, y(x), y'(x), y''(x)) = R(x, \phi_0(x), \phi_1(x), \ldots, \phi_n(x), c_1, c_2, \ldots, c_n) \tag{6.188}$$

which is a function of the independent functions and the constants $c_1, \ldots, c_n$. The Rayleigh-Ritz, Galerkin and collocation methods differ in how the residual is used to calculate the constants $c_1, c_2, \ldots, c_n$ used in the approximate solution to the differential equation.

The Rayleigh-Ritz method requires that one construct the definite integral

$$I = I(c_1, c_2, \ldots, c_n) = \int_a^b R^2(x)\, dx. \tag{6.189}$$

After integrating with respect to x and substituting in the limits of integration the result is then a function of the unknown constants $c_1, c_2, \ldots, c_n$. The problem is now treated as a maximum-minimum problem of solving for the critical point or points where I is a minimum. This requires that the system of equations

$$\frac{\partial I}{\partial c_k} = 0, \qquad k = 1, 2, \ldots, n \tag{6.190}$$

be solved for the unknown constants $c_1, c_2, \ldots, c_n$.

The Galerkin method requires that one select a set of weighting functions $\psi_k(x)$ for $k = 1, 2, \ldots, n$ and making the requirement that the residual satisfy the inner product relation

$$(R, \psi_k) = \int_a^b R\, \psi_k(x)\, dx = 0, \qquad k = 1, 2, \ldots, n \tag{6.191}$$

This gives a system of n-equations in the n-unknowns $c_1, c_2, \ldots, c_n$. The Galerkin method is called a weighted residual method. The equations (6.191) require that the residual function R be orthogonal to each of the weight functions. In many instances the weight functions are selected as the original approximating set $\phi_k(x)$ for $k = 1, 2, \ldots, n$. However, this is not a requirement of the Galerkin method.

The collocation method selects n-points x_k for $k = 1, 2, \ldots, n$ to lie in the interval $a \le x \le b$. Usually the points x_k are distributed uniformly throughout the interval (a, b) but this is not a requirement of the collocation method. One then makes the residual orthogonal to the Dirac delta functions $\delta(x - x_k)$. This gives the n-equations

$$(R, \delta(x - x_k)) = \int_a^b R \, \delta(x - x_k) \, dx = R(x_k) = 0, \qquad k = 1, 2, \ldots, n \tag{6.192}$$

which must be solved for the n-unknowns $c_1, c_2, \ldots, c_n$.

Example 6-3.   Rayleigh-Ritz Method

Use the Rayleigh-Ritz method to solve the boundary value problem

$$y'' + y + x = 0, \qquad y(0) = 0, \qquad y(1) = 0 \tag{6.193}$$

and compare the approximate solution with the exact solution

$$y(x) = \frac{\sin(x)}{\sin(1)} - x \tag{6.194}$$

Solution: Let $\phi_0(x) = 0$, since we want $y(0) = y(1) = 0$, and select the functions $\phi_1(x) = x(1 - x)$ and $\phi_2(x) = x^2(1 - x)$. Construct the approximate solution

$$y(x) = c_1 x(1 - x) + c_2 x^2(1 - x) \tag{6.195}$$

Substitute the approximate solution (6.195) into the differential equation and obtain the residual

$$R = R(x) = y'' + y + x = c_1(-2) + c_2(2 - 6x) + c_1 x(1 - x) + c_2 x^2(1 - x) + x \tag{6.196}$$

Next we calculate the integral

$$I = I(c_1, c_2) = \int_0^1 R^2(x) \, dx = \frac{1}{3} - \frac{11}{6} c_1 + \frac{101}{30} c_1^2 - \frac{19}{10} c_2 + \frac{101}{30} c_1 c_2 + \frac{131}{35} c_2^2 \tag{6.197}$$

At a stationary value

$$\frac{\partial I}{\partial c_1} = -\frac{11}{6} + \frac{101}{15} c_1 + \frac{101}{30} c_2 = 0$$

$$\frac{\partial I}{\partial c_2} = -\frac{19}{10} + \frac{101}{30} c_1 + \frac{262}{35} c_2 = 0 \tag{6.198}$$

The system of equations (6.198) has the solution

$$c_1 = \frac{46161}{246137} \quad \text{and} \quad c_2 = \frac{413}{2437} \tag{6.199}$$

We plot a graph of the approximate function $y(x)$ given by equation (6.195) together with the exact solution $y(x)$ given by equation (6.194). We also plot a graph of the error given by

$$Error = y_{approximate} - y_{exact} = c_1 x(1-x) + c_2 x^2(1-x) - \frac{\sin(x)}{\sin(1)} + x \tag{6.200}$$

These graphs are illustrated in the figure 6-8.

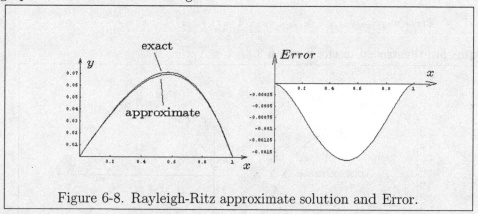

Figure 6-8. Rayleigh-Ritz approximate solution and Error.

Example 6-4.   Galerkin Method

Use the Galerkin method to solve the boundary value problem

$$y'' + y + x = 0, \qquad y(0) = 0, \qquad y(1) = 0 \tag{6.201}$$

and compare the approximate solution with the exact solution

$$y(x) = \frac{\sin(x)}{\sin(1)} - x \tag{6.202}$$

Solution: We proceed in exactly the same way as the previous example. Let $\phi_0(x) = 0$, since we want $y(0) = y(1) = 0$, and select the functions $\phi_1(x) = x(1-x)$ and $\phi_2(x) = x^2(1-x)$ which satisfy the homogeneous boundary conditions. Construct the approximate solution

$$y(x) = c_1 x(1-x) + c_2 x^2(1-x) \tag{6.203}$$

Substitute the approximate solution (6.203) into the differential equation (6.201) and obtain the same residual as in the previous example

$$R = R(x) = y'' + y + x = c_1(-2) + c_2(2-6x) + c_1 x(1-x) + c_2 x^2(1-x) + x \tag{6.204}$$

We select the weighting functions

$$\psi_1(x) = x(1-x) \qquad \text{and} \qquad \psi_2(x) = x^2(1-x) \tag{6.205}$$

and require that the following orthogonality conditions are satisfied

$$\int_0^1 R(x)\psi_1(x)\,dx = \frac{1}{12} - \frac{3}{10}c_1 - \frac{3}{20}c_2 = 0$$

$$\int_0^1 R(x)\psi_2(x)\,dx = \frac{1}{20} - \frac{3}{20}c_1 - \frac{13}{105}c_2 = 0 \tag{6.206}$$

The system of equations (6.206) has the solution

$$c_1 = \frac{71}{369} \qquad \text{and} \qquad c_2 = \frac{7}{41} \tag{6.207}$$

We plot a graph of the approximate function $y(x)$ given by equation (6.203) together with the exact solution $y(x)$ given by equation (6.202). We also plot a graph of the error given by

$$Error = y_{approximate} - y_{exact} = c_1 x(1-x) + c_2 x^2(1-x) - \frac{\sin(x)}{\sin(1)} + x \tag{6.208}$$

These graphs are illustrated in the figure 6-9.

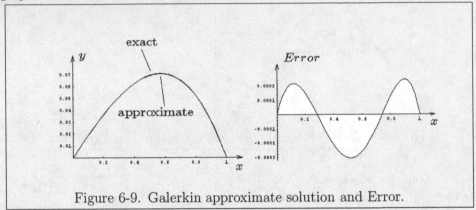

Figure 6-9. Galerkin approximate solution and Error.

Example 6-5. Collocation Method

Use the collocation method to solve the boundary value problem

$$y'' + y + x = 0, \qquad y(0) = 0, \qquad y(1) = 0 \tag{6.209}$$

and compare the approximate solution with the exact solution

$$y(x) = \frac{\sin(x)}{\sin(1)} - x \tag{6.210}$$

Solution: We proceed in exactly the same way as the previous examples. Let $\phi_0(x) = 0$, since we want $y(0) = y(1) = 0$, and select the functions $\phi_1(x) = x(1-x)$ and $\phi_2(x) = x^2(1-x)$ which satisfy the homogeneous boundary conditions. Construct the approximate solution

$$y(x) = c_1 x(1-x) + c_2 x^2(1-x) \tag{6.211}$$

Substitute the approximate solution (6.211) into the differential equation (6.209) and obtain the same residual as in the previous examples

$$R = R(x) = y'' + y + x = c_1(-2) + c_2(2 - 6x) + c_1 x(1-x) + c_2 x^2(1-x) + x \tag{6.212}$$

We select the points $x_1 = 1/4$ and $x_2 = 3/4$ and require that

$$R(1/4) = \frac{1}{4} - \frac{29}{16}c_1 + \frac{35}{64}c_2 = 0$$
$$R(3/4) = \frac{3}{4} - \frac{29}{16}c_1 - \frac{151}{64}c_2 = 0 \tag{6.213}$$

The system of equations (6.213) has the solution

$$c_1 = \frac{512}{2697} \quad \text{and} \quad c_2 = \frac{16}{93} \tag{6.214}$$

We plot a graph of the approximate function $y(x)$ given by equation (6.211) together with the exact solution $y(x)$ given by equation (6.210). We also plot a graph of the error given by

$$Error = y_{approximate} - y_{exact} = c_1 x(1-x) + c_2 x^2(1-x) - \frac{\sin(x)}{\sin(1)} + x \tag{6.215}$$

These graphs are illustrated in the figure 6-10.

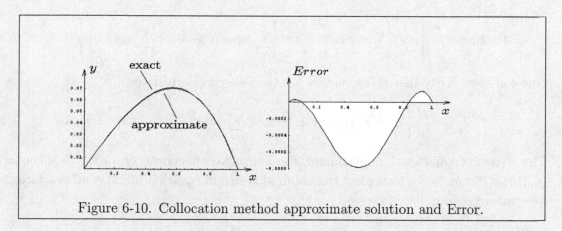

Figure 6-10. Collocation method approximate solution and Error.

Rayleigh-Ritz method and B-splines

The boundary value problem to solve

$$L(y) = -\frac{d}{dx}\left(p(x)\frac{dy}{dx}\right) + q(x)y = f(x), \qquad 0 \le x \le 1 \tag{6.216}$$

subject to the boundary conditions $y(0) = 0$ and $y(1) = 0$ is derivable from the variational problem to minimize the functional

$$I = I(y) = \int_0^1 \left\{ p(x)\left(\frac{dy}{dx}\right)^2 + q(x)y^2 - 2f(x)y \right\} dx \tag{6.217}$$

Assume that $p(x), q(x)$ and $f(x)$ are such that a unique solution to this boundary value problem exists. To obtain an approximate solution to this problem one can select a set of linearly independent basis functions $\psi_i(x)$, satisfying $\psi_i(0) = 0$ and $\psi_i(1) = 0$ for $i = 1, 2, \ldots, n$. The basis functions are used to construct an approximate solution to the differential equation (6.216) in the form of a finite series

$$y(x) = \sum_{i=1}^{n} c_i \psi_i(x) \tag{6.218}$$

where $c_1, c_2, \ldots, c_n$ are constants. The problem then reduces to determining the constants c_i, $i = 1, 2, \ldots, n$ such that $I(y) = I\left(\sum_{i=1}^{n} c_i \psi_i(x)\right)$ is minimized. A necessary condition for a minimum to occur is that

$$\frac{\partial I}{\partial c_j} = 0, \qquad \text{for each } j = 1, 2, \ldots, n \tag{6.219}$$

This gives a system of equations for determining the constants c_i for $i = 1, 2, \ldots, n$. Substitute the approximate solution given by equation (6.218) into the functional given by equation (6.217) to obtain

$$I = I(y) = \int_0^1 \left\{ p(x) \left[\sum_{n=1}^{n} c_i \psi_i'(x)\right]^2 + q(x) \left[\sum_{i=1}^{n} c_i \psi_i(x)\right]^2 - 2f(x) \left[\sum_{i=1}^{n} c_i \psi_i(x)\right] \right\} dx \tag{6.220}$$

One can then verify that this equation has the following derivatives

$$\frac{\partial I}{\partial c_j} = \int_0^1 \left\{ 2p(x) \sum_{i=1}^{n} c_i \psi_i'(x) \psi_j(x) + 2q(x) \sum_{i=1}^{n} c_i \psi_i(x) \psi_j(x) - 2f(x) \psi_j(x) \right\} dx \tag{6.221}$$

The system of equations for determining the constant coefficients c_j, $j = 1, 2, \ldots, n$, of equation (6.218) is obtain by factoring out the summation sign in equation (6.221) and evaluating the integrals to obtain

$$\frac{\partial I}{\partial c_j} = \sum_{i=1}^{n} \int_0^1 \left[p(x) \psi_i'(x) \psi_j'(x) + q(x) \psi_i(x) \psi_j(x) \right] dx - \int_0^1 f(x) \psi_j(x) = 0, \quad \text{for } j = 1, 2, \ldots, n \tag{6.222}$$

Various special functions can be selected to represent the basis functions $\psi_i(x)$ in the system of equation (6.222). The choice of certain B-splines leads to a tridiagonal system of equations to be solved. Tridiagonal systems of equations can be quickly solved and therefore B-splines are often selected as the basis functions for the Rayleigh-Ritz method for solving boundary value problems of the form given by equation (6.216).

B-splines

Partition the interval $0 \le x \le 1$ with $n+2$ points $0 = x_0 < x_1 < x_2 < x_3 < \cdots < x_n < x_{n+1} = 1$ and construct the B-splines defined by

$$B_i(x) = \begin{cases} 0, & 0 \le x \le x_{i-1} \\ \frac{x - x_{i-1}}{x_i - x_{i-1}}, & x_{i-1} < x \le x_i \\ \frac{x_{i+1} - x}{x_{i+1} - x_i}, & x_i < x \le x_{i+1} \\ 0, & x_{i+1} < x \le 1 \end{cases}$$

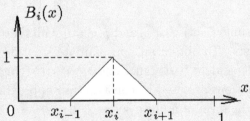

The B-spline has the value of 1 at $x = x_i$ and has the triangular shape illustrated. Hence if we select the functions $\psi_i(x) = B_i(x)$ as our basis functions the set of these functions look something like the figure 6-11.

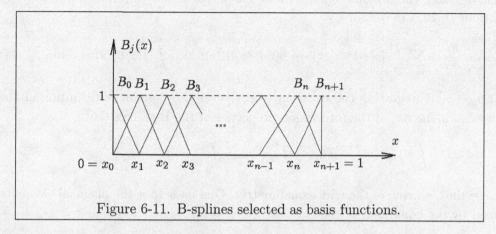

Figure 6-11. B-splines selected as basis functions.

One can verify that the B-splines have the derivatives given by

$$B_i'(x) = \begin{cases} 0, & 0 \le x \le x_{i-1} \\ \frac{1}{x_i - x_{i-1}}, & x_{i-1} < x \le x_i \\ \frac{-1}{x_{i+1} - x_i}, & x_i < x \le x_{i+1} \\ 0, & x_{i+1} < x \le 1 \end{cases}$$

The B-splines and their derivatives have the following properties that lead to the tridiagonal system of equations defining the constants of the approximate solution.

$$B_i(x)B_j(x) = 0, \quad \text{for } j \ne i-1, \ i, \ i+1$$
$$B_i'(x)B_j'(x) = 0, \quad \text{for } j \ne i-1, \ i, \ i+1 \tag{6.223}$$

Substituting the B-splines into the equations (6.222) and performing the integrations produces a tridiagonal linear system of equations having the matrix form $A\vec{c} = \vec{b}$ where A is a $n \times n$ tridiagonal matrix, $\vec{c} = \text{col}(c_1, c_2, \ldots, c_n)$ is a column vector containing the unknown coefficients and $\vec{b} = \text{col}(b_1, b_2, \ldots, b_n)$ is a column vector of right-hand sides to the linear system of equations. Recall that a tridiagonal linear system of equations having the matrix form $A\vec{c} = \vec{b}$ is characterized by a coefficient matrix A which has the form

$$A = \begin{bmatrix} a_{11} & a_{12} & & & & \\ a_{21} & a_{22} & a_{23} & & & \\ & a_{32} & a_{33} & a_{34} & & \\ & & \ddots & \ddots & \ddots & \\ & & & a_{n-1,n-2} & a_{n-1,n-1} & a_{n-1,n} \\ & & & & a_{n,n-1} & a_{nn} \end{bmatrix}$$

with elements along the main diagonal, super diagonal and subdiagonal with zero for the other elements of the matrix. This is where it gets its name tridiagonal. That is, $a_{ij} = 0$ for $i > j + 2$ and $a_{ij} = 0$ for $j > i + 2$. Let $h_i = x_{i+1} - x_i$ denote the spacing between the points x_{i+1} and x_i partitioning the domain $0 \leq x \leq 1$ and substitute $\psi_i(x) = B_i(x)$, for $i = 1, 2, \ldots, n$ into the equations (6.222) to obtain

$$\frac{\partial I}{\partial c_j} = \sum_{i=1}^{n} \int_0^1 \left[p(x) B_i'(x) B_j'(x) + q(x) B_i(x) B_j(x) \right] \, dx - \int_0^1 f(x) B_j(x) \, dx = 0 \qquad (6.224)$$

for $j = 1, 2, \ldots, n$. Integrating the equation (6.224) over the domain of definition of the B-splines we can make use of the integration properties of the B-splines that

$$\int_0^1 (\ ) \, dx = \int_0^{x_{i-1}} (\ ) \, dx + \int_{x_{i-1}}^{x_i} (\ ) \, dx + \int_{x_i}^{x_{i+1}} (\ ) \, dx + \int_{x_{i+1}}^1 (\ ) \, dx$$

to calculate the elements of the tridiagonal matrix. One finds that the diagonal elements are determined by the equations

$$a_{ii} = \frac{1}{h_{i-1}^2} \int_{x_{i-1}}^{x_i} \left[p(x) + (x - x_{i-1})^2 q(x) \right] \, dx + \frac{1}{h_i^2} \int_{x_i}^{x_{i+1}} \left[p(x) + (x_{i+1} - x)^2 q(x) \right] \, dx \qquad (6.225)$$

for $i = 1, 2, \ldots, n$. The super diagonal elements are found from the relations

$$a_{i,i+1} = \frac{1}{h_i^2} \int_{x_i}^{x_{i+1}} \left[-p(x) + (x_{i+1} - x)(x - x_i) q(x) \right] \, dx \qquad (6.226)$$

for $i = 1, 2, \ldots, n - 1$. The subdiagonal elements are found from the relations

$$a_{i,i-1} = \frac{1}{h_{i-1}^2} \int_{x_{i-1}}^{x_i} \left[p(x) + (x_i - x)(x - x_{i-1}) q(x) \right] \, dx \qquad (6.227)$$

for $i = 2, 3, \ldots, n$. The elements b_i of the column vector on the right-hand side of the linear system of equations are found from the integrations

$$b_i = \frac{1}{h_{i-1}} \int_{x_{i-1}}^{x_i} (x - x_{i-1}) f(x) \, dx + \frac{1}{h_i} \int_{x_i}^{x_{i+1}} (x_{i+1} - x) f(x) \, dx \qquad (6.228)$$

for $i = 1, 2, \ldots, n$. The verification of the above results is left as an exercise.

Introduction to the finite element method

The finite element method is similar to the Rayleigh-Ritz method used to approximate solutions to two-point boundary value problems. The finite element method is an alternative method to using difference equations to discretize partial differential equations. It has the advantage of being able to handle difficult boundary conditions associated with partial differential equations. Discretized finite difference methods requires the construction of grid points within the region of interest where a solution to the partial differential equation is to be constructed. These grid points are usually not compatible with irregular shaped boundaries and so the construction of discretized boundary conditions which involve derivatives becomes difficult if not impossible for some problems.

Consider a general second order partial differential equation which is derived from the functional

$$I = I(w) = \iint_R F(x, y, w, \frac{\partial w}{\partial x}, \frac{\partial w}{\partial y})\, dx\, dy \tag{6.229}$$

where R is the irregular region illustrated in the figure 6-12 with boundary curve ∂R. The resulting partial differential equation and associated boundary condition is referred to as a boundary value problem. There are various types of boundary conditions that can be considered.

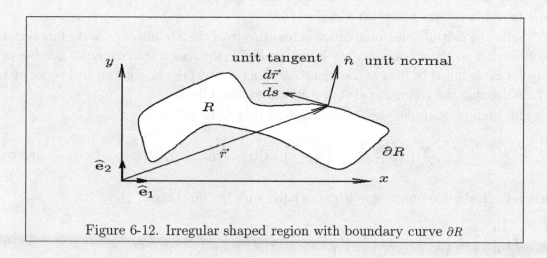

Figure 6-12. Irregular shaped region with boundary curve ∂R

Let s denote arc length measured along the boundary curve ∂R from some fixed point on the boundary. If $\vec{r} = \vec{r}(s) = x(s)\,\hat{\mathbf{e}}_1 + y(s)\,\hat{\mathbf{e}}_2$ denotes a position vector to a point on the boundary curve ∂R, then the vector $\frac{d\vec{r}}{ds}$ is a unit tangent vector to the boundary. The cross product $\hat{n} = \frac{d\vec{r}}{ds} \times \hat{\mathbf{e}}_3 = \frac{dy}{ds}\,\hat{\mathbf{e}}_1 - \frac{dx}{ds}\,\hat{\mathbf{e}}_2$ denotes a unit normal vector to the boundary curve ∂R. A Dirichlet boundary condition is said to exist whenever a value of w is specified on the boundary of the region R. This type of boundary condition is denoted

$$w(x, y)\Big|_{x, y \in \partial R} = g_1(x, y) \tag{6.230}$$

where $g_1(x, y)$ is specified function. A Neumann type boundary condition is said to exist whenever the normal derivative $\frac{\partial w}{\partial n}$ is specified along the boundary curve ∂R. This type of boundary condition is denoted

$$\frac{\partial w}{\partial n}\Big|_{x, y \in \partial R} = \nabla w \cdot \hat{n}\Big|_{x, y \in \partial R} = \left(\frac{\partial w}{\partial x}\,\hat{\mathbf{e}}_1 + \frac{\partial w}{\partial y}\,\hat{\mathbf{e}}_2 \right) \cdot \left(\frac{dy}{ds}\,\hat{\mathbf{e}}_1 - \frac{dx}{ds}\,\hat{\mathbf{e}}_2 \right)\Big|_{x, y \in \partial R} = g_2(x, y) \tag{6.231}$$

where $g_2(x,y)$ is a specified function. A Robin boundary condition is said to exist whenever some linear combination of the boundary conditions (6.230) and (6.231) are specified on the boundary curve ∂R. This type of boundary condition is denoted

$$\alpha \frac{\partial w}{\partial n} + \beta\, w \bigg|_{x,y \in \partial R} = g_3(x,y) \tag{6.232}$$

where α and β are constants and $g_3(x,y)$ is a specified function. A mixed boundary value problem is said to exist whenever different types of conditions are imposed on different sections of the boundary curve. For example, if the boundary condition $w(x,y)\Big|_{x,y\in C_1} = g_1(x,y)$ is imposed on one portion of the boundary curve C_1 and the condition $\frac{\partial w}{\partial n}\Big|_{x,y,\in C_2} = g_2(x,y)$ is imposed on another portion of the boundary curve C_2, where $C_1 \cup C_2 = \partial R$ and $C_1 \cap C_2 = \emptyset$, then a mixed boundary condition is said to exist.

To solve the partial differential equation resulting from the minimization of the functional (6.229) subject to given boundary conditions we divide the region R into a finite number of triangular elements. The lines of the triangular elements intersect at points called nodes and the lines defining the triangular elements are called nodal lines.

As an example consider the two-dimensional heat equation

$$\frac{\partial}{\partial x}\left(K_x \frac{\partial u}{\partial x}\right) + \frac{\partial}{\partial y}\left(K_y \frac{\partial u}{\partial y}\right) + Q(x,y) = 0, \qquad x,y \in R \tag{6.233}$$

which is the Euler-Lagrange equation associated with the functional

$$\iint_R \left[\frac{1}{2}\left\{ K_x \left(\frac{\partial u}{\partial x}\right)^2 + K_y \left(\frac{\partial u}{\partial y}\right)^2 \right\} - Q(x,y)\,u \right] dx dy = \iint_R f(x,y,u,u_x,u_y)\, dx dy \tag{6.234}$$

where R is a simply connected region in the x,y-plane. In the equations (6.233) and (6.234) K_x, K_y, with units $[cal/sec \cdot cm^2 \cdot C/cm]$, denote the thermal conductivity coefficients, $u = u(x,y)$, with units $[C]$, denotes the temperature at point (x,y) within the region R and $Q(x,y)$, with units $[cal/sec \cdot cm^3]$, denotes a heat source or sink at the point $(x,y) \in R$. We assume that the equation (6.233) is subject to the Dirichlet boundary conditions that

$$u(x,y) = g(x,y) \quad \text{for} \quad (x,y) \in \partial R \tag{6.235}$$

where the temperatures $g(x,y)$ are known on the boundary ∂R of the region R.

Consider for example the region R illustrated in the figure 6-13. Select points on the boundary of the region R and also select additional points interior to the region R where you wish to determine the values of the temperature. These interior nodal points are where we wish to determine the temperatures as specified by the partial differential equation (6.233).

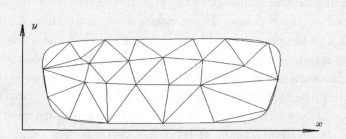

Figure 6-13. Points selected on boundary and interior to region R
connected by straight lines to form triangular elements.

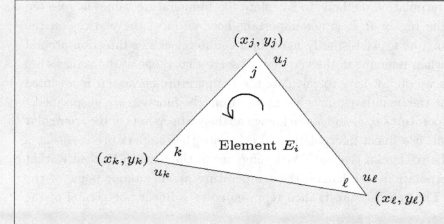

Figure 6-14. Triangular element E_i selected from within region R.

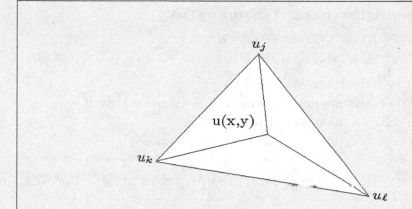

Figure 6-15. Linear interpolation used to calculate temperature $u(x, y)$ at interior point (x, y).

Construct from the points that you have selected triangular elements as illustrated in the figure 6-13. Note that in the finite element method to be developed the number of points selected within and on the boundary of the region R is related to the accuracy of the temperatures calculated at these points. If you select a small number of points, then the answers you obtain for the temperatures will be less accurate than the answers obtained by having selected a large number of points. Of course the more points selected requires more work in obtaining the answers and so some kind of compromise must be arrived at. Consider each triangle constructed within the region R as a triangular element and label these elements $1, 2, 3, \ldots, i, \ldots, N$. Let us examine the general ith triangular element illustrated in figure 6-14.

Associated with each triangular element there are vertices. For example, the ith triangular element illustrated in the figure 6-14 has the vertices labeled j, k and ℓ which have the coordinates labeled (x_j, y_j), (x_k, y_k) and (x_ℓ, y_ℓ). Our goal is to calculate the temperatures u_j, u_k and u_ℓ associated with the triangular element i for all values of $i = 1, 2, \ldots, N$. This will enable us to use interpolation methods to calculate the temperature values at selected interior points within the region R. It is not important how we label the vertices on the triangular element as long as we consistently move in a counterclockwise direction around the triangular element when referring to the vertices. Observe that if one of the vertices lies on a boundary ∂R, then we do not have to calculate the temperature because it is assumed that the temperatures at the boundary points are known from the function $g(x, y)$ specified.

Note that if the temperatures u_j, u_k and u_ℓ are known at the vertex points of the triangular element E_i, then one can use linear interpolation to calculate the temperature $u(x, y)$ at a point (x, y) interior to the triangular element. With reference to the figure 6-15 observe that one can use linear interpolation to calculate the temperature at an interior point of the triangular element E_i. The temperature is then represented as a linear polynomial of the form

$$u^{E_i} = u^{E_i}(x, y) = c_1 + c_2 x + c_3 y \tag{6.236}$$

where c_1, c_2 and c_3 are constants selected such that the linear form produces the nodal temperatures associated with the triangular element. This requires that

$$\begin{aligned} u^{E_i}(x_j, y_j) = u_j &= c_1 + c_2 x_j + c_3 y_j \\ u^{E_i}(x_k, y_k) = u_k &= c_1 + c_2 x_k + c_3 y_k \\ u^{E_i}(x_\ell, y_\ell) = u_\ell &= c_1 + c_2 x_\ell + c_3 y_\ell \end{aligned} \tag{6.237}$$

The system of equations (6.237) represents three equations in the three unknowns c_1, c_2, c_3. One can use Cramer's rule to solve for these coefficients to obtain

$$c_1 = \frac{\begin{vmatrix} u_j & x_j & y_j \\ u_k & x_k & y_k \\ u_\ell & x_\ell & y_\ell \end{vmatrix}}{2\Delta}, \qquad c_2 = \frac{\begin{vmatrix} 1 & u_j & y_j \\ 1 & u_k & y_k \\ 1 & u_\ell & y_\ell \end{vmatrix}}{2\Delta}, \qquad c_3 = \frac{\begin{vmatrix} 1 & x_j & u_j \\ 1 & x_k & u_k \\ 1 & x_\ell & u_\ell \end{vmatrix}}{2\Delta} \tag{6.238}$$

where 2Δ is the determinant

$$2\Delta = \begin{vmatrix} 1 & x_j & y_j \\ 1 & x_k & y_k \\ 1 & x_\ell & y_\ell \end{vmatrix} = (x_k y_\ell - x_\ell y_k) - (x_j y_\ell - x_\ell y_j) + (x_j y_k - x_k y_j) \tag{6.239}$$

Note that the symbol Δ represents the area of the triangular element E_i.

The equations (6.237) can also be written in the matrix form

$$\begin{bmatrix} u_j \\ u_k \\ u_\ell \end{bmatrix} = \begin{bmatrix} 1 & x_j & y_j \\ 1 & x_k & y_k \\ 1 & x_\ell & y_\ell \end{bmatrix} \begin{bmatrix} c_1 \\ c_2 \\ c_3 \end{bmatrix} \qquad (6.240)$$

and the solution can be written in the matrix form

$$\begin{bmatrix} c_1 \\ c_2 \\ c_3 \end{bmatrix} = \frac{1}{2\Delta} \begin{bmatrix} (x_k y_\ell - x_\ell y_k) & (x_\ell y_j - x_j y_\ell) & (x_j y_k - x_k y_j) \\ y_k - y_\ell & y_\ell - y_j & y_j - y_k \\ x_\ell - x_k & x_j - x_\ell & x_k - x_j \end{bmatrix} \begin{bmatrix} u_j \\ u_k \\ u_\ell \end{bmatrix} = \frac{1}{2\Delta} \begin{bmatrix} p_j & p_k & p_\ell \\ q_j & q_k & q_\ell \\ r_j & r_k & r_\ell \end{bmatrix} \begin{bmatrix} u_j \\ u_k \\ u_\ell \end{bmatrix}$$

where

$$\begin{aligned} p_j &= \frac{1}{2\Delta}(x_k y_\ell - x_\ell y_k) & p_k &= \frac{1}{2\Delta}(x_\ell y_j - x_j y_\ell) & p_\ell &= \frac{1}{2\Delta}(x_j y_k - x_k y_j) \\ q_j &= \frac{1}{2\Delta}(y_k - y_\ell) & q_k &= \frac{1}{2\Delta}(y_\ell - y_j) & q_\ell &= \frac{1}{2\Delta}(y_j - y_k) \\ r_j &= \frac{1}{2\Delta}(x_\ell - x_k) & r_k &= \frac{1}{2\Delta}(x_j - x_\ell) & r_\ell &= \frac{1}{2\Delta}(x_k - x_j) \end{aligned} \qquad (6.241)$$

Substituting these solutions for c_1, c_2 and c_3 into the equation (6.236), and rearranging terms, we can say that each triangular element E_i has associated with it a temperature function

$$u^{E_i} = u^{E_i}(x,y) = \alpha_j(x,y)u_j + \alpha_k(x,y)u_k + \alpha_\ell(x,y)u_\ell \qquad (6.242)$$

where

$$\begin{aligned} \alpha_j(x,y) &= \frac{1}{2\Delta}\left(p_j + q_j x + r_j y\right) \\ \alpha_k(x,y) &= \frac{1}{2\Delta}\left(p_k + q_k x + r_k y\right) \\ \alpha_\ell(x,y) &= \frac{1}{2\Delta}\left(p_\ell + q_\ell x + r_\ell y\right) \end{aligned} \qquad (6.243)$$

Define the functional associated with the unknown temperature u_j as

$$I = I(u_j) = \sum_{\substack{i \\ u_j \in E_i}} I_{E_i} \qquad \text{where} \qquad I_{E_i} = \iint_{E_i} f(x,y,u^{E_i},u_x^{E_i},u_y^{E_i})\,dxdy \qquad (6.244)$$

where f is defined by equation (6.234). The functional I has a stationary value when

$$\frac{\partial I}{\partial u_j} = \sum_{\substack{i \\ u_j \in E_i}} \frac{\partial I_{E_i}}{\partial u_j} = 0, \qquad j \text{ an interior node} \qquad (6.245)$$

Note that only triangular elements adjacent to the node j will contribute to the partial derivative $\frac{\partial I}{\partial u_j}$. If we calculate equation (6.245) for each unknown temperature u_j interior to the region R, then one obtains a system of equations to solve for the unknown temperatures.

Substitute the temperature function given by equation (6.242) into the functional (6.234) to obtain

$$I_{E_i} = \iint_{E_i} \left[\frac{1}{2}\left\{ K_x \left(\frac{\partial u^{E_i}}{\partial x}\right)^2 + K_y \left(\frac{\partial u^{E_i}}{\partial y}\right)^2 \right\} - Q u^{E_i} \right] dxdy \qquad (6.246)$$

Differentiate the equation (6.242) and verify the derivatives

$$\frac{\partial u^{E_i}}{\partial x} = \frac{1}{2\Delta} \left[q_j u_j + q_k u_k + q_\ell u_\ell \right], \qquad \frac{\partial u^{E_i}}{\partial y} = \frac{1}{2\Delta} \left[r_j u_j + r_k u_k + r_\ell u_\ell \right] \qquad (6.247)$$

so that the functional (6.246) can be written

$$I_{E_i} = \iint_{E_i} \left[\frac{1}{2} \left\{ \frac{K_x}{4\Delta^2} \left(q_j u_j + q_k u_k + q_\ell u_\ell \right)^2 + \frac{K_y}{4\Delta^2} \left(r_j u_j + r_k u_k + r_\ell u_\ell \right)^2 \right\} - Q u^{E_i} \right] dx dy$$

We differentiate this functional with respect to u_j to obtain

$$\frac{\partial I_{E_i}}{\partial u_j} = \iint_{E_i} \left[\frac{K_x}{4\Delta^2} \left(q_j u_j + q_k u_k + q_\ell u_\ell \right) q_j + \frac{K_y}{4\Delta^2} \left(r_j u_j + r_k u_k + r_\ell u_\ell \right) r_j - \frac{Q}{2\Delta} \left(p_j + q_j x + r_j y \right) \right] dx dy$$

This result can also be expressed in the matrix form

$$\frac{\partial I_{E_i}}{\partial u_j} = \frac{K_x}{4\Delta^2} \iint_{E_i} \left[q_j q_j \quad q_k q_j \quad q_\ell q_j \right] \begin{bmatrix} u_j \\ u_k \\ u_\ell \end{bmatrix} dx dy$$

$$+ \frac{K_y}{4\Delta^2} \iint_{E_i} \left[r_j r_j \quad r_k r_j \quad r_\ell r_j \right] \begin{bmatrix} u_j \\ u_k \\ u_\ell \end{bmatrix} dx dy - \iint_{E_i} \frac{Q}{2\Delta} \left(p_j + q_j x + r_j y \right) dx dy \qquad (6.248)$$

Observe that for Q constant the integrals in equation (6.248) can be reduced to the form

$$\iint_{E_i} dx dy = \Delta, \qquad \iint_{E_i} x \, dx dy = \bar{x} \Delta, \qquad \iint_{E_i} y \, dx dy = \bar{y} \Delta \qquad (6.249)$$

where Δ is the area of the triangular element E_i and $(\bar{x}, \bar{y})$ represent the coordinates of the centroid of the triangular element E_i. The coordinates of the centroid of a triangle is given by the relations

$$\bar{x} = \frac{1}{3}(x_j + x_k + x_\ell), \qquad \bar{y} = \frac{1}{3}(y_j + y_k + y_\ell) \qquad (6.250)$$

Consequently, one can verify that an integration of the equation (6.248), with Q constant, produces the result

$$\frac{\partial I_{E_i}}{\partial u_j} = \frac{K_x}{4\Delta} \left[q_j q_j \quad q_k q_j \quad q_\ell q_j \right] \begin{bmatrix} u_j \\ u_k \\ u_\ell \end{bmatrix}$$

$$+ \frac{K_y}{4\Delta} \left[r_j r_j \quad r_k r_j \quad r_\ell r_j \right] \begin{bmatrix} u_j \\ u_k \\ u_\ell \end{bmatrix} - \frac{Q}{2} \left[p_j + q_j \bar{x} + r_j \bar{y} \right] \qquad (6.251)$$

Associated with each triangular element E_i one can define the partial derivative matrix

$$\frac{\partial I_{E_i}}{\partial u} = \begin{bmatrix} \frac{\partial I_{E_i}}{\partial u_j} \\ \frac{\partial I_{E_i}}{\partial u_k} \\ \frac{\partial I_{E_i}}{\partial u_\ell} \end{bmatrix} \qquad (6.252)$$

This matrix can be obtained by permuting the subscripts j, k, ℓ, in a counterclockwise fashion and then rearranging terms to arrive at the result

$$
\frac{\partial I_{E_i}}{\partial u} = \begin{bmatrix} \frac{\partial I_{E_i}}{\partial u_j} \\ \frac{\partial I_{E_i}}{\partial u_k} \\ \frac{\partial I_{E_i}}{\partial u_\ell} \end{bmatrix} = \frac{K_x}{4\Delta} \begin{bmatrix} q_j q_j & q_k q_j & q_\ell q_j \\ q_j q_k & q_k q_k & q_\ell q_k \\ q_j q_\ell & q_k q_\ell & q_\ell q_\ell \end{bmatrix} \begin{bmatrix} u_j \\ u_k \\ u_\ell \end{bmatrix}
$$

$$
+ \frac{K_y}{4\Delta} \begin{bmatrix} r_j r_j & r_k r_j & r_\ell r_j \\ r_j r_k & r_k r_k & r_\ell r_k \\ r_j r_\ell & r_k r_\ell & r_\ell r_\ell \end{bmatrix} \begin{bmatrix} u_j \\ u_k \\ u_\ell \end{bmatrix} - \frac{Q}{2} \begin{bmatrix} p_j + q_j \bar{x} + r_j \bar{y} \\ p_k + q_k \bar{x} + r_k \bar{y} \\ p_\ell + q_\ell \bar{x} + r_\ell \bar{y} \end{bmatrix} \tag{6.253}
$$

In the following example we calculate the equations (6.253) for each triangular element and then form the summation given by equation (6.245). The resulting equations form a system of linear equations from which the unknowns temperature can be determined. Note in particular that in forming the equation (6.253) one need only calculate the partial derivatives in those situations where the nodes j, k or ℓ are not points belonging to the boundary ∂R. In the following example we have tried to keep the bookkeeping to a minimum and have selected only 17 points associated with the region R. These seventeen points produce 21 triangular elements and 6 nodes where the temperature is unknown. Don't expect too much accuracy from this example since the triangular elements are large and the number of unknowns is small. In practice, one can use computer codes with many more interior points and associated triangular elements to achieve a much greater accuracy. The example given is just to illustrate the finite element method with a minimum amount of effort. The limited amount of points will also allow a detailed description of each step of the finite element method (FEM).

Example 6-6.   Finite element introduction

Consider the special case where $K_x = K_y = 1$ and $Q = 2$. The equation (6.233) then becomes the Poisson equation

$$
\nabla^2 u = \frac{\partial^2 u}{\partial x^2} + \frac{\partial^2 u}{\partial y^2} + 2 = 0, \qquad x, y \in R
$$

where we select the quarter circle illustrated to represent the region R. For boundary conditions assume the Dirichlet conditions

$$
u = \begin{cases} 0, & \text{for} \quad x^2 + y^2 = 25 \\ -\frac{1}{2}(x^2 - 25), & \text{on} \quad y = 0 \\ -\frac{1}{2}(y^2 - 25), & \text{on} \quad x = 0 \end{cases}
$$

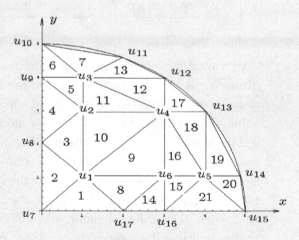

nodes and triangular elements

We selected 17 points within the region R as illustrated. Note that the nodal values $u_1, u_2, \ldots, u_6$ are unknowns and the boundary values $u_7, u_8, \ldots, u_{17}$ are known. We then constructed a set of triangular elements as illustrated. The tables 6.1 and 6.2 lists properties associated with the elements. We label both the nodes and triangular elements in some convenient way as illustrated in the figure. We begin the finite element method with the following table of values.

Table 6.1 Starting Values		
Node	Coordinates	Temperature
1	$(1,1)$	u_1 unknown
2	$(1,3)$	u_2 unknown
3	$(1,4)$	u_3 unknown
4	$(3,3)$	u_4 unknown
5	$(4,1)$	u_5 unknown
6	$(3,1)$	u_6 unknown
7	$(0,0)$	$u_7 = 12.5$
8	$(0,2)$	$u_8 = 10.5$
9	$(0,4)$	$u_9 = 4.5$
10	$(0,5)$	$u_{10} = 0.0$
11	$(2,\sqrt{21})$	$u_{11} = 0.0$
12	$(3,4)$	$u_{12} = 0.0$
13	$(4,3)$	$u_{13} = 0.0$
14	$(\sqrt{24},1)$	$u_{14} = 0.0$
15	$(5,0)$	$u_{15} = 0.0$
16	$(3,0)$	$u_{16} = 8.0$
17	$(2,0)$	$u_{17} = 10.5$

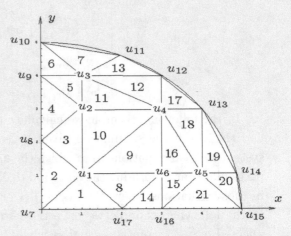

nodes and triangular elements

Table 6.2 Area and centroid of elements		
Element	Area Δ	Centroid $(\bar{x}, \bar{y})$
1	1.000	(1.000, 0.333)
2	1.000	(0.333, 1.000)
3	1.000	(0.667, 2.000)
4	1.000	(0.333, 3.000)
5	0.500	(0.667, 3.667)
6	0.500	(0.333, 4.333)
7	0.791	(1.000, 4.528)
8	1.000	(2.000, 0.667)
9	2.000	(2.333, 1.667)
10	2.000	(1.667, 2.333)
11	1.000	(1.667, 3.333)
12	1.000	(2.333, 3.667)
13	0.583	(2.000, 4.194)
14	0.500	(2.667, 0.333)
15	0.500	(3.333, 0.667)
16	1.000	(3.333, 1.667)
17	0.500	(3.333, 3.333)
18	1.000	(3.667, 2.333)
19	0.899	(4.300, 1.667)
20	0.449	(4.633, 0.667)
21	1.000	(4.000, 0.333)

We want to select the interior nodal temperatures $u_1, u_2, u_3, u_4,$ u_5 and u_6 so that the functional given by equation (6.244) has a minimum value. This requires that we select the unknown temperatures to satisfy the system of equations resulting from equation (6.245). To obtain this system of equations we first calculate the quantities $p_j, p_k, p_\ell,$ q_j, q_k, q_ℓ and r_j, r_k, r_ℓ associated with each triangular element. We calculate these quantities by moving in a counter clockwise direction around each element and then list them in table 6.3. Note that it does not matter which node of a triangular element you label j as long as you move in a counterclockwise direction around the element.

element	j	k	ℓ	p_j	p_k	p_ℓ	q_j	q_k	q_ℓ	r_j	r_k	r_ℓ

Table 6.3 Parameters associated with triangular elements

element	j	k	ℓ	p_j	p_k	p_ℓ	q_j	q_k	q_ℓ	r_j	r_k	r_ℓ
1	1	7	17	0.00	2.00	0.00	0.00	-1.00	1.00	2.00	-1.00	-1.00
2	1	8	7	0.00	0.00	2.00	2.00	-1.00	-1.00	0.00	1.00	-1.00
3	1	2	8	2.00	-2.00	2.00	1.00	1.00	-2.00	-1.00	1.00	0.00
4	8	2	9	4.00	0.00	-2.00	-1.00	2.00	-1.00	-1.00	0.00	1.00
5	2	3	9	4.00	-4.00	1.00	0.00	1.00	-1.00	-1.00	1.00	0.00
6	3	10	9	0.00	-4.00	5.00	1.00	0.00	-1.00	0.00	1.00	-1.00
7	3	11	10	10.00	-5.00	-3.42	-0.42	1.00	-0.58	-2.00	1.00	1.00
8	17	6	1	2.00	-2.00	2.00	0.00	1.00	-1.00	-2.00	1.00	1.00
9	6	4	1	0.00	-2.00	6.00	2.00	0.00	-2.00	-2.00	2.00	0.00
10	1	4	2	6.00	-2.00	0.00	0.00	2.00	-2.00	-2.00	0.00	2.00
11	4	3	2	-1.00	-6.00	9.00	1.00	0.00	-1.00	0.00	2.00	-2.00
12	4	12	3	8.00	-9.00	3.00	0.00	1.00	-1.00	-2.00	2.00	0.00
13	3	12	11	5.75	3.42	-8.00	-0.58	0.58	0.00	-1.00	-1.00	2.00
14	17	16	6	3.00	-2.00	0.00	-1.00	1.00	0.00	0.00	-1.00	1.00
15	5	6	16	-3.00	3.00	1.00	1.00	-1.00	0.00	0.00	1.00	-1.00
16	5	4	6	-6.00	-1.00	9.00	2.00	0.00	-2.00	0.00	1.00	-1.00
17	4	13	12	7.00	-3.00	-3.00	-1.00	1.00	0.00	-1.00	0.00	1.00
18	5	13	4	3.00	-9.00	8.00	0.00	2.00	-2.00	-1.00	1.00	0.00
19	5	14	13	10.70	-8.00	-0.90	-2.00	2.00	0.00	-0.90	0.00	0.90
20	15	14	5	0.90	-5.00	5.00	0.00	1.00	-1.00	-0.90	1.00	-0.10
21	15	5	16	-3.00	0.00	5.00	1.00	0.00	-1.00	-1.00	2.00	-1.00

Now for each interior node we calculate the partial derivatives given by equation (6.253). Note that we do not construct the partial derivatives for the boundary nodes as these values are fixed and are not changing. The results are listed for each interior nodal point.

Node 1

$$\frac{\partial I}{\partial u_1} = \frac{\partial I_{E_1}}{\partial u_1} + \frac{\partial I_{E_2}}{\partial u_1} + \frac{\partial I_{E_3}}{\partial u_1} + \frac{\partial I_{E_8}}{\partial u_1} + \frac{\partial I_{E_9}}{\partial u_1} + \frac{\partial I_{E_{10}}}{\partial u_1} = 0$$

which results in the equation

$$4u_1 - 0.5u_2 - 0.5u_6 = 38.833 \tag{6.254}$$

Node 2

$$\frac{\partial I}{\partial u_2} = \frac{\partial I_{E_3}}{\partial u_2} + \frac{\partial I_{E_4}}{\partial u_2} + \frac{\partial I_{E_5}}{\partial u_2} + \frac{\partial I_{E_{10}}}{\partial u_2} + \frac{\partial I_{E_{11}}}{\partial u_2} = 0$$

which results in the equation

$$-0.5u_1 + 4.25u_2 - 1.5u_3 - 0.75u_4 = 16.417 \tag{6.255}$$

Node 3

$$\frac{\partial I}{\partial u_3} = \frac{\partial I_{E_5}}{\partial u_3} + \frac{\partial I_{E_6}}{\partial u_3} + \frac{\partial I_{E_7}}{\partial u_3} + \frac{\partial I_{E_{11}}}{\partial u_3} + \frac{\partial I_{E_{12}}}{\partial u_3} + \frac{\partial I_{E_{13}}}{\partial u_3} = 0$$

which results in the equation

$$-1.5u_2 + 4.64u_3 = 7.42 \qquad (6.256)$$

Node 4

$$\frac{\partial I}{\partial u_4} = \frac{\partial I_{E_9}}{\partial u_4} + \frac{\partial I_{E_{10}}}{\partial u_4} + \frac{\partial I_{E_{11}}}{\partial u_4} + \frac{\partial I_{E_{12}}}{\partial u_4} + \frac{\partial I_{E_{16}}}{\partial u_4} + \frac{\partial I_{E_{17}}}{\partial u_4} + \frac{\partial I_{E_{18}}}{\partial u_4} = 0$$

which results in the equation

$$-0.75u_2 + 4.5u_4 - 0.75u_6 = 5.67 \qquad (6.257)$$

Node 5

$$\frac{\partial I}{\partial u_5} = \frac{\partial I_{E_{15}}}{\partial u_5} + \frac{\partial I_{E_{16}}}{\partial u_5} + \frac{\partial I_{E_{18}}}{\partial u_1} + \frac{\partial I_{E_{19}}}{\partial u_5} + \frac{\partial I_{E_{20}}}{\partial u_5} + \frac{\partial I_{E_{21}}}{\partial u_5} = 0$$

which results in the equation

$$4.65u_5 - 1.5u_6 = 7.238 \qquad (6.258)$$

Node 6

$$\frac{\partial I}{\partial u_6} = \frac{\partial I_{E_8}}{\partial u_6} + \frac{\partial I_{E_9}}{\partial u_6} + \frac{\partial I_{E_{14}}}{\partial u_6} + \frac{\partial I_{E_{15}}}{\partial u_6} + \frac{\partial I_{E_{16}}}{\partial u_6} = 0$$

which results in the equation

$$-0.5u_1 - 0.75u_4 - 1.5u_5 + 4.25u_6 = 16.583 \qquad (6.259)$$

Solving the above system of equations produces the results given in table 6.4.

Table 6.4 Comparison of finite element and exact solutions						
	u_1	u_2	u_3	u_4	u_5	u_6
Finite element solutions	11.527	7.262	3.947	3.684	3.906	7.287
Exact solutions	11.50	7.50	4.00	3.50	4.00	7.50

Using a finer mesh structure of triangular elements and more interior data points there results a larger system of equations to solve. Associated with the increase of interior points there is more additional work to obtain the system of equations for determining the unknown temperatures. The end result is that there is a much improved representation of the true temperatures. The above steps can be programmed into a digital computer which can perform the above steps fairly quickly. Finite element computational packages can be found in some selected computer software packages. Other types of elements and interpolating functions can be constructed for special one-dimensional, two-dimensional and three-dimensional problems.

> "...when new groups of phenomena compel changes in the pattern of thought... even the most eminent of physicists find immense difficulties. For the demand for change in the thought pattern may engender the feeling that the ground is to be pulled from under one's feet.... I believe that the difficulties at this point can hardly be overestimated. Once one has experienced the desperation with which clever and conciliatory men of science react to the demand for a change in the thought pattern, one can only be amazed that such revolutions in science have actually been possible at all.
>
> Werner Heisenberg (1901-1976)

Exercises Chapter 6

► 1.

Consider a surface of revolution moving with constant speed V through a fluid medium with constant fluid density ρ. Assume that the normal pressure P_n at a point (x, y) on this surface of revolution is given by

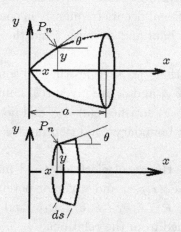

$$P_n = 2\rho V^2 \sin^2 \theta,$$

where ρ is the fluid density, V is the speed of the fluid, and θ is the tangent angle at the point (x, y) on the surface.

If P_n acts normal to the surface, it acts everywhere upon a ribbon element of width $ds = \sqrt{1 + (y')^2}\, dx$ with a force dF given by

$$dF = P_n 2\pi y\, ds = 2\rho V^2 (\sin^2 \theta)\, 2\pi y \sqrt{1 + (y')^2}\, dx.$$

The x-component of this force is the drag force and is given by

$$F_x = \int_0^a 4\pi \rho V^2 \sin^3 \theta y \sqrt{1 + (y')^2}\, dx.$$

In order to make this problem more tractable, make the following approximation. From $y' = \tan \theta$ assume that

$$\sin \theta = \frac{y'}{\sqrt{1 + (y')^2}} \approx y'$$

which implies $y' << 1$. The drag force then simplifies to

$$F_x = 4\pi \rho V^2 \int_0^a y(y')^3\, dx.$$

Find the curve $y = y(x)$ which makes the drag force F_x as small as possible, subject to the restrictions that $y(0) = 0$ and $y(a) = b$. where a and b are positive constants.

▶ **2.** Find the extremals for the functional

$$I(y, z) = \int_0^1 \left(2yz' + 2zy' + y'^2 + z'^2 \right) dx$$

which are subject to the boundary conditions $y(0) = 0$, $y(1) = 1$, $z(0) = 1$, $z(1) = 0$.

▶ **3.** Find the Euler-Lagrange equations associated with the functional

$$I = I(y) = \int_a^b \left[p(x) \left(\frac{dy}{dx} \right)^2 + q(x)y^2 + 2f(x)y \right] dx$$

where $p(x)$, $p'(x)$, $q(x)$ and $f(x)$ are given continuous functions over the interval $[a, b]$. This type of functional occurs frequently in continuum mechanics, particularly in the study of the deflection of bars and strings. Find the Euler-Lagrange equation subject to the following conditions.

(a) Assume that $y(x)$ satisfies the end conditions $y(a) = y_a$ and $y(b) = y_b$
   where y_a and y_b are given real numbers.

(b) Assume $y(x)$ satisfies $y(a) = y_a$ and no boundary conditions are specified at $x = b$.
   What boundary condition must $y(x)$ satisfy at $x = b$ if I is to be minimized?

▶ **4.** In the theory of electricity and magnetism it is known that if there exists a scalar potential $\phi = \phi(x, y, z)$ and a vector potential $\vec{A} = \vec{A}(t, x, y, z) = A_1 \widehat{e}_1 + A_2 \widehat{e}_2 + A_3 \widehat{e}_3$, then the electric force $\vec{E} = E_1 \widehat{e}_1 + E_2 \widehat{e}_2 + E_3 \widehat{e}_3$ and magnetic induction vector $\vec{B} = B_1 \widehat{e}_1 + B_2 \widehat{e}_2 + B_3 \widehat{e}_3$ are determined from the relations

$$\vec{E} = -\nabla \phi - \frac{\partial \vec{A}}{\partial t} \qquad \text{and} \qquad \vec{B} = \nabla \times \vec{A}$$

In addition the scalar potential ϕ and vector potential $\vec{A}$ can be used to calculate the potential energy V associated with a moving charge q. If the charge q is moving with a velocity $\vec{v} = \dot{x} \widehat{e}_1 + \dot{y} \widehat{e}_2 + \dot{z} \widehat{e}_3$ then the potential energy V is given by

$$V = q\phi - q\vec{A} \cdot \vec{v}$$

(a) Show that the Lagrangian for this system can be written $L = \frac{1}{2} m\vec{v} \cdot \vec{v} - q\phi + q\vec{A} \cdot \vec{v}$

(b) Show that the equations of motion can be written in the form

$$m\ddot{x} - qE_1 - q\left[\dot{y} \left(\frac{\partial A_2}{\partial x} - \frac{\partial A_1}{\partial y} \right) - \dot{z} \left(\frac{\partial A_1}{\partial z} - \frac{\partial A_3}{\partial x} \right) \right] = 0$$

$$m\ddot{y} - qE_2 - q\left[\dot{x} \left(\frac{\partial A_1}{\partial y} - \frac{\partial A_2}{\partial x} \right) + \dot{z} \left(\frac{\partial A_3}{\partial y} - \frac{\partial A_2}{\partial z} \right) \right] = 0$$

$$m\ddot{z} - qE_3 - q\left[\dot{x} \left(\frac{\partial A_1}{\partial z} - \frac{\partial A_3}{\partial x} \right) - \dot{y} \left(\frac{\partial A_3}{\partial y} - \frac{\partial A_2}{\partial z} \right) \right] = 0$$

(c) Show that by letting $x_1 = x, x_2 = y$ and $x_3 = z$ the above equations can be written in the indicial form

$$m\ddot{x}_i - q\left[\vec{E} + \vec{v} \times \vec{B} \right]_i = 0 \qquad \text{for} \quad i = 1, 2, 3$$

▶ **5.** The differential equation describing the small deflections y of a rotating shaft of length L with end load P_0 and transverse loading of $f(x)$ is

$$\frac{d^2}{dx^2}\left(EI\frac{d^2y}{dx^2}\right) + P_0\frac{d^2y}{dx^2} - \rho\omega^2 y = f(x), \qquad 0 \le x \le L \qquad (5-1)$$

where E, I,P_0, ρ and ω are constants.

(a) Multiply this equation by δy and then use integration by parts and integrate from 0 to L to show that equation (5-1), with appropriate boundary conditions, comes from the variational problem

$$\delta \int_0^L \left[\frac{1}{2}EI(y'')^2 - \frac{1}{2}P_0(y')^2 - \frac{1}{2}\rho\omega^2 y^2 - f(x)y\right]dx$$
$$+ \left[\frac{d}{dx}(EIy')\,\delta y + P_0 y'\,\delta y - EIy''\delta y'\right]_0^L = 0$$

(b) What are the essential and natural boundary conditions for this problem?
Hint: A fourth order ordinary differential equation needs four boundary conditions to specify the constants in the general solution.

▶ **6.**
(a) Use integration by parts to show that

$$\int_{x_1}^{x_2} \frac{d}{dx}(p(x)y')\,\delta y\,dx = \int_{x_1}^{x_2} -p(x)y'\,\delta y'\,dx + [p(x)y'\,\delta y]_{x_1}^{x_2}$$

(b) Show the result in part (a) implies the relation

$$\frac{d}{dx}(p(x)y')\,\delta y = -\delta\left[\frac{1}{2}p(x)(y')^2\right] + \frac{d}{dx}(p(x)y'\,\delta y)$$

▶ **7.**
(a) Use integration by parts to show that

$$\int_{x_1}^{x_2} \frac{d^2}{dx^2}(p(x)y'')\,\delta y\,dx = \left[\frac{d}{dx}(p(x)y'')\,\delta y - (p(x)y'')\,\delta y'\right]_{x_1}^{x_2} + \int_{x_1}^{x_2} p(x)y''\,\delta y''\,dx$$

(b) Show the result in part (a) implies the relation

$$\frac{d^2}{dx^2}(p(x)y'')\,\delta y = \delta\left[\frac{1}{2}p(x)(y'')^2\right] + \frac{d}{dx}\left[\frac{d}{dx}(p(x)y'')\,\delta y - (p(x)y'')\,\delta y'\right]$$

▶ **8.** Verify that the Schrödinger equation results from finding an extremum for the volume integral

$$I = \iiint_V \left[\left(\frac{\partial u}{\partial x}\right)^2 + \left(\frac{\partial u}{\partial y}\right)^2 + \left(\frac{\partial u}{\partial z}\right)^2 + \frac{8\pi^2 m}{h^2}(V-E)\,u^2\right]dxdydz$$

304

▶ **9.** Derive the variational problem associated with the deformation $u = u(x, t)$ of a string which is subjected to an external force per unit length $f = f(x, t)$.

(a) Show that the functional to be minimized is

$$I = \int_{t_1}^{t_2} \int_0^L \left[\frac{1}{2} \varrho (u_t)^2 - \frac{1}{2} \tau (u_x)^2 + fu \right] dx dt$$

(b) Show that if ϱ and τ are not constants, then the equation describing the deformation is given by

$$\frac{\partial}{\partial t} (\varrho u_t) - \frac{\partial}{\partial x} (\tau u_x) = f(x, t)$$

▶ **10.** Consider the variational problem to find y_1 and y_2 such that

$$I(y_1, y_2) = \int_{x_1}^{x_2} f(x, y_1, y_1', y_2, y_2') \, dx$$

is to be a minimum subject to a constraint condition $g(y_1, y_1', y_2, y_2') = 0$.

(a) Show

$$\delta I = \int_{x_1}^{x_2} \left(\frac{\partial f}{\partial y_1} \delta y_1 + \frac{\partial f}{\partial y_1'} \delta y_1' + \frac{\partial f}{\partial y_2} \delta y_2 + \frac{\partial f}{\partial y_2'} \delta y_2' \right) dx = 0 \qquad (10-1)$$

and

$$\delta g = \frac{\partial g}{\partial y_1} \delta y_1 + \frac{\partial g}{\partial y_1'} \delta y_1 + \frac{\partial g}{\partial y_2} \delta y_2 + \frac{\partial g}{\partial y_2'} \delta y_2' = 0 \qquad (10-2)$$

(b) Multiply equation (10-2) by λ and then integrate from x_1 to x_2 and add the result to equation (10-1). Show that after an integration by parts one obtains

$$\int_{x_1}^{x_2} \left\{ \left[\frac{\partial f}{\partial y_1} + \lambda \frac{\partial g}{\partial y_1} - \frac{d}{dx} \left(\frac{\partial f}{\partial y_1'} + \lambda \frac{\partial g}{\partial y_1'} \right) \right] \delta y_1 \right.$$
$$\left. + \left[\frac{\partial f}{\partial y_2} + \lambda \frac{\partial g}{\partial y_2} - \frac{d}{dx} \left(\frac{\partial f}{\partial y_2'} + \lambda \frac{\partial g}{\partial y_2'} \right) \right] \delta y_2 \right\} dx = 0$$

provided that $\delta y_1(x_1) = \delta y_1(x_2) = \delta y_2(x_1) = \delta y_2(x_2) = 0$.

(c) Note that δy_1 and δy_2 are not both independent because of the constraint condition $g = 0$. The equation (10-2) tells us how δy_2 is related to δy_1. If we select λ such that the coefficient of δy_2 is zero, i.e.

$$\frac{\partial f}{\partial y_2} + \lambda \frac{\partial g}{\partial y_2} - \frac{d}{dx} \left(\frac{\partial f}{\partial y_2'} + \lambda \frac{\partial g}{\partial y_2'} \right) = 0,$$

then by the fundamental lemma from chapter 3 we have

$$\frac{\partial f}{\partial y_1} + \lambda \frac{\partial g}{\partial y_1} - \frac{d}{dx} \left(\frac{\partial f}{\partial y_1'} + \lambda \frac{\partial g}{\partial y_1'} \right) = 0$$

which together with $g(y_1, y_2, y_1', y_2') = 0$ gives three equations in the three unknowns y_1, y_2 and λ.

▶ **11.**   Consider the problem of finding an extreme value for the functional

$$I = \int_{(x_1)}^{x_2} \sqrt{E(x,y) + 2F(x,y)y' + G(x,y)y'^2} \, dx$$

where E, F and G are given functions of x and y

(a) Show the Euler-Lagrange equation to be satisfied is

$$\frac{E_y + 2F_y y' + G_y y'^2}{2\sqrt{E + 2Fy' + Gy'^2}} = \frac{d}{dx}\left(\frac{F + Gy'}{\sqrt{E + 2Fy' + Gy'^2}}\right)$$

where subscripts denote partial differentiation.

(b) Show that after much algebra one can reduce the above equation to the form

$$A + By' + Cy'^2 + Dy'^3 + 2(F^2 - EG)y'' = 0$$

with $\quad A = E\,E_y - 2E\,F_x + F\,E_x \qquad\qquad C = -3F\,G_x + 2G\,E_y + 2F\,F_y - E\,G_y$

$\qquad\qquad\quad B = 3F\,E_y - 2E\,G_x - 2F\,F_x + G\,E_x \qquad D = -G\,G_x + 2G\,F_y - F\,G_y$

where the subscripts denote partial differentiation.

▶ **12.**   Derive the variational problem associated with the deformation $u = u(x,y,t)$ of a membrane which is subjected to an external force per unit area $f = f(x,y,t)$.

(a) Show that the functional to be minimized is

$$I = \int_{t_1}^{t_2} \iint_R \left[\frac{1}{2}\varrho\,(u_t)^2 - \frac{1}{2}\tau\,(\nabla u \cdot \nabla u) + fu\right] d\sigma dt$$

where $d\sigma = dxdy$ is an element of area.

(b) Find the equation describing the deformation.

▶ **13.**

(a) Find the equation describing the function which minimizes the functional

$$I = \iint_R \left[(z_{xx})^2 + (z_{yy})^2 + 2(z_{xy})^2\right] dxdy$$

(b) Find the equation describing the function which minimizes the functional

$$I = \iint_R \left[(z_{xx})^2 + (z_{yy})^2 + 2(z_{xy})^2 - 2f(x,y)z\right] dxdy$$

▶ **14.**   The Euler-buckling of a beam has the potential energy $V = \int_0^L \left[\frac{1}{2}EI(y'')^2 - \frac{1}{2}P(y')^2\right] dx$ where E, I, P are constants.

(a) Show that

$$\delta V = \left[EIy''\,\delta y' - (EIy''' + Py')\,\delta y\right]_0^L + \int_0^L (EIy'''' + Py'')\,dx$$

(b) Find the Euler-Lagrange equation and associated boundary conditions.

▶ **15.** Consider the partial differential equation

$$\frac{\partial}{\partial x}\left(\tau(x,y)\frac{\partial u}{\partial x}\right) + \frac{\partial}{\partial y}\left(\tau(x,y)\frac{\partial u}{\partial y}\right) + f(x,y)\sin(\alpha t - \phi) = \rho\frac{\partial^2 u}{\partial t^2} \qquad (15-1)$$

where $u = u(t,x,y)$ describes the displacement of a rectangular membrane over the region $x_1 \leq x \leq x_2$ and $y_1 \leq y \leq y_2$. In equation (15-1) the quantity $\tau(x,y)$ denotes the tension in the membrane, ρ is the mass density of the membrane and $f(x,y)\sin(\alpha t - \phi)$ is an external normal periodic force acting on the membrane.

(a) Assume a solution to the equation (15-1) of the form $u = u(t,x,y) = w(x,y)\sin(\alpha t - \phi)$ where α and ϕ are constants. Show that $w = w(x,y)$ must satisfy the partial differential equation

$$\frac{\partial}{\partial x}\left(\tau(x,y)\frac{\partial w}{\partial x}\right) + \frac{\partial}{\partial y}\left(\tau(x,y)\frac{\partial w}{\partial y}\right) + \rho\alpha^2 w + f(x,y) = 0 \qquad (15-2)$$

(b) Multiply equation (15-2) by $\delta w(x,y)$ and integrate over the area of the membrane to obtain

$$\int_{x_1}^{x_2}\int_{y_1}^{y_2}\left[\frac{\partial}{\partial x}\left(\tau(x,y)\frac{\partial w}{\partial x}\right) + \frac{\partial}{\partial y}\left(\tau(x,y)\frac{\partial w}{\partial y}\right)\right]\delta w\,dxdy$$
$$+ \delta\int_{x_1}^{x_2}\int_{y_1}^{y_2}\left[\frac{1}{2}\rho\alpha^2 w^2 + f(x,y)w\right]dxdy = 0 \qquad (15-3)$$

(c) Evaluate the integral $\displaystyle\int_{y_1}^{y_2}\left\{\int_{x_1}^{x_2}\frac{\partial}{\partial x}\left(\tau(x,y)\frac{\partial w}{\partial x}\right)\delta w\,dx\right\}dy$ using integration by parts.

(d) Evaluate the integral $\displaystyle\int_{x_1}^{x_2}\left\{\int_{y_1}^{y_2}\frac{\partial}{\partial y}\left(\tau(x,y)\frac{\partial w}{\partial y}\right)\delta w\,dy\right\}dx$ using integration by parts

(e) Show that equation (15-3) can be written in the form

$$\delta\int_{x_1}^{x_2}\int_{y_1}^{y_2}\left\{-\frac{1}{2}\tau\left[\left(\frac{\partial w}{\partial x}\right)^2 + \left(\frac{\partial w}{\partial y}\right)^2\right] + \frac{1}{2}\rho\alpha^2 w^2 + f(x,y)w\right\}dxdy$$
$$+ \int_{y_1}^{y_2}\left[\tau(x,y)\frac{\partial w}{\partial x}\delta w\right]_{x_1}^{x_2}dy + \int_{x_1}^{x_2}\left[\tau(x,y)\frac{\partial w}{\partial y}\delta w\right]_{y_1}^{y_2}dx = 0 \qquad (15-4)$$

(f) Show that the equation (15-4) can be written in the form

$$\delta\int_{x_1}^{x_2}\int_{y_1}^{y_2}\left[\frac{1}{2}\nabla w\cdot\nabla w - \frac{1}{2}\rho\alpha^2 w^2 - f(x,y)w\right]d\sigma - \oint_C \tau(x,y)\frac{\partial w}{\partial n}\delta w\,ds = 0$$

where $d\sigma = dxdy$ is an element of area and ds is an element of arc length along the boundary of the rectangle. Here $\dfrac{\partial w}{\partial n} = \nabla w\cdot\hat{n}$ represents the normal derivative along the boundary where $\hat{n}$ is a unit exterior normal to the boundary of the rectangle.

(g) What are the essential and natural boundary conditions for this problem?

(h) What is the variational form of equation (15-1) when

   (i) $\delta w = 0$, $f = 0$, $\alpha = 0$ and $\tau = 1$.

   (ii) $\delta w = 0$, $\alpha = 0$ and $\tau = 1$.

   (iii) $\frac{\partial w}{\partial n} = F(s)$, $\alpha = 0$, $\tau = 1$ and $f = 0$. where $F(s)$ is a given nonzero function of the arc length around the rectangle.

▶ **16.** Consider the functional $I = \int_0^1 \left[\frac{1}{2}(y')^2 + f(x)y \right] dx$ subject to the end conditions that $y(0) = y_0$ and $y(1) = y_1$ where y_0 and y_1 are constants.

(a) Show the Euler-Lagrange equation is $y'' - f(x) = 0$

(b) Select a set of basis functions $\phi_i(x)$ for $i = 0, 1, 2, \ldots, N$ where $\phi_0(x)$ satisfies the end conditions $\phi_0(0) = y_0$ and $\phi_0(1) = y_1$ with $\phi_i(0) = \phi_i(1) = 0$ for $i = 1, 2, \ldots, N$. Substitute $y(x) = \phi_0(x) + \sum_{i=1}^{N} c_i \phi_i(x)$ as an approximate solution into the given functional so that

$$I = I(c_1, c_2, \ldots, c_N) = \int_0^1 \left\{ \frac{1}{2} \left[\phi_0'(x) + \sum_{i=1}^{N} c_i \phi_i'(x) \right]^2 + f(x) \sum_{i=1}^{N} [\phi_0(x) + c_i \phi_i(x)] \right\} dx$$

Show that for I to be a minimum it is required that

$$\frac{\partial I}{\partial c_j} = \sum_{i=1}^{N} c_i \int_0^1 \phi_i'(x) \phi_j'(x)\, dx + \int_0^1 f(x) \phi_j(x)\, dx + \int_0^1 \phi_0'(x) \phi_j(x)\, dx = 0 \qquad \text{for } j = 1, 2, \ldots, N.$$

(c) Show that this leads to an approximate solution where the coefficients c_i are determined by solving the system of linear equation $A\vec{c} = \vec{b}$ where A is a $N \times N$ matrix with components $a_{ij} = \int_0^1 \phi_i'(x) \phi_j'(x)\, dx$ with $\vec{c} = \text{col}(c_1, c_2, \ldots, c_N)$ and $\vec{b} = \text{col}(b_1, b_2, \ldots, b_N)$ where $b_j = -\int_0^1 f(x) \phi_j(x)\, dx - \int_0^1 \phi_0'(x) \phi_j(x)\, dx.$

▶ **17.** Use the method outlined in problem 16 above to find an approximate solution to $y'' = f(x)$ for $0 \le x \le 1$ subject to the end point conditions $y(0) = 0$ and $y(1) = 0$. Use the basis functions $\phi_0(x) = 0$, $\phi_1(x) = x(1-x)$, $\phi_2(x) = x^2(1-x)$. Set up the matrix equation $A\vec{c} = \vec{b}$ for a general $f(x)$. Determine the approximate solution $y(x) = c_1 x(1-x) + c_2 x^2(1-x)$ and compare your approximate solution with the exact solution for the following cases.

$$(a) \quad f(x) = 1, \qquad (b) \quad f(x) = -x, \qquad (c) \quad f(x) = -x^2, \qquad (d) \quad f(x) = \sin x$$

▶ **18.** Find the function $y(x)$ which minimizes the functional $I = \int_0^\pi \left(y'^2 + y^2 \right) dx$ subject to boundary condition $y(0) = 1$. Find the minimum value for I.

▶ **19.** Consider the functional $I = \int_0^L \left[\frac{1}{2}(y')^2 + x^2 y \right] dx$

(a) Find δI

(b) Find the Euler-Lagrange equation

(c) Solve the Euler-Lagrange equation with the essential boundary conditions $y(0) = 0$ and $y(L) = 0$

(d) Use the Rayleigh-Ritz method with the basis functions $\phi_1(x) = x(x-L)$ and $\phi_2(x) = x^2(x-L)$ to find an approximate solution to the problem in part (c).

(e) Compare the approximate and exact solutions.

▶ **20.**

(a) Find the condition for the extreme value of the functional

$$I = I(y) = \int_0^{x_1} \left(y'^2 + q(x)y^2 \right) dx$$

subject to the end point conditions $y(0) = \alpha$ and $y(x_1) = \beta$ where α and β are scalar constants.

(b) Consider a generalization of the problem in part (a). Recall that the inner product of two column vectors

$$\xi = \mathrm{col}\,(\xi_1, \xi_2, \ldots, \xi_n) = \begin{pmatrix} \xi_1 \\ \xi_2 \\ \vdots \\ \xi_n \end{pmatrix} \quad \text{and} \quad \eta = \mathrm{col}\,(\eta_1, \eta_2, \ldots, \eta_n) = \begin{pmatrix} \eta_1 \\ \eta_2 \\ \vdots \\ \eta_n \end{pmatrix}$$

with real components is defined as the scalar quantity

$$(\xi^T, \eta) = (\xi_1, \xi_2, \ldots, \xi_n) \begin{pmatrix} \eta_1 \\ \eta_2 \\ \vdots \\ \eta_n \end{pmatrix} = \sum_{i=1}^{n} \xi_i \eta_i = \xi_1 \eta_1 + \xi_2 \eta_2 + \cdots + \xi_n \eta_n$$

Let $y = \mathrm{col}\,(y_1(x), y_2(x), \ldots, y_n(x))$ denote an n-dimensional column vector with derivative denoted using the notation $y' = \mathrm{col}\,(y_1'(x), y_2'(x), \ldots, y_n'(x))$ and let $A(x)$ denote a positive definite $n \times n$ symmetric matrix. Consider the problem of finding an extreme value for the functional

$$I(y) = \int_0^{x_1} \left[(y'^T, y') + (y^T, A(x)y) \right] dx$$

subject to the end point conditions

$$y(0) = \alpha = \mathrm{col}\,(\alpha_1, \alpha_2, \cdots, \alpha_n) \quad \text{and} \quad y(x_1) = \beta = \mathrm{col}\,(\beta_1, \beta_2, \ldots, \beta_n)$$

where now α and β denote vector constants. Show that the resulting Euler-Lagrange equation is the vector differential equation

$$y'' - A(x)y = 0, \qquad y(0) = \alpha, \qquad y(x_1) = \beta$$

where 0 denotes the column vector of zeros.

▶ **21.** Consider the variational problem

$$I = \iiint_V \left[\nabla u \cdot \nabla u - 2\psi(x,y,z)u \right] d\tau + \iint_S \left[2uf(x,y,z) + u^2 g(x,y,z) \right] d\sigma$$

where $d\tau$ is an element of volume and $d\sigma$ is an element of surface area. Here V is the volume enclosed by the surface S. Let $\hat{n}$ denote the unit exterior normal to the closed surface S which encloses V.

(a) Show that the function u, which produces an extremum, must satisfy the equations

$$\nabla^2 u + \psi(x,y,z) = 0 \quad \text{for } x,y,z \in V$$

$$\nabla u \cdot \hat{n} + f(x,y,z) + u\,g(x,y,z) = 0 \quad \text{for } x,y,z \text{ on the surface } S$$

(b) Interpret the resulting equations and boundary conditions if u represents temperature.

▶ **22.** Construct an approximate solution to the partial differential equation $\frac{\partial^2 u}{\partial x^2} + \frac{\partial^2 u}{\partial y^2} = -1$
for $(x, y) \in R = \{(x, y) \mid -a \le x \le a, -b \le y \le b\}$ where a and b are positive constants.
The solution is to be subjected to the boundary condition that $u = \sin \frac{\pi}{2a}(x + a)$ for $y = -b$,
$-a \le x \le a$ and $u = 0$ for points (x, y) on the other edges of the rectangular region.
Hint: Consider the functional $I = \iint_R \left[(u_x)^2 + (u_y)^2 - 2u \right] dx dy$

▶ **23.** Consider the functional $I = \int_0^1 \left[x^2(y'')^2 + xy^2 - 10xy \right] dx$ which is to be an extremal,
where y is subjected to the end point conditions $y(1) = y'(1) = 0$.
(a) Construct an approximate solution of the form $y = y_0(x) = c_0(x - 1)^2$
(b) Construct an approximate solution of the form $y = y_1(x) = (x - 1)^2(c_0 + c_1 x)$
(c) Construct an approximate solution of the form $y = y_2(x) = (x - 1)^2(c_0 + c_1 x + c_2 x^2)$
State the condition or conditions that you are imposing upon the constants to obtain
your approximate solution.

▶ **24.** Find the Euler-Lagrange equations and boundary conditions associated with the
variational problem $I = \iiint_V \left[\nabla u \cdot \nabla u - 2\psi(x, y, z) u \right] d\tau + \iint_S \left[2uf(x, y, z) + u^2 g(x, y, z) \right] d\sigma$ where
$d\tau$ is an element of volume and $d\sigma$ is an element of surface associated with the surface S
enclosing the volume V.

▶ **25.** Verify the following statements.
(a) The Schrödinger equation in spherical coordinates (r, θ, ϕ) for the hydrogen atom has
solutions of the form $u = u(r, \theta, \phi) = F(r)G(\theta)H(\phi)$ obtained by the method of separation of variables. The separated equations for F, G and H are Sturm-Liouville ordinary
differential equations.
(i) Solutions for $F = F(r)$ exist if and only if $n = 1, 2, 3, \dots$ Here n is called the principle
quantum number.
(ii) Solutions for $G - G(\theta)$ exist if and only if $\ell = 0, 1, 2, 3, \dots, n - 1$. Here ℓ is called the
orbital quantum number.
(iii) Solutions for $H = H(\phi)$ exist if and only if $m_\ell = -\ell, -\ell+1, \dots, -1, 0, 1, 2, \dots, \ell-1, \ell$. Here
m_ℓ is called the magnetic quantum number.
(b) The principle quantum number n is sometimes called the shell quantum number and the
values of n equal to $1, 2, 3, \dots$ are sometimes replaced by the shell labels $K, L, M, \dots$.
(c) The orbital quantum number ℓ is sometimes referred to as the subshell quantum number.
(i) The n-th shell contains n subshells.
(ii) The subshell numbers ℓ equal to $0, 1, 2, 3, 4, 5, \dots$ are sometimes replaced by the labels
$s, p, d, f, g, \dots$.
(d) The magnetic quantum number is m_ℓ and the ℓ-th subshell contains $2\ell + 1$ orbitals.
(e) Show that there are n^2 orbitals in the n-th shell for the values $n = 1, 2, 3, 4$. Is this true for
all values of n?
(f) Show that $1 + 3 + 5 + 7 + \dots + 2n - 1 = n^2$ for $n = 1, 2, 3, \dots$.

Bibliography

- Akiezer, N.I., *The Calculus of Variations*,
    Blaisdell Publishing Company, New York, 1962.

- Arfken, G., *Mathematical Methods for Physicists*,
    Academic Press, New York, 1970.

- Arthurs, A.M., *Complementary Variational Principles*,
    Clarendon Press, Oxford 1970.

- Birkhoff, G., Rota, G., *Ordinary Differential Equations*,
    Ginn and Company, Boston, 1962.

- Bolza, O., *Lectures on the Calculus of Variations*,
    Dover Publications, New York, 1961.

- Bliss, G.A., *Lectures on the Calculus of Variations*,
    University of Chicago Press, Chicago, 1946.

- Courant, R., *Calculus of Variations*, The Courant Institute of Mathematical Sciences,
    New York University Press, 1957.

- Courant, R., Hilbert, D., *Methods of Mathematical Physics*, Vol. 1.,
    Interscience Publishers, Inc., 1961.

- Courant, R., Robbins, H., *What is Mathematics?*, Second Edition,
    New York: Oxford University Press, 1943.

- Denn, M.M, *Optimization by Variational Methods*,
    McGraw Hill Book Company, New York, 1969.

- Edwards, J., *A Treatise on the Integral Calculus*, Vol. II,
    Chelsea Publishing Company, New York, 1922.

- Elsgolc, L.E., *Calculus of Variations*, Translated from the Russian,
    Addison-Wesley Publishing Co., Reading, Massachusetts, 1962.

- FinLayson, B.A., *The Method of Weighted Residuals*,
    Academic Press, Orlando Florida, 1972.

- Forsyth, A.R., *Calculus of Variations*, Dover Publications, New York, 1960.

- Fox, C., *An Introduction to the Calculus of Variations*,
    Oxford University Press, New York, 1950.

- Fox, R.L., *Optimization Methods for Engineering Design*,
    Addison-Wesley Publishing Company, Reading Massachusetts, 1971.

- Gelfand, I.M., Fomin, S.V., *Calculus of Variations*,
    Prentice-Hall, Inc., Englewood Cliffs, New Jersey, 1963.

- Hildebrand, F.B., *Methods of Applied Mathematics*,
    Prentice-Hall, Inc., Englewood Cliffs, New Jersey, 1960.

- Hsu, J.C., Meyer, A.U., *Modern Control principles and Applications*,
   McGraw-Hill Book Company, 1968.

- Lanczos, C., *The Variational Principles of Mechanics*,
   Dover Publications, Inc., New York, 1970.

- Lebedev, N.N., Skalskaya, I.P., Uflyand, Y.S., *Worked Problems in Applied Mathematics*,
   Dover Publications, Inc., New York, 1965.

- Milnes, H.W., *Calculus of Variations*,
   Texas Technological College, Lubbock, Texas, 1970.

- Mura, T., Koya, T., Variational Methods in Mechanics,
   Oxford University Press, New York, 1992.

- Murnaghan, F.D., *The Calculus of Variations*,
   Spartan Book, Washington, D.C., 1962.

- Pars, L.A., *Introduction to Calculus of Variations*,
   John Wiley and Sons Inc., New York, 1962.

- Pauling, L., Wilson, E.B., *Introduction to Quantum Mechanics
   with Applications to Chemistry*, McGraw-Hill book Company, New York, 1935.

- Reddy, J.N., *Energy and Variational Methods in Applied Mechanics*,
   John Wiley & Sons, New York, 1984.

- Rektorys, K., *Variational Methods in Mathematics, Science and Engineering*,
   D. Reidel Publishing Company, Dordrecht, Holland, 1975.

- Sagen H., *Introduction to the Calculus of Variations*,
   McGraw Hill Book Company, New York, 1969.

- Schechter, R.S., *The Variational Method in Engineering*,
   McGraw Hill Book Company, New York, 1967.

- Todhunter, I., *History of the Calculus of Variations*,
   Chelsea Publishing Company, New York, N.Y., 1962.

- Todhunter, I., *Researches in the Calculus of Variations*,
   G.E. Strechert & Company, New York, 1924.

- Van Wylen, G.J., Sonntag, R.E., *Fundamentals of Classical Thermodynamics*,
   John Wiley and Sons, New York, 1965.

- Weinstock, R., *Calculus of Variations with Applications to Physics and Engineering*,
   McGraw Hill Book Company, New York, 1952.

- Young, L.C., *Calculus of Variations and Optimal Control Theory*,
   W.B. Saunders Company, Philadelphia, 1969.

Bibliography

APPENDIX A
Units of Measurement

The following units, abbreviations and prefixes are from the Système International d'Unitès   (designated SI in all Languages.)

Prefixes.

Abbreviations		
Prefix	Multiplication factor	Symbol
tera	10^{12}	T
giga	10^9	G
mega	10^6	M
kilo	10^3	K
hecto	10^2	h
deka	10	da
deci	10^{-1}	d
centi	10^{-2}	c
milli	10^{-3}	m
micro	10^{-6}	μ
nano	10^{-9}	n
pico	10^{-12}	p

Basic Units.

Basic units of measurement		
Unit	Name	Symbol
Length	meter	m
Mass	kilogram	kg
Time	second	s
Electric current	ampere	A
Temperature	degree Kelvin	$^\circ$K
Luminous intensity	candela	cd

Supplementary units		
Unit	Name	Symbol
Plane angle	radian	rad
Solid angle	steradian	sr

Appendix A

DERIVED UNITS		
Name	Units	Symbol
Area	square meter	m^2
Volume	cubic meter	m^3
Frequency	hertz	Hz (s^{-1})
Density	kilogram per cubic meter	kg/m^3
Velocity	meter per second	m/s
Angular velocity	radian per second	rad/s
Acceleration	meter per second squared	m/s^2
Angular acceleration	radian per second squared	rad/s^2
Force	newton	N $(kg \cdot m/s^2)$
Pressure	newton per square meter	N/m^2
Kinematic viscosity	square meter per second	m^2/s
Dynamic viscosity	newton second per square meter	$N \cdot s/m^2$
Work, energy, quantity of heat	joule	J $(N \cdot m)$
Power	watt	W (J/s)
Electric charge	coulomb	C $(A \cdot s)$
Voltage, Potential difference	volt	V (W/A)
Electromotive force	volt	V (W/A)
Electric force field	volt per meter	V/m
Electric resistance	ohm	Ω (V/A)
Electric capacitance	farad	F $(A \cdot s/V)$
Magnetic flux	weber	Wb $(V \cdot s)$
Inductance	henry	H $(V \cdot s/A)$
Magnetic flux density	tesla	T (Wb/m^2)
Magnetic field strength	ampere per meter	A/m
Magnetomotive force	ampere	A

Physical Constants:

- $4 \arctan 1 = \pi = 3.14159\,26535\,89793\,23846\,2643\ldots$
- $\lim_{n \to \infty} \left(1 + \frac{1}{n}\right)^n = e = 2.71828\,18284\,59045\,23536\,0287\ldots$
- Euler's constant $\gamma = 0.57721\,56649\,01532\,86060\,6512\ldots$
- $\gamma = \lim_{n \to \infty} \left(1 + \frac{1}{2} + \frac{1}{3} + \cdots + \frac{1}{n} - \log n\right)$
- Speed of light in vacuum $c = 2.997925(10)^8\,m\;s^{-1}$
- Electron charge $= 1.60210(10)^{-19}\,C$
- Avogadro's constant $= 6.02252(10)^{23}\,mol^{-1}$
- Plank's constant $= 6.6256(10)^{-34}\,J\,s$
- Universal gas constant $= 8.3143\,J\,K^{-1}\,mol^{-1} = 8314.3\,J\,Kg^{-1}\,K^{-1}$
- Boltzmann constant $= 1.38054(10)^{-23}\,J\,K^{-1}$
- Stefan–Boltzmann constant $= 5.6697(10)^{-8}\,W\,m^{-2}\,K^{-4}$
- Gravitational constant $= 6.67(10)^{-11}\,N\,m^2kg^{-2}$
- μ_0 permeability of free space $= 4\pi\,(10)^{-7}\,H/m$
- ϵ_0 permittivity of free space $= \frac{1}{\mu_0 c^2}\,F/m$

Appendix A

APPENDIX B

Gradient, Divergence, Curl and Laplacian in Cartesian, Cylindrical and Spherical Coordinates

Cartesian Coordinates (x, y, z)

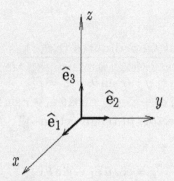

Gradient

For $u = u(x, y, z)$ a scalar function of position, then the gradient of u is represented

$$\operatorname{grad} u = \nabla u = \frac{\partial u}{\partial x} \, \widehat{\mathbf{e}}_1 + \frac{\partial u}{\partial y} \, \widehat{\mathbf{e}}_2 + \frac{\partial u}{\partial z} \, \widehat{\mathbf{e}}_3$$

Divergence

If $\vec{V} = \vec{V}(x, y, z) = V_1 \, \widehat{\mathbf{e}}_1 + V_2 \, \widehat{\mathbf{e}}_2 + V_3 \, \widehat{\mathbf{e}}_3$ is a vector function of position, then the divergence of $\vec{V}$ is represented

$$\operatorname{div} \vec{V} = \nabla \cdot \vec{V} = \frac{\partial V_1}{\partial x} + \frac{\partial V_2}{\partial y} + \frac{\partial V_3}{\partial z}$$

Curl

If $\vec{V} = \vec{V}(x, y, z) = V_1 \, \widehat{\mathbf{e}}_1 + V_2 \, \widehat{\mathbf{e}}_2 + V_3 \, \widehat{\mathbf{e}}_3$ is a vector function of position, then the curl of $\vec{V}$ is represented

$$\operatorname{curl} \vec{V} = \nabla \times \vec{V} = \begin{vmatrix} \widehat{\mathbf{e}}_1 & \widehat{\mathbf{e}}_2 & \widehat{\mathbf{e}}_3 \\ \frac{\partial}{\partial x} & \frac{\partial}{\partial y} & \frac{\partial}{\partial z} \\ V_1 & V_2 & V_3 \end{vmatrix} = \left(\frac{\partial V_3}{\partial y} - \frac{\partial V_2}{\partial z} \right) \widehat{\mathbf{e}}_1 - \left(\frac{\partial V_3}{\partial x} - \frac{\partial V_1}{\partial z} \right) \widehat{\mathbf{e}}_2 + \left(\frac{\partial V_2}{\partial x} - \frac{\partial V_1}{\partial y} \right) \widehat{\mathbf{e}}_3$$

Laplacian

For $u = u(x, y, z)$ a scalar function of position, then the Laplacian of u is represented

$$\Delta u = \nabla^2 u = \frac{\partial^2 u}{\partial x^2} + \frac{\partial^2 u}{\partial y^2} + \frac{\partial^2 u}{\partial z^2}$$

Cylindrical Coordinates (r, θ, z)

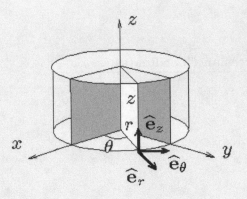

$x = r\cos\theta, \quad y = r\sin\theta, \quad z = z$

$\vec{r} = x\,\widehat{\mathbf{e}}_1 + y\,\widehat{\mathbf{e}}_2 + z\,\widehat{\mathbf{e}}_3 = r\cos\theta\,\widehat{\mathbf{e}}_1 + r\sin\theta\,\widehat{\mathbf{e}}_2 + z\,\widehat{\mathbf{e}}_3$

Unit vectors in the directions $\frac{\partial \vec{r}}{\partial r}, \ \frac{\partial \vec{r}}{\partial \theta}, \ \frac{\partial \vec{r}}{\partial z}$ are given by

$$\widehat{\mathbf{e}}_r = \cos\theta\,\widehat{\mathbf{e}}_1 + \sin\theta\,\widehat{\mathbf{e}}_2$$
$$\widehat{\mathbf{e}}_\theta = -\sin\theta\,\widehat{\mathbf{e}}_1 + \cos\theta\,\widehat{\mathbf{e}}_2$$
$$\widehat{\mathbf{e}}_z = \widehat{\mathbf{e}}_3$$

Gradient

For $u = u(r, \theta, z)$ a scalar function of position, then the gradient of u is represented

$$\operatorname{grad} u = \nabla u = \frac{\partial u}{\partial r}\,\widehat{\mathbf{e}}_r + \frac{1}{r}\frac{\partial u}{\partial \theta}\,\widehat{\mathbf{e}}_\theta + \frac{\partial u}{\partial z}\,\widehat{\mathbf{e}}_z$$

Divergence

If $\vec{V} = \vec{V}(r, \theta, z) = V_r\,\widehat{\mathbf{e}}_r + V_\theta\,\widehat{\mathbf{e}}_\theta + V_z\,\widehat{\mathbf{e}}_z$ is a vector function of position, then the divergence of $\vec{V}$ is represented

$$\operatorname{div} \vec{V} = \nabla \cdot \vec{V} = \frac{1}{r}\frac{\partial (rV_r)}{\partial r} + \frac{1}{r}\frac{\partial V_\theta}{\partial \theta} + \frac{\partial V_z}{\partial z}$$

Curl

If $\vec{V} = \vec{V}(r, \theta, z) = V_r\,\widehat{\mathbf{e}}_r + V_\theta\,\widehat{\mathbf{e}}_\theta + V_z\,\widehat{\mathbf{e}}_z$ is a vector function of position, then the curl of $\vec{V}$ is represented

$$\operatorname{Curl} \vec{V} = \nabla \times \vec{V} = \frac{1}{r}\begin{vmatrix} \widehat{\mathbf{e}}_r & r\,\widehat{\mathbf{e}}_\theta & \widehat{\mathbf{e}}_z \\ \frac{\partial}{\partial r} & \frac{\partial}{\partial \theta} & \frac{\partial}{\partial z} \\ V_r & rV_\theta & V_z \end{vmatrix}$$

$$\nabla \times \vec{V} = \left(\frac{1}{r}\frac{\partial V_z}{\partial \theta} - \frac{\partial V_\theta}{\partial z}\right)\widehat{\mathbf{e}}_r + \left(\frac{\partial V_r}{\partial z} - \frac{\partial V_z}{\partial r}\right)\widehat{\mathbf{e}}_\theta + \frac{1}{r}\left(\frac{\partial}{\partial r}(rV_\theta) - \frac{\partial V_r}{\partial \theta}\right)\widehat{\mathbf{e}}_z$$

Laplacian

For $u = u(r, \theta, z)$ a scalar function of position, then the Laplacian of u is represented

$$\Delta u = \nabla^2 u = \frac{\partial^2 u}{\partial r^2} + \frac{1}{r}\frac{\partial u}{\partial r} + \frac{1}{r^2}\frac{\partial^2 u}{\partial \theta^2} + \frac{\partial^2 u}{\partial z^2}$$

Appendix B

Spherical Coordinates (ρ, θ, ϕ)

$$x = \rho\sin\theta\cos\phi, \quad y = \rho\sin\theta\sin\phi, \quad z = \rho\cos\theta$$
$$\vec{r} = x\,\widehat{\mathbf{e}}_1 + y\,\widehat{\mathbf{e}}_2 + z\,\widehat{\mathbf{e}}_3 = \rho\sin\theta\cos\phi\,\widehat{\mathbf{e}}_1 + \rho\sin\theta\sin\phi\,\widehat{\mathbf{e}}_2 + \rho\cos\theta\,\widehat{\mathbf{e}}_3$$

Unit vectors in the directions $\frac{\partial \vec{r}}{\partial \rho}$, $\frac{\partial \vec{r}}{\partial \theta}$, $\frac{\partial \vec{r}}{\partial \phi}$ are given by

$$\widehat{\mathbf{e}}_\rho = \sin\theta\cos\phi\,\widehat{\mathbf{e}}_1 + \sin\theta\sin\phi\,\widehat{\mathbf{e}}_2 + \cos\theta\,\widehat{\mathbf{e}}_3$$
$$\widehat{\mathbf{e}}_\theta = \cos\theta\cos\phi\,\widehat{\mathbf{e}}_1 + \cos\theta\sin\phi\,\widehat{\mathbf{e}}_2 - \sin\theta\,\widehat{\mathbf{e}}_3$$
$$\widehat{\mathbf{e}}_\phi = -\sin\phi\,\widehat{\mathbf{e}}_1 + \cos\phi\,\widehat{\mathbf{e}}_2$$

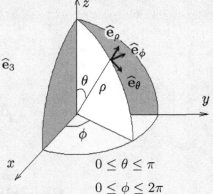

$$0 \leq \theta \leq \pi$$
$$0 \leq \phi \leq 2\pi$$

Gradient

For $u = u(\rho, \theta, \phi)$ a scalar function of position, then the gradient of u is represented

$$\operatorname{grad} u = \nabla u = \frac{\partial u}{\partial \rho}\,\widehat{\mathbf{e}}_\rho + \frac{1}{\rho}\frac{\partial u}{\partial \theta}\,\widehat{\mathbf{e}}_\theta + \frac{1}{\rho\sin\theta}\frac{\partial u}{\partial \phi}\,\widehat{\mathbf{e}}_\phi$$

Divergence

If $\vec{V} = \vec{V}(\rho, \theta, \phi) = V_\rho\,\widehat{\mathbf{e}}_\rho + V_\theta\,\widehat{\mathbf{e}}_\theta + V_\phi\,\widehat{\mathbf{e}}_\phi$ is a vector function of position, then the divergence of $\vec{V}$ is represented

$$\operatorname{div} \vec{V} = \nabla \cdot \vec{V} = \frac{1}{\rho^2}\frac{\partial}{\partial \rho}\left(\rho^2 V_\rho\right) + \frac{1}{\rho\sin\theta}\frac{\partial}{\partial \theta}\left(\sin\theta V_\theta\right) + \frac{1}{\rho\sin\theta}\frac{\partial V_\phi}{\partial \phi}$$

Curl

If $\vec{V} = \vec{V}(\rho, \theta, \phi) = V_\rho\,\widehat{\mathbf{e}}_\rho + V_\theta\,\widehat{\mathbf{e}}_\theta + V_\phi\,\widehat{\mathbf{e}}_\phi$ is a vector function of position, then the curl of $\vec{V}$ is represented

$$\operatorname{curl} \vec{V} = \nabla \times \vec{V} = \frac{1}{\rho^2\sin\theta}\begin{vmatrix} \widehat{\mathbf{e}}_\rho & \rho\,\widehat{\mathbf{e}}_\theta & \rho\sin\theta\,\widehat{\mathbf{e}}_\phi \\ \frac{\partial}{\partial \rho} & \frac{\partial}{\partial \theta} & \frac{\partial}{\partial \phi} \\ V_\rho & \rho V_\theta & \rho\sin\theta V_\phi \end{vmatrix}$$

$$\nabla \times \vec{V} = \frac{1}{\rho\sin\theta}\left(\frac{\partial}{\partial \theta}\left(\sin\theta V_\phi\right) - \frac{\partial V_\theta}{\partial \phi}\right)\widehat{\mathbf{e}}_\rho + \left(\frac{1}{\rho\sin\theta}\frac{\partial V_\rho}{\partial \phi} - \frac{1}{\rho}\frac{\partial}{\partial \rho}\left(\rho V_\phi\right)\right)\widehat{\mathbf{e}}_\theta + \frac{1}{\rho}\left(\frac{\partial}{\partial \rho}\left(\rho V_\theta\right) - \frac{\partial V_\rho}{\partial \theta}\right)\widehat{\mathbf{e}}_\phi$$

Laplacian

For $u = u(\rho, \theta, \phi)$ a scalar function of position, then the Laplacian of u is represented

$$\nabla^2 u = \frac{1}{\rho^2}\frac{\partial}{\partial \rho}\left(\rho^2\frac{\partial u}{\partial \rho}\right) + \frac{1}{\rho^2\sin\theta}\frac{\partial}{\partial \theta}\left(\sin\theta\frac{\partial u}{\partial \theta}\right) + \frac{1}{\rho^2\sin^2\theta}\frac{\partial^2 u}{\partial \phi^2}$$

Appendix B

APPENDIX C

Solutions to Selected Exercises

Selected Solutions Chapter 1

▶ 1. (a)

$$y = \sin^{-1} x$$
$$\sin y = x$$
$$\cos y \frac{dy}{dx} = 1$$
$$\frac{dy}{dx} = \frac{1}{\cos y} = \frac{1}{\sqrt{1 - x^2}}$$
$$-1 \le x \le 1$$

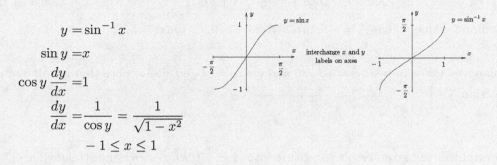

▶ 1. (c) $xy^2 - y - x^3 = 0$ defines two functions

$$y_1 = y_1(x) = \frac{1 + \sqrt{1 + 4x^4}}{2x} \quad \text{and} \quad y_2 = y_2(x) = \frac{1 - \sqrt{1 + 4x^4}}{2x}$$

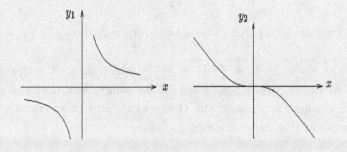

$$\frac{dy_1}{dx} = \frac{4x^2}{\sqrt{1 + 4x^4}} - \frac{1 + \sqrt{1 + 4x^4}}{2x^2} \quad \text{and} \quad \frac{dy_2}{dx} = \frac{-4x^2}{\sqrt{1 + 4x^4}} - \frac{1 - \sqrt{1 + 4x^4}}{2x^2}$$

▶ 3.

$$\frac{dy}{dx} = \frac{2t - 1}{2t}, \qquad \frac{d^2y}{dx^2} = \frac{-1}{4t^3}$$

▶ 4. (b)

$$\frac{\partial w}{\partial x} = \frac{-y}{x^2 + y^2}, \quad \frac{\partial w}{\partial y} = \frac{2y}{x^2 + y^2}, \quad \frac{\partial^2 w}{\partial x^2} = \frac{2xy}{(x^2 + y^2)^2}, \quad \frac{\partial^2 w}{\partial x \partial y} = \frac{y^2 - x^2}{(x^2 + y^2)^2}, \quad \frac{\partial^2 w}{\partial y^2} = \frac{-2xy}{(x^2 + y^2)^2}$$

▶ 5.

(a) $Area = bh$,   (b) $Area = \frac{1}{2}bh$

► 6.

(a) $\frac{13}{2}\sqrt{2}$

(b) Let $\widehat{\mathbf{e}}_\alpha = \cos\alpha\,\widehat{\mathbf{e}}_1 + \sin\alpha\,\widehat{\mathbf{e}}_2$, then $I = \frac{df}{ds} = \nabla f \cdot \widehat{\mathbf{e}}_\alpha\Big|_{x=1,y=2} = 8\cos\alpha + 5\sin\alpha$. I has a maximum value when $\frac{dI}{d\alpha} = -8\sin\alpha + 5\cos\alpha = 0$, or $\tan\alpha = 5/8$.

This gives the values $\sin\alpha = 5/\sqrt{89}$ and $\cos\alpha = 8/\sqrt{89}$. Also note that a unit vector in the direction $\nabla f\Big|_{x=1,y=2}$ is $\widehat{\mathbf{e}}_\alpha = \frac{8}{\sqrt{89}}\,\widehat{\mathbf{e}}_1 + \frac{5}{\sqrt{89}}\,\widehat{\mathbf{e}}_2$

► 7.

(a) To establish the given result substitute into $A = \frac{1}{2}\int_c x\,dy - y\,dx$ the straight lines

$$y - y_1 = \frac{y_3 - y_1}{x_3 - x_1}(x - x_1)$$

$$y - y_2 = \frac{y_1 - y_2}{x_1 - x_2}(x - x_2)$$

$$y - y_2 = \frac{y_3 - y_2}{x_3 - x_2}(x - x_2)$$

and break the line integral into three parts integrating counterclockwise around the triangle.

(b) If $A = 0$, then one can show $\frac{y_2-y_1}{x_2-x_1} = \frac{y_3-y_2}{x_3-x_2} = \frac{y_3-y_1}{x_3-x_1}$ which shows the slopes of the lines connecting each combination of the points are all the same. Alternatively, construct the equation of the line through any one of the three points and show the other two points are on the line because of the above slope relationship.

► 9.

(c) $\dfrac{dy}{dx} = \dfrac{2x + \sin(x+y) + \frac{y}{x^2+y^2}}{\frac{x}{x^2+y^2} - \sin(x+y)}$, (d) $\dfrac{dy}{dx} = \dfrac{2x - \frac{2x}{x^2+y^2}}{\frac{2y}{x^2+y^2} - 2y}$

► 10. $f(u,v) = 0$, $u = \alpha x + \beta y + \gamma z$ $v = x^2 + y^2 + z^2$ Assume $z = z(x,y)$, then

differentiate with respect to x and show

$$\frac{\partial f}{\partial u}\frac{\partial u}{\partial x} + \frac{\partial f}{\partial v}\frac{\partial v}{\partial x} = 0$$

$$\frac{\partial f}{\partial u}\left(\alpha + \gamma\frac{\partial z}{\partial x}\right) + \frac{\partial f}{\partial v}\left(2x + 2z\frac{\partial z}{\partial x}\right) = 0$$

$$\frac{\partial z}{\partial x} = \frac{-2xf_v - \alpha f_u}{\gamma f_u + 2z f_v}$$

differentiate with respect to y to show

$$\frac{\partial f}{\partial u}\frac{\partial u}{\partial y} + \frac{\partial f}{\partial v}\frac{\partial v}{\partial y} = 0$$

$$\frac{\partial f}{\partial u}\left(\beta + \gamma\frac{\partial z}{\partial y}\right) + \frac{\partial f}{\partial v}\left(2y + 2z\frac{\partial z}{\partial y}\right) = 0$$

$$\frac{\partial z}{\partial y} = \frac{-2yf_v - \beta f_u}{\gamma f_u + 2z f_v}$$

Substitute $\frac{\partial z}{\partial x}$ and $\frac{\partial z}{\partial y}$ into the given partial differential equation and show the coefficients of the partial derivative terms f_u and f_v are zero.

Selected Solutions Chapter 1

▶ 11.

$$(a) \quad \left.\frac{df}{ds}\right|_{(1,1,1)} = -24, \qquad (b) \quad \left.\nabla f\right|_{(1,1,1,)} = 3\,\widehat{\mathbf{e}}_1 - 7\,\widehat{\mathbf{e}}_2 + \widehat{\mathbf{e}}_3$$

▶ 12. Unit tangent vector in the direction of the tangent is $\widehat{\mathbf{e}} = \dfrac{\widehat{\mathbf{e}}_1 + y'(x)\,\widehat{\mathbf{e}}_2}{\sqrt{1+[y'(x)]^2}}$

▶ 13. If $f = f(x, y + \epsilon\eta, y' + \epsilon\eta')$, then

$$\frac{\partial f}{\partial \epsilon} = \frac{\partial f}{\partial(y+\epsilon\eta)}\frac{\partial(y+\epsilon\eta)}{\partial\epsilon} + \frac{\partial f}{\partial(y'+\epsilon\eta')}\frac{\partial(y'+\epsilon\eta')}{\partial\epsilon}\bigg|_{\epsilon=0} = \frac{\partial f}{\partial y}\eta + \frac{\partial f}{\partial y'}\eta'$$

$$\frac{\partial^2 f}{\partial\epsilon^2} = \eta\left[\frac{\partial^2(y+\epsilon\eta)}{\partial\eta^2} + \frac{\partial^2 f}{\partial(y+\epsilon\eta)\partial(y'+\epsilon\eta')}\eta'\right] + \eta'\left[\frac{\partial^2 f}{\partial(y'+\epsilon\eta')\partial(y+\epsilon\eta)}\eta + \frac{\partial^2 f}{\partial(y'+\epsilon\eta')^2}\eta'\right]$$

$$\frac{\partial^2 f}{\partial\epsilon^2}\bigg|_{\epsilon=0} = \eta^2\frac{\partial^2 f}{\partial y^2} + 2\eta\eta'\frac{\partial^2 f}{\partial y\partial y'} + (\eta')^2\frac{\partial^2 f}{\partial y'^2}$$

▶ 16.

$$(b) \quad \frac{\partial f}{\partial x} = \frac{x}{\sqrt{x^2+y^2}}, \qquad \frac{\partial f}{\partial y} = \frac{-y}{\sqrt{x^2+y^2}}$$

$$\frac{\partial^2 f}{\partial x^2} = \frac{1}{f} - \frac{x^2}{f^3}, \quad \frac{\partial^2 f}{\partial x\partial y} = \frac{xy}{f^3}, \quad \frac{\partial^2 f}{\partial y^2} = -\frac{1}{f} - \frac{y^2}{f^3} \quad \text{where} \quad f = \sqrt{x^2+y^2}$$

▶ 17.

$$(b) \quad \frac{\partial z}{\partial x} = -\frac{x}{z}, \qquad \frac{\partial z}{\partial y} = -\frac{y}{z}$$

▶ 18.

(a) The directional derivative $\frac{dF}{ds}$ is a maximum at the points

   (i) $(1,1)$ in the direction $(3\,\widehat{\mathbf{e}}_1 + \widehat{\mathbf{e}}_2)$ and

   (ii) $(-1,-1)$ in the direction $(-3\,\widehat{\mathbf{e}}_1 - \widehat{\mathbf{e}}_2)$

Let s denote arclength along the perimeter of the square starting at the point $(1,-1)$, then one can plot $|\nabla F|$ as one moves counterclockwise around the square to obtain the figure illustrated.

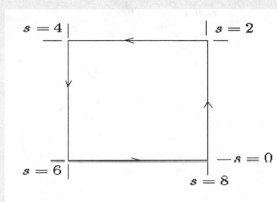
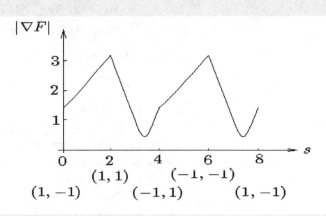

(b) I=0

Selected Solutions Chapter 1

► 19.

(a) $J = \dfrac{\partial(x,y)}{\partial(u,v)} = \alpha_0\beta_1 - \beta_0\alpha_1$

(b) $u = \dfrac{\beta_1(x - \gamma_0) - \beta_0(y - \gamma_1)}{J}, \qquad v = \dfrac{\alpha_0(y - \gamma_1) - \alpha_1(x - \gamma_0)}{J}$

(d) $\beta_1 x - \beta_0 y = \beta_1\gamma_0 - \beta_0\gamma_1$

► 20. Let $z_0 = \alpha x_0^2 + \beta y_0^2$, then equation of tangent plane is given by

$$-2\alpha x_0(x - x_0) - 2\beta y_0(y - y_0) + (z - z_0) = 0$$

► 21.

$$\nabla_\alpha u = \frac{\partial u}{\partial x}\cos\alpha + \frac{\partial u}{\partial y}\sin\alpha$$

$$\nabla_\alpha v = \frac{\partial v}{\partial x}\cos\alpha + \frac{\partial v}{\partial y}\sin\alpha$$

$$\nabla_{\alpha+\frac{\pi}{2}} v = \frac{\partial v}{\partial x}\cos\left(\alpha + \frac{\pi}{2}\right) + \frac{\partial v}{\partial y}\sin\alpha = -\frac{\partial v}{\partial x}\sin\alpha + \frac{\partial v}{\partial y}\cos\alpha$$

But $\dfrac{\partial v}{\partial x} = -\dfrac{\partial u}{\partial y}$ and $\dfrac{\partial v}{\partial y} = \dfrac{\partial u}{\partial x}$

Hence $\nabla_{\alpha+\frac{\pi}{2}} v = \dfrac{\partial u}{\partial y}\sin\alpha + \dfrac{\partial u}{\partial x}\cos\alpha = \nabla_\alpha u$

► 22.

(a)

$$\nabla_\theta U = \frac{\partial U}{\partial x}\cos\theta + \frac{\partial U}{\partial y}\sin\theta$$

$$\frac{\partial U}{\partial r} = \frac{\partial U}{\partial x}\frac{\partial x}{\partial r} + \frac{\partial U}{\partial y}\frac{\partial y}{\partial r} = \frac{\partial U}{\partial x}\cos\theta + \frac{\partial U}{\partial y}\sin\theta$$

so that $\nabla_\theta U = \dfrac{\partial U}{\partial r}$

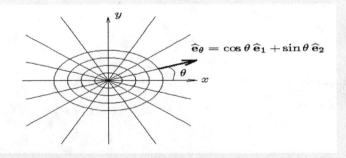

Selected Solutions Chapter 1

▶ 22.

(b)

$$\frac{\partial U}{\partial \theta} = \frac{\partial U}{\partial x}\frac{\partial x}{\partial \theta} + \frac{\partial U}{\partial y}\frac{\partial y}{\partial \theta} = -\frac{\partial U}{\partial x} r\sin\theta + \frac{\partial U}{\partial y} r\cos\theta$$

$$\nabla_{\theta + \frac{\pi}{2}} U = \frac{\partial U}{\partial x}\cos(\theta + \frac{\pi}{2}) + \frac{\partial U}{\partial y}\sin(\theta + \frac{\pi}{2})$$

$$\nabla_{\theta + \frac{\pi}{2}} U = -\frac{\partial U}{\partial x}\sin\theta + \frac{\partial U}{\partial y}\cos\theta = \frac{1}{r}\frac{\partial U}{\partial \theta}$$

▶ 23.

(a) $\quad \frac{dw}{dt} = \frac{\partial w}{\partial x}\frac{dx}{dt} + \frac{\partial w}{\partial y}\frac{dy}{dt}$

(c) $\quad w = w(x_1, x_2, \ldots, x_n), \quad x_i = x_i(u_1, u_2, \ldots, u_m) \quad$ for $i = 1, 2, \ldots, n$

$\quad \frac{\partial w}{\partial u_j} = \frac{\partial w}{\partial x_1}\frac{\partial x_1}{\partial u_j} + \frac{\partial w}{\partial x_2}\frac{\partial x_2}{\partial u_j} + \cdots + \frac{\partial w}{\partial x_n}\frac{\partial x_n}{\partial u_j} \quad$ for $j = 1, 2, \ldots, m$

▶ 24. $\quad \dfrac{dz}{dt} = \dfrac{-4x(t+1)\sin y}{2x + \cos x} + \dfrac{(2t - y)x^2\cos y}{t + 1 + 2y}$

▶ 25.

(c) If $p = p(T, U)$ and $V = V(T, U)$, then

$$\frac{\partial p}{\partial T} dT + \frac{\partial p}{\partial U} dU = dp, \quad \text{and} \quad \frac{\partial V}{\partial T} dT + \frac{\partial V}{\partial U} dU = dV$$

Now solve the second equation for dU in terms of dV and dT to obtain

$$dU = \frac{1}{\frac{\partial V}{\partial U}} dV - \frac{\frac{\partial V}{\partial T}}{\frac{\partial V}{\partial U}} dT$$

This result is substituted into the first equation for dp to obtain

$$\frac{\partial p}{\partial T} dT + \frac{\partial p}{\partial U}\left[\frac{1}{\frac{\partial V}{\partial U}} dV - \frac{\frac{\partial V}{\partial T}}{\frac{\partial V}{\partial U}} dT\right] = dp$$

which simplifies to

$$\left(\frac{\frac{\partial V}{\partial U}\frac{\partial p}{\partial T} - \frac{\partial p}{\partial U}\frac{\partial V}{\partial T}}{\frac{\partial V}{\partial U}}\right) dT + \frac{\frac{\partial p}{\partial U}}{\frac{\partial V}{\partial U}} dV = dp$$

which implies that

$$\frac{\partial p}{\partial T} = \frac{\frac{\partial(P,V)}{\partial(T,U)}}{\frac{\partial V}{\partial U}}, \quad \text{and} \quad \frac{\partial U}{\partial V} = \frac{1}{\frac{\partial V}{\partial U}}$$

Therefore one can convert the second law in the form $\frac{\partial U}{\partial V} - T\frac{\partial p}{\partial T} + p = 0$ to the form

$$\frac{1}{\frac{\partial V}{\partial U}} - T\frac{\frac{\partial(P,V)}{\partial(T,U)}}{\frac{\partial V}{\partial U}} + p = 0, \quad \text{or} \quad -1 + T\frac{\partial(P,V)}{\partial(T,U)} - p\frac{\partial V}{\partial U} = 0$$

Selected Solutions Chapter 1

▶ 26. $Area = \pi ab$

▶ 27.

$$P = P(V,T) \quad \text{with} \quad dP = \frac{\partial P}{\partial V} \, dV + \frac{\partial P}{\partial T} \, dT$$

(i) If P is held constant, then $dP = 0$ and $\left(\dfrac{dV}{dT}\right)_P = -\dfrac{\frac{\partial P}{\partial T}}{\frac{\partial P}{\partial V}} = \alpha V$

(ii) If T is held constant, then $dT = 0$ and $\left(\dfrac{dP}{dV}\right)_T = \dfrac{\partial P}{\partial V} = -\dfrac{E}{V}$

(iii) If V is held constant, then $dV = 0$ and
$$\left(\frac{dP}{dt}\right)_V = \frac{\partial P}{\partial T} = -\alpha V \left(\frac{\partial P}{\partial V}\right) = -\alpha V \left(-\frac{E}{V}\right) = \alpha E$$

▶ 28.

(a)

We have $\vec{r} = x\,\widehat{\mathbf{e}}_1 + y\,\widehat{\mathbf{e}}_2 = \vec{r}(u,v)$ and the vector $d\vec{r}$ has the components $\frac{\partial \vec{r}}{\partial u}\,du$ and $\frac{\partial \vec{r}}{\partial v}\,dv$ which make up the sides on an area element dA having area

$$dA = \left| \frac{\partial \vec{r}}{\partial u}\,du \times \frac{\partial \vec{r}}{\partial v}\,dv \right| = \left| \frac{\partial \vec{r}}{\partial u} \times \frac{\partial \vec{r}}{\partial v} \right| du\,dv$$

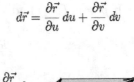

where

$$\frac{\partial \vec{r}}{\partial u} \times \frac{\partial \vec{r}}{\partial v} = \begin{vmatrix} \widehat{\mathbf{e}}_1 & \widehat{\mathbf{e}}_2 & \widehat{\mathbf{e}}_3 \\ \frac{\partial x}{\partial u} & \frac{\partial y}{\partial u} & 0 \\ \frac{\partial x}{\partial v} & \frac{\partial y}{\partial v} & 0 \end{vmatrix} = \widehat{\mathbf{e}}_3 \left(\frac{\partial x}{\partial u}\frac{\partial y}{\partial v} - \frac{\partial x}{\partial v}\frac{\partial y}{\partial u} \right) = \widehat{\mathbf{e}}_3 \frac{\partial(x,y)}{\partial(u,v)}$$

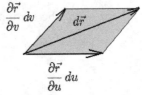

so that $dA = \left| \dfrac{\partial \vec{r}}{\partial u}\,du \times \dfrac{\partial \vec{r}}{\partial v}\,dv \right| du\,dv = \left| \dfrac{\partial \vec{r}}{\partial u} \times \dfrac{\partial \vec{r}}{\partial v} \right| du\,dv = \dfrac{\partial(x,y)}{\partial(u,v)}\,du\,dv$

(b) If $\vec{r} = \vec{r}(u,v,w)$, then $d\vec{r} = \frac{\partial \vec{r}}{\partial u}\,du + \frac{\partial \vec{r}}{\partial v}\,dv + \frac{\partial \vec{r}}{\partial w}\,dw$ produces a volume element $d\tau$ having the vector sides $\frac{\partial \vec{r}}{\partial u}\,du$, $\frac{\partial \vec{r}}{\partial v}\,dv$, and $\frac{\partial \vec{r}}{\partial w}\,dw$. The element of volume is calculated from the triple scalar product

$$d\tau = \left| \left(\frac{\partial \vec{r}}{\partial u}\,du \times \frac{\partial \vec{r}}{\partial v}\,dv \right) \cdot \frac{\partial \vec{r}}{\partial w}\,dw \right| = \left| \left(\frac{\partial \vec{r}}{\partial u} \times \frac{\partial \vec{r}}{\partial v} \right) \cdot \frac{\partial \vec{r}}{\partial w} \right| du\,dv\,dw = \frac{\partial(x,y,z)}{\partial(u,v,w)}\,du\,dv\,dw$$

Here we have used the triple scalar product result

$$\vec{A} \cdot (\vec{B} \times \vec{C}) = \begin{vmatrix} A_1 & A_2 & A_3 \\ B_1 & B_2 & B_3 \\ C_1 & C_2 & C_3 \end{vmatrix}$$

▶ 29.

$$\vec{F} \cdot \hat{n} = \frac{2x^2 y + 2y^2 z + x^2}{\sqrt{1 + 4x^2 + 4y^2}}, \qquad d\sigma = \frac{dx dy}{|\hat{n} \cdot \widehat{e}_3|} = \sqrt{1 + 4x^2 + 4y^2}\, dx dy$$

so that

$$I = \iint_S \vec{F} \cdot \hat{n}\, d\sigma = \iint_R (2x^2 y + 2y^2(1 - x^2 - y^2) + x^2)\, dx dy$$

where R is the region bounded by the unit circle $x^2 + y^2 = 1$ Make the change of variables $x = r \cos \theta$ and $y = r \sin \theta$ with $dx dy = r dr d\theta$ and show

$$I = \int_0^{2\pi} \int_0^1 \left[2r^3 \cos^2 \theta \sin \theta + 2r^2 \sin^2 \theta (1 - r^2 \cos^2 \theta - r^2 \sin^2 \theta) + r^2 \cos^2 \theta \right] r dr d\theta = \frac{\pi}{6}$$

▶ 30.

(a) The equations $x(u, v) - x = 0$ and $y(u, v) - y = 0$ are implicit equations which define u, v in terms of x and y. The differential of these equations give

$$dx = \frac{\partial x}{\partial u} du + \frac{\partial x}{\partial v} dv, \qquad dy = \frac{\partial y}{\partial u} du + \frac{\partial y}{\partial v} dv$$

Solve for du and dv and show

$$du = \frac{\frac{\partial y}{\partial v} dx - \frac{\partial x}{\partial v} dy}{J}, \qquad dv = \frac{-\frac{\partial y}{\partial u} ddx + \frac{\partial v}{\partial y} dy}{J}$$

where $J = \frac{\partial(x,y)}{\partial(u,v)}$ is the Jacobian of the transformation. Compare the above results with

$$du = \frac{\partial u}{\partial x} dx + \frac{\partial u}{\partial y} dy \quad \text{and} \quad dv = \frac{\partial v}{\partial x} dx + \frac{\partial v}{\partial y} dy$$

to show

$$\frac{\partial u}{\partial x} = \frac{1}{J} \frac{\partial y}{\partial v}, \qquad \frac{\partial u}{\partial y} = \frac{-1}{J} \frac{\partial x}{\partial v}, \qquad \frac{\partial v}{\partial x} = \frac{-1}{J} \frac{\partial y}{\partial u}, \qquad \frac{\partial v}{\partial y} = \frac{1}{J} \frac{\partial x}{\partial u}$$

(b) For $x = r \cos \theta$ and $y = r \sin \theta$ we replace u, v in part (a) by r, θ to obtain

$$\frac{\partial x}{\partial r} = \cos \theta, \qquad \frac{\partial x}{\partial \theta} = -r \sin \theta, \qquad \frac{\partial y}{\partial r} = \sin \theta, \qquad \frac{\partial y}{\partial \theta} = r \cos \theta$$

with Jacobian $J = \begin{vmatrix} \cos \theta & -r \sin \theta \\ \sin \theta & r \cos \theta \end{vmatrix} = r$ Consequently,

$(i) \quad \dfrac{\partial r}{\partial x} = \dfrac{1}{J} \dfrac{\partial y}{\partial \theta} = \cos \theta, \qquad (iii) \quad \dfrac{\partial \theta}{\partial x} = \dfrac{-1}{J} \dfrac{\partial y}{\partial r} = \dfrac{-1}{r} \sin \theta$

$(ii) \quad \dfrac{\partial r}{\partial y} = \dfrac{-1}{J} \dfrac{\partial x}{\partial \theta} = \sin \theta \qquad (iv) \quad \dfrac{\partial \theta}{\partial y} = \dfrac{1}{J} \dfrac{\partial x}{\partial r} = \dfrac{1}{r} \cos \theta$

Check

$$r^2 = x^2 + y^2 \qquad\qquad \tan \theta = \frac{y}{x}$$

$$2r \frac{\partial r}{\partial x} = 2x \Rightarrow \frac{\partial r}{\partial x} = \frac{x}{r} = \cos \theta \qquad \sec^2 \theta \frac{\partial \theta}{\partial x} = \frac{-y}{x^2} \Rightarrow \frac{\partial \theta}{\partial x} = \frac{-y}{x^2(1 + \tan^2 \theta)} = -\frac{\sin \theta}{r}$$

$$2r \frac{\partial r}{\partial y} = 2y \Rightarrow \frac{\partial r}{\partial x} = \frac{y}{r} = \sin \theta \qquad \sec^2 \theta \frac{\partial \theta}{\partial y} = \frac{x}{x^2} \Rightarrow \frac{\partial \theta}{\partial y} = \frac{x}{x^2(1 + \tan^2 \theta)} = \frac{\cos \theta}{r}$$

Selected Solutions Chapter 1

▶ 31. $x = x(u,v,w), \qquad y = y(u,v,w), \qquad z = z(u,v,w)$ with Jacobian given by

$$J = \frac{\partial(x,y,z)}{\partial(u,v,w)} = \begin{vmatrix} \frac{\partial x}{\partial u} & \frac{\partial x}{\partial v} & \frac{\partial x}{\partial w} \\ \frac{\partial y}{\partial u} & \frac{\partial y}{\partial v} & \frac{\partial y}{\partial w} \\ \frac{\partial z}{\partial u} & \frac{\partial z}{\partial v} & \frac{\partial z}{\partial w} \end{vmatrix}$$

Calculate the differentials dx, dy and dz

$$dx = \frac{\partial x}{\partial u}\,du + \frac{\partial x}{\partial v}\,dv + \frac{\partial x}{\partial w}\,dw$$
$$dy = \frac{\partial y}{\partial u}\,du + \frac{\partial y}{\partial v}\,dv + \frac{\partial y}{\partial w}\,dw$$
$$dz = \frac{\partial z}{\partial u}\,du + \frac{\partial z}{\partial v}\,dv + \frac{\partial z}{\partial w}\,dw$$

Use Cramer's rule and solve for du, dv and dw to obtain

$$du = \frac{1}{J}\begin{vmatrix} dx & \frac{\partial x}{\partial v} & \frac{\partial x}{\partial w} \\ dy & \frac{\partial y}{\partial v} & \frac{\partial y}{\partial w} \\ dz & \frac{\partial z}{\partial v} & \frac{\partial z}{\partial w} \end{vmatrix}, \quad dv = \frac{1}{J}\begin{vmatrix} \frac{\partial x}{\partial u} & dx & \frac{\partial x}{\partial w} \\ \frac{\partial y}{\partial u} & dy & \frac{\partial y}{\partial w} \\ \frac{\partial z}{\partial u} & dz & \frac{\partial z}{\partial w} \end{vmatrix}, \quad dw = \frac{1}{J}\begin{vmatrix} \frac{\partial x}{\partial u} & \frac{\partial x}{\partial v} & dx \\ \frac{\partial y}{\partial u} & \frac{\partial y}{\partial v} & dy \\ \frac{\partial z}{\partial u} & \frac{\partial z}{\partial v} & dz \end{vmatrix}$$

where J is the Jacobian of the transformation.

This gives

$$du = \frac{1}{J}\frac{\partial(y,z)}{\partial(v,w)}\,dx + \frac{1}{J}\frac{\partial(z,x)}{\partial(v,w)}\,dy + \frac{1}{J}\frac{\partial(x,y)}{\partial(v,w)}\,dz$$
$$dv = \frac{1}{J}\frac{\partial(y,z)}{\partial(w,u)}\,dx + \frac{1}{J}\frac{\partial(z,x)}{\partial(w,u)}\,dy + \frac{1}{J}\frac{\partial(x,y)}{\partial(w,u)}\,dz$$
$$dw = \frac{1}{J}\frac{\partial(y,z)}{\partial(u,v)}\,dx + \frac{1}{J}\frac{\partial(z,x)}{\partial(u,v)}\,dy + \frac{1}{J}\frac{\partial(x,y)}{\partial(u,v)}\,dz$$

▶ 32.

$$r = \sqrt{x^2 + y^2}$$

(a) $\quad r^2 \qquad\qquad$ (b) $\quad \theta = \arctan(y/x) \qquad\qquad$ (c)

$$z = z$$

$$\frac{\partial r}{\partial y} = \sin\theta$$
$$\frac{\partial \theta}{\partial x} = \frac{-1}{r}\sin\theta$$
$$\frac{\partial \theta}{\partial y} = \frac{1}{r}\cos\theta$$

▶ 33.

$$\rho = \sqrt{x^2 + y^2 + z^2}$$

(a) $\rho^2 \sin \theta$ (b) $\theta = \arctan\left(\dfrac{\sqrt{x^2 + y^2}}{z}\right)$ (c) $\dfrac{\partial \rho}{\partial y} = \sin \theta \sin \phi$

$$\phi = \arctan(y/x)$$

$$\frac{\partial \theta}{\partial z} = \frac{-\sin \theta}{\rho}$$

$$\frac{\partial \phi}{\partial x} = \frac{-\sin \phi}{\rho \sin \theta}$$

▶ 34. Use the law of cosines to show $r^2 = x^2 + y^2 - 2xy \cos \theta$, then

$$\frac{dr}{dt} = \frac{1}{r}\left(xV_B + yV_A - 2(xV_A + yV_B)\cos \theta\right)$$

▶ 35.

(a) $I_a = 8$,   (b) $I_b = 0$,   (c) $I_c = 3$

▶ 36. $\nabla \times \vec{F} \cdot \hat{n} = x(x-1) - y^2 + z(2x+1)$ with $d\sigma = \dfrac{dxdy}{|\hat{n} \cdot \hat{\mathbf{e}}_3|} = \dfrac{dxdy}{z}$ Convert to polar coordinates $x = r\cos\theta$, $y = r\sin\theta$, $dxdy = rdrd\theta$ to obtain $I = \pi$

▶ 37. (a) $I_a = 3/2$,   (b) $I_b = 89/60$,   (c) $I_c = 1$

▶ 38. 4π

▶ 39. Also let $\vec{F} = \psi \nabla \phi$ in divergence theorem and subtract results to obtain answer.

▶ 40.

(b) $\vec{r} = x\,\hat{\mathbf{e}}_1 + y\,\hat{\mathbf{e}}_2 + z(x,y)\,\hat{\mathbf{e}}_3$ with $\dfrac{\partial \vec{r}}{\partial x} = \hat{\mathbf{e}}_1 + \dfrac{\partial z}{\partial x}\hat{\mathbf{e}}_3$ and $\dfrac{\partial \vec{r}}{\partial y} = \hat{\mathbf{e}}_2 + \dfrac{\partial z}{\partial y}\hat{\mathbf{e}}_3$ so that a normal

vector to the surface is $\mathcal{N} = \dfrac{\partial \vec{r}}{\partial x} \times \dfrac{\partial \vec{r}}{\partial y} = \begin{vmatrix} \hat{\mathbf{e}}_1 & \hat{\mathbf{e}}_2 & \hat{\mathbf{e}}_3 \\ 1 & 0 & \frac{\partial z}{\partial x} \\ 0 & 1 & \frac{\partial z}{\partial y} \end{vmatrix} = -\dfrac{\partial z}{\partial x}\hat{\mathbf{e}}_1 - \dfrac{\partial z}{\partial y} + \hat{\mathbf{e}}_3$ and a unit

normal to the surface is

$$\hat{n} = \frac{-\frac{\partial z}{\partial x}\hat{\mathbf{e}}_1 - \frac{\partial z}{\partial y}\hat{\mathbf{e}}_+ \hat{\mathbf{e}}_3}{\sqrt{\left(\frac{\partial z}{\partial x}\right)^2 + \left(\frac{\partial z}{\partial y}\right)^2 + 1}}$$

▶ 41. $\displaystyle\oiint_S \vec{r} \cdot \hat{n}\, d\sigma = \iiint_V \text{div}\,(\vec{r})\, d\tau = 3 \iiint_V d\tau = 3V$ since div$(\vec{r})$=3. Hence $V = \dfrac{1}{3}\displaystyle\oiint_S \vec{r} \cdot \hat{n}\, d\sigma$

▶ 42.

(a) From (42c) $da = du - Tds - sdT = Tds - Pdv - Tds - sdT = -Pdv - sdT$

(b) From (42d) $dg = dh - Tds - sdT = vdP - sdT$

Selected Solutions Chapter 1

► 42.

(c)

$$(42a) \ du = Tds - Pdv \ \Rightarrow \ \left(\frac{\partial u}{\partial s}\right)_v = T, \quad \left(\frac{\partial u}{\partial v}\right)_s = -P$$

$$(42b) \ dh = Tds + vdP \ \Rightarrow \ \left(\frac{\partial h}{\partial s}\right)_P = T, \quad \left(\frac{\partial h}{\partial P}\right)_s = v$$

$$(42e) \ da = -Pdv - sdT \ \Rightarrow \ \left(\frac{\partial a}{\partial v}\right)_T = -P, \quad \left(\frac{\partial a}{\partial T}\right)_v = -s$$

$$(42f) \ dg = vdP - sdT \ \Rightarrow \ \left(\frac{\partial g}{\partial P}\right)_T = v, \quad \left(\frac{\partial g}{\partial T}\right)_P = -s$$

(d)

$$\frac{\partial^2 u}{\partial s \partial v} = \left(\frac{\partial T}{\partial v}\right)_s = \frac{\partial^2 u}{\partial v \partial s} = -\left(\frac{\partial P}{\partial s}\right)_v$$

$$\frac{\partial^2 h}{\partial s \partial P} = \left(\frac{\partial T}{\partial P}\right)_s = \frac{\partial^2 h}{\partial P \partial s} = \left(\frac{\partial v}{\partial s}\right)_P$$

$$\frac{\partial^2 a}{\partial v \partial T} = -\left(\frac{\partial P}{\partial T}\right)_v = \frac{\partial^2 a}{\partial T \partial v} - \left(\frac{\partial s}{\partial v}\right)_T$$

$$\frac{\partial^2 g}{\partial P \partial T} = \left(\frac{\partial v}{\partial T}\right)_P = \frac{\partial^2 g}{\partial T \partial P} = -\left(\frac{\partial s}{\partial P}\right)_T$$

(e)

$$(i) \quad h = g + Ts = g - T\left(\frac{\partial g}{\partial T}\right)_P$$

$$(ii) \quad a = u - Ts = h - Pv - Ts = g - Pv = g - P\left(\frac{\partial g}{\partial P}\right)_T$$

► 43.

(a) $\quad h = C_p T = u + RT = C_v T + RT \ \Rightarrow \ C_p = C_v + R$

(b) $\quad dq = C_v dT + Pdv = C_v d\left(\frac{Pv}{R}\right) + Pdv$ since $T = \frac{Pv}{R}$

(c) If $dq = 0$ (adiabatic process), then

$$-Pdv = \frac{C_v P}{R} dv + \frac{C_v v}{R} dP$$

$$-\frac{C_v v}{R} dP = \frac{C_v P}{R} dv + Pdv = \frac{(C_v + R)}{R} Pdv = \frac{C_p}{R} Pdv$$

$$\text{or} \quad -\gamma \frac{dv}{v} = \frac{dP}{P}$$

(d) Integrate part (c) above to obtain $-\gamma \ln v + \ln c = \ln P$ where $\ln c$ is a constant of integration. Simplify this result to the form $Pv^\gamma = constant$.

Selected Solutions Chapter 1

Selected Solutions Chapter 2

▶ 1.

(a)

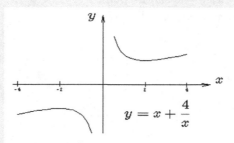

$x = 2, \quad y = 4$ local minimum
$x = -2, y = -4$ local maximum

$y = x + \dfrac{4}{x}$

(b)

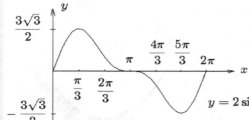

$x = \pi, \quad y = 0$ inflection point
$x = \pi/3, \quad y = 3\sqrt{3}/2$ maximum
$x = 5\pi/3, \quad y = -3\sqrt{3}/2$ minimum

$y = 2\sin x + \sin 2x$

(c)

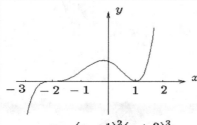

$x = 1, \quad y = 0$ local minimum
$x = -2, \quad y = 0$ inflection point
$x = -1/5, \quad y = 8.39808$ local maximum

$y = (x - 1)^2 (x + 2)^3$

▶ 2.

(a)

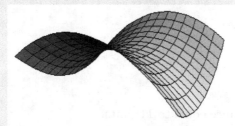

$x = 1, \quad y = 1, \quad z = 2$ saddle point
$A = z_{xx} = 4,$
$B = z_{xy} = -1,$
$C = z_{yy} = -6,$

$AC - B^2 < 0$

$z = 2x^2 - xy - 3y^2 - 3x + 7y$

► 2.

(b)

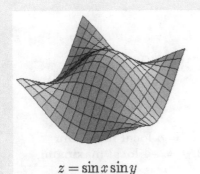

$x = \frac{\pi}{2}\ y = \frac{\pi}{2}$ maximum
$x = 0,\ y = 0$ saddle point
$x = 0,\ y = \pi$ saddle point
$x = \pi,\ y = 0$ saddle point
$x = \pi,\ y = \pi$ saddle point

$z = \sin x \sin y$

► 3.

(a) $x = y = \sqrt{A_0}$ square
(b) $x = y = \sqrt{A_0}$ square

► 4.

(a)

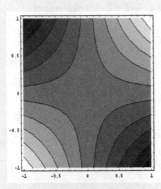

$x = 0,\ y = 0$ saddle point

(b)

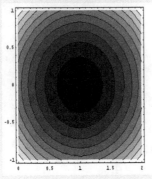

$x = 1\ y = 0$ minimum

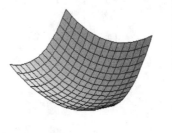

► 5. $\lambda = 2\pi\sqrt{\dfrac{\tau}{\rho g}}, \qquad V_{min}^2 = 2\sqrt{\dfrac{\tau g}{\rho}}$

► 6. Minimum occurs at $t = 0$ with minimum distance equal to 11 units.

► 7. Minimize $d^2 = f = \sum_{i=1}^{n}(x - x_i)^2 + (y - y_i)^2 + (z - z_i)^2$. If

$$\frac{\partial f}{\partial x} = \sum_{i=1}^{n}2(x - x_i) = 0, \quad \frac{\partial f}{\partial y} = \sum_{i=1}^{n}2(y - y_i) = 0, \quad \frac{\partial f}{\partial z} = \sum_{i=1}^{n}2(z - z_i) = 0$$

then

$$x = \frac{1}{n}\sum_{i=1}^{n} x_i, \quad y = \frac{1}{n}\sum_{i=1}^{n} y_i, \quad z = \frac{1}{n}\sum_{i=1}^{n} z_i$$

Here (x, y, z) represents the center of gravity for a system of particles all having a mass of 1 unit.

▶ 8. Minimize $f = d^2 = x^2 + y^2 = x^2 + (1-x)^2$. We have $\frac{\partial f}{\partial x} = 2x + 2(1-x)(-1) = 0$ implies $x = 1/2, y = 1/2$

▶ 9. Consider extreme values of $f = x^2 + 24xy + 8y^2 + \lambda(x^2 + y^2 - 25)$ we have

$$\frac{\partial f}{\partial x} = 2x + 24y + 2\lambda x = 0$$

$$\frac{\partial f}{\partial y} = 24x + 16y + 2\lambda y = 0$$

$$\frac{\partial f}{\partial \lambda} = x^2 + y^2 - 25 + 0$$

For a nonzero solution to this system of equations one must require

$$\begin{vmatrix} 2 + 2\lambda & 24 \\ 24 & 16 + 2\lambda \end{vmatrix} = 4\lambda^2 + 36\lambda - 544 = 4(\lambda + 17)(\lambda - 8) = 0$$

This gives $\lambda = 8$ and $\lambda = -17$. Examine these cases to show

$$(3, 4) \text{ and } (-3, -4) \text{ maximum,} \qquad (-4, 3) \text{ and } (4, -3) \text{ minimum}$$

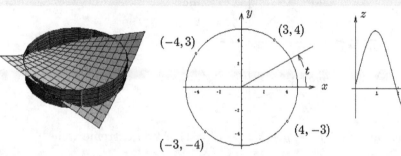
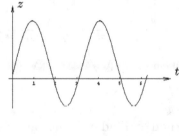

▶ 10. Consider the extreme values of $f = xyz + \lambda_1(x^2 + y^2 - 1) + \lambda_2(x - z)$ Show that at an extreme value

$$\frac{\partial f}{\partial x} = yz + 2\lambda_1 x + \lambda_2 = 0$$

$$\frac{\partial f}{\partial y} = xz + 2\lambda_1 y = 0$$

$$\frac{\partial f}{\partial z} = xy - \lambda_2 = 0$$

$$\frac{\partial f}{\partial \lambda_1} = x^2 + y^2 - 1 = 0$$

$$\frac{\partial f}{\partial \lambda_2} = x - z = 0$$

Selected Solutions Chapter 2

Show the above equations can be combined to obtain the equations

$$x(2y^2 - x^2) = x(\sqrt{2}y - x)(\sqrt{2}y + x) = 0, \quad \text{and} \quad x^2 + y^2 - 1 = 0$$

Consider the cases (i) $x = 0$, (ii) $y = x/\sqrt{2}$, and (iii) $y = -x/\sqrt{2}$.

x	y	z	w	Comment
0	1	0	0	local minimum
0	-1	0	0	local maximum
$+\sqrt{2/3}$	$+1/\sqrt{3}$	$+\sqrt{2/3}$	$+2/3\sqrt{3}$	local maximum
$-\sqrt{2/3}$	$-1/\sqrt{3}$	$-\sqrt{2/3}$	$-2/3\sqrt{3}$	local minimum
$+\sqrt{2/3}$	$-1/\sqrt{3}$	$+\sqrt{2/3}$	$-2/3\sqrt{3}$	local minimum
$-\sqrt{2/3}$	$+1/\sqrt{3}$	$-\sqrt{2/3}$	$+2/3\sqrt{3}$	local maximum

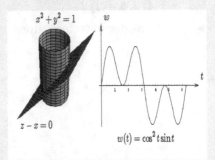

▶ 11.

Method 1: Let $x = \cos t$ $y = \sin t$ and find extreme values for $z = \cos t \sin t$

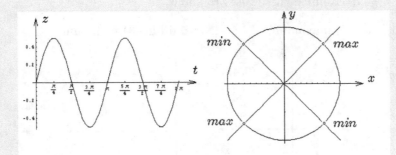

Method 2: Find extreme values for $f = xy + \lambda(x^2 + y^2 = 1)$. At an extreme value

$$\frac{\partial f}{\partial x} = y + 2\lambda x = 0$$

$$\frac{\partial f}{\partial y} = x + 2\lambda y = 0$$

$$\frac{\partial f}{\partial \lambda} = x^2 + y^2 - 1 = 0$$

Show that equations on left imply that $x^2 - y^2 = (x - y)(x + y) = 0$. The case $y = x$ produces $x = \pm 1/\sqrt{2}$, $y = \pm 1/\sqrt{2}$, $z = 1/2$ which produces a local maximum.

The case $y = -x$ produces $x = \pm 1/\sqrt{2}$, $y = \mp 1/\sqrt{2}$, $z = -1/2$ which produces a local minimum.

▶ 12.

Method 1: Minimize $\ell = x^2 + y^2 + z^2$ where $z = 8/xy$, then $\ell = x^2 + y^2 + 64/x^2y^2$ with

$$\frac{\partial \ell}{\partial x} = 2x - 128/x^2 y^2 = 0 \quad \text{and} \quad \frac{\partial \ell}{\partial y} = 2y - 128/x^2 y^3 = 0$$

Solve these equations and show $y^2 x^4 = 64$, $x^2 y^4 = 64$ which implies $x^2 y^2(y^2 - x^2) = 0$

Selected Solutions Chapter 2

Along $y = x$ one finds $x = 2$, $y = 2$, $z = 2$ and $x = -2$, $y = -2$, $z = 2$. Along $y = -x$ one finds $x = 2$, $y = -2$, $z = -2$ and $x = -2$, $y = 2$, $z = -2$ with $d = \sqrt{\ell} = 2\sqrt{3}$.

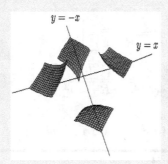

Method 2: Find extreme values for $f = x^2 + y^2 + z^2 + \lambda(xyz - 8)$. At an extreme value

$$\frac{\partial f}{\partial x} = 2x + \lambda yz = 0$$

$$\frac{\partial f}{\partial y} = 2y + \lambda xz = 0$$

$$\frac{\partial f}{\partial z} = 2z + \lambda xy = 0$$

$$\frac{\partial f}{\partial \lambda} = xyz - 8 = 0$$

One finds $\lambda = -1$ and $y^2 - x^2 = 0$ and $z^2 - x^2 = 0$ which will lead to the same results as Method 1 above.

▶ 13.

Method 1: $f = xy + 2xz + 2yz$ is to be minimized subject to the constraint $xyz = 32$. Substitute $z = 32/xy$ into f to obtain $f = f(x,y) = xy + 64/y + 64/x$. At a minimum one finds $x = y = 4$ and $z = 2$.

Method 2: Find extreme values of $f = xy + 2xz + 2yz + \lambda(xyz - 32)$ to obtain the same answers as above.

▶ 14.

$$(a) \quad d_{min} = \left| \frac{c}{\sqrt{a^2 + b^2}} \right| \qquad (b) \quad x = \frac{ac}{a^2 + b^2}, \ y = \frac{bc}{a^2 + b^2}$$

$$(c) \quad d_{min} = \left| \frac{d}{\sqrt{a^2 + b^2 + c^2}} \right| \qquad (d) \quad x = \frac{ad}{a^2 + b^2 + c^2}, \ y = \frac{bd}{a^2 + b^2 + c^2}, \ z = \frac{cd}{a^2 + b^2 + c^2}$$

One can show the general case that

(a) Distance from point (x_1, y_1) to line $ax + by - c = 0$ is given by $\ell = \left| \frac{ax_1 + by_1 - c}{\sqrt{a^2 + b^2}} \right|$

(c) Distance from point (x_1, y_1, z_1) to the plane $ax + by + cz - d = 0$ is given by $\ell = \left| \frac{ax_1 + by_1 + cz_1 - d}{\sqrt{a^2 + b^2 + c^2}} \right|$.

Selected Solutions Chapter 2

▶ 15. $A_\ell = \pi(x + R)\sqrt{h^2 + (R - x)^2}$ Note that when $x = R$ we have $A_\ell = 2\pi Rh$ the surface area of a cylinder and when $x = 0$ we have $A_\ell = \pi R\sqrt{h^2 + R^2}$ the surface area of a cone.

$$\frac{dA_\ell}{dx} = 2\pi \left[\frac{x^2 - Rx + \frac{h^2}{2}}{\sqrt{h^2 + (R - x)^2}} \right]$$

complete the square of numerator $\quad \dfrac{dA_\ell}{dx} = 2\pi \left[\dfrac{\left(x - \frac{R}{2}\right)^2 + \frac{1}{4}(2h^2 - R^2)}{\sqrt{h^2 + (R - x)^2}} \right]$

If $2h^2 > R^2$, then $\frac{dA_\ell}{dx} > 0$ so that the surface area increases from $x = 0$ to $x = R$

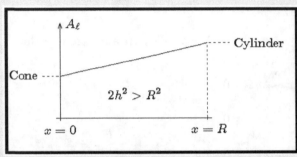 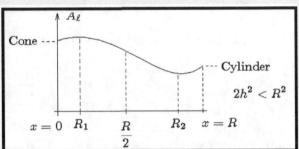

If $2h^2 < R^2$, then setting $\frac{dA_\ell}{dx} = 0$ one finds the two roots, $x = R_1 = \frac{R}{2} - \frac{\sqrt{R^2 - 2h^2}}{2}$ and $x = R_2 = \frac{R}{2} + \frac{\sqrt{R^2 - 2h^2}}{2}$ where R_1 gives maximum A_ℓ and R_2 gives minimum A_ℓ.

▶ 16. Minimum distance is 5 units.

▶ 17. $x = a/3$, $y = b/3$

▶ 18. Let $\vec{r}_0 = x_0\,\widehat{\mathbf{e}}_1 + y_0\,\widehat{\mathbf{e}}_2 + z_0\,\widehat{\mathbf{e}}_3$ and $\vec{r}_1 = \frac{x_0\,\widehat{\mathbf{e}}_1 + y_0\,\widehat{\mathbf{e}}_2 + z_0\,\widehat{\mathbf{e}}_3}{\sqrt{x_0^2 + y_0^2 + z_0^2}}$ then

   (a) $\ell_{min} = |\vec{r}_0 - \vec{r}_1|$ and $\ell_{max} = |\vec{r}_0 - \vec{r}_1| + 2$

   (b) $\ell_{min} = |\vec{r}_1 - \vec{r}_0|$ and $\ell_{max} = 2 - \ell_{min}$

▶ 19. Maximum at $\left(1 - \sqrt{2}, \frac{1 + \sqrt{2}}{2}\right)$ and minimum at $\left(1 + \sqrt{2}, \frac{1 - \sqrt{2}}{2}\right)$

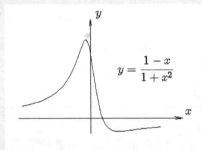

▶ 20. If $h > a/\sqrt{2}$ then place light at height $a/\sqrt{2}$.
If $h < a/\sqrt{2}$, then place light at height h.

Selected Solutions Chapter 2

► 21.

$$I = k\frac{I_A}{x^2} + k\frac{I_B}{(\ell - x)^2} \qquad k > 0, \ I_A > 0, \ I_B > 0, \ \ell > 0$$

$$\text{and} \quad \frac{dI}{dx} = k\left\{-\frac{2I_A}{x^3} + \frac{2I_B}{(\ell - x)^3}\right\} = 0 \quad \text{when}$$

$$\frac{I_A}{x^3} = \frac{I_B}{(\ell - x)^3} \Rightarrow \frac{x}{\ell - x} = \left(\frac{I_A}{I_B}\right)^{1/3} \text{ Solve for } x$$

$$\text{Minimum exists because} \quad \frac{d^2I}{dx^2} = k\left\{\frac{6I_A}{x^4} + 6\frac{I_B}{(\ell - x)^4}\right\} > 0$$

► 22.

$$y^3 = 6xy - x^3 - 1$$

$$3y^2\frac{dy}{dx} = 6x\frac{dy}{dx} + 6y - 3x^2$$

$$\frac{dy}{dx} = \frac{6y - 3x^2}{3y^2 - 6x} \text{ and } \frac{dy}{dx} = 0 \text{ when } y = \frac{1}{2}x^2$$

When $y = \frac{1}{2}x^2$ the equation $y^3 = 6xy - x^3 - 1$ becomes $(\frac{1}{2}x^2)^3 = 6x(\frac{1}{2}x^2) - x^3 - 1$ which simplifies to the quadratic equation $(x^3)^2 - 16(x^3) + 8 = 0$ with roots $x^3 = 8 \pm \sqrt{56} = 8 \pm 2\sqrt{14}$ Also one can calculate

$$\frac{d^2y}{dx^2} = \frac{(3y^2 - 6x)(6\frac{dy}{dx} - 6x) - (6y - 3x^2)(6y\frac{dy}{dx} - 6)}{(3y^2 - 6x)^2}$$

Then when $x^3 = 8 \pm 2\sqrt{14}$ we have $\frac{dy}{dx} = 0$ with

$$\left.\frac{d^2y}{dx^2}\right|_{x^3 = 8 + 2\sqrt{14}} < 0 \Rightarrow \text{ which implies a local maximum exists}$$

$$\left.\frac{d^2y}{dx^2}\right|_{x^3 = 8 - 2\sqrt{14}} > 0 \Rightarrow \text{ which implies a local minimum exists}$$

► 23.

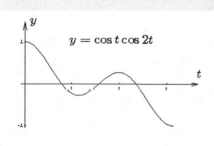

$y = \cos t \cos 2t$

Show $\frac{dy}{dt} = 0$ when $t = 0, t = \pi$ and where $\tan t = \pm\sqrt{5}$ and verify that $y = 1$ is a maximum, $y = -1$ is a minimum, $y = \sqrt{6}/9$ is a local maximum and $y = -\sqrt{6}/9$ is a local minimum.

Selected Solutions Chapter 2

► 24.

(a)

$$y = \frac{x(x-1)}{(x+1)(x-5)} \text{ with } \frac{dy}{dx} = \frac{(x+1)(x-5)(2x-1) - x(x-1)2x - 4)}{(x+1)^2(x-5)^2}$$

$\frac{dy}{dx} = 0$ when $(x+1)(x-5)(2x-1) - x(x-1)2x - 4) = -3x^2 - 10x + 5 = 0$ which has the roots $x = -5/3 \pm 2\sqrt{10}/3$

$$y \Big|_{x=\frac{-5+2\sqrt{10}}{3}} = \frac{1}{18}(7 - 2\sqrt{10}) \text{ maximum}$$

$$y \Big|_{x=\frac{-5-2\sqrt{10}}{3}} = \frac{1}{18}(7 + 2\sqrt{10}) \text{ minimum}$$

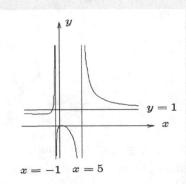

► 24.

(b)

$$y = \frac{x(x-1)}{x-2} \text{ with } \frac{dy}{dx} = \frac{(x-2)(2x-1) - x(x-1)}{(x-2)^2}$$

$\frac{dy}{dx} = 0$ when $x^2 - 4x + 2 = 0$ which gives the roots $x = 2 \pm \sqrt{2}$.

$$y \Big|_{x=2+\sqrt{2}} = 3 + 2\sqrt{2}$$

$$y \Big|_{x=2-\sqrt{2}} = 3 - 2\sqrt{2}$$

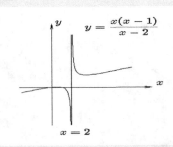

Selected Solutions Chapter 2

▶ 25.

$$f = x^3 - y^3 + 3x^2 - 9x$$

$$\frac{\partial f}{\partial x} = 3x^2 + 6x - 9 = 3(x^2 + 2x + 3) = 3(x+3)(x-1) = 0$$

$$\frac{\partial f}{\partial y} = -3y^2 + 6y = -3y(y-2) = 0$$

gives the roots $x = -3$, $x = 1$, $y = 0$ and $y = 2$.

$$f_{xx} = 6x + 6, \quad f_{xy} = 0, \quad f_{yy} = -6y + 6$$

At $(-3, 0)$, $H_1 < 0$, $H_2 < 0$ gives saddle point.

At $(-3, 2)$, $H_1 < 0$, $H_2 > 0$ gives local maximum.

At $(1, 0)$, $H_1 > 0$, $H_2 > 0$ gives local minimum.

At $(1, 2)$, $H_1 > 0$, $H_2 < 0$ gives saddle point.

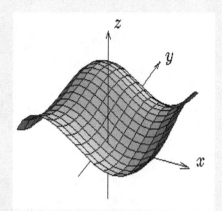

▶ 26.

(a) Point (x, c) is where x has the value $x^* = \frac{x_0 + \alpha x_1}{1 + \alpha}$, where $\alpha = \frac{c - y_0}{c - y_1}$ is a constant.
In expanded form $x^* = \frac{(c - y_1)x_0 + (c - y_0)x_1}{2c - y_0 - y_1}$

(b) Let $L = d_0 + d_1 = \sqrt{(x - x_0)^2 + (c - y_0)^2} + \sqrt{(x_1 - x)^2 + (c - y_1)^2}$ with
$\frac{\partial L}{\partial x} = \frac{1}{2}\frac{2(x - x_0)}{d_0} - \frac{1}{2}\frac{2(x_1 - x)}{d_1}$. If $\frac{\partial L}{\partial x} = 0$, then

$$\frac{x - x_0}{d_0} = \frac{x_1 - x}{d_1} \quad \Rightarrow \quad \sin\theta_1 = \sin\theta_2 \quad \Rightarrow \quad \theta_1 = \theta_2$$

(c) Equation of the straight line through the points (x_1, y_1) and (x^*, c) is obtained from the point-slope formula and has the form $y - y_1 = -\left(\frac{c - y_1}{x_1 - x^*}\right)(x - x_1)$. When $x = x_0$, one can use the result form part (a) to write

$$y = y_1 - \frac{(c - y_1)}{(x_1 - x^*)}(x_0 - x_1) = \frac{y_1(x_1 - x^*) - (c - y_1)x_0 + (c - y_1)x_1}{x_1 - x^*}$$

$$y = = \frac{y_1(x_1 - x^*) - [(2c - y_0 - y_1)x^* - (c - y_0)x_1] + (c - y_1)x_1}{x_1 - x^*} = 2c - y_0.$$

▶ 27.

(a)

$$\left|\frac{d\vec{F}}{ds}\right| = |\nabla f| |\hat{e}| \cos 0 \text{ is a maximum when } 0 = 0.$$

(b) $V = \frac{\cos\theta}{\rho^2}$, $\quad \operatorname{grad} V = \frac{\partial V}{\partial \rho}\hat{\mathbf{e}}_\rho + \frac{1}{\rho}\frac{\partial V}{\partial \theta}\hat{\mathbf{e}}_\theta + \frac{1}{\rho\sin\theta}\frac{\partial V}{\partial \phi}\hat{\mathbf{e}}_\phi = -\frac{2\cos\theta}{\rho^3}\hat{\mathbf{e}}_\rho - \frac{\sin\theta}{\rho^3}\hat{\mathbf{e}}_\theta$

Selected Solutions Chapter 2

▶ 28. Find extreme values of $f = xy + yz + \lambda_1(x^2 + y^2 - 2) + \lambda_2(yz - 2)$

$$(i) \qquad f_x = y + 2x\lambda_1 = 0$$
$$(ii) \qquad f_y = x + z + 2\lambda_1 y + \lambda_2 z = 0$$
$$(iii) \qquad f_z = y + \lambda_2 y = 0$$
$$(iv) \qquad f_{\lambda_1} = x^2 + y^2 - 2 = 0$$
$$(v) \qquad f_{\lambda_2} = yz - 2 = 0$$

For $\lambda_2 = -1$ the equations (ii) and (i) become

$$x + 2\lambda_1 y = 0$$
$$2x\lambda_1 + y = 0$$

For a nonzero solution we require $\begin{vmatrix} 1 & 2\lambda_1 \\ 2\lambda_1 & 1 \end{vmatrix} = 0$ which requires $\lambda_1 = \pm 1/2$ For $\lambda_1 = 1/2$ one finds $(1, -1, 2)$ gives $f = 1$ a minimum and $(-1, 1, 2)$ with $f = 1$ a minimum. For $\lambda_1 = -1/2$ one finds $(1, 1, 2)$ gives $f = 3$ a maximum and $(-1, -1, -2)$ with $f = 3$ a maximum.

▶ 29. If $y = ax^2 + bx + c$, then
(a)

$$y(-2) = 4a - 2b + c \qquad\qquad e_1 = y(-2) - y_1 = 4a - 2b + c - y_1$$
$$y(-1) = a - b + c \qquad\qquad e_2 = y(-1) - y_2 = a - b + c - y_2$$
$$y(0) = c \qquad\qquad e_3 = y(0) - y_2 = c - y_3$$
$$y(1) = a + b + c \qquad\qquad e_4 = y(1) - y_4 = a + b + c - y_4$$
$$y(2) = 4a + 2b + c \qquad\qquad e_5 = y(2) - y_5 = 4a + 2b + c - y_5$$

The sum of squares of the errors is given by $E(a, b, c) = \sum_{i=1}^{5} e_i^2$ and at a minimum $\frac{\partial E}{\partial a} = 0$, $\frac{\partial E}{\partial b} = 0$, and $\frac{\partial E}{\partial c} = 0$. This requires that

$$34a + 10c = 4y_1 + y_2 + y_4 + 4y_5$$
$$10b = -2y_1 - y_2 + y_4 + 2y_5$$
$$10a + 5c = y_1 + y_2 + y_3 + y_4 + y_5$$

and solving for a, b, c one finds $a = \frac{1}{14}(2y_1 - y_2 - 2y_3 - y_4 + 2y_5)$,
$b = \frac{1}{10}(-2y_1 - y_2 + y_4 + 2y_5)$, $\quad c = \frac{1}{35}(-3y_1 + 12y_2 + 17y_3 + 12y_4 - 3y_5)$
(b) $y = \frac{8}{14}x^2 - \frac{1}{10}x + \frac{95}{70} = 0.571429x^2 - 0.1x + 1.35714$

▶ 30.
(a) $x = 5$, $y = 5$ a square.
(b) $r = \left(\frac{V}{2\pi}\right)^{1/3}$, $\quad h = 2\left(\frac{V}{2\pi}\right)^{1/3}$

Selected Solutions Chapter 2

▶ 31.

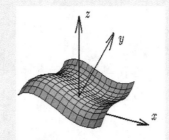

$(2, 3)$	local minimum
$(2, -3)$	saddle point
$(-2, 3)$	saddle point
$(-2, -3)$	local maximum

▶ 32. $x = \dfrac{ab^2}{2(a^2 + b^2)}, \qquad y = \dfrac{a^2 b}{2(a^2 + b^2)}$

▶ 33. $x = 4, \quad y = 4, \quad z = 2$

▶ 34. $p = \sqrt{p_1 p_2}$

▶ 35. $x = 6, \quad y = 6, \quad z = 3$

▶ 37. $x = C, \quad y = C, \quad f = 2C^2$

▶ 38. $\dfrac{\partial E}{\partial c_i} = \displaystyle\int_a^b w(x) 2\Big(f(x) - \sum_{n=1}^{\infty} c_n \phi_n(x) \Big) \phi_i(x)\, dx = 0, \quad i = 1, 2, 3, \ldots$

$\int_a^b w(x) f(x) \phi_i(x)\, dx - \sum_{n=1}^{\infty} c_n \int_a^b w(x) \phi_n(x) \phi_i(x)\, dx = 0$ By hypothesis the set of functions $\{\phi_m(x)\}$ are orthogonal over the interval (a, b) so that $(\phi_n, \phi_i) = \begin{cases} 0, & \text{if } n \neq i \\ \| \phi_i \|^2, & \text{if } n = i \end{cases}$ consequently $(f, \phi_i) - c_i(\phi_i, \phi_i) = 0$ or $c_i = \dfrac{(f, \phi_i)}{\|\phi\|^2}$

▶ 42.

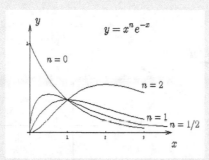

$y = x^n e^{-x}$

▶ 44.   Minimum distance is 1 unit.

▶ 45.   For ℓ_1 and ℓ_2 representing the length of the sides, then $\ell_1 = 2a/\sqrt{2}, \quad \ell_2 = 2b/\sqrt{2}$

▶ 46. $y = \dfrac{1}{4}\Big[h + \sqrt{h^2 + 8R^2} \Big] \quad x^2 = R^2 - y^2 = \tfrac{1}{8}\Big[4R^2 - h^2 - h\sqrt{h^2 + 8R^2} \Big]$

▶ 47.

(a) Use $\cos 2\theta = \cos^2 \theta - \sin^2 \theta$ with $\theta = \frac{x - \xi}{2}$ and show

$$y = \left(\frac{1 - \alpha}{1 + \alpha} \right) \frac{(1 + \alpha)^2}{1 + \alpha^2 - 2\alpha\big(\cos^2 \theta - \sin^2 \theta \big)} = \left(\frac{1 - \alpha}{1 + \alpha} \right) \frac{(1 + \alpha)^2}{(1 + \alpha^2)(\cos^2 \theta + \sin^2 \theta) - 2\alpha(\cos^2 \theta - \sin^2 \theta)}$$

which simplifies to the desired result.

Selected Solutions Chapter 2

▶ 47.

(b) Set $\frac{dy}{dx} = 0$ to show this requires $\sin(x - \xi) = 0$ or $x - \xi = 0, \pi, 2\pi, 3\pi, \ldots$
When $x - \xi = 0, 2\pi, 4\pi, \ldots$, then $y = \frac{1+\alpha}{1-\alpha}$ is a maximum
When $x - \xi = \pi, 3\pi, 5\pi, \ldots$, then $y = \frac{1-\alpha}{1+\alpha}$ is a minimum.

▶ 48.   Minimum distance is 5 units.

▶ 49.   Minimum distance is 6 units.

▶ 50. $x = 1$ is local maximum, $x = 3$ give local minimum, $x = 6$ is inflection point.

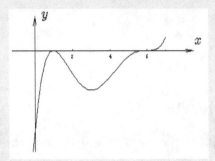

▶ 51.

(a) Local minimum at $(\sqrt{C}, 2\sqrt{C})$ and local maximum at $(-\sqrt{C}, -2\sqrt{C})$

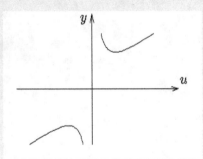

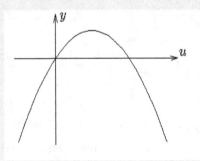

(b) Maximum at $(C/2, C^2/4)$

▶ 52. $H/R = \sqrt{2}\pi/\alpha$ produces a minimum value.

▶ 53. $F = Area = (\ell - 2x + x\cos\theta)x\sin\theta$ with $\frac{\partial F}{\partial x} = 0$ and $\frac{\partial F}{\partial \theta} = 0$ when $\theta = \pi/3$ and $x = \ell/3$
Here $F_{xx}F_{\theta\theta} - (F_{x\theta})^2 > 0$ with $F_{xx} < 0$ implies F is a maximum.

Selected Solutions Chapter 2

Selected Solutions Chapter 3

▶ 1.

(a) Green's theorem is $\oint_C P\,dx + Q\,dy = \iint_R \left(\frac{\partial Q}{\partial x} - \frac{\partial P}{\partial y}\right) dxdy$ and in the special case $P = 0$ and $Q = f(x,y)g(x,y)$, it reduces to the form

$$\oint_C fg\,dy = \iint_R \left(f\frac{\partial g}{\partial x} + g\frac{\partial f}{\partial x}\right) dxdy$$

The unit normal to the boundary curve C is given by $\hat{n} = \frac{dy}{dx}\hat{e}_1 - \frac{dx}{ds}\hat{e}_2$ with $\hat{n}\cdot\hat{e}_1 = \frac{dy}{ds}$, hence the Green's theorem can also be written as

$$\oint_C fg(\hat{n}\cdot\hat{e}_1)\,ds = \iint_R \left(f\frac{\partial g}{\partial x} + g\frac{\partial f}{\partial x}\right) dxdy$$

which can also be written in the form given on page 131, number 1(a).

▶ 3.

(a) Show $y - y_0 = m(x - x_0)$ is straight line with $m = \frac{y_1 - y_0}{x_1 - x_0}$.

(b) Substitute y from part (a) into $\Delta = \begin{vmatrix} x & y & 1 \\ x_0 & y_0 & 1 \\ x_1 & y_1 & 1 \end{vmatrix}$ and do the algebra to calculate Δ.

▶ 5.

(a) $\frac{d}{dx}\left(\frac{\partial f}{\partial y'}\right) = \frac{\partial^2 f}{\partial y\partial x} + \frac{\partial^2 f}{\partial y\partial y'}y' + \frac{\partial^2 f}{\partial y'^2}y'' + \frac{\partial^2 f}{\partial y'\partial y''}y'''$

▶ 7.

(a) $\frac{\partial f}{\partial y} - \frac{d}{dx}\left(\frac{\partial f}{\partial y'}\right) = 0 \Rightarrow \frac{\partial\alpha}{\partial y} + \frac{\partial\beta}{\partial y}y' + \frac{\partial\gamma}{\partial y}(y')^2 - \frac{d}{dx}\left(\beta + 2\gamma y'\right) = 0$

▶ 9.

(b) If $f = f(y, y', y'', y''')$, then a first integral of the Euler-Lagrange equation is given by

$$f + \alpha = \frac{\partial f}{\partial y'}y' + \frac{\partial f}{\partial y''}y'' + \frac{\partial f}{\partial y'''}y''' - \frac{d}{dx}\left(\frac{\partial f}{\partial y''}\right)y' - \frac{d}{dx}\left(\frac{\partial f}{\partial y'''}\right)y'' + \frac{d^2}{dx^2}\left(\frac{\partial f}{\partial y'''}\right)y'$$

▶ 11. $dV = \pi y^2\,dx$ with $V = V(\alpha) = \int_0^1 \pi\left[x + \alpha x(1-x)\right]^2 dx$. Expand and integrate to show $V - V(\alpha) - \pi\left(\frac{1}{3} + \frac{\alpha}{6} + \frac{\alpha^2}{30}\right)$. At an extreme value $\frac{dV}{d\alpha} = \pi\left(\frac{1}{6} + \frac{\alpha}{15}\right) = 0$ or $\alpha = -15/6$. The second derivative is $\frac{d^2V}{d\alpha^2} = \pi/15 > 0$ which implies $V(-15/6) = \pi/8$ is a minimum value.

▶ 14. $\frac{d}{dx}(p(x)y') - q(x)y = 0$ with $y(x_0) = y_0$ and $y(x_1) = y_1$

▶ 16. $z = x$ and $y = \dfrac{\sinh(x)}{\sinh(1)}$

▶ 18. $y = \left(\dfrac{\pi - \cosh \pi}{\sinh \pi}\right) \sinh x + \cosh x - x$

▶ 21. $u_{xx} + u_{yy} = q(x,y)$ is the Helmholtz equation with boundary condition

$$u = u(x,y)\Big|_{(x,y) \in \partial R} = g(x,y)$$

▶ 23.

(b) $\nabla u = \dfrac{\partial u}{\partial r}\,\widehat{\mathbf{e}}_r + \dfrac{1}{r}\dfrac{\partial u}{\partial \theta}\,\widehat{\mathbf{e}}_\theta + \dfrac{\partial u}{\partial z}\,\widehat{\mathbf{e}}_z$ with $d\tau = r\,dr\,d\theta\,dz$

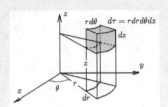

(c) $\nabla u = \dfrac{\partial u}{\partial \rho}\,\widehat{\mathbf{e}}_\rho + \dfrac{1}{\rho}\dfrac{\partial u}{\partial \theta}\,\widehat{\mathbf{e}}_\theta + \dfrac{1}{\rho \sin \theta}\dfrac{\partial u}{\partial \phi}\,\widehat{\mathbf{e}}_\phi$

$$d\tau = \rho^2 \sin \theta\, d\rho\,d\theta\,d\phi$$

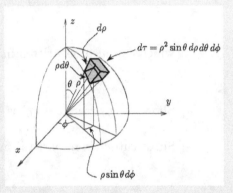

▶ 25. Let $u = \displaystyle\int_{x_1}^{x} y(\xi)\,d\xi$ with $u' = y$ and $u'' = y'$ and write $I = \displaystyle\int_{x_1}^{x_2} f(x,u,u',u'')\,dx$

▶ 27. $y = y_1 \sin x + y_0 \cos x$

▶ 29. Let $Y = y(x) + \epsilon \eta(x)$ and calculate $I(\epsilon) = \dfrac{\int_{x_0}^{x_1} F(x,Y,Y')\,dx}{\int_{x_0}^{x_1} G(x,Y,Y')\,dx}$ together with $I'(\epsilon) = \dfrac{dI}{d\epsilon}$. Then evaluate $I'(0)$ and put the result in a form where you can use the fundamental lemma.

▶ 32. Let $Y(x) = y(x) + \epsilon \eta(x)$ and then calculate $I(\epsilon)$ and $I'(\epsilon) = \dfrac{dI}{d\epsilon}$. Make use of the symmetric kernel $G(x,\xi) = G(\xi,x)$ to examine the integrals

$$\int_{x_1}^{x_2} G(\xi,x)y(\xi)\,d\xi \qquad \text{and} \qquad \int_{x_1}^{x_2} G(\xi,x)y(x)\,dx$$

and then reduce $I'(0)$ to the form

$$2\int_{x_1}^{x_2}\left[\int_{x_1}^{x_2} \lambda G(\xi,x)y(\xi)\,d\xi + H(x) - y(x)\right]\eta(x)\,dx = 0$$

Then use fundamental lemma to obtain the desired result.

▶ 37. $z_{xxxx} + 2z_{xxyy} + z_{yyyy} = 0$

Selected Solutions Chapter 3

Selected Solutions Chapter 4

▶ 1. $y = c_1 \sinh x + c_2 \cosh x + c_3 \sin x + c_4 \cos x$ where

$$c_1 = \frac{1}{2}\cosh\frac{\pi}{2}, \quad c_2 = \frac{1}{2}(\sinh\frac{\pi}{2} - 1), \quad c_3 = -\frac{1}{2}\cosh\frac{\pi}{2}, \quad c_4 = \frac{1}{2}(1 - \sinh\frac{\pi}{2})$$

▶ 2. $z = \left(\dfrac{z_2 - z_1}{\theta_2 - \theta_1}\right)(\theta - \theta_1) + z_1 \qquad s = \sqrt{r^2 + \left(\dfrac{z_2 - z_1}{\theta_2 - \theta_1}\right)^2}\,(\theta_2 - \theta_1)$

▶ 3. $ds = \sqrt{dv^2 + v^2 \sin^2\alpha\, d\theta^2}$ with $s = \int_{\theta_1}^{\theta_2}\sqrt{(v')^2 + v^2\sin^2\alpha}\,d\theta$ Show the Euler-Lagrange differential equation becomes $vv'' - 2(v')^2 - v^2\sin^2\alpha = 0$ Make the substitution $\dfrac{dv}{d\theta} = u$ with $\dfrac{d^2v}{d\theta^2} = \dfrac{du}{dv}\dfrac{dv}{d\theta} = u\dfrac{du}{dv}$ and perform an integration to show there results $\dfrac{dv}{v\sqrt{v^2 - A^2}} = c_1\,d\theta$ where c_1 is a constant of integration and $A = \dfrac{\sin\alpha}{c_1}$. Another integration gives the final result $v = \dfrac{\sin\alpha}{c_1}\sec(\theta\sin\alpha + c_2)$ where c_2 is another constant of integration.

▶ 4. $v = \dfrac{2\alpha r}{\sqrt{a^2 - r^2}}\arctan\left[\dfrac{\sqrt{a^2 - r^2}\tan\frac{u}{2}}{a + r}\right] + \beta$ where α and β are constants and $a > r$.

▶ 5. $y = 3 - 2/x$ The natural boundary conditions would be $y'(1) = -1/2$ and $y'(2) = -1/8$

▶ 6. $y = -\sin x$ The natural boundary conditions would be $y'(0) = 0$ and $y'(\pi) = 0$

▶ 7. $y = ax - (2a + b)x^2 + (a + b)x^3$ The natural boundary conditions would be
$y''(0) = 0,\ y''(1) = 0,\ y'''(0) = 0,\ y'''(1) = 0$

▶ 10.

$$x = x(t) = \left(\frac{-\sqrt{2}\sinh\sqrt{2}L - \cosh\sqrt{2}L}{\sinh\sqrt{2}L - \sqrt{2}\cosh\sqrt{2}L}\right)\sinh\sqrt{2}t + \cosh\sqrt{2}t$$

$$y = y(t) = x(t) + \frac{dx(t)}{dt}$$

▶ 11. $y = y(x) = \dfrac{1}{2}e^x - \dfrac{1}{2}e^{-x} + \dfrac{1}{2}\sin x$

▶ 14. $w = w(x) = \cos x, \quad y = y(x) = \sin x, \quad z = y = \sin x, \quad I = \dfrac{\pi}{4}$

▶ 16. $\left(\dfrac{\partial g}{\partial y} - \dfrac{\partial f}{\partial y'}\right)\delta y\,\bigg|_{x=x_1} = 0, \qquad \dfrac{\partial f}{\partial y'}\delta y\,\bigg|_{x=x_2} = 0$

▶ 19. $\left(\dfrac{\partial f}{\partial y'} + \dfrac{\partial h}{\partial y}\right)\delta y\,\bigg|_{x=x_1} = 0, \qquad \left(\dfrac{\partial f}{\partial y'} + \dfrac{\partial g}{\partial y}\right)\delta y\,\bigg|_{x=x_2} = 0$

▶ 21. $\dfrac{\partial f}{\partial w} - \dfrac{\partial}{\partial x}\left(\dfrac{\partial f}{\partial w_x}\right) - \dfrac{\partial}{\partial y}\left(\dfrac{\partial f}{\partial w_y}\right) = 0$ For $\dfrac{dx}{ds} = -\sin\theta$ and $\dfrac{dy}{ds} = \cos\theta$, then boundary condition can be written

$$\oint_C\left[\left(\frac{\partial f}{\partial w_y} - \frac{\partial g}{\partial w}\right)\sin\theta + \left(\frac{\partial f}{\partial w_x} - \frac{\partial h}{\partial w}\right)\cos\theta\right]\delta w\,ds = 0$$

▶ 23.　$\nabla^2 u + \psi(x, y, z) = 0$ for $x, y, z \in V$. If $\delta u = 0$ on S, then u is specified on S so that there results a Dirichlet boundary condition $u(x, y, z)\Big|_{x,y,z \in S} = h(x, y, z)$. Otherwise one must require $\nabla u \cdot \widehat{n} + f + gu = 0$ for $x, y, z \in S$ this represents either a Neumann or Robin boundary condition. The partial differential equation represents steady state temperature within a region with a heat source.

▶ 24.　$(3, 4)$ gives $I_{min} = 5$ and $(-3, -4)$ gives $I_{max} = 15$

▶ 26.　$y = 2\alpha x - x^2$

▶ 27.　$\sqrt{5}$

▶ 28.　$3\sqrt{2}/8$

▶ 30.　$\gamma u + f - \dfrac{\partial}{\partial x}\left(\alpha \dfrac{\partial u}{\partial x}\right) - \dfrac{\partial}{\partial y}\left(\beta \dfrac{\partial u}{\partial y}\right) = 0$

▶ 31.　$r^2 = \sec 2\theta$

▶ 32.　$y = \frac{3}{7}x^2 - \frac{12}{7}x + 1$

▶ 33.　Consider the cases $\lambda = \omega^2 > 0$, $\lambda = 0$ and $\lambda = -\omega^2 < 0$ General solution is $y = C\sin\frac{n\pi x}{\ell}$ and $\int_0^\ell \left(\frac{dy}{dx}\right)^2 dx$ is a minimum when $n = 1$. This gives the solution $y = y(x)C\sin\frac{\pi x}{\ell}$.

▶ 35.　$y = 12x^5 + 10x^4 + 1$

▶ 36.　(a)　$\sqrt{5}$,　　(b)　$\ell^2 = 5$

▶ 37.　$y = x$

▶ 38.　$2\sqrt{5}$

▶ 40.　Let $f = (\dot{y}^2 + \dot{x}^2 - 4t\dot{x} - 4x) + \lambda(\dot{y}^2 - t\dot{y} - \dot{x}^2)$

$$\lambda = -10/11 \qquad\qquad \lambda = -12/11$$
$$y = -\frac{5}{2}t^2 + \frac{7}{2}t \qquad\qquad y = 3t^2 - 2t$$
$$x = t \qquad\qquad\qquad x = t$$
$$I_{min} = 1/12 \qquad\qquad I_{max} = 1$$

▶ 41.　$y = -\dfrac{3}{4}x - \dfrac{1}{4}x^2 + 1$　　$I_{min} = 23/48$

▶ 45.　$\theta = c_1 \displaystyle\int \dfrac{\sqrt{1 + (f')^2}}{r\sqrt{r^2 - c_1^2}}\, dr + c_2$

Selected Solutions Chapter 4

Selected Solutions Chapter 5

▶ 2.   (b) $\frac{d^2y}{dt^2} = g$. If at $t = 0$ we have $y(0) = 0$ and $\dot{y}(0) = 0$, then $y = \frac{1}{2}gt^2$

▶ 5.   (c)   Using Newton's second law

$$I\frac{d^2\theta}{dt^2} = \underbrace{-k_\theta\theta}_{restoring\ torque} \quad \underbrace{-c_\theta\frac{d\theta}{dt}}_{damping\ torque} \quad \underbrace{+F_\theta(t)}_{external\ torque}$$

▶ 6.   *tautochrone* — The time taken for a particle moving under the action of gravity to arrive at its lowest point on the brachistrochrone curve is the same no matter what the starting point on the curve.

▶ 7.   (a)

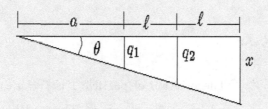

$$\tan\theta = \frac{q_1}{a} = \frac{q_2}{a+\ell} = \frac{x}{a+2\ell} \quad \Rightarrow \quad x = 2q_2 - q_1$$

▶ 10.

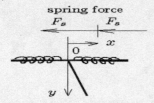

The position of the mass m is given by the vector $\vec{r} = x\,\hat{\mathbf{e}}_1 + \ell\cos\theta\,\hat{\mathbf{e}}_2 + \ell\sin\theta\,\hat{\mathbf{e}}_1$ where $\hat{\mathbf{e}}_1$ and $\hat{\mathbf{e}}_2$ are unit vectors in the directions of the x and y axes. Here $T = \frac{1}{2}mv^2$ where $v^2 = \frac{d\vec{r}}{dt}\cdot\frac{d\vec{r}}{dt}$ simplifies to the result given. When the spring is extended a distance x the work done is $dW = -2kx\,dx$ so that $W = -kx^2$ and the potential energy is then $V = -W = kx^2$

► 13. $y = \dfrac{q_0}{24\,EI}\left(x^4 - 2Lx^3 + L^3x\right)$ with $y\Big|_{x=L/2} = \dfrac{5q_0}{384\,EI}$

► 14. $y = \dfrac{q_0}{24\,EI}\left(x^4 - 4Lx^3 + 6L^2x^2\right)$ with $y\Big|_{x=L} = \dfrac{q_0 L^4}{8\,EI}$

► 16.   The position of the particle is $\vec{r} = R\sin\theta\,\widehat{\mathbf{e}}_1 + R\cos\theta\,\widehat{\mathbf{e}}_2 + z\,\widehat{\mathbf{e}}_3$. One can calculate $\dfrac{d\vec{r}}{dt}$ and $v^2 = \dfrac{d\vec{r}}{dt} \cdot \dfrac{d\vec{r}}{dt}$. Note also that $h = R\sin\theta$ so that $mgh = mgR\sin\theta$

► 17.   The position of the particle is $\vec{r} = (r_0\cos\omega t + \ell\sin\theta)\,\widehat{\mathbf{e}}_1 + (r_0\sin\omega t - \ell\cos\theta)\,\widehat{\mathbf{e}}_2$ Calculate $\dfrac{d\vec{r}}{dt}$ and $v^2 = \dfrac{d\vec{r}}{dt} \cdot \dfrac{d\vec{r}}{dt}$ in order to calculate $T = \frac{1}{2}mv^2$.

► 19. $f(E) = E_0 e^{-E/E_0}$

► 20.

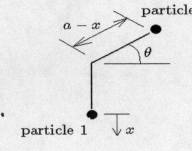

The position of particle 1 is $\vec{r} = x\,\widehat{\mathbf{e}}_3$ with $\dfrac{d\vec{r}}{dt} = \dot{x}\,\widehat{\mathbf{e}}_3$ and $v_1^2 = \dfrac{d\vec{r}}{dt} \cdot \dfrac{d\vec{r}}{dt} = \dot{x}^2$ The position of particle 2 is $\vec{r} = (a - x)\cos\theta\,\widehat{\mathbf{e}}_1 + (a - x)\sin\theta\,\widehat{\mathbf{e}}_2$ with $v_2^2 = \dfrac{d\vec{r}}{dt} \cdot \dfrac{d\vec{r}}{dt} = (a-x)^2\dot{\theta}^2 + \dot{x}^2$, then $T = \frac{1}{2}mv_1^2 + \frac{1}{2}mv_2^2$ and $V = -mgx$

     (f) $\ddot{x} + \alpha^2 x = 0$ with $\alpha^2 = \dfrac{3g}{2a}$

► 22.   (b) $\vec{r}_1 = \ell_1\cos\theta_1\,\widehat{\mathbf{e}}_2 + \ell_1\sin\theta_1\,\widehat{\mathbf{e}}_1$ with $v_1^2 = \dfrac{d\vec{r}_1}{dt} \cdot \dfrac{d\vec{r}_1}{dt} = \ell_1^2\dot{\theta}_1^2$

Similarly, $\vec{r}_2 = (\ell_1\sin\theta_1 + \ell_2\sin\theta_2)\,\widehat{\mathbf{e}}_1 + (\ell_1\cos\theta_1 + \ell_2\cos\theta_2)\,\widehat{\mathbf{e}}_2$ with $v_2^2 = \dfrac{d\vec{r}_2}{dt} \cdot \dfrac{d\vec{r}_2}{dt} = \ell_1^2\dot{\theta}_1^2 a + 2\ell_1\ell_2\dot{\theta}_1\dot{\theta}_2\cos(\theta_1 - \theta_2) + \ell_2^2\dot{\theta}_2^2$, then $T = \frac{1}{2}m_1v_1^2 + \frac{1}{2}m_2v_2^2$.

Selected Solutions Chapter 6

▶ 1. To solve the differential equation $3yy'' + y'^2 = 0$ make the substitutions $y' = p$ and $y'' = \dfrac{dp}{dy}p$

and then separate the variables to obtain $\dfrac{3dp}{p} = -\dfrac{dy}{y}$, then an integration gives the first integral $yp^3 = c_0^3$ where c_0^3 is an arbitrary constant. This leads to the differential equation $y^{1/3}\,dy = c_0 dx$ and a second integration produces the result $y^{4/3} = \dfrac{4}{3}c_0 x + c_1$ where c_1 is an arbitrary constant. The boundary conditions $y(0) = 0$ and $y(a) = b$ produce the final solution $y = b\left(\dfrac{x}{a}\right)^{3/4}$

▶ 2. $y(x) = x$ and $z = z(x) = 1 - x$

▶ 5.

Essential Boundary Conditions	Natural Boundary Conditions

Essential Boundary Conditions

$y(0) = 0, \qquad y(L) = 0$

$y'(0) = 0, \qquad y'(L) = 0$

Natural Boundary Conditions

$\dfrac{d}{dx}\left(EIy''\right)\Big|_{x=0} = 0, \quad \dfrac{d}{dx}\left(EIy''\right)\Big|_{x=L} = 0$

$EIy''\Big|_{x=0}, \qquad EIy''\Big|_{x=L} = 0$

▶ 13. (a) The biharmonic equation $\nabla^4 z = \dfrac{\partial^4 z}{\partial x^4} + 2\dfrac{\partial^4 z}{\partial^2 x \partial^2 y} + \dfrac{\partial^4 z}{\partial y^4} = 0$

▶ 17. (c)

For general $f(x)$

$$\frac{1}{3}c_1 + \frac{1}{6}c_2 = -\int_0^1 f(x)x(1-x)\,dx$$

$$\frac{1}{6}c_1 + \frac{2}{15}c_2 = -\int_0^1 f(x)x^2(1-x)\,dx$$

For $f(x) = -x^2$

$$\Rightarrow \quad \frac{1}{3}c_1 + \frac{1}{6}c_2 = \frac{1}{20}$$

$$\Rightarrow \quad \frac{1}{6}c_1 + \frac{2}{15}c_2 = \frac{1}{30}$$

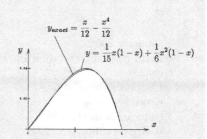

This gives $c_1 = 1/15$ and $c_2 = 1/6$ with $y_{exact} = \dfrac{x}{12} - \dfrac{x^4}{12}$ and $y = \dfrac{1}{15}x(1-x) + \dfrac{1}{6}x^2(1-x)$

▶ 18. $y(x) = c_1 \sinh(x) + \cosh(x)$ substituted into I gives

$$I = I(c_1) = \frac{1}{2}c_1^2 \sinh(2\pi) + c_1(\cosh(2\pi) - 1) + \frac{1}{2}\sinh(2\pi)$$

and $I(c_1)$ is a minimum when $c_1 = \dfrac{1}{\sinh(2\pi)}\dfrac{\cosh(2\pi)}{}$

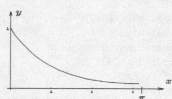

This gives $y = y(x) = \left(\dfrac{1 - \cosh(2\pi)}{\sinh(2\pi)}\right)\sinh x + \cosh x$ with $y(\pi) = \operatorname{sech}\pi$ and $I_{min} = \tanh\pi$

Index

A

absolute maximum minimum 56
algebraic constraint condition 130
allowed quantum states 266
amplitude of u 19
anti-derivative 10
arc length 6
area parallelogram 9
associated Legendre function 262

B

basic lemma 97
beam deformation 247
Beltrami's identity 103
binormal vector 7
boundary 3
brachistrochrone 195
B-splines 288

C

catenary 200
chain rule differentiation 16, 24
change of variable 107
Clairaut's theorem 5
closed region 3
collocation method 283
comparison function 100
composite function 16
concavity 56
conic section 216
conjugate point 149
connected open set 3
conservation of energy 206
conservative force 205
constraint condition 72
constraint functions 75
continuity 1, 3
contour plot 4, 62
convex function 75, 76
coordinate curves 8
coordinate transformation 33
Coulomb's law 259
critical points 55
curvature 7

curvilinear coordinates 8
cycloids 197

D

delta neighborhood 3, 60
dependent variable absent 103
derivative 1
derivative of implicit function 25
derivative test 64, 65
differential constraint condition 130
differentials 24
directional derivative 36, 39, 61
Dirichlet boundary value problem 246
discontinuity 1
dissipative forces 209
domain 1
Du Bois-Reymond form 135

E

eccentricity 216
eigenvalues and eigenfunctions 254
elastic beam theory 228
elliptic integrals 18
end point condition 157
equations of motion 205
Euler-Lagrange equation 102, 114
Euler-Poisson equation 114
Euler's theorem 40
exact or total derivative 103
extreme values 55, 99

F

Fibonacci search 275
finite element 290, 297
first derivative test 57
first mean value theorem 11
first variation 111
fixed end conditions 111, 250
Fletcher-Powell gradient search 275
flux per unit volume 15
Foucault pendulum 223
free boundary condition 250
free end conditions 162
function 1, 55

functionals 95
Functions of two variables 60

G

Galerkin method 280
Gauss divergence theorem 15, 124
general variation 153
generalized bachistrochrone problem 251
geodesic curves 175
gradient vector 37
Green's theorem 13, 121, 125, 131

H

Hamilton integral 204, 244
Hamilton's principle 203, 245
hanging cable 199
heat equation 135
Hessian functions 66
Hessian matrix 64
higher derivatives 113
holonomic system 251
homogeneous function 40
hydrogen atom 259

I

implicit form 10
implicit function theorem 21
implicit functions 20
independent variable absent 103
integral used to define function 18
integration by parts 11
invariance 107
inverted cycloid 198
irrotational field 205
isoperimetric problem 199
J

Jacobi differential equation 146
Jacobi elliptic functions 18
Jacobi test 150
Jacobian determinant 28, 33

K

kinetic energy 204

L

Lagrange multipliers 71
Lagrangian 206

Laplace equation 33, 122, 126, 135,
Law of reflection 59
Legendre and Jacobi analysis 142
Legendre's necessary conditions 146
Leibnitz rule 18
level curves 37, 69
line integral 11, 121
linear interpolation 293
linear programming 82
local maximum or minimum 55, 60

M

mathematical programming 81
maxima and minima 55, 79
Maxwell-Boltzmann distribution 85
Maxwell's equations 271
mean value theorem 2
method of steepest ascent 275
mixed end condition 138
monotone function 76
movable boundaries 155
multiply-connected regions 13

N

natural boundary conditions 138, 141
Navier equations 226
negative definite 65
Neumann boundary value problem 246
nonconservative system 209
nonholonomic system 251
normal plane 7
numerical methods 273

O

objective function 75
open interval 57
open set 3
orbitals 265
osculating plane 7
other functionals 111

P

parametric equations 5, 8, 11, 17, 30, 108
partial derivatives 3, 121
pattern search 275
Pauli exclusion principle 266
piecewise smooth functions 164
plane curve 13
planimeter 14

points of inflection 56
Poisson equation 122
polar coordinates 33
positive definite 64
positive sense 13
potential 196
potential function 204
power series 40
principal submatrics 66
principal unit normal 7
product rule 16
projection 37

Q

quadratic forms 64
quantum numbers 264
quantum states 266

R

radius of curvature 7
random search 275
range 1
Rayleigh-Ritz method 276, 287
rectifying plane 7
relative extreme 56
rheonomic system 251
Robin boundary value problem 246
Rolle's theorem 2
rolling cylinders 252

S

saddle point 63, 66
scalar field 61
Schrödinger equation 257
scleronomic system 251
search techniques 273
second derivative test 57
second derivatives 112
second directional derivative 39, 63
second law of thermodynamics 35
second variation 111, 145
separation of variables 260
shortest distance 105
simple closed curve 13
simple pendulum 218
simply connected region 13
smooth surface 4
Snell's law of refraction 59
soap film problem 201
solenoidal field 16

space curve 4
space curves 5
special cases 103
special cases 114
spherical harmonics 267
spherical pendulum 222
spring-mass system 210
stationary points 55
stationary solutions 111
stationary value 99
Stoke's theorem 14
straight line 96
strong variation 100
Sturm-Liouville system 256
subscript notation 121
surface 8
surface of revolution 95
system of equations 23

T

tangent vector 6
Taylor series 40, 58
Taylor series expansion 64
test for maxima and minima 57, 142
torsion 7
transformations 32
transversality condition 157
triangular element 294
tridiagonal system 288

U

unit normal vector 8

V

variable end point 157
variational notation 109
variational problem 100
vector field 11
vibrating membrane 247
vibrating string 243

W

wave equation 135
wave functions 259
weak variation 100
Weierstrass criticism 130
Weierstrass theorem 57
work done 11
work-energy theorem 195

Printed in the United States
By Bookmasters